10th Research in Engineering Education Symposium

Connecting Research-Policy-Practice for Transforming Engineering Education

KLE Technological University,
Hubballi, Karnataka, India
4 – 6th January, 2024

Editors
Dr. Preethi Baligar
Mr. Kaushik Mallibhat
Dr. Rohit Kandakatla
Ms. Radhika Amashi

RESEARCH IN ENGINEERING EDUCATION NETWORK REEN

KLE Technological University | Creating Value, Leveraging Knowledge

Indo Universal Collaboration for Engineering Education IUCEE

Printed from e-media with permission by:

Curran Associates, Inc.
57 Morehouse Lane
Red Hook, NY 12571

Some format issues inherent in the e-media version may also appear in this print version.

Printed with permission by Curran Associates, Inc. (2024)

Print ISBN: 978-1-7138-9381-3
e-ISBN: 978-1-7138-9440-7

Additional copies of this publication are available from:

Curran Associates, Inc.
57 Morehouse Lane
Red Hook, NY 12571 USA
Phone: 845-758-0400
Fax: 845-758-2633
Email: curran@proceedings.com
Web: www.proceedings.com

About REES

The Research in Engineering Education Symposium, REES 2024, was held from Jan 4-6, 2024, at KLE Technological University, Hubballi, Karnataka, India, in collaboration with the Research in Engineering Education Network (REEN) and Indo-Universal Collaboration for Engineering Education (IUCEE).

The theme of REES 2024 was "Connecting Research-Policy-Practice for Transforming Engineering Education." The conference aspired to galvanize the efforts of leaders, academia, researchers, practitioners, and policy, regulatory, and monitoring agencies in engineering education to contribute to and nurture the engineering education research ecosystem.

REES 2024 invited contributions for the following themes and tracks

Sub-theme 1: Research
1. Engineering education problems that require inquiry through research
2. Engaging engineering faculty in educational research
3. Capacity building models for engineering education research in relevant contexts
4. Pathways to transition from traditional engineering to engineering education research
5. Engineering education research leading to the development of theories and frameworks
6. Global quality benchmarks for engineering education research

Sub-theme 2: Policy
1. Recognition of engineering education research as a viable scholarly field of inquiry at institutional and national levels
2. Identify and overcome barriers that discourage engineering education research
3. Incentives and support systems to engage engineering faculty in engineering education research
4. Funding engineering education research
5. Creating engineering education ecosystems: Moving from regulator mindset to enabler mindset

Sub-theme 3: Practice
1. Impact evaluation of engineering education research on practice
2. Effective models of disseminating engineering education research to practice
3. Designing rigorous, inclusive, and sustainable learning environments for engineering faculty and students
4. Transnational education and policy
5. Scholarship of teaching and learning

About the Hosts

Research in Engineering Education Network

The Research in Engineering Education Network (REEN) is a community of scholars interested in conducting high-quality work in and advancing engineering education research. REEN differs from other networks and communities of engineering educators in its primary focus on research in engineering education worldwide. The mission of REEN is to provide an independent, international, and inclusive forum to advance scholarly discourse on research in engineering education. REEN's vision is to establish and maintain an independent, international, vibrant, and inclusive community that supports, discusses, and disseminates scholarly research on engineering education. The community is intended to nurture new and veteran researchers through collaboration and sound methodological approaches to influence policy at all levels and teaching practices around the globe.
Website:https://reen.co/

KLE Technological University

KLE Technological University (KLE Tech) is rooted in one of Karnataka's premier engineering institutions, B. V. Bhoomaraddi College of Engineering and Technology (BVB), a prestigious engineering college in Hubli. The founding organization, KLE Society, Belgaum, established BVB College in 1947 to create an institution that would lay the foundation of modern engineering education in the northern region of Karnataka. Over the years, it evolved to reach and hold a unique position of pride in India's technical education system. As we entered the 21st century, the college undertook a comprehensive reform process to adapt to the challenging global engineering education scenario. In pursuit of academic excellence, the college attained academic autonomy from the University Grant Commission (UGC) in 2007. As an autonomous college, BVB established its distinctive character in the educational space through its curriculum and outstanding student experience. In 2014, the Government of Karnataka recognized the college as a state-private university. The rich heritage of BVB College as one of the best engineering colleges in Hubli, combined with the brand equity of KLE Society, are the starting points for KLE Technological University to emerge as a university with a national distinction. Over time, it gained tremendous credibility with industries and employers and emerged as a brand to reckon with. The institute's alumni have done exceedingly well in all spheres of life at national and international levels, bringing name and fame to themselves and their alma mater.
Website: www.kletech.ac.in

Indo-Universal Collaboration for Engineering Education

The vision of the Indo-Universal Collaboration for Engineering Education (IUCEE) is to improve the quality and global relevance of engineering education in India. Its mission is to build an ecosystem for transforming engineering education in India with the assistance of engineering education experts and industry worldwide. IUCEE operates through two formal organizations: IUCEE Foundation, a Section 8 company in India, and IUCEE Inc., a 501c3 company in the US. A team of Indian and international experts from academia and industry guides IUCEE. They are passionate about the mission of IUCEE and devote time and energy to offering courses and workshops, facilitating collaborations, organizing and speaking at events, and mentoring faculty, students, and college leaders. A fundamental strength of IUCEE is its network of engineering educators from around the world who are passionate about the quality of engineering education. It also has ties with global organizations such as the International Federation of Engineering Education Societies (IFEES), the Global Engineering Deans Council (GEDC), the American Society for Engineering Education (ASEE), the Accreditation Board for Engineering and Technology (ABET), the International Society for Engineering Pedagogy (IGIP), and Engineering Projects in Community Service (EPICS).

Website:www.iucee.org

Patron and Chairs

Dr. Prabhakar Kore
Chairman, KLE Society, Belagavi, Chancellor, KLE Technological University (KLE Tech), BVB Campus, Hubballi, Karnataka, India

Dr. Ashok Shettar
Conference General Chair
Vice Chancellor, KLE Technological University, Hubballi

Dr. Gopalkrishna Joshi
Conference General Chair
Vice Chancellor, MIT Vishwaprayag University, Solapur

Dr. Vijayalakshmi M
Chair, Organizing Committee
Professor & Head, School of Computer Science and Engineering, KLE Tech, BVB Campus, Hubballi, Karnataka, India

Dr. Sohum Ashok Sohoni
Co-Chair, Organizing Committee
Professor and Program Director Software Engineering
Milwaukee School of Engineering Milwaukee, Wisconsin, USA

Dr. Preethi Baligar
Chair, Technical Committee
Centre for Engineering Education Research, KLE Tech, BVB Campus, Hubballi, Karnataka, India

Dr. Rohit Kandakatla
Co-Chair, Technical Committee
Director – KG Reddy College of Engineering and Technology, India

Mr. Kaushik Mallibhat
Member Secretary, Technical Committee, KLE Tech, BVB Campus, Hubballi, Karnataka, India

Technical Committee Members

(in alphabetical order of first name)

Dr. Ashish Agrawal

Assistant Professor, College of Engineering Technology, Rochester Institute of Technology, NY, USA.

Dr.Avneet Hira

Boston College, United States of America

Dr Buddha Chandrashekhar

Chief Coordinating Officer, AICTE, Ministry of Education, Government of India

Prof. Faris Tarlochan

Qatar University, Qatar

Dr. Helen Inglis

University of Pretoria, South Africa

Dr. Homero Murzi

Virginia Tech, United States of America

Prof. John Mitchell

Centre for Engineering Education, University College London,

United Kingdom

Mr. Keith Raymond Fernandes

St Joseph Engineering College, Mangaluru, Karnataka, India

Dr. Mitra Mohd Addi

Universiti Teknologi Malaysia, Malaysia

Dr. Preeti Basavaraj Patil

Centre for Engineering Education Research

KLE Technological University, Hubballi, India

Dr. Prithvi Sekhar Pagala

Specialist, Global Engineering Academy

L&T Technology Services Limited, Bengaluru, India

Dr. Rajlaxmi Chouhan

Assistant Professor, Department of Electrical Engineering

Indian Institute of Technology Jodhpur, India

Dr. Rucha V. Joshi

Faculty, Plaksha University, Mohali, Punjab, India

Mr. Sanjeev M Kavale

Arizona State University, United States of America

Prof. Sivakumar Krishnan

Director, Learning and Innovation Vishnu Educational Development and Innovation Centre (VEDIC), Sri Vishnu Educational Society, Hyderabad, India

Dr. Scott Daniel

University of Technology Sydney, Australia

Dr. Teresa Hattingh

University of Johannesburg, South Africa

Program Agenda

Keynotes

Inaugural Keynote Theme - Future of Engineering Education in the World of Artificial Intelligence	Dr Viraj Kumar Professor, Kotak IISc AI-ML Centre.
Theme - Evolution of Engineering Education Research as a recognized field of inquiry across the globe	Dr. Teresa Hattingh - Associate Professor and Engineering Education Specialist, University of Johannesburg
Theme - Methodological Requirements for Engineering Education Research	Dr. Johannes Strobel - Associate Dean of Research & Graduate Studies, University of Texas at El Paso
Theme - Emerging and Future Trends in Engineering Education Research	Dr. Aditya Johri, Professor of Information Sciences & Technology at George Mason University

Panel Discussions

Building a community of Engineering Education Research at the regional, national, and international level	**Panel Members:** Dr. Gopalkrishna Joshi - Founding Vice-Chancellor, Vishwaprayag University, Solapur Dr. Sohum Sohoni - Professor and Program Director of Software Engineering, Milwaukee School of Engineering Dr. Helen Inglis, Senior Lecturer, University of Pretoria Spandhana Gonuguntla - Senior Customer Success Engineer, Mathworks **Moderator** Mr. Sanjeev Kavale - Graduate Research Assistant, Arizona State University
Transitioning from engineering to engineering education research	**Panel Members:** Dr. Rucha Joshi - Assistant Professor, Plaksha University Prof. William Oakes - Director of EPICS, Purdue University Dr. Shannon Chance - Honorary Professor at Center for Engineering Education, University College London Dr. Siva Krishnan - Director for Learning and Innovation, Vishnu Educational Development and Innovation Centre (VEDIC) **Moderator** Mr. Kaushik Mallibhat - Assistant Professor, KLE Technological University
Enablers and Barriers to Promote Engineering Education Research	**Panel Members:** Dr. Adam R Carberry, Head. Department of Engineering Education, Ohio State University Dr. Rohit Kandakatla - Director, KG Reddy College of Engineering and Technology Dr. Preethi Baligar, Director, Centre for Engineering Education Research, KLE Technological University Dr. Helen Inglis, Senior Lecturer, University of Pretoria **Moderator** Dr. Sohum Sohoni Professor and Program Director of Software Engineering, Milwaukee School of Engineering

Plenary Sessions

Theme - Impactful Research: Curricula, Practices, and Shaping Policies	**Speakers:** Dr. Ashok Shettar - Vice-Chancellor, KLE Technological University Prof. Edward Berger - Interim Head for School of Engineering Education, Purdue University Dr. Xiangyun Du - Director, UNESCO Center for PBL at Aalborg University

Keynote Speaker

Future of Engineering Education in the World of Artificial Intelligence

Dr Viraj Kumar
Professor, Kotak IISc AI-ML Centre, India

Dr Viraj Kumar is a Visiting Professor at the Kotak IISc AI-ML Centre. He holds a PhD in Computer Science from the University of Illinois at Urbana-Champaign and he conducts research in Computer Science education (with an emphasis on India-specific challenges). Prof Kumar serves as an elected member of the ACM India Council, and chairs its Education Committee. He also serves on a Ministry of Education Committee to examine the implications of Generative AI on Engineering Education. Previously, Prof Kumar served as a consultant to both the Kasturirangan Committee (MHRD) to draft the National Education Policy and the Committee to frame the National Curricular Frameworks (MoE). In addition to teaching at IISc, he contributes to faculty and school-teacher training initiatives, and is a co-author of IISc's institutional policy on the usage of Generative AI.

Keynote Speaker

Evolution of Engineering Education Research as a recognized field of inquiry across the globe

Dr. Teresa Hattingh - Associate Professor and Engineering Education Specialist, University of Johannesburg, South Africa

Teresa Hattingh is an Associate Professor and Engineering Education specialist at the University of Johannesburg in South Africa. In this role, she provides support to enhance student access and success and build scholarly approaches to teaching and learning in the Faculty. Teresa completed her Bachelors and Master's degrees in engineering and her PhD in Engineering Education. She is registered as a professional engineer and has extensive teaching experience in higher education and industry experience in project engineering, operations and supply chain management in the metallurgical and gas industries. She is actively involved in engineering education research and is the current chair of the Global Research in Engineering Education (REEN) board and is the current President of the Southern African Institute for Industrial Engineering.

Keynote Speaker

Methodological Requirements for Engineering Education Research

Dr. Johannes Strobel, Associate Dean of Research & Graduate Studies, University of Texas, El Paso, USA

Dr. Johannes Strobel is Associate Dean for Research and Graduate Studies at the University of Texas at El Paso, USA. He joint UTEP from SRI International where he was Director, Policy Research, Center for Innovation Strategy and Policy (CISP). For the past 20 years, Dr. Strobel has been actively shaping the K-16 engineering education research landscape in the United States, as exemplified through serving as the founding editor of the Journal of Pre-College Engineering Education Research. His research and teaching focuses on engineering 'habits of mind' and worldviews, empathy and care in engineering and integrated STEM competencies. He published more than 180 papers in proceedings, journals and book chapters, co-edited five books and was on successful STEM/engineering education research grants with over US$35MM. He is the lead author of "Hands-on Standards STEM in Action", a nationally and internationally available set of learning modules for grades preK - 5th grade published by hand2mind.

Keynote Speaker

Emerging and Future Trends in Engineering Education Research

Dr. Aditya Johri, Professor of Information Sciences & Technology
George Mason University, USA

Aditya Johri is Professor of Information Sciences & Technology at the College of Engineering and Computing at George Mason University (GMU), USA. He studies how technology shapes learning across formal and informal settings and the ethical implications of using technology. He publishes broadly in the fields of engineering and computing education, educational technology, and computer-supported collaborative work and learning. His research has been recognized with several best paper awards (CSCW, IEEE ETHICS, ICWL) and his co-edited volume, the Cambridge Handbook of Engineering Education Research (CHEER), received the 2015 Best Book Publication Award from Division I of AERA. His edited volume International Handbook of Engineering Education Research (IHEER) was published by Routledge in 2023. He served as a Fulbright-Nokia Distinguished Chair in ICT at Aalto University, Finland (2021) and is a past recipient of the NSF CAREER Award (2009). His teaching and mentoring have been recognized with the University Teaching Excellence Award (2002) and Mentoring Excellence Award (2022) for undergraduate research. He was awarded a Ph.D. in Learning Sciences & Technology Design (2007) from Stanford University, Palo Alto, CA. More information is available at:http://mason.gmu.edu/~johri

Panel Discussion 1

Building a community of Engineering Education Research at the regional, national, and international level

Prof.Gopalkrishna Joshi,
Founding Vice Chancellor, MIT Vishwaprayag University,
Solapur, Maharashtra (India).

Prof.Gopalkrishna Joshi is the Founding Vice Chancellor of MIT Vishwaprayag University, Solapur, Maharashtra (India). Earlier, Prof.Joshi served as the Executive Director of Karnataka State Higher Education Council, Bengaluru (India). During his tenure at KSHEC, he was involved in the implementation of the National Education Policy 2020 and international collaborations in the space of higher education. Prior to joining KSHEC, he was a Professor of Computer Science and Dean of Curriculum Innovation & Program Assessment) at KLE Technological University, Hubballi, India. He was also the Director of the Centre for Engineering Education Research (CEER) at KLE Tech. His areas of research interest include data engineering and engineering education. He is involved in practice and research in outcome-based education. He has contributed to innovations in first-year engineering curriculum and PBL practices. He has 53 publications to his credit. He is a member of the International Advisory Body of the UNESCO Centre for PBL in Engineering at Aalborg University, Denmark.

Prof.Joshi is a member of professional bodies, including the Institution of Engineers (India), the Association of Computing Machinery, and the Indian Society for Technical Education. He is on the Board of Indo Universal Collaboration for Engineering Education(IUCEE), a not-for-profit organization working to improve the quality of engineering education in India. He is also on the Board of IT Education Standards (BITES), another Bengaluru-based not-for-profit organization.

Dr. Sohum Sohoni,
Professor, Program Director of Software Engineering,
Department of Electrical Engineering and Computer Science, Milwaukee School of Engineering USA

Dr. Sohum Sohoni is a Professor and Program Director of Software Engineering in the department of Electrical Engineering and Computer Science at the Milwaukee School of Engineering. Prior to joining MSOE, he was faculty at Arizona State University and Oklahoma State University. His research is in Computer Engineer and Engineering Education. He is interested in the adoption and translation of effective teaching techniques, and has established a vision for networks of discipline-based education research (DBER) ambassadors to facilitate the adoption and evaluation of research-based instructional strategies. A computer architecture visualization platform that he and his students designed has been used by over 1000 students at 3 universities in the US.

He has published over 40 peer-reviewed papers in journals and conferences including papers in ACM SIGMETRICS, IEEE Transactions on Computers, the International Journal of Engineering Education, and Advances in Engineering Education. He has received a best paper award for his work in computer engineering from the IETE Technical Review journal, and three best paper awards for his work in engineering education, from the American Society for Engineering Education (ASEE). He is a popular and well-respected instructor, and has received many teaching awards including the Regents Distinguished Teaching Award in 2010 at Oklahoma State University.

Dr. Sohoni is a member of ACM (Association for Computing Machinery) and ASEE (American Society for Engineering Education), and a Senior Member of IEEE (Institute of Electrical and Electronics Engineers). He has reviewed for various journals and conferences, and served as session chair at a number of conferences. He is a member of the board of the Indo Universal Collaboration for Engineering Education (IUCEE) and the Research in Engineering Network (REEN) and serves as the Co-Editor-in-Chief of the Journal of Engineering Education Transformations. He also serves as a Program Evaluator for the Accreditation Board for Engineering and Technology (ABET).

Dr. Spandhana Gonuguntla
Senior Customer Success Engineer at MathWorks India

Spandhana is a Senior Customer Success Engineer at MathWorks India. She works with education customers to enable effective use of MATLAB and Simulink in engineering curriculum and research. She specializes in Data analytics, mathematical modelling, Machine Learning and Deep Learning. She has presented at numerous conferences including the Grace-Hopper Conference and SWE WE Local 2022. She has a PhD from National University of Singapore and a B.Tech from National Institute of Technology, Surathkal.Network (REEN) and on the executive committee of the Global Engineering Deans Council (GEDC).

Helen Inglis, Senior Lecturer, Department of Mechanical and Aeronautical Engineering University of Pretoria in South Africa

Dr Helen Inglis is an engineering education researcher focusing on student success.

Helen is a Senior Lecturer in the Department of Mechanical and Aeronautical Engineering at the University of Pretoria in South Africa. Her masters and doctoral degrees, from the University of Illinois at Urbana-Champaign, were in the field of solid mechanics, but her passion for teaching and for social justice has led her to grow her research activities in engineering education since joining the University of Pretoria. Helen is part of the organising team for a South African project to develop innovative engineering curricula, and is a REEN board member. Dr. Helen Inglis, Senior Lecturer, University of Pretoria

Panel Discussion 2

Transitioning from engineering to engineering education research

Dr. Rucha Joshi – Tamaskar, Associate Professor at Plaksha University, Punjab, India

Dr. Rucha Joshi – Tamaskar, Ph.D., is an Associate Professor at Plaksha, formerly an Assistant Professor at the University of California, Davis; and a postdoctoral fellow at Purdue University, where she worked on tissue engineering as well as engineering education research, specifically on biomaterials, drug delivery and instructional innovation using active learning in biomedical engineering. Rucha received a BS in biotechnology engineering from Shivaji University, India, in 2009, and MS in biomedical engineering from Vanderbilt University in 2011, a Ph.D. in Biomedical Engineering in 2016. Rucha has two patents, on making low calorie biscuits from banana peel pulp, and second on "collagen based therapeutic delivery systems". Her current research focuses on silk-biomaterials, project-based learning, personalizing learning, and entrepreneurially minded engineers. She is also a director for Grand Challenges Scholars Program at Plaksha, creating experiential learning through a combined curricular, co-curricular, and extra-curricular program.

Rucha is actively working to promote the inclusive maker movement by serving learning needs of visually impaired high school children. She has served as a BME Liaison for the Committee for Diversity, Equity, and Inclusion (CDEI) of the American Society for Engineering Education (ASEE) in 2021-22, and as a secretary of New Engineering Educators division of ASEE in 2019-20. On extra-curriculars, Rucha is a NAUI certified SCUBA diver and 2023 Punjab state badminton winner in 35+ Women category, representing Punjab in 45th National Badminton

Veterans Championship. Her book in Marathi tells a story of her journey of representing India at Japan in class 12th, being honored by Dr. APJ Abdul Kalam, the late president of India, for her invention of "low calorie biscuits from banana peel pulp". The book won a literary award from the State Government of Maharashtra in 2008. Her latest book on collagen based drug delivery was published by Springer in 2021.

Professor Shannon Chance, PhD, SFHEA, LEED-AP
Programme Chair (Honours BSc in Digital Construction), TU Dublin & Visiting Professor, UCL Centre for Engineering Education

Professor Shannon Chance is an advocate for excellence in engineering education with a notable track record in securing international research funding, including prestigious Fulbright and Marie Curie fellowships. With her cross-disciplinary expertise spanning architecture and education, she offers workshop participants practical insights into obtaining research grants and fellowships. Her work reflects a deep commitment to global educational practices and the advancement of research in engineering education. Shannon's session, "Applying for International Research Grants and Fellowships," is designed to empower engineering educators with the knowledge to secure funding, supporting research that spans continents and cultures. Attend this workshop to discover strategies that align with funders' goals and enhance your capacity to contribute to engineering education's global dialogue.

Dr. Sivakumar Krishnan, Director of Learning and Innovation with Vishnu Educational Development and Innovation Center (VEDIC), Hyderabad

Dr. Sivakumar Krishnan is currently Director of Learning and Innovation with Vishnu Educational Development and Innovation Center (VEDIC) at Sri Vishnu Educational Society (SVES). He has a B.Tech. in Aerospace Engineering (1995) from IIT, Madras and a Ph.D. in Aerospace Engineering (2000) from University of Michigan, Ann Arbor.

He is a Certified International Engineering Educator by IGIP, Austria (International Society for Engineering Pedagogy). Since 2013, he has developed and offered various faculty development initiatives throughout India, with the Indo-Universal Collaboration for Engineering Education (IUCEE). Since 2017 he has continued to offer programs on teaching and learning, accreditation, curriculum development and research, so far covering about 2500 faculty in 30 unique workshops till now.

Between 2002 and 2010, as Assistant Professor, he taught at the Department of Mechanical Engineering, Purdue University School of Engineering at IUPUI and has authored journal and conference publications as well as holds patents in the areas of combustion, radiative heat transfer, infrared sensors and engineering education. His 2009 ASEE (American Society for Engineering Education) Conference paper on Project-Based Learning in Introductory Thermodynamics was recognized as an 'Outstanding Contribution to Mechanical Engineering Education'. He believes strongly that the Indian demographic dividend needs to be effectively capitalized through focus on holistic student mentoring by faculty as well as courageous and thoughtful institutional leadership.

From 2011 to 2016, as Technical Specialist with Fiat Chrysler Automobiles, he developed innovative technologies to improve the fuel economy and performance of a high volume V6 gasoline engine. Between 2016 and 2017, he was Chief Product Officer with IntelliEd Innovations, a visual media education technology startup. He is also a certified Heartfulness meditation trainer (since 2018) and practitioner.

Prof. William (Bill) Oakes is the Assistant Dean for Experiential Learning, Director of the EPICS Program, Professor of Engineering Education, Purdue University

William (Bill) Oakes is the Assistant Dean for Experiential Learning, a 150th Anniversary Professor, Director of the EPICS Program, Professor of Engineering Education at Purdue University, and a registered professional engineer. He is one of the founding faculty in the School of Engineering Education having courtesy appointments in Mechanical, Environmental and Ecological Engineering and Curriculum and Instruction. He was the first engineer to receive the U.S. Campus Compact Thomas Ehrlich Faculty Award for Service-Learning and a co-recipient of the U.S. National Academy of Engineering's Bernard Gordon Prize for Innovation in Engineering and Technology Education. He is a fellow of NSPE and ASEE and elected to the ASEE Hall of Fame.

Panel Discussion 3

Enablers and Barriers to promote Engineering Education Research

Dr. Adam R. Carberry, Professor and Chair, Department of Engineering Education, The Ohio State University, USA

Dr. Adam R. Carberry is Professor and Chair in the Department of Engineering Education at The Ohio State University (OSU). He joined OSU after having served as an Associate Professor in The Polytechnic School within Arizona State University's Fulton Schools of Engineering (FSE). There he served as the Graduate Program Chair for the Engineering Education Systems & Design (EESD) Ph.D. Program. His engineering education research has been supported by the National Science Foundation, Kern Family Foundation, Lemelson Foundation, and Helmsley Foundation. He is currently a Deputy Editor for the Journal of Engineering Education and co-maintains the Engineering Education Community Resource wiki. Additional career highlights include serving as Chair of the Research in Engineering Education Network (REEN), visiting École Nationale Supérieure des Mines in Rabat, Morocco as a Fulbright Specialist, receiving an American Society for Engineering Education (ASEE) Educational Research and Methods Division Apprentice Faculty Award, receiving a Frontiers in Education New Faculty Award, and being named an ASEE Fellow.

Dr Preethi Baligar, Associate professor and Director, Centre for Engineering Education Research, KLE Technological University, Hubballi, India

Dr Preethi Baligar is an associate professor and director for the Centre for Engineering Education Research (CEER) at KLE Technological University, Hubballi. Her core contribution lies in the design of instruction and faculty development for interdisciplinary, design-based, PBL, and blended learning courses in first-year undergraduate engineering. She has engaged with the professional community to develop institutional best practices, outcome-based education, cooperative and collaborative learning, and design student-centered learning environments. She has served as a technical chair, session chair, and in other capacities for conferences like REES 2024 and RRSPBL 2019-2021. She has several publications related to engineering education research and is also a co-author of the International Handbook for Engineering Education Research. She is also a reviewer for several engineering education conferences and Journals. She holds a Master's in Computer Science and a PhD in Engineering Education from KLE Tech.

Dr. Rohit Kandakatla, Director for Strategy, International Affairs, and Human Resource Development, KG Reddy College of Engineering and Technology, Hyderabad, India

Dr. Rohit Kandakatla is currently serving as the Director for Strategy, International Affairs, and Human Resource Development at KG Reddy College of Engineering and Technology. He also has an adjunct faculty appointment with the Center for Engineering Education Research at KLE Technological University. He has completed his Bachelors and Masters of Engineering, Ph.D. in Engineering Education from Purdue University, and post-doctorate research from Indian Institute of Information Technology (IIIT) Hyderabad. His research interests include service-learning, education technology, understanding organizational development in higher education, innovation and entrepreneurship in engineering, and education policy. He has been involved in the global engineering education community for more than a decade starting with his role as a student leader with Student Platform for Engineering Education Development (SPEED) where he has led many student-based initiatives to help solve engineering education issues at the local and global level. He got elected as the Global President in 2014 at the 9th Global Student Forum held in Dubai and served in the role for 2 years when he helped increase SPEED's global presence through locally organized student forums that are hosted to voice local issues. He got elected as the Vice-President for Student Engagement of the International Federation for Engineering Education Societies (IFEES) in 2015 to represent students' voices at the global level. He was awarded Young Engineering Educator Scholarship by National Science Foundation (NSF) in 2018, IUCEE Young Leader Award in 2015, and IGIP Young Scientist Award in 2014. He currently serves on the governing body of the Research in Engineering Education Network (REEN) and on the executive committee of the Global Engineering Deans Council (GEDC).

Helen Inglis, Senior Lecturer, Department of Mechanical and Aeronautical Engineering University of Pretoria in South Africa

Dr Helen Inglis is an engineering education researcher focusing on student success. Helen is a Senior Lecturer in the Department of Mechanical and Aeronautical Engineering at the University of Pretoria in South Africa. Her masters and doctoral degrees, from the University of Illinois at Urbana-Champaign, were in the field of solid mechanics, but her passion for teaching and for social justice has led her to grow her research activities in engineering education since joining the University of Pretoria. Helen is part of the organising team for a South African project to develop innovative engineering curricula, and is a REEN board member. Dr. Helen Inglis, Senior Lecturer, University of Pretoria

Impactful Research: Curricula, Practices, and Shaping Policies

Dr. Edward Berger, Associate Vice Provost for Learning Innovation, and Director of the Innovation Hub, Purdue University, USA

Edward Berger is the Associate Vice Provost for Learning Innovation, and Director of the Innovation Hub, as well as Professor of Engineering Education and Mechanical Engineering at Purdue University. Starting on February 1, 2023, he accepted the role of Interim Head of the School of Engineering Education at Purdue. He is also the Executive Director of the Mechanical Engineering Education Research Center at Purdue (MEERCat Purdue), which pursues fundamental research and research-to- practice projects in student success, teaching with technology, and institutional culture and change. He has been a tenured engineering professor at three insti- tutions (University of Cincinnati 1996-2004, University of Virginia 2005-2014, Purdue University 2014-present). He began his faculty career with a research portfolio broadly in the area of tribomechanics, including dynamics of interfaces and structures with friction. From 2005-2010, his research portfolio underwent a transformation into engineering education research, and since then he has been a PI or co-PI on $12M+ in federally-funded projects for engineering education research. He has 60+ journal publications across mechanical engineering and engineering education topics. In 2008, he won a Commonwealth of Virginia state-wide teaching award for his integration of technology into undergraduate education. He was recently a program officer at the National Science Foundation (August 2019 - December 2020) in the Engineering Directorate, Division of Engineering

Education and Centers, where he oversaw a $15M annual budget for engineering education research and collaborated across the agency on pro- grams including ERC, AI Institutes, and Advanced Manufacturing. He recently concluded a two-year term as the faculty representative to the Indiana Commis- sion for Higher Education, during which he served as the chair of the Academic Affairs and Quality Committee.

Dr. Xiangyun Du, Professor and Director at UNESCO Centre for Problem and Project-Based Learning, Aalborg University, Denmark

Xiangyun Du, Ph.D., is a professor and Director at UNESCO Center for Problem and Project-Based Learning, Aalborg University, Denmark. Prof. Du has been committed to research in educational transformation through pedagogical change using Problem-Based and Project-Based Learning methodology) – in diverse social, cultural, and educational contexts. Having worked in fields of development ranging from engineering, STEM, teaching preparation, language teaching, and business to health (medicine, dental, and public health sciences) education, etc.), her research topics included change from an inter/cross-cultural perspective, curriculum and pedagogy development, faculty/staff/ teacher development, intercultural learning and teaching, and gender studies. Having won multiple teaching and learning prizes herself, Prof. Du has also been engaged with educational institutions n over 30 institutions across countries doing substantial work on pedagogy development. Prof. Du has over 200 relevant international publications, including 10 monographs, over 100 journal papers (SCOPUS, Web of Science, and SSCI), 15 edited books, and over 40 book chapters as well as over 60 conference contributions. She has also been actively involved in several international academic programs, networks, and editorial works for journals.

Paper Presentations

Review Process

For authors, REES 2024 followed a two-level submission procedure: abstracts and full papers. The review process was double-blind, and the criteria for review are included below.

Review criteria for abstracts:

1. Focus of the Paper: The paper clearly describes the research question OR hypothesis and explains the implications of the project to research, policy, and/or practice.
2. Relevance: The paper relates the work undertaken to relevant discussions in the literature and other disciplinary literature as required and describes its contribution to these discussions.
3. Approach: The paper clearly describes and justifies the appropriateness of the overall approach, which could include designs, methods, theories, and analytic processes, and discusses the study's limitations.
4. Argument: The paper presents novel ideas or results of significance to others supported by convincing evidence and reasoning, illustrating the connection between claims and evidence.
5. Writing Quality: The paper is written in English with a sufficient standard to enable the reader to make sense of it.

List of Reviewers

We thank all the 101 reviewers who ensured the quality and rigor of the submissions. The list of the reviewers is included in the Appendix.

Paper Presentations

The Statistics for papers received are as follows, with participation from 6 countries, including the USA, UK, Ghana, South Africa, Ireland, and India. Around 100+ reviewers from 18 countries supported REES in the review process.

Total number of abstracts received	174
Number of abstracts accepted	143
Total number of full papers received	83
Total number of accepted papers	30

Review criteria for Full paper

Content	Originality	Content contains highly original treatment of, or new perspective on, the topic.	Content contains moderately original treatment of, or new perspective on, the topic.	Content contains minimal original treatment of, or new perspective on, the topic.
	Scholarship	Content reviews and builds on appropriate prior work to a significant extent.	Content reviews and builds on appropriate prior work to a limited extent.	Content does not review and build on appropriate prior work.
	Relevance	The paper makes a highly significant contribution to the field of engineering education.	The paper makes a moderate contribution to the field of engineering education.	The paper makes a minimal contribution to the field of engineering education.
	Approach	The approach is novel and/or sophisticated and appropriate for the purpose of the paper and is consistent with the perspective (quantitative, qualitative, mixed, or more specific).	The approach is basic but still appropriate for the purpose of the paper and is consistent with the perspective (quantitative, qualitative, mixed, or more specific).	The approach is inadequate and/or not appropriate for the purpose of the paper.
	Results	Data collection and assessment results are very clear and logical, strongly supporting the goals of the paper.	Data collection and assessment results are somewhat clear and logical, moderately supporting the goals of the paper.	Data collection and assessment results need improvement.
Focus	Goals	The goals are strongly developed and explicitly stated	The goals are not fully developed and/or stated.	The goals are not developed and/or stated.
	Order	The order in which ideas are presented is explicitly and consistently clear, logical and effective.	The order in which ideas are presented is reasonably clear, logical and effective but could be improved.	There is little apparent structure to the flow of ideas, causing confusion.
	Conclusions	The conclusions are very well formulated and are strongly supported by the data.	The conclusions are moderately effective and are only partially supported by the data.	The conclusions are minimally effective and do not appear to be supported by the data.
Language	Style	The paper is clear, concise, and consistent. It is easily understandable and a pleasure to read.	The paper is mostly understandable, with occasional inconsistencies that could be improved.	Multiple sections of the paper are difficult to read/understand. The paper could be better structured or more clearly explained.
	Mechanics	The writing is near perfect, with little to no grammar or spelling errors.	Minor grammar or spelling errors are present but do not detract from the content. Content is clear	Some grammar or spelling errors are significant and detract from the content. The paper requires further editing.

Session Chairs and Session moderators

We gratefully acknowledge the contribution of the session chairs and moderators, who were critical to the success of the paper presentation sessions.

Session No	Session Chairs	Session moderator
1	Prof. Johannes Strobel	Mrs. Jayanti Shinge
2	Prof. Maartje Van den Bogaard	Ms. Christina J Rebello
3	Prof. Shannon Massie Chance	Ms. Vishakha M
4	Dr. Rohit Kandakatla	Mr Naveenkumar Aigol
5	Dr. Soumya Narayanan	Mrs Jayanti Shinge
6	Dr. Helen Inglis	Ms. Christina J Rebello
7	Prof. Adam Carberry	Ms. Vishakha M
8	Prof. Sivakumar Krishnan	Mr Naveenkumar Aigol

Guidelines for paper presentation

Presentation

1. The time for the author's presentation is 05 minutes. The authors are free to use their own template for presentation.
2. The authors shall bring a one-page handout (15-20 copies) on their study, ending with 1 or 2 questions for the audience. The handouts are given to authors at the start of the presentations. The handout contains the research objectives, design, and findings.
3. The presentations will be held back-to-back.
4. The participants can jot down questions on the handout itself.

Post presentation

1. The participants will be split into two groups, with 2/3 authors in each group.
2. The author will raise their questions first, followed by the participant's questions.
3. The authors will then be swapped between the groups.
4. The authors/participants will share key takeaways from the small group discussion.

Time split up for a Session of 90 minutes

1. Paper presentations (5 * 05) = 25 minutes
2. Post-presentation discussions = 30 minutes3.Share insights = 20 minutes
4. Buffer 15 minutes

Duncan Fraser Award Winner

KLE Technological University | Creating Value Leveraging Knowledge

REEN | Indo Universal Collaboration for Engineering Education IUCEE FOUNDATION

Research in Engineering Education Symposium-2024
Jan 4-6, 2024
KLE Technological University, Hubballi, India

DUNCAN FRASER AWARD

SPONSORED BY KLE TECHNOLOGICAL UNIVERSITY

2024 AWARD WINNER

Lyndsay Ruane

Ph.D. Scholar, University of Colorado Boulder
Title of the paper : Using PALAR to Formalize Informal Education

Duncan Fraser Award Certificate

KLE Technological University

Indo Universal Collaboration for Engineering Education
IUCEE FOUNDATION

Research in Engineering Education Symposium - 2024

Jan 4-6, 2024 - Hubballi, India

Duncan Fraser award

(Sponsored by KLE Technological University)

The award honors the memory of Professor Duncan Fraser, a leader in engineering education research and an inspiration to many young educators and researchers.

This is to certify that **Lyndsay Ruane**
from the **University of Colorado Boulder**
has presented a paper titled **Using PALAR to Formalize Informal Education**
co-authored by **Hannah Sanders, Laura MacDonald, Jessica Rush Leeker**
has been awarded the Duncan Fraser Award.

This award recognizes 'emerging scholars' and is presented to the student-author whose work focuses on equity, inclusion, student engagement, and/or curriculum development.

Dr. Ashok Shettar
Vice Chancellor,
KLE Technological University, Hubballi, India

Workshops and Workshop Presenters

1. Failing Up: Opportunities for Growth in Inter-institutional Collaborations

About Workshop:

PALAR, an acronym for participatory action learning and action research, pioneers a dynamic fusion of theories, creating a comprehensive structure for community engagement beyond the ordinary. This workshop is not just about theories; it's a hands-on exploration that empowers participants to wield PALAR as a versatile philosophy, a robust methodology, a theory of learning, and a facilitation process for community engagement. We'll share real-world insights gained from a PALAR-based model developed for an exciting undergraduate-focused community engagement project. Join us as we unlock the potential of PALAR and equip you with the skills to drive transformative community engagement.

About Presenters

Presenter Name: Dr. Jessica Rush Leeker

Designation: Professor, Engineering Management Program

Affiliation: University of Colorado Boulder

When expanding her impact within her community— and encouraging others to do the same— there is no shortage of inspiration and devotion within Dr. Jessica Rush Leeker. Fueled by her desire to deepen her knowledge and understand how she can leave a lasting impact on the world and the people around her, Dr. Rush Leeker has cultivated a rich educational background. Equipped with her undergraduate degree in Supply Chain and Information Systems from Penn University and her Ph.D. in Engineering Education and MBA in Sustainability and Operations from Purdue University, she is proud to currently share her expertise as an Engineering Professor at CU Boulder. As the founder of RL Strategies, Dr. Rush Leeker supports organizations in project areas such as goal setting and employee development and DEI-centered workshops and coaching. By applying her extensive educational and professional experience to address each business's unique needs, she strives to devise long-term, customizable corporate strategies for creating diversity, equity, and inclusion for all. Through her coaching business, Dr. Rush Leeker has worked with several successful companies and private coaching clients looking to improve their business practices to find professional and personal growth. Today's youth will soon be tomorrow's leaders. As a children's book author and Yoga teacher, Dr. Rush Leeker hopes to mentor the next generation as she encourages students of all ages to explore their creativity, always choose kindness, and be brave enough to pursue their passions.

Presenter Name: Dr. Laura MacDonald

Designation: Managing Director, Mortenso Center in Global Engineering & Resilience

Affiliation: University of Colorado Boulder

Dr. Laura MacDonald is the Managing Director of the Mortenson Center in Global Engineering & Resilience and the Global Engineering Residential Academic Program. She holds a Ph.D. in Geography and Environmental Engineering from Johns Hopkins University and a BS in Environmental Engineering from Northwestern University. As Managing Director, Dr. MacDonald manages the Mortenson Center's large-scale research program in Rwanda and supports the Mortenson Center's broader research portfolio in sub-Saharan Africa and the western US. Dr. MacDonald also provides administrative, programming, and financial oversight of the Mortenson Center's graduate and undergraduate education programs. She is currently a CU Dialogues Faculty Fel is pursuing a micro-credential in Just and Equitable Teaching from CU Boulder.

Presenter Name: Lyndsay Ruane

Designation: PhD Student, Aerospace Engineering Department

Affiliation: University of Colorado Boulder

Lyndsay Ruane is a PhD student in aerospace engineering at the University of Colorado Boulder. She completed her bachelor's degree in physics in 2019 at Hastings College in Hastings, NE. From there, she completed a master's degree in aerospace engineering at CU Boulder, focusing on astrodynamics and satellite navigation. This research focused on position determination with limited signal access and secure verifiability of those signals received. Lyndsay continues to study aerospace at CU Boulder, centering her research on diversity, accessibility, and retention in aerospace engineering education. Her current research project partners CU Boulder with an HBCU in the southern United States, combining qualitative and quantitative methods and bringing diverse inspiration and knowledge-sharing opportunities to her work.

2. Optimizing Engineering Student Learning Outcomes: Proficiency Alignment with Assessment Questions Difficulty Index

About Workshop:

In the realm of education, assessment serves as a pivotal tool to gauge the progress of learning. This workshop delves into the intricate world of assessment, exploring its multifaceted role in evaluating students; performance and informing teaching strategies. Emphasizing the significance of Rasch model psychometrics techniques, the workshop aims to enhance the quality of formative and summative assessments, aligning them with a learning-oriented perspective. This hands-on workshop seeks to empower participants with the knowledge and skills to refine the quality of test items in both formative and summative assessments by applying Rasch model psychometrics techniques. The presenter will elucidate Rasch's concepts, demonstrating simple dichotomous and polytomous analyses applied to typical assessments. By the end of the workshop, participants will possess the capability to optimize formative or summative assessments. This involves analyzing assessments to ascertain if students; performance aligns with the expectations of the paper setter and investigating the uniform distribution of test items within an assessment. The workshop will feature a hands-on session, guiding participants to construct data files of students' scores for dichotomous and polytomous analysis using the Rasch model. Practical application will be facilitated through the open-access Ministeps software. Participants must bring a laptop with internet access or access to a computer lab with internet connectivity.

About Presenters

Presenter Name: Dr. Rashpinder Kaur
Designation: Assistant Professor, Department of Electronics and Communication Engineering
Affiliation: Chitkara University, Punjab Campus

Dr. Rashpinder Kaur has 15+ years of commendable experience as an engineering educator at the NAAC A+ accredited Chitkara University. She has demonstrated a steadfast commitment to advancing engineering education. Having successfully completed her Ph.D. in engineering education in 2023, she developed a two-tier concept inventory assessment instrument integral to her Ph.D. dissertation titled "QUALITATIVE STUDY OF CONCEPTUAL KNOWLEDGE AND CHARACTERISATION OF EXPERTISM IN ENGINEERING STUDENTS." This instrument drew inspiration from the Signals and System Concept Inventory (SSCI) created by Kathleen E. Wage, Electrical and Computer Engineering Professor at the Volgenau School of Engineering, George Mason University. The outcomes of Dr. Kaur`s dissertation provide valuable insights into effective teaching strategies that foster conceptual understanding. In a leadership role, Dr. Kaur initiated the establishment of the Teaching and Learning Centre (TLC) at Chitkara University, where she serves as the coordinator. She actively organizes various workshops, Faculty Development Programs (FDPs), and symposia focused on improving engineering educators' and students' teaching and learning processes. Beyond her administrative and research contributions, Dr. Kaur has authored and co-authored 22 research articles in SCI/SCIE/Scopus-indexed journals and international conferences. Her substantial citation index of 199, as per Google Scholar, attests to the impact of her scholarly work. Dr. Kaur`s expertise extends beyond engineering education to signal processing and communication systems and proficiency in MATLAB and LABVIEW programming.

3. Applying for Research Grants and Fellowships

About Workshop:

It is often tricky for engineering teachers in higher education to find support and secure resources needed to learn engineering education research methods, build networks, and conduct studies. This workshop will help participants identify and envision possibilities for securing grants and fellowships at the international/global level. This interactive workshop will focus on participant interaction and learning, and participants will be asked to bring wifi-enabled digital devices with them to permit online searching during focused periods of the workshop. At the end of this workshop, participants will understand and be able to explain (1) basic processes for obtaining grants and fellowships; (2) what fellowship programs exist to facilitate study in the US and Europe, where EER is well established; (3) where to find databases listing research jobs and fellowships; (4) how to identify funders' objectives; (5) how to contact prospective hosts; and (6) how to locate helpful resources to support grant writing. The workshop's target audience is people interested in traveling abroad to do EER. They need to be interested in EER and curious about visiting other parts of the world.

About Presenters

Presenter Name: Dr. Shannon Chance
Designation: Lecturer & Program Chair, School of Architecture, Building & Environment
Affiliation: Technological University Dublin

Professor Shannon Chance advocates for excellence in engineering education and has a notable track record of securing international research funding, including prestigious Fulbright and Marie Curie fellowships. With her cross-disciplinary expertise spanning architecture and teaching, she offers workshop participants practical insights into obtaining research grants and fellowships. Her work reflects a deep commitment to global educational practices and the advancement of research in engineering education. Shannon's session, "Applying for International Research Grants and Fellowships," is designed to empower engineering educators with the knowledge to secure funding and support research that spans continents and cultures. Attend this workshop to discover strategies that align with funders' goals and enhance your capacity to contribute to engineering education's global dialogue.

4. Finding and developing true personal stories in engineering

About Workshop:

Student persistence in engineering programs is linked to students' sense of belonging and identification with their major. Lack of professional identification and lack of belonging exacerbate the departures of underrepresented minority students from engineering programs, for whom exclusionary department cultures and biased policies have amplified the impact. This workshop provides an overview of the science and craft of storytelling developed with The Story Collider, an international, national non-profit focused on storytelling in the sciences. We will demonstrate how a storytelling model can be integrated into an engineering classroom.

About Presenters

Presenter Name: Dr. Krishna Pakala
Designation: Associate Professor, Department of Mechanical & Biomedical Engineering
Affiliation: Boise State University

Krishna Pakala, PhD, is an Associate Professor in the Department of Mechanical and Biomedical Engineering at Boise State University (Boise, Idaho). He was the director of the Industrial Assessment Center at Boise State University. He was the Faculty in Residence for the Engineering and Innovation Living Learning Community (2014 - 2021). He was the inaugural faculty associate for mobile learning and the faculty associate for accessibility and universal design for learning. He received the Foundation Excellence Award, the David S.

Taylor Service to Students Award, and the Thelden Apple Award from Boise State University. He also received the 2023 National Outstanding Teacher Award, the ASEE PNW Outstanding Teaching Award, the ASEE Mechanical Engineering division's Outstanding New Educator Award, and several course design awards. He serves as the campus representative and past chair for the ASEE PNW Section. His academic research interests include innovative teaching and learning strategies, emerging technologies, and mobile teaching and learning strategies.

Presenter Name: Dr. Anne Hamby
Designation: Associate Professor, Department of Marketing
Affiliation: Boise State University

Anne Hamby, PhD, is an Associate Professor in the Department of Marketing in the College of Business at Boise State University (Boise, Idaho). Her research focus is in the area of consumer psychology. Specifically, she studies how emotional and structural aspects of stories engage their audiences and how engagement in stories influences beliefs and behavior in a consumer-related context. She is also interested in issues related to consumer well-being and examines the psychological, social, and cultural factors that influence risky consumption practices and prosocial behavior. She received the College of Business and Economics Advisory Council Research Excellence award in 2023 and is the College of Business and Economics Distinguished Research Professor for 2023-2025. Her work has been published in top-tier journals such as the Journal of Marketing Research, the Journal of the Academy of Marketing Science, and the Journal of Consumer Psychology. She is an associate editor at the Journal of Advertising.

Presenter Name: Dr. Eric Jankowski
Designation: Associate Professor, Micron School of Materials Science & Engineering.
Affiliation: Boise State University

Dr. Eric Jankowski is an Associate Professor in the Micron School of Materials Science and Engineering at Boise State University. He received his Ph.D. in Chemical Engineering from the University of Michigan, was a Director's Fellow at the National Renewable Energy Laboratory, and is an NSF Career and ASEE Young Pacesetter awardee. In addition to advancing scientific software and training around molecular simulations for organic materials through his research programs, he serves as Board President for The Story Collider. Their curricular collaboration was recognized this year with a Silver Medal for Diversity, Equity, and Inclusivity Partnerships by the Anthem Awards.

Presenter Name: Dr. Sara Hagenah
Designation: Associate Professor, Micron School of Materials Science & Engineering.
Affiliation: Boise State University

Sara Hagenah, Ph.D., is an Associate Professor in the Department of Curriculum, Instruction, and Foundational Studies at Boise State University. Her research deeply engages with informal and formal P-20 school-community partnerships and aims to advance equitable science teaching and learning collaboratively. She has expertise in designing P-20 STEM curricula, identity research, and qualitative research methodologies. She designs, leads, and studies job-embedded professional development, focusing on rigorous and responsive teaching and learning opportunities.

5. Leveraging Mobile Devices to Improve Engineering Learning Thermal Fluids: Ways to Measure the Impact

About Workshop:

Mobile technology, when purposefully integrated into teaching, can impact student engagement, enhancement, and extension of learning. In this workshop, participants will engage in reflection and small group peer discussions to identify how mobile devices (smartphones, tablets) can enhance the teaching and learning of engineering content. Participants will identify ways in which the impact of mobile device use on the student's learning experience can be measured using qualitative and quantitative methods.

About Presenters

Presenter Name: Dr. Krishna Pakala

Designation: Associate Professor, Department of Mechanical & Biomedical Engineering

Affiliation: Boise State University

Krishna Pakala, Ph.D., is an Associate Professor in the Department of Mechanical and Biomedical Engineering at Boise State University (Boise, Idaho). He was the director of the Industrial Assessment Center at Boise State University. He was the Faculty in Residence for the Engineering and Innovation Living Learning Community (2014 - 2021). He was the inaugural faculty associate for mobile learning and the faculty associate for accessibility and universal design for learning. He received the Foundation Excellence Award, the David S. Taylor Service to Students Award, and the Golden Apple Award from Boise State University. He also received the 2023 National Outstanding Teacher Award, the ASEE PNW Outstanding Teaching Award, the ASEE Mechanical Engineering division's Outstanding New Educator Award, and several course design awards. He serves as the campus representative and past chair for the ASEE PNW Section. His academic research interests include innovative teaching and learning strategies, emerging technologies, and mobile teaching and learning strategies.

Presenter Name: Dr. Devshikha Bose

Designation: Senior Educational Development Specialist, Center for Teaching and Learning

Affiliation: Boise State University

Devshikha Bose is a senior educational development consultant at Boise State University, USA. Her expertise includes the scholarship of teaching and learning, the integration of technology in teaching and learning, academic development workshops, learning community facilitation, and grant implementation support.

Presenter Name: Dr. Diana Bairaktarova

Designation: Associate Professor, Department of Engineering Education

Affiliation: Virginia Polytechnic Institute and State University

Diana Bairaktarova is an associate professor in the Department of Engineering Education, an affiliate faculty member in the Department of Mechanical Engineering, and the director of the Abilities, Creativity, and Ethics in Design [acedvt.com] Research Lab at Virginia Tech. Her current research projects investigate how aptitudes and abilities, interest, and manipulation of physical and virtual objects influence learning and performance in engineering. Diana was awarded the 2020 Virginia Tech XCaliber Award for extraordinary contributions to the innovative use of technology to improve student learning.

6. Methodological Requirements for Engineering Education

About Workshop:

This workshop offers a comprehensive learning experience for three distinct groups within the engineering community. Aimed at (1) engineering faculty seeking to integrate engineering education research into their repertoire, (2) those collaborating with education faculty on education-centric studies, and (3) those utilizing social science methods like human-centric design, the workshop equips participants with valuable skills. Throughout the workshop, participants will gain the ability to navigate the entire process of designing an educational research study, from conceptualization to reporting. They will learn to discern the most suitable research methodologies for their specific research questions and distinguish between various qualitative research frameworks and statistical tests. Structured in two parts, the workshop begins with an in-depth introduction to educational research, methodologies, and research study design. Small group activities enhance understanding. The second part involves hands-on application, with facilitators guiding participants in applying the concepts to their unique research questions through individualized support. Recognizing the importance of sustained engagement, workshop facilitators will transition all materials to an online platform. Participants are encouraged to extend their planning and research in this digital space for an additional two weeks—an optional yet valuable opportunity for continued collaboration and learning beyond the physical workshop. In essence, this workshop promises to empower engineering faculty with the knowledge and practical skills needed to make meaningful contributions to the field of engineering education research.

About Presenters

Presenter Name:Dr. Johannes Strobel

Designation: Associate Dean of Research & Graduate Studies, Full Professor, STEM Education

Affiliation: University of Texas at El Paso

Dr. Johannes Strobel is Associate Dean for Research and Graduate Studies at the University of Texas at El Paso, USA. He joint UTEP from SRI International where he was Director, Policy Research, Center for Innovation Strategy and Policy (CISP). For the past 20 years, Dr. Strobel has been actively shaping the K-16 engineering education research landscape in the United States, as exemplified through serving as the founding editor of the Journal of Pre-College Engineering Education Research. His research and teaching focuses on engineering 'habits of mind' and worldviews, empathy and care in engineering and integrated STEM competencies. He published more than 180 papers in proceedings, journals and book chapters, co-edited five books and was on successful STEM/engineering education research grants with over US$35MM. He is the lead author of "Hands-on Standards STEM in Action", a nationally and internationally available set of learning modules for grades preK - 5th grade published by hand2mind.

Presenter Name: Dr. Maartje van den Bogaard

Designation: Associate Professor, Teacher Education

Affiliation: University of Texas at El Paso

Dr. Van den Bogaard is an associate professor in STEM education at the University of Texas at El Paso. Prior to coming to UTEP Dr. Van den Bogaard was appointed tenured assistant professor in STEM Education and Head of Program for the Delft University of Technology (TU Delft) Science Education and Communication program where she was successful in doubling enrollment in the teacher training program and led the program's accreditation renewal. Under her leadership, the program was voted best master's program at TU Delft in a national student survey (2020). She earned a PhD in Technology, Policy and Management from TU Delft for which she received the Outstanding Dissertation Award by the International Society for Educational Planning. She holds an MSc in Education Sciences from the University of Groningen – both in the Netherlands. Dr. Van den Bogaard's research focuses on student success, complexity and curriculum innovation, with a particular interest in research methodology, and Discipline-Based Education Research (DBER). She is an associate editor for the European Journal of Engineering Education, the International Journal of Design and Technology Education, and Springer Nature Social Sciences. Dr. Van den Bogaard was inducted as a Fellow in the European Society of Engineering Education in the fall of 2023.

7. Tools and Strategies for Disseminating Knowledge via the Scholarship of Teaching and Learning

About Workshop:

Many disciplinary engineering Ph.D. programs (e.g., non-engineering Education programs) do not adequately prepare graduates for teaching and/or disseminating best teaching practices via the scholarship of teaching and learning (SOTL). As we look to broaden participation in engineering, the limited scholarly training of disciplinary engineering educators causes a potential challenge for implementing effective pedagogical approaches critical for improving representation among minoritized and underrepresented populations. This workshop aims to provide participants with tools and strategies to implement innovative curricula in the engineering classroom and share lessons learned via SoTL. By the end of the workshop, participants will learn how to apply six different tools and strategies as it relate to curriculum development, assessment, and SoTL knowledge dissemination. Participants will receive six mini-lectures (and associated active learning assignments) summarizing specific tools and strategies related to curriculum development, curriculum assessment, and the dissemination of knowledge via SoTL.

About Presenters

Presenter Name:Dr. Lisa Bosman

Designation: Associate Professor

Affiliation: Purdue University

Dr. Lisa Bosman, PhD in Industrial Engineering, is an Associate Professor at Purdue University in West Lafayette, IN, USA. She is the founding director of iAGREE Labs (Inclusive, Applied, and Grounded Research in Entrepreneurially-Minded Education), which aims to empower action through real-world solutions and evidence-based practices. While working at Purdue University (2018- present), Dr. Bosman was awarded 15 grants totaling $1,166,589. Her research centers around three themes related to developing the entrepreneurial mindset in future leaders and innovators: (1) Teaching and curriculum development; (2) educator Professional Development, and (3) Real-World applied learning (Academic Entrepreneurship Research). However, she consistently employs the "20% time rule", a concept popularized by Google, in which employees spend about 20% of their time working on collaborative projects outside their core research area. Dr. Bosman has disseminated research through more than 75 peer-reviewed publications. In addition, she has published two teaching-oriented books: Teaching the Entrepreneurial Mindset to Engineers (Springer, 2018) and Teaching the Entrepreneurial Mindset Across the University – An Integrative Approach (Springer, 2021). In addition, she has been a workshop facilitator and invited speaker for over 20 international engagements most recently including Israel, Colombia, South Africa, Sweden, Germany, Indonesia, Tanzania, Lebanon, Palestine, South Korea, and Portugal.

8. Writing an engineering education research paper

About Workshop:

This workshop, titled "Writing an Engineering Education Research (EER) Paper," is designed for participants actively engaged in EER but seeking to enhance their proficiency in crafting impactful research papers. Aimed at individuals with some prior experience in the field, the workshop is structured to improve existing papers within the realm of engineering education research or the scholarship of teaching and learning.

The first hour of the workshop will feature an interactive lecture focusing on the mechanics of writing EER papers. Emphasis will be placed on aligning research questions with appropriate methodologies and data analysis techniques. Participants will gain insights into structuring their papers effectively, ensuring clarity, coherence, and methodological rigor.

The second hour will be dedicated to hands-on, collaborative activities. Participants will engage in peer-review sessions, offering constructive feedback on each other's existing papers. This interactive component aims to cultivate a supportive environment where individuals can draw upon collective expertise to enhance the quality and impact of their work. Through this collaborative approach, participants will refine their individual projects and develop a deeper understanding of effective strategies for presenting research in the domain of engineering education.

Overall, this workshop provides a comprehensive and practical opportunity for EER practitioners to refine their writing skills, foster collaboration within the community, and contribute to advancing knowledge in engineering education research and the scholarship of teaching and learning.

About Presenters

Presenter Name: Dr. Sohum Sohoni

Designation: Professor and Program Director, Software Engineering, Department of Electrical Engineering and Computer Science

Affiliation: Milwaukee School of Engineering

Dr. Sohum Sohoni is a professor and program director of software engineering in the Department of Electrical Engineering and Computer Science at the Milwaukee School of Engineering. Before joining MSOE, he was a faculty at Arizona State University and Oklahoma State University. His research is in Computer Engineering and Engineering Education. He is interested in the adoption and translation of effective teaching techniques and has established a vision for networks of discipline-based education research (DBER) ambassadors to facilitate the adoption and evaluation of research-based instructional strategies. A computer architecture visualization platform that he and his students designed has been used by over 1000 students at three universities in the US.

He has published over 40 peer-reviewed papers in journals and conferences, including ACM SIGMETRICS, IEEE Transactions on Computers, the International Journal of Engineering Education, and Advances in Engineering Education. He has received the best paper award for his work in computer engineering from the IETE Technical Review journal and three best paper awards for his work in engineering education from the American Society for Engineering Education (ASEE). He is a popular and well-respected instructor and has received many teaching awards, including the Regents Distinguished Teaching Award in 2010 at Oklahoma State University.

Dr. Sohoni is a member of ACM (Association for Computing Machinery) and ASEE (American Society for Engineering Education) and a senior member of IEEE (Institute of Electrical and Electronics Engineers). He has reviewed for various journals and conferences and served as session chair at many conferences. He is a member of the board of the Indo-Universal Collaboration for Engineering Education (IUCEE) and the Research in Engineering Network (REEN). He serves as the Co-Editor-in-Chief of the Journal of Engineering Education Transformations. He also serves as a program evaluator for the Accreditation Board for Engineering and Technology (ABET).

Presenter Name:Dr. Prathiba Nagabhushan

Designation: Faculty of Education and Arts, School of Education

Affiliation: Australian Catholic University

Dr. Prathiba Nagabhushan, an accomplished educator and psychologist, pursued her undergraduate and postgraduate studies in Education in Bangalore before dedicating seventeen years to a distinguished career at a secondary school in the same city. Relocating to Australia with her husband for their daughter's higher education, she seized the opportunity to embark on a new academic journey. Dr. Nagabhushan successfully completed her PhD in Educational Psychology at the Australian National University in 2014. Currently holding positions at St. Mary MacKillop College and the Australian Catholic University, she imparts knowledge in Psychology and Educational Psychology to postgraduate students. Dr. Nagabhushan's association with IUCEE since 2014 has been marked by significant contributions. As an integral member, she instructs the IIEECP certificate program, conducts workshops in engineering education research, and reviews articles for JEET. Her specialization in motivation and engagement, coupled with a focus on foundational concepts, has left a lasting impact on students across various age groups. Her research has identified 11 factors crucial in learning situations, shaping educational practices in Engineering and Medical courses.

Presenter Name: Dr. Helen Inglis

Designation: Engineering Education Specialist, Faculty of Engineering & the Built Environment

Affiliation: University of Johannesburg

Dr Helen Inglis is an engineering education researcher with a focus on student success. Helen is a Senior Lecturer in the Department of Mechanical and Aeronautical Engineering at the University of Pretoria in South Africa. Her masters and doctoral degrees, from the University of Illinois at Urbana-Champaign, were in the field of solid mechanics, but her passion for teaching and for social justice has led her to grow her research activities in engineering education since joining the University of Pretoria. Helen is part of the organising team for a South African project to develop innovative engineering curricula, and is a REEN board member.

Presenter Name: Dr. Teresa Hattingh

Designation: Senior Lecturer

Affiliation: University of Pretoria

Teresa Hattingh is an Associate Professor and Engineering Education specialist at the University of Johannesburg in South Africa. In this role, she provides support to enhance student access and success and build scholarly approaches to teaching and learning in the Faculty. Teresa completed her Bachelors and Master's degrees in engineering and her PhD in Engineering Education. She is registered as a professional engineer and has extensive teaching experience in higher education and industry experience in project engineering, operations and supply chain management in the metallurgical and gas industries. She is actively involved in engineering education research and is the current chair of the Global Research in Engineering Education (REEN) board and is the current President of the Southern African Institute for Industrial Engineering.

TABLE OF CONTENTS

Engineering Education Research Collaborations: Use of Futures, Values, Systems, and Strategic Thinking

Medha Dalal[a] and Adam Carberry[b]
Arizona State University[a], The Ohio State University[b]
Corresponding Author Email: medha.dalal@asu.edu

Abstract

Context

Countries, like the United States, invest millions of taxpayers' dollars to support engineering education research (EER). Many calls for research look to bring novel ways of thinking through interdisciplinary collaborations between engineering researchers and social scientists. However, the use of different ways of thinking is often implicit or taken for granted.

Purpose or Goal

The purpose of this research study is to better understand how ways of thinking are applied in collaborative EER projects. The following research question is explored: *In what ways do collaborating engineering and social sciences researchers use futures, values, systems, and strategic thinking in their EER projects?*

Methods

A survey was distributed to examine the use of four specific ways of thinking in EER – futures, values, systems, and strategic thinking. The participant sample included awardees of one specific National Science Foundation program that required collaboration with a social scientist with the goal of designing revolutionary novel approaches to engineering education. A sample of 84 researchers were contacted with 48 responses received. The open-ended survey responses were analyzed qualitatively for emergent themes to examine use of ways of thinking in EER.

Outcomes

Results uncovered deeper themes behind researchers' enactments of ways of thinking, such as workforce development, pedagogical innovation, inclusion and social justice, weaving a tapestry, and stakeholder engagement. Results highlight how ways of thinking are enacted in EER and influence engineering education practice in order to drive innovation and transformation in the field.

Conclusion

This study contributes to the broader conversation on transforming engineering education through a ways of thinking lens. These thinking approaches, when integrated and applied purposefully, empower stakeholders to anticipate, address, and transcend the complex challenges facing the field, ultimately advancing engineering education in response to the evolving needs of society.

Keywords—ways of thinking, interdisciplinary collaboration, transformation.

I. INTRODUCTION

A well-established body of literature shows the benefits of interdisciplinary collaborations between engineering and social sciences researchers for the improvement of education in engineering colleges (Carr et al., 2017; McKenna et al., 2009; Olds et al., 2005). Such collaborative research typically involves drawing on theories and research methods from learning sciences, instructional design, or educational psychology and applying them to the teaching, learning, and other related activities within engineering education and research. Collaborating researchers share their domain-specific knowledge and skills, engage in meaning-making, evaluate multiple perspectives, and work together to solve the problems (Borrego & Newswander, 2008; Dalal et al., 2017).

The underlying notion behind such collaborations is to foster innovation in the engineering education system. The United States invests millions of taxpayers' funds in engineering education research (EER) via National Science Foundation (NSF), with the goal that resulting research will lead to improved engineering education. Many NSF calls require an interdisciplinary collaboration between engineering faculty and social scientists to bring novel ways of thinking about educational research in the engineering domain (NSF, 2017; Wankat et al., 2002).

Adopting new ways of thinking is seen as one necessary means to bring about change and inform the existing practices within the global engineering ecosystem (ASEE, 2014; NSF, 2017). A necessary first step is to better understand what ways of thinking are currently used in EER. Numerous activities associated with EER collaborations are not well documented. These include problem solving approaches, ways of thinking, vision, values, and strategies toward transformation of the field. The use of different ways of thinking is often implicit or taken for granted. Recent publications have brought this issue to the forefront, including a proposed framework for applying four specific ways of thinking in EER – futures, values, systems, and strategic thinking (Dalal et al., 2021, 2023).

This study aimed to better understand how ways of thinking are applied in collaborative EER projects. The following

https://doi.org/10.52202/073963-0001

research question is explored: *In what ways do collaborating engineering and social sciences researchers use futures, values, systems, and strategic thinking in their EER projects?* The following sections describe the ways of thinking framework, methods, and results. The results are then discussed in the context of current challenges in EER and potential use in informing future research practices.

II. WAYS OF THINKING FRAMEWORK AND LITERATURE

The term *ways of thinking* is often associated with a systematic thought process (Sousa, 2016). Different ways of thinking facilitate different strategies and subsequent actions to innovate. The definition of ways of thinking used by different fields varies depending on the context. For example, the field of learning sciences considers ways of thinking as an approach to solving complex problems through coherent patterns in reasoning (Harel & Sowder, 2005). Business and finance view ways of thinking as combination of intuition and rules that inform decisions (Douglas, 2000). Sustainability education equates ways of thinking to a lens that addresses complex challenges regarding sustainability literacy (Warren et al., 2014). This study operationalizes ways of thinking as a systematic thought process that informs decision-making to address complex engineering education challenges. It is not a heuristic, but rather an approach used by researchers to think, act, and engage with their research. More specifically, the study is guided by the Framework for Applying Ways of Thinking in Engineering Education Research (FAWTEER), that proposed four ways of thinking including futures, values, systems, and strategic thinking to address complex engineering education challenges (Dalal et al., 2021).

Futures thinking focuses on working to address tomorrow's problems today with anticipatory approaches to understand and prepare for future changes, problems, and solutions (Dalal et al., 2023). Values thinking is about recognizing the concepts of ethics, equity, and social justice (Warren et al., 2014). It involves understanding these concepts in the context of varying cultures and accordingly making decisions. Systems thinking involves considering holistic approaches to problem-solving that understand and analyze the complexity of various elements and their interrelationships in the overall ecosystem (McKenna et al., 2014). Strategic thinking is the ability to create a plan of action to achieve the desired vision and act upon the other ways of thinking (Wiek et al., 2011).

Futures, values, systems, and strategic thinking can be implemented in conjunction with one another or used individually depending on the problem under consideration. When used in a networked fashion, they link topics that may seem disconnected and build capacity problem solving capacity with respect to complex engineering education challenges. This study was designed to better understand how these four ways of thinking are used by engineering education researchers in collaborative EER.

III. METHODS

A. Instrument Development

The survey instrument was developed through iterative construction and validation over a three-month period. The survey included Likert scale items to measure importance of various activities associated with different ways of thinking as well as open-ended items focused on examples of ways of thinking enactments. Discussion of the Likert-scale items and scale results are outside the scope of this study. This study is focused on the analysis and results of the open-ended items for understanding how different ways of thinking are applied in collaborative EER projects.

The open-ended items asked participants the following questions: i) In your [NSF Award Name] project, do you believe you have used futures thinking? [yes, maybe, no] ii) If yes, please describe an example from your [NSF Award Name] project that you believe involved futures thinking. iii) If maybe, please describe an example from your [NSF Award Name] project that you think may have involved futures thinking. iv) If no, why do you think you have not? These questions repeated for values, systems, and strategic thinking. A definition of the specific way of thinking was provided before the questions for clarity. The instrument was validated through expert reviews and think aloud pilot sessions (Dalal & Carberry, 2019).

B. Sample and Participants

The potential survey participants were selected from among awardees listed in the public database on the NSF website (https://www.nsf.gov/awardsearch). The database search was limited to one specific programs within the Division of Engineering Education and Centers that stated a required collaboration with a social scientist and the goal of designing revolutionary novel approaches to engineering education. Listserv created within the program was also used to reach other researchers who may not be listed on the NSF site.

A total population of 84 researchers resulted from these processes who were asked to participate in the survey. We received 48 responses (57% response rate) which included 25% researchers from engineering disciplines, 18% from social sciences, 42% from both including engineering education, and 15% did not disclose their discipline.

C. Data Collection & Analysis

The survey was deployed over a five-week period in October and November 2018 using the Tailored Design Method (Dillman et al., 2014) of web-based surveys for attaining higher response rates. A pre-notification was sent three days ahead of the survey link. Three reminders were sent once a week while the survey was open to increase the response rate.

The scope of this study is limited to the open-ended survey responses which were analyzed qualitatively for thematic analysis. coded inductively by the study team following procedures recommended by Saldaña (2009). First-order coding followed the inductive, open coding method (Corbin & Strauss, 2015). Second-order, axial coding was then used to understand the relationship among the previously identified open codes, informed by FAWTEER and focusing on a specific way of thinking. Finally, open codes, and second-order codes were configured into themes to answer the research question.

IV. RESULTS

Table 1 captures the responses to the survey questions: *In your [NSF Award Name]* project, do you believe you have used futures/vales/systems/strategic thinking? *[no, maybe, yes]*

TABLE I
SUMMARY OF USE OF A SPECIFIC WAY OF THINKING

Use	Futures Thinking	Values Thinking	Systems Thinking	Strategic Thinking
Yes	22	30	31	30
Maybe	13	8	6	5
No	6	4	3	4
N/A	7	6	8	9

Qualitative results are presented for each ways of thinking with relevant themes and illustrative text in a narrative synthesis and Table 2. Participant quotations are embedded in the narrative as evidence and to enhance contextual understanding.

A. Futures Thinking

Futures thinking in EER emerged as a multi-dimensional approach, encompassing themes that drive innovation and transformation in the field. Two major themes included workforce development and pedagogical innovation. Participants seemed acutely aware of the need to equip students with skills and knowledge that will not only serve them well in their current academic pursuits but also make them agile and adaptable professionals for the ever-evolving job market. As one participant aptly put it,

> "All of our innovations are oriented toward producing engineers that are appropriate for the changing social and economic system."

This sentiment underscores the driving force behind the efforts to prepare students as future-ready engineers not only possessing technical expertise but also as individuals who understand the broader societal implications of their work.

The second pivotal theme in the application of futures thinking in EER centered around pedagogical innovation. Participants wrote about reimagining engineering education by incorporating novel approaches to teaching and learning. These approaches extended beyond traditional disciplinary boundaries, encouraging students to think holistically and consider the broader contexts in which their engineering work would take place. Innovative teaching methods such as technical writing in a cross-disciplinary way, active learning, sustainability topics, and the incorporation of future-focused content like fairness in algorithms were mentioned in the survey responses. A few participant responses (n=6) also highlighted the need for engineering faculty professional development for pedagogical innovation. Those who stated not using futures thinking (n=6) indicated that "it was not relevant to the grant" or "Never heard of it before."

TABLE 2
RESULTING THEMES FOR WAYS OF THINKING

	Theme	Illustrative Quote
Futures thinking	Workforce development	"...doing lots of research into trends in the field and what future employment looks like."
	Pedagogical innovation	"We are encouraging faculty participants to think about a future state of pedagogical innovation for their teaching in the classroom."
Values thinking	Diversity, equity, and inclusion	"We are trying to change the department culture to value students who might come into the program with different professional goals than our current "typical" student (or at least, what the faculty consider to be the typical student)."
	Social justice in engineering education	"...revamping the curriculum [...] addressing social justice and empathy as a key factor of design."
Systems thinking	Holistic approach	"Holism is core to our research perspective. Our research questions, data, and publications reflect the engineering school's past-present-future worldview and activities, as well as the context of the engineering school within the broader university setting and academia generally."
	Weaving a tapestry	"We are developing vertical integration of topics...across the curriculum in close collaboration with industry partners. The goal is to change the culture in the field."
Strategic thinking	Project management	"We are always thinking strategically to get the best outcome and optimize our effort. We also think carefully about personnel and how to get the right people to fulfill the right roles."
	Collaboration and stakeholder engagement	"A needs assessment is being conducted that includes the voices from faculty, students, [PROJECT NAME] team, and external stakeholders to identify program strengths and areas for improvement."
	Adaptation and continuous improvement	"We are continuously revisiting and refining our project plan with all team members to ensure that we reach our goals and consider alternative solutions when we meet road blocks."

B. Values Thinking

Two themes of diversity, equity, and inclusion (DEI) and social justice in engineering education emerged from the survey

https://doi.org/10.52202/073963-0001

data on values thinking examples. The majority of the statements covered examples wherein values thinking was enacted in relation to the concepts of DEI. Enactment examples included faculty and project teams actively engaged in creating environments where participants can authentically express themselves and feel valued for their uniqueness. This statement below coneys the emphasis on DEI:

> "Our aim is to create more inclusive learning and work environments where participants feel both connected and are valued for being their authentic self."

This commitment also extended to students, with a strong emphasis on ensuring that the learning experience was inclusive and considered diversity and equity as evident from the following statement,

> "We have extensively discussed what the values of our department are, how to best serve all of our students, with equity in mind, not just equality. We are mindful of the different cultures present on our campus and are working to create a feeling of inclusiveness in all of our students."

Another aspect of values thinking focused on social justice in engineering education. This emergent theme illuminated that for the participants the curricular focus was not solely on technical knowledge but also on fostering graduates who are socially conscious, responsible global citizens. One participant highlighted this by writing,

> "Our project explicitly includes finding ways to include discussions of social justice in engineering classes. That meant that we talked about the scope of the topics that we thought could be included and the kinds of issues we might like to see addressed through values thinking."

Overall, the values thinking examples underscore the importance of not only technical knowledge but also the broader societal and ethical dimensions of engineering education. Those who stated not using values thinking (n=4) indicated that it was not part of the scope of what they were trying to accomplish.

C. Systems Thinking

Two themes of holistic approach and weaving a tapestry emerged from the survey data on systems thinking examples. Participants emphasized the importance of taking a holistic approach to engineering education reform. This involves considering the entire ecosystem of engineering education, including curriculum, faculty, students, and the broader institutional context as evident from this statement:

> "Our approach to achieving the goals of the project is holistic and multipronged - for example, we provide direct support to students, integrate new content in classes, provide faculty development for inclusive pedagogy, partner with other colleges to leverage expertise, establish seed funding grants to bring in more faculty, and are developing a certificate program that counts toward T&P."

Several participants (n=9) specifically mentioned the application of systems thinking in the redesign of engineering curricula. They emphasized the importance of considering how changes in curriculum affect students, faculty, departments, colleges, and the university as a whole. One participant noted:

> "During new curriculum development, we considered the impact on the students, department, college, university, and we involved faculty, students, staff, and faculty from other departments in the development."

Participants also emphasized the value of working with different departments, faculty members, and external industry partners to weave a tapestry and bring synergy and alignment in the engineering education initiatives using systems thinking. One participant mentioned,

> "We are developing a new degree (BA in CS with a minor in education) that has resulted from a systems thinking approach and involvement of folks: teachers, HE educators, researchers, non-profits, industry..."

Multiple participants (n=7) mentioned considering institutional context and goals and aligning their research goals accordingly to bring synergy and potentially greater impact. Overall, the examples of systems thinking highlighted the interconnectedness of different elements within the system and the need to address multiple aspects of engineering education simultaneously to bring about comprehensive change. Those who stated not using systems thinking (n=3) indicated that the work was just starting, and they would know better about systems thinking later.

D. Strategic Thinking

Three themes of project management, collaboration and stakeholder engagement, and adaptation and continuous improvement emerged from the survey data on strategic thinking examples. Project management was a prevalent theme in the responses highlighting the importance of careful planning and strategic resource allocation to achieve project goals. Participants described the need to set clear objectives, develop timelines, and allocate resources effectively to overcome project implementation challenges. A few statements below capture the emphasis on planning with strategic thinking:

> "We have made attempts to set goals and timelines...we need to do more of this to be optimally effective."
>
> "We created a logic model...planned how to intervene in our curriculum...creating a communication plan...to carry out the planned work."
>
> "Curriculum and program changes that affect and depend upon multiple factor such as human resources, funding, space and lab resources ..."

One participant mentioned teaching project planning to students by embedding strategic thinking in the curriculum.

Collaboration and engagement with stakeholders, both within and outside the department or institution, were highlighted as key aspects of strategic thinking. Respondents stressed the importance of involving key offices, faculty, staff, and students strategically to garner support and drive change:

> "We have worked hard to ensure that the project is not viewed as a disciplinary endeavour...members were

https://doi.org/10.52202/073963-0001

chosen strategically to ensure participation by key offices."

This also involved a needs assessment involving all stakeholders as evident from the following quotation: "A needs assessment is being conducted...to identify program strengths and areas for improvement."

Collaboration and engagement with stakeholders also covered thinking strategically about scaling up and sustaining the impact in the future. While this theme is similar to the theme of weaving a tapestry under systems thinking, there are nuances. Systems thinking about collaborations with industry and other departments was about bringing in alignment and synergy in the engineering education efforts, whereas strategically thinking about collaborations was more about long term sustainability of initiatives that aim to transform engineering education.

Participants also emphasized the need for adaptability and continuous improvement in strategic thinking. They mentioned the importance of revisiting and refining project plans, monitoring progress, and adjusting strategies based on changing circumstances and stakeholder feedback:

"We are always thinking strategically...carefully about personnel (and personalities) and how to get the right people to fulfill the right roles."

"Looking back at goals in proposal and checking to see where we are...checking in with stakeholders' views vs. our own."

A few participants (n=4) mentioned that strategic thinking encapsulated everything as "This is central to our five-year project that we hope will lead to permanent change." Overall, the survey responses highlighted the application of strategic thinking in EER through planning and resource allocation, collaboration and stakeholder engagement, and a focus on adaptability and continuous improvement. Some of those who stated not using strategic thinking (n=4) stated: "While [we] see the importance of strategic thinking, we have not applied this approach well."

V. DISCUSSION

In the realm of (EER), the adoption of multifaceted thinking approaches plays a crucial role in addressing complex challenges and fostering innovation (JEE 2006; Dalal et al., 2023). Four distinctive ways of thinking—futures, values, systems, and strategic thinking—have been identified as integral components of EER, each offering unique perspectives and methodologies (Dalal et al., 2021). This discussion explores the emergent themes within each of these thinking approaches and underscores their interplay in the pursuit of advancing engineering education.

Futures thinking in EER encourages scholars to explore uncharted territories, anticipate evolving trends, and embrace uncertainty (Warren et al., 2014). The emergent themes underscore the importance of preparing the future workforce and innovation. Participants in our survey emphasized the significance of staying attuned to technological and social advancements and emerging pedagogical paradigms to prepare future-ready engineers. Futures thinking in EER embodies the spirit of proactivity, acknowledging that the engineering landscape is constantly evolving (Dalal & Carberry, 2018). By incorporating this perspective, educators and researchers can proactively anticipate shifts in the engineering field, align curricula with emerging needs, and prepare students to thrive in a rapidly changing environment.

Findings for Values thinking illuminate the moral compass that guides EER endeavors. Participants in our survey emphasized the importance of embedding shared values, such as inclusivity, social justice, and diversity, into the fabric of engineering education. These values often serve as guiding principles for curriculum design, faculty development, and decision-making processes. The emergent themes within values thinking underscore the commitment of educational stakeholders to create inclusive and equitable learning environments. By prioritizing values such as social justice and DEI, engineering education can become a more holistic and empathetic endeavor, instilling these principles in the next generation of engineers (Leydens & Lucena, 2017; Swan et al., 2014).

Systems thinking in EER fosters a holistic understanding of educational ecosystems (McKenna et al., 2014). Participants stressed the importance of recognizing the interdependencies between various components of the engineering education system. Participants also recognized the interconnectedness of educational institutions and stakeholders. Systems thinking encourages a shift from isolated problem-solving to systemic change, facilitating a more comprehensive approach to addressing persistent challenges. By considering the broader context in which engineering education operates, researchers and educators can develop interventions that create cascading effects and leverage sub-systems to drive change.

Strategic thinking in EER emphasizes deliberate planning, resource allocation, and stakeholder engagement (Wiek et al., 2011). Survey participants stressed the importance of setting clear objectives, developing timelines, and collaborating strategically with key offices, faculty, staff, and students. The emergent themes within strategic thinking underscore the need for structured planning and the flexibility to adjust strategies in response to changing circumstances. By adopting a strategic approach, researchers and educators can enhance the efficiency and efficacy of interventions aimed at improving engineering education. Strategic thinking also extends to sustainability planning, ensuring that the impact of initiatives endures beyond the project's duration.

While each of these thinking approaches—futures, values, systems, and strategic thinking—bring its unique strengths to the field of engineering education research, they are not mutually exclusive. Instead, they complement and enrich one another. For instance, values thinking informs the ethical underpinnings of strategic planning, ensuring that educational strategies align with overarching principles of inclusivity and social justice. Systems thinking, on the other hand, aids in the

https://doi.org/10.52202/073963-0001

identification of strategic partners and key stakeholders whose collaboration is essential for implementing meaningful change.

Our purpose in sharing these results, particularly from one specific program, was to show how cross-disciplinary partnerships and ways of thinking could cross-fertilize ideas that work broadly to bring a cultural change in engineering education. Prior studies have concluded that when professionals from different disciplines come together for a common goal, they often deconstruct traditional disciplinary ways of thinking, change their beliefs, values, and attitudes, and "assimilate new ways of thinking into new approaches to practice" (Borrego & Newswander, 2008; Frodeman et al., 2010; McCallin, 2004, p. 38). The need for re-conceptualizing how we think about engineering education necessitates research that identifies novel ways of thinking and how they are applied. It was also surprising to see statements that indicated not using strategic or values thinking or not knowing about futures thinking. Such statements further highlight the need to create awareness about ways of thinking and their explicit use in EER endeavors.

It should be noted that the study's sample size was limited to one particular NSF award and hence small. The scope of the qualitative study was intentionally limited to get a preliminary sense and deepen our understanding of the futures, values, systems, and strategic ways of thinking used in EER projects. This effort to qualitatively assess ways of thinking was not intended for generalizability. With the initial findings on hand, future research could explore each way of thinking in further detail through broader surveys. We intentionally refrained from collecting gender data from participants for this study believing that this demographic would not likely have an impact on responses. Future research could investigate differences in ways of thinking among various groups (e.g., experience, gender, or discipline). Replication of this empirical investigation with other samples would help strengthen the evaluation of futures, values, systems, and strategic thinking for EER in different contexts.

We believe this study contributes to the broader conversation on transforming engineering education through a ways of thinking lens. Results highlight how futures, values, systems, and strategic thinking are enacted in EER and influence engineering education practice. Integration of thinking approaches equips researchers and educators with a comprehensive toolkit to effect positive change in engineering education.

VI. CONCLUSION

In conclusion, the four ways of thinking—futures, values, systems, and strategic thinking—play pivotal roles in shaping the trajectory of engineering education research. These thinking approaches, when integrated and applied purposefully, empower stakeholders to anticipate, address, and transcend the complex challenges facing the field, ultimately advancing engineering education in response to the evolving needs of society. As the field continues to evolve, the fusion of these thinking approaches will remain instrumental in fostering innovation and promoting inclusivity in engineering education.

REFERENCES

Creswell, J. W. (2012). Educational research: Planning, conducting, and evaluating quantitative and qualitative research (4th ed). Pearson.

Crouch, C. H., & Mazur, E. (2001). Peer Instruction: Ten years of experience and results. American Journal of Physics, 69(9), 970–977.

Esakia, A., & McCrickard, D. S. (2016). An adaptable model for teaching mobile app development. 2016 IEEE Frontiers in Education Conference (FIE), 1–9.

Fellah, A., & Bandi, A. (2018). The Essence of Recursion: Reduction, Delegation, and Visualization. Journal of Computing Sciences in Colleges, 33(5), 115–123.

Guzdial, M., & du Boulay, B. (2019). The History of Computing Education Research. In S. A. Fincher & A. V. Robins (Eds.), The Cambridge Handbook of Computing Education Research (1st ed., pp. 11–39). Cambridge University Press.

American Society for Engineering Education (ASEE). (2014). Transforming undergraduate education in engineering (TUEE) phase I: Synthesizing and integrating Industry perspectives. American Society for Engineering Education, Arlington, VA.

Borrego, M., & Newswander, L. K. (2008). Characteristics of successful cross-disciplinary engineering education collaborations. Journal of Engineering Education, 97(2), 123-134.

Carr, G., Blanch, A. R., Blaschke, A. P., Brouwer, R., Bucher, C., Farnleitner, A. H., … Blöschl, G. (2017). Emerging outcomes from a cross-disciplinary doctoral programme on water resource systems. Water Policy, 19(3), 463–478. http://doi.org/10.2166/wp.2017.054

Corbin, J., & Strauss, A. (2015). Basics of qualitative research, (4th ed.). Los Angeles: Sage.

Dalal, M. & Carberry, A. (2018). Work in progress: 'ways of thinking' of interdisciplinary collaborators. Proceedings of the 2018 American Society for Engineering Education (ASEE) Annual Conference & Exposition, Salt Lake City, UT. doi: 10.18260/1-2—31317

Dalal, M. & Carberry, A. (2019). Work in progress: Exploring 'ways of thinking' of interdisciplinary collaborators. Proceedings of the 2019 American Society for Engineering Education (ASEE) Annual Conference & Exposition. Tampa, FL. https://doi.org/10.18260/1-2--33614

Dalal, M., Carberry, A. & Archambault, L. (2021). Developing a ways of thinking framework for engineering education research. Studies in Engineering Education, 1(2), 108-125. https://doi.org/10.21061/see.38

Dalal, M., Carberry, A., & Archambault, L. (2023). Exploring the use of futures, values, systems, and strategic thinking among engineering education research collaborators. Journal of Engineering Education, 112(2) 382-402. https://doi.org/10.1002/jee.20511

Dalal, M., Larson, J., Zapata, C., Savenye, W., Hamdan, N. & Kavazanjian, E. (2017). An interdisciplinary approach to developing an undergraduate module on biogeotechnical engineering. In Proceedings of Society for Information Technology & Teacher Education International Conference 2017 (pp. 2074-2079), Chesapeake, VA: Association for the Advancement of Computing in Education (AACE).

Dillman, D. A., Smyth, J. D., & Christian, L. M. (2014). Internet, phone, mail, and mixed-mode surveys: the tailored design method. 4th Ed. Hoboken, NJ: Wiley & Sons

Douglas, M. J. (2000). Trading in the zone: master the market with confidence, discipline and a winning attitude. Paramus, NJ: New York Institute of Finance.

Frodeman, R., Klein, J. T., & Mitcham, C. (Eds.). (2010). The Oxford handbook of interdisciplinarity. Oxford University Press.

Harel, G., & Sowder, L. (2005). Advanced mathematical-thinking at any age: Its nature and its development. Mathematical Thinking and Learning, 7(1), 27-50.

Journal of Engineering Education. (2006). Special report: The research agenda for the new discipline of engineering education. Journal of Engineering Education, 95(4), 259-261.

Leydens, J. A., & Lucena, J. C. (2017). Engineering justice: Transforming engineering education and practice. Hoboken, NJ: John Wiley & Sons.

McCallin, A. (2004). Pluralistic dialoguing: A theory of interdisciplinary teamworking. The Grounded Theory Review, 4(1), 25-42.

McKenna, A. F., Froyd, J., & Litzinger, T. (2014). The complexities of transforming engineering higher education: Preparing for next steps. Journal of Engineering Education, 103(2), 188–192. http://doi.org/10.1002/jee.20039

McKenna, A. F., Yalvac, B. & Light, G. J. (2009). The role of collaborative reflection on shaping engineering faculty teaching approaches. Journal of Engineering Education, 98(1), 17-26.

National Science Foundation. (2017). PFE: Research initiation in engineering formation (PFE: RIEF) Retrieved from https://www.nsf.gov/pubs/2017/nsf17514/nsf17514.pdf

Olds, B.M., B.M. Moskal, & R.L. Miller. (2005). Assessment in engineering education: Evolution, approaches and future collaborations. Journal of Engineering Education 94(1), 13-25.

Saldaña, J. (2009). The coding manual for qualitative researchers. London, UK: SAGE Publications Ltd.

Sousa, D. A. (2016). How the brain learns (5th ed). Singapore: Sage Publications Asia-Pacific Pte. Ltd.

Swan, C., Paterson, K., & Bielefeldt, A. R. (2014). Community engagement in engineering education as a way to increase inclusiveness. In Johri, A., & Olds, B. M. (Eds.). Cambridge handbook of engineering education research (pp 357-372). New York, NY: Cambridge University Press.

Wankat, P. C., Felder, R. M., Smith, K. A., & Oreovicz, F. S. (2002). The scholarship of teaching and learning in engineering. In Huber and Morrealle (Eds), Disciplinary styles in the scholarship of teaching and learning: Exploring common ground (pp 217-238), Washington, D.C.: AAHE publications.

Warren, A., Archambault, L., & Foley, R. W. (2014). Sustainability education framework for teachers: Developing sustainability literacy through futures, values, systems, and strategic thinking. The Journal of Sustainability Education, 6(1), 1-14.

Wiek, A., Withycombe, L., & Redman, C. L. (2011). Key competencies in sustainability: A reference framework for academic program development. Sustainability Science, 6(2), 203-218.

Copyright Statement

 https://doi.org/10.52202/073963-0001

Using PALAR to Formalize Informal Education

Lyndsay Ruane, Hannah Sanders, Laura MacDonald, Jessica Rush Leeker
University of Colorado Boulder
Corresponding Author Email: Jessica.Leeker@colorado.edu

Abstract

Context

This paper analyzes a self-developed, STEM-focused community engagement project undertaken by faculty, staff, graduate and undergraduate students from a predominately white and a minority-serving institution. The inter-institutional project uses the Participatory Action Learning and Action Research (PALAR) theoretical framework for community engagement to examine tenets of anti-racism and decolonization within higher education.

Purpose or Goal

The central hypothesis is that an inter-institutional approach to educational transformation centered on democratizing innovation across institutional boundaries will prepare next-generation innovators to address systemic and institutional racism within STEM by challenging higher educational norms. The PALAR approach provides a robust framework through which researchers can simultaneously participate in "action learning" and analyze the effectiveness of the informal educational setting they have created.

Methods

The PALAR framework is a process-based knowledge, research, and development paradigm incorporating emotions, communication, logical problem-solving, critical thinking, and social experiences. The research team collected multiple forms of qualitative data, including quarterly interviews, group meeting observations, and weekly student journals, to investigate the development of participant attitudes and relationships.

Outcomes

One way PALAR-framed pedagogies differ from traditional classrooms is in the role of faculty and staff researchers, serving primarily as guidance rather than authority. The ambiguity and lack of a formal classroom format challenged researchers to critically self-reflect but also acted as an initial hindrance for student participants. Accustomed to traditional classrooms, the students reported discomfort and confusion while they navigated an unfamiliar level of control over their learning.

Conclusion

Informal education through PALAR allowed researchers and students to reflect critically on learning and education assumptions. The process-built subjectivity inherent to PALAR led to improved knowledge sharing compared to traditional learning methods. This subjectivity also allowed researchers and community members to present themselves as resources and consultants, rather than authority figures, making those involved more comfortable with the new (informal) learning process.

Keywords— Undergraduate students, informal learning, institutional change, PALAR framework

I. INTRODUCTION

PALAR (Participatory Action Learning and Action Research) is a paradigm designed to confront complex and dynamic social problems. By incorporating research subjects as active participants, PALAR challenges typical relationship dynamics and power structures to support broadened perspectives, increased agency, and personal growth. The framework facilitates community engagement with cyclical and reflective processes that are intentionally adaptable and self-motivated.

This project is fundamentally focused on the collaborative creation of a Living Learning Laboratory in the southern United States. The Laboratory will concentrate on education, sustainability, and community service while also studying and accounting for the racial and socio-historical influences of the land. Through its creation, community engagement will be established as a partnership, actively involving and recognizing perspectives and expertise from local populations. The Laboratory will support sustainable infrastructure and climate resiliency research in a uniquely versatile and informal learning environment while deliberately incorporating local culture and history. These research topics will provide a platform to explore educational norms in higher education, addressing systemic and institutional racism within STEM fields through an increased understanding of existing institutional boundaries.

To accomplish this, one cohort of students from two different undergraduate institutions will work together with the support of faculty and community members. The institutions consist of a predominately white institution (PWI) and Historically Black Colleges and Universities (HBCU), combining knowledge and resources to facilitate profound change in undergraduate education by understanding and enacting tenets of anti-racism and decolonization. Each year, over four years, a cohort of ten students from each institution is selected through an application and interview process. Student collaboration is mainly remote, with occasional in-person site visits. A team of faculty, staff, graduate and undergraduate student researchers, and community members supports them.

The structure of that support team provides a significant distinction from other projects, as PALAR delegates faculty researchers to serve primarily as guidance rather than authority members. This unusual authority structure was a defining factor of the PALAR framework and its implementation in inter-institutional informal education.

II. PALAR Overview

Participatory Action Learning and Action Research combine multiple theories of action research to establish a comprehensive and dynamic structure for community engagement. Created by Ortrun Zuber-Skerritt (2011), PALAR combines participatory action research and action learning concepts in a project- and process-based paradigm and learning theory. By design, the framework pulls pieces from existing action research practices to serve as an adaptable "philosophy, a methodology, a theory of learning, and as a facilitation process for community engagement" (Zuber-Skerritt, 2015, p. 5).

PALAR is best described as a sum of its parts. The "AL" portion refers to action learning (AL), a problem-solving method involving taking action and reflecting on results afterward. In "learning by doing," AL typically focuses on collaboration and critical reflection to generate fresh understandings (Marquardt, 1999; McGill & Brockbank, 2007; Zuber-Skerritt, 2015). In PALAR, action learning works with action research (AR), a more systematic method that seeks to solve social problems via transformative change. AR utilizes a repetitive cycle: taking action, observing, reflecting on those results, and then retaking action with reflection-based reevaluations. Together, AR and AL actively collect knowledge and facilitate an involved, dynamic, and cumulative method of inquiry.

Finally, the "P" in PALAR refers to "participatory" research. This paradigm requires the deliberate involvement of research subjects, all working toward "inclusion, social justice, and equality of participants" (Zuber-Skerritt, 2015, p. 7). Ultimately, PALAR designates that participants observe and reflect on action results and are personally invested in project outcomes, granting a uniquely dynamic and observational perspective on the project's creation.

Our project was designed to utilize PALAR's four standard recurring stages: plan-act-observe-reflect. These four stages comprise the cycle consistently repeated at all levels throughout each project year. At the end of each cycle, reflections are utilized to plan the next round of action steps, and those constant reevaluations are key for PALAR projects to respond "effectively to complex issues in rapidly changing contexts" (Zuber-Skerritt, 2011, p. 1).

In its deliberate design, the framework's informal, interdisciplinary, and learner-centered approach differs vastly and fundamentally from a traditional classroom setting. PALAR aims to find meaningful solutions to social justice problems with dynamic collaboration and project-based development (Teare & Zuber-Skerritt, 2013). This approach's self-initiated and self-directed nature is essential for meaningful personal growth and sustainable social change.

Structurally, the research team in a PALAR project is established drastically differently from a traditional classroom. The research subjects, the undergraduate cohort in this context, contribute to the project and the research as active participants. The lead investigator, on the other hand, joins the subjects and contributes to conversations more similarly to a peer than an authority figure. That dynamic supports the investigator "researching with, rather than on, community members while perceiving them as co-participants rather than mere informants and/or recipients of knowledge" (Kearney et al., 2013, p. 118). This unconventional structure introduces an informal relationship between co-researchers, encouraging humanized, personal perspectives and meaningful context.

When researchers offer guidance instead of acting with authority, PALAR also opens meaningful growth opportunities among participants. For societal change to be meaningful and withstand time, participants must be willing and able to maintain progress through self-motivation, confidence, and agency. Without an authority figure to dominate advancement, PALAR "allows academic researchers to partner with people to help them learn how to improve their situation, drawing on their lived experience and intimate knowledge of the challenges they face" (Wood, 2015, pp. 79-80). When successful, PALAR provides contextualized solutions to social problems and enables participants to continue meaningful and transformational work in their communities.

III. Methods

To follow and understand PALAR, student and faculty researchers regularly reevaluated and reorganized the methods and structures of the project. We regularly collected qualitative data throughout the project, analyzing dynamics, relationships, and attitudes over time. This qualitative data included quarterly interviews, surveys, group meeting recordings, weekly journals, and field notes.

Over the first year, faculty researchers performed 30-45 minute quarterly interviews with each undergraduate student in the cohort. To promote honest feedback, the researchers who performed these interviews were not the same faculty members with whom the cohort is in regular contact. Interviews were conducted with open-ended questions, allowing space for free expression and authentic reactions. After the first round of interviews, we decided to implement surveys to accompany all subsequent interviews, which provided additional structure and allowed interviewers to prepare better to ask about topics most relevant to individual students.

Additionally, the undergraduate cohort met remotely as a group every week for one hour. During this hour, the students

https://doi.org/10.52202/073963-0002

facilitated their discussions and held each other accountable for completing deliverables. A graduate student researcher attended each weekly meeting to take field notes and provide support, and recordings of each meeting were transcribed for observation and trend identification. Students often utilized these meetings for work that required high levels of collaboration, providing a platform for researchers to study their interactions.

Along with weekly meetings, student participants must submit a weekly journal. These journals were available only to the research team, not to the other cohort students. That semi-privacy initiated an additional level of anonymity so that students could voice sensitive concerns, such as complaints about particular relationship dynamics, without fear of retribution. The journals also allowed students to reflect on their experiences individually.

Finally, during the first year, the student cohort gathered in person twice. One of those in-person gatherings was at the future site of the Living Learning Laboratory, which was close to home for students from the HBCU, and the second meeting took place in a different state from both institutions, on relatively neutral ground. These meetings were rich in information gathering, collaboration, and relationship-building. Two graduate student observers collected qualitative data from these in-person meetings through field notes.

In the undergraduate cohort, reflections included thoughts on data collection, architectural methods, community involvement, and project and team structure. Concurrently, faculty researchers focused more on the informal learning structure and inter-institutional and community collaboration while considering student perspectives and individual developments.

The research team members range from professional social scientists to third-party graduate student researchers, allowing for the evaluation of various forms of qualitative data on multiple levels. We analyzed the results in the context of a PALAR framework and how it adapted to fit the needs of this multi-dimensional, contextually complex, and highly dynamic project.

IV. Results

PALAR is explicitly designed to identify and solve nuanced, dynamic, and multifaceted societal problems. With research subject participation and consistent reflection and reevaluation, the paradigm allows problem-solving processes to adapt alongside solutions still in development. In particular, student participation facilitates purposeful individual advancement, such as confidence, self-advocacy, and self-agency.

The role of faculty and staff researchers proved to be a significant distinction from the traditional classroom structure. Rather than operating from a position of authority, researchers acted primarily as guidance for the student cohort, challenging typical relationship expectations. Researchers were forced to reflect critically on interactions and intentional influence, taking care to give the students control of their work.

The undergraduate cohort initially struggled with the project's ambiguity. PALAR provided each student with a level of control over unfamiliar learning, contrasting their accustomed experiences in traditional classrooms. Journals, meeting notes, and interview transcriptions identified structure-based struggles by nearly every student participant, amplified by difficulty with the fully remote format. Likewise, faculty encountered difficulty in maintaining consistent motivation among the students. However, over time, analysis of these same data sources showed improvement via in-person meetings, notable personal development, and recognition of anti-racism perspectives.

A. *Structure Struggles*

Almost every student reported varying levels of discomfort and confusion in the ambiguity at the beginning of this project. During the second group meeting, a student asked the graduate researcher whether to regard graduate students as "the teachers" or if they should report to the faculty lead. Accustomed to traditional classrooms, they expected some authority to provide assignments, discipline, and general order. This situation reflects the uncertainty sometimes associated with informal learning and emphasizes the importance of individual agency and leadership.

Through journals, more students expressed their initial confusion and desire for more structure. After the first month, one student recognized that without formalized consequences, "many [messages] resulted in poor follow through, or poor results." In the same time frame, another student echoed that sentiment about ambiguity, "having such a loose setup made it hard for me to be fully invested/dedicate enough time to the project," and "not everyone is on the same page" regarding effort levels.

Even in the first round of interviews, which took place about three months into the project, students articulated frustrations, "at first, days weren't as productive as we thought they'd be, or we couldn't...get certain details done." Encouragingly, that same student associated this discomfort with "growing pains," articulating later that "now everyone chimes into [discussions] to some extent, and when they do they sound more relaxed... [it] definitely feels more organic."

Faculty and graduate students felt the pressure of these struggles with structure. Without rubrics or grades to enforce consistent expectations and consequences, the faculty members were met with uneven effort levels among student participants. Students with strong personal motivation or with particularly relevant skills were often forced to make up for incomplete work from their fellow cohort. Therefore, faculty was required to delay several deadlines and readjust expectations repeatedly, which further added to student confusion and frustration.

Eventually, the ambiguity and lack of formal consequences emerged from personal thoughts in journals into the group discussion. Around the end of the third month, students self-directed a group conversation about "a few concerning issues regarding expecting every member to fully participate and submit assignments ahead of time to prevent delays." This discussion encouraged the students to advocate for themselves, as they began actively participating in the project's structure and asking researchers for the support they needed.

This self-advocacy also emerged in journal writings, where participants brainstormed solutions to their concerns. Those concerns and solutions led directly to changes in the students' self-created structure. For example, one student suggested in their journal, "I wonder if there should be some sort of student leadership to guide everyone along." This situation led to the establishment of weekly facilitators designated to lead a group meeting ahead of time. Multiple students expressed a desire to "organize communication throughout the week so that everyone is up to date on what we should be working on," which was addressed via scheduled weekly check-ins. Students advocated for "a student-led meeting earlier in the week in addition to our Thursday meetings," leading to the establishment of small groups that meet according to expertise and scheduling, separate from the large group meeting.

After this advocacy, faculty still observed unequal workloads and a few more failures to meet deadlines, but on a decreased scale and with an increased understanding and optimism for the future. The student participants were directly involved in identifying issues, searching for answers, and implementing solutions. This structure allowed them to build confidence, agency, and problem-solving skills that would not be present in a traditional classroom. While some struggled initially, individual journals reflected personal growth by the end of the fifth month, "I could see the growth in my communication skills as well as my fellow cohort members."

B. In-Person Benefit

Another notable theme in the qualitative results addressed the difficulty of remote collaboration. A significant focus for this project is the establishment of genuine inter-institutional relationships. In reflections, a staff researcher identified the importance of everyone's presence at the large group meetings, "Time spent in full company is vital in the establishment of meaningful relationships. Group activities, like icebreakers and team-building exercises, are vital in identifying common ground and building mutual respect."

Fairly early on, during the second month, the students met for the first time in person, attending a two-day site visit at the future Laboratory building site in the southern United States. Then, during the fifth month, they met again for a more extended, five-day retreat in Taos, New Mexico, serving as a relatively neutral location for participants from both institutions. After each visit, students reflected and recognized the importance of these meetings, noting that they could complete more work and build more personal relationships over just a few hours of in-person collaboration.

After the first few weeks of ambiguity and confusion, students seemed to find a slight clarity immediately following the first site visit. One student reported in a journal entry the week after that site trip, "I feel like last week's trip was the real start of this project because now we all have a feeling of the site and a lot of valuable information that we can use moving forward." In a group meeting, another student told their peers, "I feel like I have a better understanding of the cultural aspect of the project and what we're actually trying to accomplish," echoed in that week's student meeting.

Completing the five-day retreat in the fifth month diminished some concerns regarding dedication and investment levels. Journal entries from that week stated, "I feel inspired and enriched after our trip to Taos...I gained so much insight. I think that as a group, we are all very invested in this, and at the time and energy to dedicate." Participants even noted personal impacts, "this trip has helped my social [abilities] in a tremendous way. I'm able to translate this to my relationships outside," and another noted within the group, "the trip became an internal and external advantage being that the connection with the team grew stronger and the memories we made are forever with us."

Beyond relationship building and personal growth, students responded positively to the in-person gathering because of increased productivity. One journal entry, written after the second site visit, noted that "when [we] met a few weeks ago, it took no less than an hour to accomplish the same thing it took three weeks to convey to the others."

These reflections from the participants emphasize the importance of meeting in person, even briefly. The two-day and five-day gatherings increased motivation, understanding, and meaningful relationship building.

C. Anti-racism and Decolonization

A key goal for this project was to utilize PALAR to examine tenets of anti-racism and decolonization. At first, students did not pursue this path of discussion. When faculty researchers questioned this at the Taos retreat, students mentioned in the discussion that they "felt like equals" and believed "we are making history with this project."

In encouraging conversations on race and racism, faculty split the student cohort by institution. In those discussions, students were asked to reflect and share their experiences with the complexity of racism. This forced the students to confront unwanted "friction," which was met with resistance. They discussed the importance of recognizing the impacts of racism and colonization, particularly within this project, located in the American Deep South, which is steeped in violent, triumphant, and meaningful history.

Participants mentioned resisting this conversation for several reasons. They did not like being separated and felt it would be more meaningful as a group conversation. They also wanted to "focus on the design," not on any "painful" context. Through this process, faculty observed a lack of deep understanding

https://doi.org/10.52202/073963-0002

among the students, some of them failing to grasp the anti-racism tenet of the project while focusing on technical and historical aspects.

It is important to note that these young people, particularly those young people who are relative strangers to each other, were extremely hesitant to discuss uncomfortable topics like prevalent and systemic racism in existence today. The participants discussed historical racism in their weekly meetings but did not touch on modern impacts until faculty researchers breached the subject.

Promisingly, once the subject was introduced, students seemed more comfortable speaking about racism and decolonization independently. During a group meeting in the sixth month after the Taos trip, participants facilitated their conversation on affirmative action, later described in journal entries as a "rich discussion [that] got everyone's gears turning." Following the same meeting, another journal entry read:

> This incident allowed us students to share additional perspectives on other social issues that are actively occurring in our society and generation. The conversations then led to the topic of systemic racism and recognizing the different areas in society where racism serves as a disadvantage, specifically towards African Americans. This issue ranges from the education system to the workforce of employees.

This student-led conversation and meaningful following reflections indicate a willingness to consider systemic problems in conversation. While it needed some researcher encouragement to get started, the resulting discussions could have a lasting impact on the project development and individual participants.

D. *Personal Development*

Finally, a fundamental characteristic of PALAR is the potential impact on participants. By pushing students to assume leadership and control of their education, we encouraged the development of skills like communication, leadership, initiative, agency, self-advocacy, and broadened understanding. Rather than an "objective, impersonal ideal of scientific detachment," PALAR researchers recognize that "personal transformation is as much an outcome of the process as practical change and theory generation" (Wood, 2015, p. 81) (Polanyi, 1958).

In journals and interviews, students reported new perspectives, recognizing that this "research project so far has broadened my thought process more than I would have imagined." They also began to relate with the community and account for local needs, "I have to consider how things could be replicated in the community, so I try to be mindful of what community members have [access to]."

Students also reflected on increased communication and meaningful conversational skills, "I found all the suggestions and advice [from faculty guidance] to be extremely helpful. I think it...will be helpful in opening up the conversation or deepening it." They also journaled about the impact of their fellow cohort members, especially praising the discussion leaders for specific weeks, "the discussion leader...enhanced the discussion of positionality statements greatly."

Finally, after only five months with the project, or about one school semester, student participants reported increased initiative and self-advocacy. This report included acknowledging when help was necessary, first recognizing that "I need to practice the utilization of these sources in the future." That same student later noticed about themselves, "I am also learning the value of reaching out to others for help or knowledge."

Faculty recognized this personal growth through participation levels in group projects and meetings. While still encountering uneven workload frustrations and lack of motivation to meet deadlines, researchers observed deep growth in understanding and personal participation by each student. This helped to contextualize perceived effort levels and provided opportunities to attempt various methods of engagement for future projects.

Ultimately, the structural issues and reevaluations demonstrate the students' progression into self-motivated problem-solvers. More than half of the participants expressed at some point their desire to create a lasting impact beyond the scope of this project. In one interview, one of the students defined their idea of success: "I think if I can just leave a mark on it… that would make me really proud of, like, being in there. My idea of success is kind of in longevity."

V. Conclusions

In the first half-year of this project, faculty and staff researchers immediately struggled to balance their preexisting ideas of zero control and complete control of a class. Simultaneously, student participants struggled to comprehend and take advantage of their atypical levels of control. Along the way, most students acknowledged the benefits of in-person meetings for remote collaborations, the significance of systemic racism and colonization, and the opportunities for personal development that will impact participants for years to come.

These findings reflect lessons learned in designing a PALAR project and provide guidelines for future researchers. For example, if roles and responsibilities are articulated, participants may adjust more quickly to an ambiguous, informal structure. The unusual power dynamic and lack of authority caused students to feel confused and unmotivated, and they needed some initiation structure. Future projects should consider establishing roles and expectations within the first few meetings to avoid a "slow start." Additionally, if possible, opportunities for in-person collaboration should be prioritized, as participants value these meetings, even if only for a few days. That introductory meeting creates a foundation for solid relationships, communication, and emotional buy-in and can introduce a vital kickstart to the project's progress.

The AL/AR portion of PALAR was really put to the test by

https://doi.org/10.52202/073963-0002

faculty, expressly through their attempts to inspire self-motivation and deep introspection. Without negative consequences typically seen in formal classrooms for incomplete work or missed deadlines, it seemed easy for students to fall into relaxed roles. Faculty addressed this in several reassessment cycles, attempting accountability through fellow students, graduate leads, and finally faculty intervention. The most success was gained via expectations that were articulated from the beginning. Future projects will take this understanding to articulate necessary requirements from the beginning of the project and to enforce deadlines with more tangible consequences, such as a three-strike removal system. Another example of AL/AR cycle learning was when students hesitated to discuss what they considered to be controversial topics in anti-racism and decolonialization. Faculty realized that kind of deep introspection and confrontation were never articulated as expectations. In future projects, researchers should ensure that students understand the objectives of the project itself, including requirements for addressing social issues for transformative impact.

Other lessons learned also affect the speed of a project. In this case, participants needed an in-person initiation to feel comfortable discussing issues like systemic racism. If future projects are on a faster timeline, they may need to consider introducing sensitive topics early to allow students time to feel comfortable. However, if future projects have sufficient time and flexibility, the added freedom of making mistakes and self-correcting along the way encouraged students to advocate for themselves and develop intrinsic motivation that they will carry beyond this project's scope.

The PALAR framework implemented in this project required critical self-analysis at all levels, introducing and establishing informal learning as a powerful and legitimate education tool and compelling students to take their education into their own hands. The flexibility allowed by PALAR facilitated knowledge sharing and learning between institutions, researchers, and community members of varying backgrounds.

Throughout the first five months of this project (one semester), student participants were able to learn and grow in an informal environment. By leading themselves and using faculty/staff researchers as guides, the undergraduate cohort independently recognized, addressed, and suggested solutions for various complex and multifaceted problems. This process encouraged problem-solving that was inclusive of a wide range of community members and also encouraged individual improvement. The informal learning process facilitated discussions, personal development, and critical reflection that would have been unavailable in traditional classrooms, and PALAR provided a designated framework for analysis amid intentional ambiguity.

REFERENCES

Kearney, J., Wood, L., & Zuber-Skerritt, O. (2013). Community–University Partnerships: Using Participatory Action Learning and Action Research (PALAR). *Gateways: International Journal of Community Research and Engagement, 6*, 113–130. https://doi.org/10.5130/ijcre.v6i1.3105

Marquardt, M. J. (1999). *Action Learning in Action: Transforming Problems and People for World-Class Organizational Learning* (1st edition). Davies-Black Publishing.

McGill, I., & Brockbank, A. (2007). *Facilitating Reflective Learning in Higher Education.*

Teare, R., & Zuber-Skerritt, O. (2013). *Lifelong Action Learning for Community Development: Learning and Development for a Better World.* Brill. https://brill.com/display/title/37217

Wood, L. (2015). Reflecting on Reflecting: Fostering Student Capacity for Critical Reflection in an Action Research Project. *South African Journal of Higher Education, 4*, 79–93.

Zuber-Skerritt, O. (2011). *Action Leadership: Towards a Participatory Paradigm* (2011th edition). Springer. https://link.springer.com/book/10.1007/978-90-481-3935-4

Zuber-Skerritt, O. (2015). Participatory Action Learning and Action Research (PALAR) for Community Engagement: A Theoretical Framework. *Educational Research for Social Change, 4*(1), 5–25.

Post-Colonial Agenda for Engineering Education Research in India

Rohit Kandakatla[a], Rucha Joshi[b,*], and Ashish Agrawal [c,*]
KG Reddy College of Engineering and Technology, Hyderabad, Telangana, India [a], Plaksha University, Mohali, Punjab, India[b], Rochester Institute of Technology, Rochester, New York, USA[c]
Corresponding Author Email: rohit.kandakatla@kgr.ac.in

Abstract

Context

India in 2020 announced the National Education Policy (NEP), which aimed at decolonizing and revamping all aspects of the education structure, including its regulation and governance. NEP 2020 was designed with the principle to nurture India's next generation into human beings who are capable of rational thought and action, who are compassionate and emphatic, who are courageous and resilient, and who possess scientific temper, creative imagination, and sound ethical values. Engineering education in India, therefore, is at a critical juncture to transform itself in alignment with the vision of NEP 2020.

Purpose or Goal

This work aims to systematically explore the existing efforts and future directions for aligning engineering education in India within the framework of the NEP. The broader goal of this study is to provide evidence-based suggestions to policymakers to establish a research agenda that helps engineering education align with the NEP 2020.

Methods

We employed a qualitative approach in the study and get the insights and perspectives of different stakeholders on critical challenges in Indian engineering education that need to be solved through research inquiry. To achieve this goal, semi-structured interviews were conducted with engineering education researchers, leaders, and practitioners in India, and the data were analyzed through thematic analysis to present the results in specific themes as they emerged from the data.

Outcomes

The results are reported through a proposed research agenda that is categorized into various themes identified to guide engineering education research in India along with specific questions that can be taken up for inquiry.

Conclusion

The NEP 2020 represents a crucial milestone in decolonizing the Indian education system, and engineering education must align with its vision. The findings from the study can guide engineering educators in India toward a systematic transformation of their education system.

Keywords—Engineering Education Research; National Education Policy (NEP) 2020; Research Agenda.

*Authors Joshi and Agrawal contributed equally to the paper.

I. INTRODUCTION

India is now in its 77th year of independence and has since then taken a significant leap with respect to engineering education in the country. The set up of five Indian Institute of Technology (IITs) established in the period of 1951 to 1961, and sixth set up in 2001 helped establish quality of engineering education. Today, India is a home to more than 2500 engineering colleges producing approximately 1.5 million engineering graduates annually. A significant amount of this growth in engineering colleges can be attributed to liberalization. The liberalization of the Indian economy in 1991 led many multinational companies to enter the Indian markets. This coupled with the internet and software revolution a few years later led to a large number of IT services companies established in the country. The government's foresight to prepare India's youth for these emerging opportunities led to change in policies to allow privatization of higher education. India witnessed an exponential growth in number of engineering colleges since early 2000's, with most of them being affiliated to a local government-funded university, which prescribes the academic regulations, curriculum structure, course syllabus, and assessment process (Pratik et al., 2013).

However, the mushrooming of colleges across the country happened at the expense of good quality of education. The total number of colleges increased in engineering as the government established new institutions like the Indian Institutes of Information Technology (IIITs), as well as 23 IITs and 32 National Institute of Technologies (NITs), in an effort to produce a larger number of world-class engineers. Despite these efforts, they were still insufficient to meet the demand for technical and engineering education in India. The expansion of higher education took place along with loosely monitored regulatory systems that failed to put in check the necessary measures to ensure quality of education. As a result, many new engineering colleges began sprouting (table 1), many of which are private colleges monetizing education. Although some of them are accredited by All India Council for Technical Education (AICTE), or other bodies, many lack proper

https://doi.org/10.52202/073963-0003

autonomy, quality infrastructure, have insufficient number of qualified faculty, and fail to equip students with required skills for employability in engineering. As reported by the "National Employability Report for Engineers 2019", 80% of these engineering graduates are not suitable for employment after graduation, as they lack the required technical and professional skills and mindset to work in the industry (Minds, 2019).

Table 1: Total No. of Universities in the Country offering engineering degree as on 25.01.2023	
Universities	Total No.
State Universities	460
Deemed to be Universities	128
Central Universities	56
Private Universities	430
Total	1074

A major factor that impacts engineering education in India is outdated curriculum, taught through traditional teaching practices, and mostly assessed at the basic level of rote learning (Choudhury, 2016). Most of Indian education system focuses currently on the reproduction of knowledge which is long lasting imprint left by the British from their colonization of India for over two centuries. The Indian education system was originally designed by the British to create a subservient workforce. Since independence, its core principles have largely remained unchanged (Chaudhary, 2007). The Indian education system in the last many decades lacks avenues to prepare students for skills such as critical thinking, problem-solving, teamwork, communication, etc. The emphasis on rote learning continues to limit the learning capacity and development of Indian students even in engineering education, as most students are inadequately prepared to apply their technical knowledge to analyse and solve problems in the industry and society (Mohanty & Dash, 2016).

While curriculum reform, interdisciplinary education, teacher training, incorporating technology, and focus on soft skills seem to be increasingly discussed to change this situation, what is missing is systematic research reporting the effectiveness of such measures and identification of new measures that are needed and how they should be implemented. What factors have what effect, which factors need to be prioritized, and how various factors might be acting to make impact in India's engineering education, needs to be explored systematically using research. It's important to note that changes required to revamp India's engineering education quality cannot happen overnight and require a collective effort from educational institutions, government, industry, and society at large. However, with sustained effort, it's possible to transform the education system to better prepare students for the challenges of the 21st century. Thus, there is a need to identify research questions for transforming India's engineering education so that the education system can come out of the shackles of age-old processes, beliefs and systems. It is unique opportunity to deliver for the country's post-colonial engineering education needs by engagement of deliberations with all stakeholders including policy leaders, institutional decision makers, faculty, and students.

It is noteworthy to mention that India in 2020 announced in National Education Policy (NEP) which aims to produce engaged and productive citizens who would contribute to the development of an equitable, inclusive, and plural society as envisaged by our constitution (Kumar, 2021). NEP 2020 opens opportunities for engineering education in India to transform itself through systematic research, practice, and reflection on how Indian engineering students can be trained to think critically, solve problems, engage in higher-order and lifelong learning, develop care and empathy for the community, and communicate and work in teams to solve societal problems. However, effective implementation of NEP still requires work as each institute is figuring out their own ways to overcome challenges in engineering education. Research in this direction by engineering education researchers, will help create possible solutions to initiate the transformation of engineering education in India. This paper aims to empower engineering education researchers, India's practitioners, leaders, faculty, and policymakers for this transformation, by finding the right questions to focus on, through semi-structured interviews with stakeholders performed to discuss and deliberate a research agenda for post-colonial engineering education in India. Through various themes identified to guide engineering education research in India, and by formulating specific questions that can be taken up for inquiry, this paper seeks to provide the starting point that can be used to transform engineering education in India as envisioned by NEP 2020.

II. LITERATURE REVIEW

A. Pre-Colonial Education System in India

Higher education in pre-colonial India was characterized by a vibrant ecosystem of learning, intellectual exchange, and pursuit of knowledge across diverse disciplines. It nurtured a deep respect for wisdom, scholarship, and holistic development, providing a solid foundation for intellectual pursuits and the advancement of society. India had renowned centers of learning and universities such as Nalanda, Takshashila, Vikramashila, and Valabhi universities, which were instrumental in the dissemination of knowledge (Pandya, 2014). These institutions attracted scholars and students from across the Indian subcontinent and beyond, offering education in a wide range of disciplines. Indian higher education placed significant emphasis on the liberal arts and sciences. The curriculum covered diverse subjects such as philosophy, mathematics, science, astronomy, grammar, literature, music, and arts. This multidisciplinary approach aimed to cultivate a well-rounded education and foster critical thinking and intellectual development.

There was a great emphasis on holistic development of students as it aimed to nurture physical, intellectual, moral, and

https://doi.org/10.52202/073963-0003

spiritual aspects (Tilak, 2020). Students were encouraged to cultivate virtues, practice self-discipline, and lead a balanced life as the goal was to create well-rounded individuals capable of contributing to society. Pre-colonial Indian education encouraged critical thinking, questioning, and independent inquiry (Crozet, 2012). Students were encouraged to debate, analyze, and interpret various texts and philosophical ideas to develop a deep understanding of the subject matter and promote intellectual growth. Overall, education in pre-British India emphasized spiritual, moral, and intellectual growth, aiming to produce individuals who were not only knowledgeable but also ethical and responsible members of society.

B. British Colonization of Indian Education System

During the British rule in India between 1858 and 1947, the education system underwent significant changes that reflected the colonial objectives and policies. The British introduced English education in India with the aim of creating a class of Indians who could serve as intermediaries between the British administration and the local population (Cohn, 2009). English became the medium of instruction in schools and colleges, and it was given more prominence than traditional Indian languages. The curriculum focused on English language proficiency and subjects relevant to colonial administration, such as law, administration, and engineering. The British administration showed little interest in promoting traditional Indian education systems and often viewed them as inferior (Ellis, 2020). Traditional educational institutions were marginalized and received little support from the colonial government. This led to a decline in traditional knowledge systems and practices. The British education system focused on promoting English literature, culture, and values, and the curriculum aimed to create a sense of admiration and loyalty towards British culture among the educated Indians. The education policies implemented by British were aligned with their colonial interests and the curriculum was designed to produce clerks, administrators, and professionals who could assist in the functioning of the British administration.
(Ghosh, 1993).

The British education system in India had several drawbacks and negative impacts on the country. The emphasis on English education and the imposition of Western values and cultural norms led to a disconnection from Indian cultural roots. Students were often alienated from their own traditions, languages, and heritage, resulting in cultural displacement and a sense of inferiority regarding their own culture and identity (Fischer-Tiné & Mann, 2004). The curriculum of the British education system had little relevance to the local Indian context. The subjects taught were often disconnected from the needs and realities of Indian society, failing to address the pressing issues and challenges faced by the people. The education system aimed to produce a class of Indians who could serve the interests of the British colonial administration, which hindered the development of independent thought and critical analysis among Indian students (Chaudhary, 2007). The British education system prioritized theoretical knowledge and academic subjects while neglecting vocational and practical education. This limited the development of practical skills and hindered the emergence of a skilled workforce capable of meeting the needs of the Indian society.

C. Review of Education Policies in India post-Independence

India achieved its Independence in 1947 and became a democratic republic a few years later when the Constitution of India came into effect on 26 January 1950. For the first two decades, education was given less of a priority by the Indian government, and little changes were made to decolonize the Indian education system. The most comprehensive education policy was first announced in the year 1968 and based on the recommendations of the Kothari Commission Report (Mukhopadhyay, 2017). One of the main focus of the National Education Policy (NEP) 1968 was to emphasize the need for universal primary education across the country especially to achieve national integration and social justice. The policy also provided a caution by highlighting the poor state of education across most schools and colleges due to lack of basic infrastructure and unavailability of skilled and passionate teachers. The next major education policy, the National Policy on Education (NPE) 1986, emphasized the need to provide equal opportunities to all in education as a way to reduce disparities and promote social justice (Majid & Kouser, 2020). The focus of NPE 1986 expanded its goal from universal primary education (as envisioned by the Kothari Commission Report) to secondary and tertiary education in India. The NPE 1986 also indicated the need to strengthen research in higher education. Multiple national and state research focused institutions were established to enhance research infrastructure in the country (Ramamurthy & Pandiyan, 2017).

D. De-Colonization of Indian Education System – the National Education Policy (NEP) 2020

India witnessed it first major intention to decolonize its education system from beliefs and influence of the British colonial rule with the announcement of the National Education Policy (NEP) in 2020 2020. Until NEP 2020, most of the previous education policies focused on improving the accessibility and quality of the education system that was mired by poor infrastructure and lack of skilled human resource (Rao, 2022). There were hardly any attempts made to fundamentally question the educational philosophy, the vision and aspirations of the education system, the local relevance of education curriculum, and the process followed teaching and assessment. The NEP 2020 was one of the first comprehensive policy framework to promote holistic development, foster creativity, and critical thinking skills and instill principles of ethics, empathy, and integrity among students (Kumar, 2021). It advocated for a flexible and multidisciplinary curriculum that allows students to choose from a wide range of subjects, encouraging a holistic and well-rounded education. The policy also aims to transform the assessment system to promote

https://doi.org/10.52202/073963-0003

formative assessment and reduce the emphasis on rote learning (Malakar, 2022).

The NEP 2020 in its policy highlighted their intent to addresses many essential issues deeply rooted in the education system since post-independence (Jain et al., 2021). While the NEP 2020 incorporated certain elements from the ancient Indian education system, it is important to also note that it also took inspiration from global best practices to create a relevant and comprehensive policy for the modern educational landscape in India. The policy aims to build on India's rich educational heritage while also adapting to the changing needs of the 21st-century world. One of the major challenges expected in its successful implementation of NEP2020 is to critically examine and question colonial legacies, paradigms, and structures. Decolonization calls all stakeholders in the Indian education system to re-evaluate and restructure the curriculum, pedagogical approaches, and assessment practices to ensure they are inclusive, relevant, and representative of the local context.

In the last few years, higher education in India is undergoing several reformations in line with the recommendations of NEP 2020. However, successful transformation would require systemic disciplinary focused efforts from various stakeholders (teachers, students, parents, policymakers) to be involved in a dialogue to critically examine deeply rooted colonial influences on India's education system. As engineering is one of the most sought-after choices for higher education in India, we believe the transformation of engineering education should be led by research-driven inquiry to investigate and explore how the discipline can go through a post-colonial transformation by drawing the best from both ancient Indian systems and global practices. This research study aims to engage multiple stakeholders in the Indian engineering education ecosystem to examine, deliberate, and collectively suggest a post-colonial agenda for engineering education research in India. We believe such a research agenda will guide the systematic transformation of engineering education in India (Vijaylakshmi et al., 2022), in alignment with the vision and aspirations of NEP 2020.

III. METHODS

The authors employed a qualitative approach to address the research objectives of the study. Semi-structured interviews were conducted with the various stakeholders that play a key role in designing and implementing curricula at engineering. These stakeholders included engineering education leaders (i.e., deans and vice chancellors of engineering colleges and universities), heads of teaching and learning centers at engineering colleges, engineering education researchers and practitioners (i.e., people who either conduct research in engineering education or have been implementing research-informed pedagogies in classrooms). Cognitive interviews were conducted with few individuals to test the reliability of the semi-structured interview protocol and changes were made to a few questions based on the feedback received. A total of 18 participants (involving 6 engineering education leaders (EEL), 8 engineering education researchers (EER), 2 engineering education practitioners (EEP), and 2 heads of Teaching Learning Centers (TLC) were interviewed. It should be noted that all participants had some experience of working in the Indian engineering education context, i.e., they were working at an institution in India or, if they were employed by universities outside India, they had conducted research that directly pertained to issues related to engineering education in India. The interview questions asked participants about their prior experience with engineering education research and/or practice, current state of engineering education research in India, and their thoughts on the potential directions of research in this field. Prior to the start of each interview, verbal consent was taken from each participant for analyzing the data collected. The interview data were transcribed using Microsoft Teams' in-built transcription feature. The names of all the interview participants have been later anonymized during the data analysis process.

The transcribed interviews were qualitatively analyzed using the thematic analysis process (Braun & Clarke, 2006). The first step involved the authors reading through transcripts to identify instances where participants talked about either the challenges faced stakeholders in engineering teaching and learning or the potential directions for working in this space. This process involved inducting coding where each of the instances were assigned codes that were descriptive of the instance. In the next step, the authors explored the emergence of the themes by grouping the various inductive codes together. This step was an iterative process as different codes were grouped into different themes that described distinct potential areas for engineering education research in India. The analysis resulted in seven themes.

Trustworthiness was established during both steps of the analysis. The first step of assigning codes to interview excerpts was divided among the three authors. However, most interviews from this step were then reviewed by either one or both co-authors to identify points of disagreement. Each disagreement was resolved through discussions that resulted in the modification of the code. At the second stage of analysis that involved collapsing codes into themes, the authors again reviewed the final themes and the codes that were identified as part of a particular theme that were mutually agreed upon.

IV. FINDINGS

In this section, we present the themes that emerged from the data analysis. Exemplar quotes are provided as evidence for each theme. We provide pseudonyms (EER, EEL, EEP, and

https://doi.org/10.52202/073963-0003

TLC) to all the participate and present their quotes by referring to the category of sample.

A. Theme 1 - Transformation of engineering curriculum based on the needs of the industry and society

Interviews with the participants revealed the growing need to transform the curriculum of engineering education in India and better align it with the local contexts. For example, EER1 emphasized the need to look at engineering education as a way to improve the socio-political aspects at both national and regional levels in India – *"Like the socio-political situation in India within a state, within a city, and within a town would be completely different. And so what can engineering education bring there? I think there's some kind of like assessment to make it contextualized, to make it community-based, would be really important."* This participant believed that it is important for engineering education to look inward, recognize local needs and problems, and empower engineering students with the knowledge and skills to address those problems in the society. Engaging students in meaningful experiences with the society has been reported significantly boost their preparation for personal and professional life after graduation (Kandakatla et al., 2023).

Another major transformation of engineering curriculum in India is required to increase its relevance to the needs of the industry. Engineering curriculum offered in most of the engineering institutions was designed several decades ago and is known to be outdated and irrelevant with the requirements of the industry. Participants believed there should be more extensive involvement and engagement of the industry in the design of engineering curricula so that the outgoing graduates are better prepared for their future roles. For example, EEL4 mentioned – *"I think that having industry involved in engineering curriculum setting and having a dialogue with them is very important? To let them know what we think and to accept what their requirements are, and find the middle ground is very important. This is an aspect of engineering education that we should look at seriously."* He indicated the need for the Indian government and associated policy makers to encourage autonomy and flexibility to adapt the curriculum with the continuous and rapid changes in the industry and how engineering institutions should explore different ways to integrate industry-oriented problems into the Indian curriculum.

B. Theme 2 - Adoption of evidence-based and student-centric teaching practices to enable higher-order learning.

One of the major post-colonial shifts needed in the Indian education system is to reduce the emphasis on rote learning in India. TLC 1 spoke in the interview about the continued emphasis on rote learning in India and need to promote original thinking which was at the core of Indian education system before the British colonization – "*One of the things I'm constantly reminded about in the context of India is the need to promoting original thinking. I can say analyzing we don't do enough of that. Encouraging and supporting questioning. And you know that that needs to be embedded back into the culture. I don't understand how we lost track of that tradition of questioning that we had always, but we need to bring that element into the fold and sort of move it, take it forward*". EEL2 spoke of the negative implications of the rote learning on development of the Indian students - "*Rote learning is something you know, which makes a child's brain limited*" and the TLC1 mentioned need to differentiate between surface and deep learning *"I have some superficial understanding of the subject just by reading through some things and I think that I have learned something. I think that is a very important topic for India to be able to understand between surface learning and deep learning."* EEL6 believed it is important to enable students to have learning opportunities think and engage at the higher-order cognitive levels - *"We need to really look at the pedagogy and the assessment to ensure that we consciously engage students in higher order learning. There need to be experiences where student need to solve open-ended problems."*

TLC1 encouraged the need for engineering faculty in India to build an understanding of evidence-based teaching and learning practices to enable students to engage in higher-order learning - "*We have to position the student in the context of solving a problem, work in teams as it needs to do critical thinking, problem solving, and you know all these things enable higher order thinking skills. If you refer to Bloom's Taxonomy, apply, analyze, evaluate, and create, those cognitive levels require one to be able to solve complex problems. You have pedagogies like PBL, service learning which essentially embed higher-order learning experiences*".

EEL3 emphasized a need for engineering education India to shift from teacher-centric to student-centric learning with a focus to improve students' motivation and increase their active engagement in the classroom "*Students need to be engaged in the class to promote higher level thinking. They should be made to think and express in the classroom. Because it's a 50-minute session and there are sixty such sessions in a semester. Each faculty member should at least spend a couple of classes where students are fully engaged through group discussions etc. How can we increase the number of faculty doing this?*"

Integration of experiential learning experiences in Indian engineering education was reported by EEL1 as another way to engage students in higher-order learning – "*Lots of experiential learning with problem-based learning approach. There are many approaches which we can learn and bring into the practice to ensure higher order learning*". He further elaborated the need to understand how and to what extent experiential learning can be embedded into students learning experiences – "*How we can bring in more experiential, contextual, situated learning into our student experiences? How pedagogy can be designed to challenge the student?*" EEL3 also emphasized that

https://doi.org/10.52202/073963-0003

the assessment practices followed in Indian engineering education must also be aligned to evaluate higher-order learning, which would require a paradigm shift in how Indian educators perceive the purpose and value of assessment in educational process. As she noted – "*How do Indian students learn? That's an interesting question at a broad level because we have so much of this emphasis on exams, marks, coaching, etc. Do they really assess learning of students? And if so, how much of it is happening? So, I think those are all important questions to answer*". EEP1 also mentioned the need to increase focus formative assessment to provide students feedback on their learning "*There are so many exams that are held and assessments that are done. What is the quality of the feedback that is given to the student and how many students get that personalized feedback? We don't know. But there is a way, through formative assessments, how do you increase that?*". Evidence-based assessment methods must therefore be explored, analyzed, and adopted in the Indian context.

C. Theme 3 - Breaking disciplinary boundaries to promote inter and multidisciplinary engineering.

EEL6 recognized that industry in the current contexts require engineering professional to work on problems which are multidisciplinary in nature thereby emphasizing its importance among engineering graduates "*Complex problems of today's world cannot be solved by disciplinary knowledge alone. While the way I understand that the purpose of higher education is to equip the learners with disciplinary competencies, we need to also focus on students' ability to work in multidisciplinary teams to solve the problems, of today's world. So, these are the perspectives through which we need to look at multidisciplinary education and its need in the context of engineering education*". The boundaries between engineering disciplines are converging in the modern context "*Disciplinary boundaries are blurring as today's complex problems require multidisciplinary skills*" and academic institutions must therefore evolve from the existing disciplinary silos. EEL1 said "*How the traditional institutions which have lived for a very long time in silos, an engineering institution, a law institution, a liberal arts institution, a science institution, must now break their boundaries, develop capability to bring multidisciplinary experience. Because I think the way forward, not just for engineering education but for higher education in general is to break down silos.*"

As there is a shift towards inter and multidisciplinary learning, there is also a need for exploring innovative approaches to embed multidisciplinary learning experiences among students (Amashi et al., 2021). EEL6 said there is a lack of understanding currently on how multidisciplinary learning as a concept can be put to practice – "*We need to really understand multidisciplinary [learning]. How we can bring multidisciplinary learning into our curriculum. We are not yet very clear about it. [A] lot of research needs to happen in Indian context. It needs to be embedded strategically and we still are not very clear on how. There is lot of literature but in Indian context, these experiments need to happen*". Similarly, EER3 also discussed about interdisciplinarity as an approach to provide students with varied career pathways after graduation "*I always thought that it probably, you know, it would help to have more interdisciplinary approach for students who want to probably explore other new areas to do something that they want to do.*"

D. Theme 4 - Reimagining engineering education through technology

Participants believed it is important reimagine engineering education using technology, so that engineering educators can understand the different ways in which technology can serve as an enabler (Deepika et al., 2021). The sudden shift to online learning due to the disruptions caused by COVID19 pandemic in India has opened avenues to experiment with technology in engineering education (Syed et al., 2021). One of the prominent approaches recommended by participants was blended learning, which combines both classroom and online learning (Kandakatla et al., 2020). Through technology tools, students can now have access to learning resources from renowned faculty both inside and outside of their institutions (Kandakatla et al., n.d.). EEL5 said - "*You can have blended learning and learn online. For us as educational institutes, the best professors can record their lecture and all interested students can learn from them. Because everyone cannot get admission in the premier institutions which have the best faculty. However, I can access some of their teachings through technology*". TLC2 mentioned how the use of technology could provide opportunities for engineering institutions in low resource contexts to engage their students in experiential learning through virtual labs – "*Most engineering schools in India are suffer from lack of infrastructure. They don't really have good laboratories. So virtual labs could be a really good complement for actual experience.*" How does specific technology affect or change engineering education in India, should also be studied.

Participants noted the tremendous potential of emerging technologies such as Artificial Intelligence (AI), Augmented Reality (AR), and Virtual Reality (VR) and reflected on how they could be integrated into engineering education in India. EEL5 noted the advances in AI on it could disrupt current practices in engineering education – "*The way AI is penetrating the system, the way data analytics is maturing, I'm very sure we will have well informed approaches on self-regulation and self-motivation of learners [and] support for the teacher in engaging the learner effectively in the teaching learning process through the use of AI technologies*". EER6 mentioned opportunities for research in the use of generative AI tools in engineering education – "*a lot of research required at how AI and ChatGPT is being used for engineering education*". EEL1 indicated other emerging technologies like AR and VR to have huge scope for engineering education in India as they provide opportunities to simulate industry setting in the engineering

https://doi.org/10.52202/073963-0003

institutions – "*You know AR, VR kind of environment or metaverse kind of environment will be a good surrogate for real hands-on work. This requires quite a bit of research to understand students can experiment in such virtual environment.*"

E. Theme 5 – Nurturing innovation and entrepreneurship for self-reliant India

The COVID19 pandemic has pushed governments across the globe to recognize the important of robust supply chains that are self-reliant in terms of consumption and production in the country (Kandakatla et al., 2021). EEL6 believed innovation and entrepreneurial efforts in India must be aligned to the local needs of the diverse population in India – "*how can typical Indian institution particularly in Tier 2, Tier 3 cities can build an innovation and entrepreneurial ecosystem. To drive innovation and entrepreneurial ecosystem, we must understand that each of us should influence the socioeconomic scenario of the region in which we live. Where our institutions are present, we need to go beyond teaching and research and should participate in a transformation of the region. For us to play that role, we need to find that model of driving innovation and entrepreneurship.*" EER1 said there is immense potential for entrepreneurship in India if we're able to embed innovation thinking and mindset among the upcoming younger generation – "*It's about the mindset. So, in that sense to me it's absolutely important, at least for a country like India where it's a huge country with diverse culture, and one size does not fit all here. Our engineering education system should formally recognize the importance of both innovation and entrepreneurship and create adequate space in our curricular settings to promote and formally rewarded students. Then it becomes part of one's DNA.*"

However, the nurturing of innovation and entrepreneurship in engineering institutions requires a complete ecosystem. EEL1 noted, "*You see, innovation and entrepreneurship need an ecosystem. It does not come from one of subject on innovation or one of subject on entrepreneurship. The institutional ecosystem is very essential. For you to drive innovation, the first question we need to ask is how? Learning experience in our institution that challenges the student to solve the real problems, to connect to the real world. Because always the innovation happens when we connect to the real world and understand the issues and challenges through observation, through engagement. Without that, innovation does not happen just like that. So how well can we engage our students to with the larger ecosystem which really challenges him*". EEL3 said the ecosystem should also align student's innovative ideas with entrepreneurial knowledge "*How well our ecosystem helps them to move forward by making them to build his own solution process and see whether this problem is a business opportunity or this problem has a greater social impact, or it has both. So, all these things we need to build without building the required ecosystem in the institution.*"

F. Theme 6 – Role of faculty as change agents in the transformation of engineering education

Engineering faculty in India are considered as one of the key stakeholders as they're the change agents who will lead the transformation of engineering education. EEL6 said – "*You know, to me, one of the major concerns is the teacher. He is the change agent. So how do I enhance the productivity of the teacher in the field? Everything that you do, you know, all issues related to teachers are to me absolutely important because they are the change agents. So, to me, engineering, education, research, focusing on the changing role of the teachers in today's context, the challenges that they have, how do we empower them? How do we keep him on the pedestal of learner? So that he continues to learn and thereby continues to contribute to the engineering education space*". However, the EEL3 considers a change in engineering faculty beliefs and mindset as one of the major challenges that could impede any transformation – "*Transformation of faculty toughest because most of us teachers, we come from a legacy. We have our own practices, 30-40 years of our experiences have shaped us in a particular way. I mean particular way of working, particular way of thinking. NEP 2020 really challenges us. It wants a very different approach from us. How well can we transform ourselves at individual level*".

EEL6 believed there is growing acceptance among Indian faculty on the change in expectations from Indian engineering institutions – "*But today I'm seeing there is a growing recognition among engineering educators themselves that. They also need to know pedagogical skills they need to have. They also need to be aware of technological skills, you know, so disciplinary knowledge, pedagogical skills and technological awareness.*" However, large scale capacity building efforts needs to be taken up to change the beliefs of faculty from being present in the past to being futuristic "*I think it will have really positive impact if faculty are able to understand why it is done that. Any old institution will have a heterogeneity of faculty who are very seniors, who are in the middle, who are the youngsters. So how do you manage these people in getting them trained? Because generally we will be talking about the past only, but we should come out from that past, talk about present in the future.*" The use of faculty development programs and communities of practice have been reported to enable large scale change among faculty towards their instructional practices (Kandakatla & Palla, 2020). TLC1 reported how faculty should be instilled with feelings of care towards students – "*I think is very important that faculty are curious and empathetic enough that they want to see their student do well. Am I able to mentor students throughout the four years because I have a personal interest in their growth?*"

G. Theme 7 – Change in engineering students' attitudes, aspirations, and learning preferences.

The last theme of the study focused another important and the most essential stakeholder that is the engineering students. Engineering is the one of most sought-after discipline for higher

https://doi.org/10.52202/073963-0003

education in India, with over a million students graduating every year. EEL2 said majority of the students often lack the interest and motivation as most of them join engineering education due to parental or peer pressure "*We have many students who have joined a relevant branch not with their interest but because of parental and peer pressure*". TLC1 believed students motivation in classroom as a major challenge - "*Motivation is a big part of it, but how do you bring day-to-day classroom? So there's a 50 hours of classroom. How much time and how many students are attentive? So what is the connection between attention and so more into the ability to self-regulate the mind and be focused? Of course meditation is an important part of our culture and it is, but it's not perceived like that among the students. So it is, it is an important aspect because attention is a big problem today.*"

The current students who are part of the GenZ (students born in and after early 2000s) particularly have challenges with low attention span and engagement in the classroom. EEL4 believes engineering institutions must therefore explore innovative methods to engage with these learners - "*Handling GenZ students is not so easy according to me. It needs a different kind of pedagogy because the retention rates are very, very less. They're all born with a mobile phone. A lot of innovation is required, lot of education research is required for us to understand how exactly all these dimensions can be brought into the student experience as per the unique aspirations.*" EEL5 emphasized the need to encourage self-regulated learning among students, to help them become life-long learners and keep up with the rapid changes in the industry – "*Another change that I foresee is, no skill will stay beyond the three to five years with anybody. If you want to remain valid, you must upgrade yourself. So lifelong learning is going to be a major thing probably in one's lifetime. One will have to visit, you know, institutions multiple times to reskill and upskill himself, you know.*" EER2 said efforts must also be taken to instill compassion, ethics, and empathy among students with a goal to make them socially responsible - "*So we have to train our students in that way in the four years they are with us, they learn about sustainability, SDGS, ethics etc. I think those are the changes which must be taken in into account when we are actually nurturing students into good human beings*".

V. DISCUSSION

We now present a research agenda for engineering education in India through various focus areas along some guiding requestions for engineering education researchers in India. It is important to note that while some of the research questions proposed were directly suggested by the participants (specifically the engineering education researchers), the other research questions are being proposed by the authors (as engineering education researchers themselves) based on the themes and discussions presented in the results section.

A. Discipline-based educational research

- How can student-centered teaching and learning practices be adopted to address issues of motivation, attention, and retention among GenZ learners?
- What evidence-based teaching practices and assessment methods could be designed to engage Indian engineering students in higher-order learning?
- How can the horizontal and vertical alignment of engineering curriculum be improved in India?
- What are the disciple-specific skills expected from Indian engineering graduates to overcome the barriers of transition to workspace?
- How can discipline-specific collaborations models could be implemented in engineering institutions to bridge the gap between industry and academia?

B. Multidisciplinary engineering

- What models of experiential learning can be adopted by Indian engineering institutions to promote multidisciplinary learning?
- How can strategic collaborations be established among different disciplines in an institution to promote multidisciplinary engineering?
- How can students be provided with informal opportunities to engage in multidisciplinary learning experiences?
- What skills are required by engineering faculty to engage students in multidisciplinary learning experiences?
- How can computer science be integrated into core engineering disciplines (civil, electrical, and mechanical) to improve students' interest and align with needs of the industry 4.0?

C. Technology-enhanced Learning

- How can blended learning be adopted in Indian engineering institutions to promote student engagement and learning outside the classroom?
- How can virtual laboratories be used to provide students with experiential learning opportunities especially in low resource contexts?
- How can data analytics be used to assess individual learning needs, preferences,
- How can AR/VR be developed and embedded to provide engineering students with authentic learning experiences?
- How can artificial intelligence based used to predict the enablers and barriers in the learning pathways of engineering students?

D. Community engagement through Service-Learning

- How can the engineering curriculum be localized to the needs of the Indian society?

 https://doi.org/10.52202/073963-0003

- How can Indian engineering graduates be nurtured into ethical, empathetic, and socially responsible citizens of the country?
- How can engineering institutions increase the relevance towards the socio-economic development of nearby regions?
- What are the enablers and barriers for engineering faculty and students to engage in service-learning?

E. Innovation and Entrepreneurship in Engineering Education

- How can engineering students be engaged in design and development of innovations through curricular and in-formal learning experiences?
- How can entrepreneurial thinking be nurtured among engineering students in India?
- What are the effective models engineering institutions foster an ecosystem for innovation and entrepreneurship?
- What are the formal and in-formal mode for engineering faculty to engage in innovation and entrepreneurship?
- How can entrepreneurial mindset be inculcated among engineering faculty to increase potential for monetization of research activities?

F. Facilitating Change in Engineering Institutions

- What are the barriers to change among engineering institutions to transform themselves in line with recommendations of NEP2020?
- How can engineering faculty in India be supported towards the pedagogical transformation of their classroom practices?
- How can engineering faculty's perceptions of assessment be changed to encourage increase in use of formative practices?
- How can engineering faculty be inculcated with feelings of care and compassion towards the holistic development of their students?
- How can engineering faculty and students in India be enabled to become life-long learners?
- What models can engineering institutions adopt to promote diversity and inclusion among their stakeholders?

VI. CONCLUSION

Research in engineering education in India can transform our education, bringing students out of rote learning mode and inbreeding innovative mindset and ownership of learning. It is necessary to change faculty and institutional landscape and mindset for shaping today's engineers, and furthermore mould our thought processes to deliver the quality of education with intentionality for preparing the workforce and tomorrow's leaders capable of addressing post-colonial India's grand challenges. The research agenda presented in this study is an attempt to guide engineering education researchers to answer critical questions that are essential to be answered for post-colonial transformation of engineering education in India.

ACKNOWLEDGMENT

The authors declare that they have no conflict of interest. Authors would like to first thank Prof. William Oakes from Purdue University for providing the inspiration and being the catalyst to start this study. We next thank Prof. Gopal Joshi and Prof. Sohum Sohoni for their guidance and for providing suggestions to recruit the various participants for the study. Lastly, we would like to thank all the participants including the engineering education leaders, teaching and learning heads, engineering education researchers, and practitioners for giving their insights and time through the interviews.

REFERENCES

Amashi, R., Joshi, G., & Kandakatla, R. (2021). The Influence of Sustainable Development Module on the Values and Beliefs of First-Year Students in Undergraduate Engineering Education. *Journal of Engineering Education Transformations*, *34*(0), Article 0. https://doi.org/10.16920/jeet/2021/v34i0/157112

Braun, V., & Clarke, V. (2006). Using thematic analysis in psychology. *Qualitative Research in Psychology*, *3*(2), 77–101. https://doi.org/10.1191/1478088706qp063oa

Chaudhary, L. (2007). Essays on Education and Social Divisions in Colonial India. *The Journal of Economic History*, *67*(2), 500–503. https://doi.org/10.1017/S0022050707250181

Choudhury, P. K. (2016). Growth of Engineering Education in India: Status, Issues and Challenges. *Higher Education for the Future*, *3*(1), 93–107. https://doi.org/10.1177/2347631115610223

Cohn, B. (2009). Colonialism and Its Forms of Knowledge: The British in India. In *The New Imperial Histories Reader*. Routledge.

Crozet, C. (2012). The core tenets of education in ancient India, inspirations for modern times. *International Journal of Pedagogies and Learning*, *7*(3), 262–265. https://doi.org/10.5172/ijpl.2012.7.3.262

Deepika, A., Kandakatla, R., Saida, A., & Reddy, V. B. (2021). Implementation of ICAP Principles through Technology Tools: Exploring the Alignment between Pedagogy and Technology. *Journal of Engineering Education Transformations*, *34*(0), Article 0. https://doi.org/10.16920/jeet/2021/v34i0/157210

Ellis, C. (2020). History of Colonial Education: Key Reflections. In P. M. Sarangapani & R. Pappu (Eds.), *Handbook of Education Systems in South Asia* (pp. 1–27). Springer. https://doi.org/10.1007/978-981-13-3309-5_70-1

Fischer-Tiné, H., & Mann, M. (2004). *Colonialism as Civilizing Mission: Cultural Ideology in British India*. Anthem Press.

Ghosh, S. C. (1993). 'English in taste, in opinions, in words and intellect': Indoctrinating the Indian through textbook, curriculum and education. In *The Imperial Curriculum*. Routledge.

Jain, S. B., Choudhary, S., & Philip, J. (2021). Revival and Resurgence of Our Lost GEM -- The Ancient Indian

Education System. *Shanlax International Journal of Education*, *9*(4), 382–388.
Kandakatla, R., Aluvalu, R., Devireddy, S., Kulkarni, N., & Joshi, G. (2021). Role of Indian Higher Education Institutions towards Aatmanirbhar India: Government Policies and Initiatives to promote Entrepreneurship and Innovation. *2021 World Engineering Education Forum/Global Engineering Deans Council (WEEF/GEDC)*, 8–14. https://doi.org/10.1109/WEEF/GEDC53299.2021.9657261
Kandakatla, R., Berger, E. J., Rhoads, J. F., & DeBoer, J. (2020). Student Perspectives on the Learning Resources in an Active, Blended, and Collaborative (ABC) Pedagogical Environment. *International Journal of Engineering Pedagogy (iJEP)*, *10*(2), 7. https://doi.org/10.3991/ijep.v10i2.11606
Kandakatla, R., Berger, E., Rhoads, J. F., & DeBoer, J. (n.d.). *Technology-Mediated Resources as a Substitute to Human Resources: Experiences of International Undergraduate Students in an Active, Blended, and Collaborative (ABC) Classroom in the USA*. 83.
Kandakatla, R., Dustker, S. M., Bandi, S., & Oakes, W. (2023). Achieving Indian National Board of Accreditation Engineering Graduate Attributes through Project-Based Service-Learning: Conceptual Analysis. *International Journal for Service Learning in Engineering, Humanitarian Engineering and Social Entrepreneurship*, *18*(1), Article 1.
Kandakatla, R., & Palla, A. (2020). Role of Community of Practice (CoP) to Facilitate Change in STEM Instructional Practices through Faculty Development Programs. *2020 IFEES World Engineering Education Forum - Global Engineering Deans Council (WEEF-GEDC)*, 1–5. https://doi.org/10.1109/WEEF-GEDC49885.2020.9293660
Kumar, A. (2021). New education policy (NEP) 2020: A roadmap for India 2.0. *University of South Florida (USF) M3 Publishing*, *3*(2021). https://www.doi.org/10.5038/9781955833042
Majid, I., & Kouser, S. (2020). *Indian Higher Education from the Lens of National Policy of Education (1986) and National Policy on Education (2016): A Comparative Study. IX*, 1152–1166.
Malakar, S. (2022). National Education Policy 2020: Reform's In Schools And Higher Education: An Analytical Study. *Global Journal of Applied Engineering in Computer Science and Mathematics (GJAECSMA)–Special Edition*.
Minds, A. (2019). National Employability Report: Engineers. 2019. *Gurugram, India*.
Mohanty, A., & Dash, D. (2016). Engineering Education in India: Preparation of Professional Engineering Educators. *Journal of Human Resource and Sustainability Studies*, *4*(2), Article 2. https://doi.org/10.4236/jhrss.2016.42011
Mukhopadhyay, D. (2017). Revisiting the Kothari Commission (1964–66) Report from the perspective of strengthening our science education and research enterprise. *Current Science*, *113*(12), 2258–2261.
Pandya, D. R. N. (2014). *Indian Education System- A Historical Journey*. *3*(3).
Pratik, A., Radhakrishnan, D. B., & Kandakatla, R. (2013, August 30). Innovation in Indian Engineering Educational System: Study and Analysis. *WEEF 2013 Cartagena*. WEEF 2013 Cartagena. https://www.acofipapers.org/index.php/acofipapers/2013/paper/view/201
Ramamurthy, S., & Pandiyan, K. (2017). National Policy on Education 2016: A Comparative Critique with NPE 1986. *Economic and Political Weekly*, *52*(16), 46–53.
Rao, L. P. (2022). National Educational Policies in India. *Sumedha Journal of Management*, *11*(2), 56–62.
Syed, K., Kandakatla, R., Madhulita, Yadav, R. I., & Himasagarika, R. (2021). Responding to COVID-19 and Transitioning to Online Learning: Evaluation of an Institution wide Capacity Building Efforts on Technology-Enhanced Learning. *Journal of Engineering Education Transformations*, *34*(0), Article 0. https://doi.org/10.16920/jeet/2021/v34i0/157232
Tilak, J. B. G. (2020). What can we learn from Universities of Ancient India? *Journal of Contemporary Educational Research*, *4*(1), Article 1. https://doi.org/10.26689/jcer.v4i1.925
Vijaylakshmi, M., Kandakatla, R., Baligar, P., Joshi, G., & Shettar, A. (2022). Design of Research Canvas to align Research Efforts at Engineering Education Research Centre in India. *2022 IEEE Global Engineering Education Conference (EDUCON)*, 844–849. https://doi.org/10.1109/EDUCON52537.2022.9766458

Analyzing Open Mindedness and Critical Thinking in Research Mindset: A Quantitative Approach for Engineering Doctoral Students

Sanjeev Kavale[a], Shawn Jordan[a], and Adam Carberry[b].
Arizona State University[a], The Ohio State University[b]
Corresponding Author Email: skavale@asu.edu

Abstract

Context

A research mindset corresponds to cognitive procedures, mental filters, or beliefs that help in performing a successful research task. A previous qualitative study examining doctoral students' research mindsets revealed six attributes: (1) open mindedness, (2) believing in oneself and the research, (3) persistence, (4) honesty, (5) being critical, and (6) a writing mindset.

Purpose or Goal

This current study uses a quantitative approach to: (1) assess the research mindset attributes of open mindedness and being critical, and (2) explore possible gender and engineering disciplinary differences within these attributes.

Methods

A total of 89 doctoral students belonging to different engineering disciplines participated in the survey. An exploratory factor analysis was performed to reveal an initial factor structure for 'open mindedness' and 'being critical.' Kruskal Wallis tests and multiple linear regression analyses were performed to further understand the differences in these attributes across the demographic characteristics of gender and engineering discipline.

Outcomes

The exploratory factor analysis revealed a two-factor solution in line with the research mindset's attributes of 'open mindedness' and 'being critical.' Further analysis revealed no significant evidence to claim that gender differences exist when examining 'open mindedness' and 'being critical' scores. Some differences were observed for 'being critical' across disciplines for the doctoral student sample.

Conclusion

The study contributes valuable insights related to research mindset of doctoral students. The two attributes, open mindedness and being critical, are gender neutral but have differences in engineering disciplines leading to potential implication on how doctoral education can be designed and delivered. This also necessitates further research to gain a deeper understanding of the research mindset.

Keywords**—**Critical thinking; Doctoral students; Mindset; Open mindedness; Research Mindset.

I. INTRODUCTION

MINDSET has been defined in a variety of ways. Cognitive psychology identifies mindset as the sum total of the activated cognitive procedures that consist of the cognitive orientation most conducive to successful task performance (French II, 2016; Gollwitzer & Bayer, 1999). Positive psychology views mindset as a set of beliefs that shape how one perceives this world and themselves (Brooks et al., 2012; Dweck, 2011; French II, 2016). Social psychology considers mindset as a cognitive filter or a frame of reference (French II, 2016; Gupta & Govindarajan, 2002). Each of these definitions of mindset suggest mindsets are crucial for the performance of any task.

Literature is abundant with studies related to many mindsets. The studies related to growth and fixed mindset (Dweck, 2006) can be considered pivotal in the research related to mindsets and have inspired many other researchers to investigate new spaces of mindsets. Having a certain mindset to suit specific task performances is beneficial (Eilers et al., 2022; Fang et al., 2022; Zingoni & Corey, 2017) and could be the reason for studies on different mindsets in the literature. Many non-discipline specific mindsets such as global mindset (Gupta & Govindarajan, 2002) and developmental mindset (Thurbon, 2016) and discipline-specific mindsets such as maker mindset (Dougherty, 2013) and entrepreneurial mindset (Naumann, 2017) have emerged in recent years.

Research can also be considered as a cognitive task or set of cognitive tasks as explained by the previous definitions of mindset. Conducting research may also need cognitive orientation, beliefs or mental filters that can be termed as 'research mindset' for the successful conduct of research tasks. The concept of 'research mindset' has gained attention in different contexts, and many scholars have explored its significance in multiple ways. For example, Kveven et al. (2014) have considered research mindset as a transformative process, empowering students to become critical thinkers who are adept at asking pertinent questions, conducting scientific research, and navigating complex data. On a similar note, Clark

https://doi.org/10.52202/073963-0004

& Johnstone (2018) have explored undergraduate music students for research mindset seeking to uncover not only their information-seeking behaviors but also their attitudes, comfort levels, and approaches towards research and writing within their academic journey. Conversely, Lee Chuen et al. (2019) advocate the cultivation of research mindset by actively engaging students in inquiry-based learning, promoting hands-on experience, interdisciplinary collaboration, and innovation. McEachern & Horton (2016) extend the concept's reach by explaining the necessity of research mindset for the development of researcher identity among faculty and students. Moreover, within the context of undergraduate engineering research experiences, other scholars, such as Branch et al. (2015) and Prasad and Bhat (2021), have also recognized the need for a research mindset. Despite the considerable attention given to the concept of a research mindset in the literature, it is noteworthy that an explicit and universally accepted definition of what constitutes a research mindset remains elusive. This gap in the literature underscores the complexity and evolving nature of this concept, necessitating further exploration and clarification.

Prior work undertaken by the research team revealed six different attributes of the research mindset held by doctoral engineering students: (1) open-mindedness, (2) believing in oneself and the research, (3) persistence, (4) honesty, (5) being critical, and a (6) writing mindset (Kavale & Carberry, 2023). It is interesting to see that some of the attributes of research mindset are attributes that have been cited as attributes of other mindsets (e.g., open-mindedness or be open minded is an attribute of entrepreneurial mindset (Brunhaver et al., 2018) and design thinking mindset (Maier et al., 2017), while truth-seeking, analyticity, systematicity, and inquisitiveness noted in critical thinking mindset (Bramhall et al., 2012; Facione et al., 2016) are closely related to the being critical attribute of the research mindset.

The current study is an extension of previous qualitative studies aimed to assess the elements of research mindset. The current study also examines potential differences across gender and engineering disciplines for two attributes of the research mindset: open-mindedness and critical thinking. Specifically, the current study's research questions are:

1. How do open mindedness and critical thinking manifest among doctoral students?
2. What, if any, gender differences exist for open-mindedness and critical thinking aspects of the research mindset held by doctoral students?
3. What, if any, engineering disciplinary differences exist for open-mindedness and critical thinking aspects of the research mindset held by doctoral students?

Recent scholarly articles mention the need for a research mindset (Branch et al., 2015; Prasad & Bhat, 2021). Such efforts are part of a larger effort happening in the field of mindsets. The topic resonates with engineering education researchers as a core capability of researchers. This study lays the foundation for understanding research mindset at a larger scale in various domains of STEM education that can become useful for the scientific community.

II. Conceptual Framing

The six attributes of research mindset identified by (Kavale & Carberry, 2023) - open-mindedness, believing in oneself and the research, persistence, honesty, being critical, and a writing mindset - were used to form a conceptual framework for the current study (Figure 1). The conceptual framework was used to create an instrument that informed the study's research questions. This study explores the attributes of 'open mindedness' and 'being critical.'

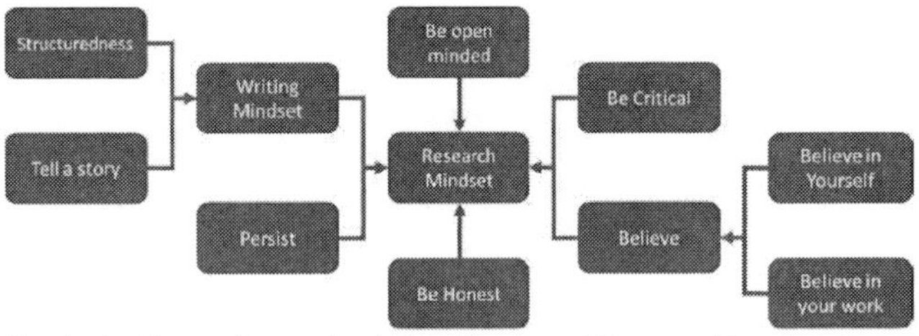

Fig. 1. Attributes of research mindset (Conceptual Framework).

Numerous studies within the field of engineering education have explored and measured open-mindedness among undergraduate engineering students. Some studies suggest that open mindedness improves doctoral education (Albertyn, 2022; Boud & Lee, 2005; Ortwein, 2015), but investigations of open-mindedness among engineering doctoral students, considering gender and disciplinary differences, is notably scarce.

A similar pattern is observed in the examination of critical thinking. A substantial body of research exists examining and assessing critical thinking among undergraduate engineering students (Ahern et al., 2019; Caratozzolo et al., 2019; Douglas, 2012). Developing and enhancing critical thinking skills is a fundamental responsibility of any educational program, particularly in doctoral education. A noticeable gap exists in the literature regarding graduate students. This is somewhat surprising considering the deliberate focus that has been placed in STEM doctoral education in the U.S. on fostering critical thinking throughout the Ph.D. journey (Golde, 2005; Leshner & Scherer, 2018).

III. Methods

A survey-based study was undertaken with engineering doctoral students. The study was approved by the Institutional Review Board (IRB) at the first author's institution. An exploratory factor analysis was conducted to address the first research question. Kruskal Wallis tests and multiple regression analyses were performed to address the remaining two research questions. The following subsections explain the methods in

https://doi.org/10.52202/073963-0004

detail.

A. Researcher Positionality

The authors of this paper all hold degrees in engineering and have substantial experience in the field of engineering education. The first author, currently pursuing a doctoral degree in engineering education, acknowledges a personal perspective that believes in the existence of a research mindset. It is important to recognize that this belief may have influenced the deliberations presented in this paper.

B. Participants and Data Collection

Doctoral students belonging to different schools of engineering at a research-intensive public university in the Southwestern region of the United States were surveyed to capture their research mindsets. The engineering college at the chosen university has an average enrolment of 1194 Ph.D. students per year for the last 5 years. Participant recruitment was undertaken through advising offices within the college. Personal email invitations were also shared with all students whose information was publicly available through lab or other university websites. It was not possible to assess the response rate of the participants because the total number of students receiving the invite is unknown. Ten participants were randomly selected to receive a $10 gift card as an incentive.

TABLE I
PARTICIPANT SAMPLE DEMOGRAPHICS

Gender	Female	26 (29.21%)
	Male	61 (68.53%)
	Genderqueer	1 (1.12%)
	Preferred not to say	1 (1.12%)
Racial and ethnic groups	Asians	53 (59.55%)
	Middle Eastern or North African	7 (7.86%)
	White	14 (15.73%)
	Hispanic, Latino or Spanish origin	4 (4.49%)
	Jewish	1 (1.12%)
	Multiracial	7 (7.86%)
	Preferred not to say	3 (3.37%)
Ph.D. major	Chemical engineering, biomedical engineering, and biotechnology	11 (12.35%)
	Computer or information technology engineering	36 (40.44%)
	Electrical engineering	4 (4.49%)
	Mechanical engineering, aeronautical engineering, civil engineering, and material science engineering	16 (17.97%)
	Human systems engineering	12 (13.48%)
	Engineering education	10 (11.23%)
International Student status	Yes	59 (66.29%)
	No	30 (33.70%)
Current year in Ph.D.	1st or 2nd year	38 (42.69%)
	3rd or higher	51 (57.30%)
No. of articles published	2 or less	40 (44.94%)
	3 or more	49 (55.05%)

A total of 114 responses were obtained. Responses from graduate students not currently enrolled in a Ph.D. program were removed from the data set (13 responses). Incomplete responses were also eliminated (12 responses). A total of 89 responses were included in the final analysis, which is approximately 7.5% of all engineering Ph.D. students enrolled at the institution. The demographic information of the sample is presented in Table I.

C. Measures

A total of 25 items were included in the survey. Sixteen items captured the independent variables of open mindedness (9 items) and being critical (7 items) using a 5-point Likert scale. The remaining items captured student demographics – gender, Ph.D. major, year in Ph.D., number of articles published, and international student status, which were used as dependent variables. The measures of the two constructs were created based on the codes found in the work by (Kavale & Carberry, 2023).

D. Validation of the instrument

Validity testing of the instrument was performed using recommendations provided by the *Encyclopedia of Social Measurement* (McGartland Rubio, 2005). Four engineering education research faculty examined the content embedded in the instrument. A spreadsheet containing all items was shared with each faculty reviewer. The faculty rated each item for representativeness and clarity. An option was also given to provide additional comments. The feedback led to the removal of 5 items.

The modified instrument was then pilot tested by two engineering doctoral students. Pilot testing was done using think aloud session to allow participants to voice their opinions in real time about the items in the instrument (Ericsson & Simon, 1993; Someren et al., 1994). The students were also asked to comment on overall relevance, number of items, response alternatives, wording, or additional comments. The instrument was further refined based on the inputs from the students. The total number of items remained at 16.

E. Exploratory Factor Analysis

An exploratory factor analysis was conducted to further reduce the number of items in the survey and to address the first research question. The analysis was performed using R version 4.2.2. Responses of all items were checked for means, standard deviations, Kurtosis, and skewness. Then, the Kaiser-Meyer-Olkin measure of sampling adequacy and Bartlett's test for sphericity were conducted to ensure that the sample was adequate for further analysis (McCoach et al., 2013). An exploratory factor analysis using the principal axis factoring method (McCoach et al., 2013) was performed on the data set. Scree plots (Cattell, 1966; Horn & Engstrom, 1979), parallel analysis (Slocum-Gori & Zumbo, 2011), and MAP test

(Velicer, 1976) were used to determine the appropriate number of latent factors,. The Oblimin rotation method (Clarkson & Jennrich, 1988) was employed to anticipate that the factors would be somewhat correlated. The factor structure was considered acceptable based on the following criteria (McCoach et al., 2013). Items were retained if they had a minimum factor loading value of 0.40 and a value less than 0.30 on all other factors. Cross-loaded items were not included in the factors. Inter-item correlations for all items were checked to be less than 0.85, and each factor had at least three loaded items. The reliability of the items within the factors was checked using Cronbach's alpha for a minimum value of 0.7 (Cronbach, 1951). Lastly, the factor correlations were examined, and a maximum value of 0.85 was deemed acceptable (McCoach et al., 2013).

F. Kruskal Wallis test and Regression Analysis

The second and third research questions were addressed using multiple regression analysis (Kutner et al., 2005) and Kruskal Wallis test (Theodorsson-Norheim, 1986; Vargha & Delaney, 1998). The composite scores of the constructs 'open mindedness' and 'being critical' were calculated based on the weighted averages using the loadings obtained from the exploratory factor analysis. Before further analysis, diagnostic tests were performed to check the assumptions of normality, linearity, and homoscedasticity. Histograms, scatter plots, and quantile-quantile plots were used to confirm these assumptions. The generalized Variance Inflation Factor (VIF) was calculated to identify multicollinearity issues in the regression models for a cutoff of 10 (Kutner et al., 2005). It is observed that the generalized variance inflation factor values for all dependent variables were within the cut-off value of 10. A visual inspection of residual plots suggested that there exists heteroscedasticity in the given data. Also, quantile-quantile plots suggested that the data is non normal and there exist a few outliers. The dependent variables were suitably dummy coded as needed by the multiple regression analysis.

IV. RESULTS

The results section is divided into two sections. The first section addresses the first research question on generalizing the attributes of the research mindset. The second section addresses the second and third research questions exploring potential impacts of demographic differences for research mindset.

A. Section 1: Exploratory Factor Analysis

Simple descriptive statistics of the responses for all items were checked first. Skewness and Kurtosis were evaluated for the normality of the data. The thresholds used to evaluate normality were ± 2 for skewness and Kurtosis (Aron et al., 2013). The skewness of all items varied between -4.74 to -1.13, 8 items failed to fall within the threshold. Similarly, the Kurtosis of all items varied between 3.61 to 29.71; 9 items failed to fall within the threshold. These findings indicate the

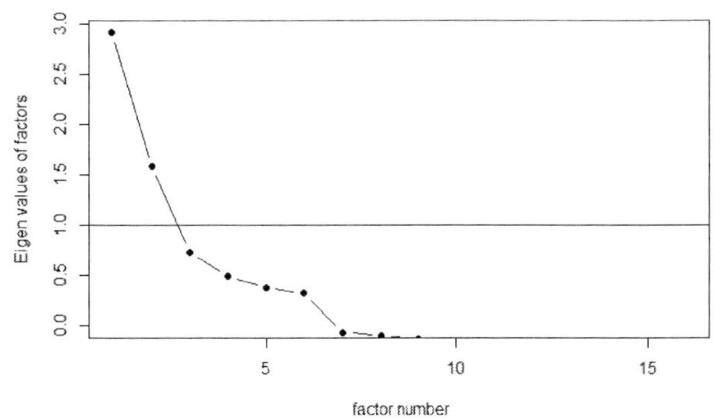

Fig. 2. Scree plot.

TABLE II
EXTRACTED FACTORS WITH ITEMS AND THEIR LOADINGS

Sl. No	Item	Factor Loading
	Factor 1: Open Mindedness	
1	I believe being open to other researchers' suggestions during any research activity is important.	0.429
2	I believe that my peers' opinions, criticisms, suggestions, and feedback are important for my research.	0.386
3	I am willing to learn from others to do improve my research activities.	0.677
4	I am willing to learn new things needed to do my research.	0.500
5	I believe my advisors' opinions, criticisms, suggestions, and feedback are important for my research	0.421
6	I believe getting feedback on the research activities is very crucial.	0.325
7	I believe that taking other researchers' opinions will give me different perspectives to work on my research.	0.613
8	Being open-minded while doing any research related activity is important.	0.612
9	Being open to feedback is an important aspect of research.	0.631
	Factor 2: Being Critical	
10	It is crucial for me to critically evaluate every step of the research process by questioning each task performed.	0.803
11	It is essential for me to validate each task performed during the research process.	0.578
12	It is important for me to have a clear understanding of my research tasks.	0.427
13	It is important for me to pay attention and be observant while conducting research tasks.	0.302
14	It is important for me to critically evaluate my views and perspectives while performing research tasks.	0.463
15	It is important for me to critically examine the views and perspectives of others while performing research tasks.	0.473
16	I believe that staying focused is crucial in performing research-related tasks.	0.477

https://doi.org/10.52202/073963-0004

presence of non-normally distributed data in this study. Exploratory factor analysis was conducted using the principal against violations of the assumption of multivariate normality in the data (Fabrigar et al., 1999). Therefore, no additional steps were taken to address this issue. Inter-item correlations were checked for all items, and no items were found to have correlations beyond 0.85. The sample can be considered adequate (n = 89) because it meets the minimum of 5 to 10 participants per variable or item. Bartlett's test of sphericity was significant (p = 0.000). The KMO measure obtained was 0.60, which meets the minimum threshold of 0.60 to determine sample adequacy, indicating a sufficient correlation between variables to proceed with the analyses.

The number of factors that can be extracted based on eigenvalues (Kaiser criterion) from a visual inspection of the Scree plot and the original MAP test suggest two factors. The Scree plot is shown in Figure 2. Parallel analysis suggested six factors. For verification, models with one, two, and three factors were created. The one-factor and three-factor models had multiple cross-loadings, suggesting the two-factor model best fit the data (Table II). This result also aligns with the expected number of constructs.'

Three items (items 2, 6 and 13) were removed due to low factor loadings. Cronbach's alpha for the factors 'open mindedness' and 'being critical' were 0.756 and 0.701, respectively. The correlation between the two factors was 0.13.

B. *Section 2: Kruskal Wallis Test and Regression Analysis*

Considering the non-normal nature of the data, the Kruskal Wallis test was performed to check if there were any differences in 'open mindedness' and 'being critical' scores based on gender or engineering discipline. There were no significant differences found (open mindedness based on gender: H(3) = 2.801, p = 0.423; being critical based on gender: H(3) = 3.086, p = 0.378; open mindedness based on Ph.D. Major: H(5) = 2.290, p = 0.807; being critical based on Ph.D. Major: H(5) = 8.132, p = 0.149).

Multiple linear regression analysis was also performed to address the second and third research questions. Table III represents the regression model for predicting the two attributes 'open mindedness' and 'being critical' of the research mindset. The created models with included demographic variables explain close to zero variance in 'open mindedness' and 'being critical' attributes of the research mindset. There is no sufficient evidence to say that there exist gender differences in the 'open mindedness' and 'being critical' attributes of the research mindset. This indicates a potential need for a study with a larger sample size to make statistical inferences. Despite this, an effort was made to investigate the results to inform future research.

TABLE III
EXTRACTED FACTORS WITH ITEMS AND THEIR LOADINGS

Variables	Open mindedness	Being critical
Intercept	4.847 ***	4.786 ***
Gender (base: Female)		
Genderqueer	-0.083	-0.483
Male	-0.006	0.077
Preferred not to say	-0.553	-0.257
Racial and Ethnic background (base: Asian)		
Hispanic, Latino or Spanish origin	-0.081	0.115
Jewish	0.139	0.411
Middle Eastern or North African	-0.081	-0.139
Multiracial	-0.154	-0.048
Preferred not to say	0.160	0.244
White	0.024	0.031
Ph.D. Major (base: Chem, Biotech, BioMed Engg)		
Computer or Information technology engineering	0.082	-0.227
Electrical Engineering	0.073	-0.176
Human Systems Engineering	0.020	-0.275
Mech, Civil, Materials and allied Engg	0.054	-0.473**
Engineering Education	0.150	-0.487**
International Student (base: No)		
Yes	-0.028	0.127
Number of articles published (base: 2 or less articles)		
3 or more articles	-0.051	0.123
Current year in Ph.D. (base: 1st or 2nd year)		
3rd or higher	-0.045	-0.241*
Adjusted R squared	-0.115	0.017
F Test	0.480	1.092
n	89	89

Note: All terms are standardized regression coefficients. *p<0.1.; **p<0.05.

No significant differences were observed for the individual item scores of 'open mindedness' based on participant majors. Some significant differences were observed for 'being critical' items based on Ph.D. majors. Comparing the scores of 'being critical' among students from different engineering disciplines (Note: Chemical, Biotechnology, and Biomedical Engineering were used as the baseline), it was found that students from Mechanical, Civil, Materials Science, and allied engineering disciplines had significantly lower scores (-0.473 points, p = 0.018) after controlling for other variables. Similarly, the students belonging to engineering education also had significantly lower scores (-0.487, p = 0.040). It is important to remember that the study had a limited sample size of 89 students, and that the measure was based on a self-reported survey data only. These findings suggest that there may be differences in critical thinking skills among engineering disciplines, but further research is needed to confirm these results and explore potential factors that may explain these differences.

V. Discussion and Implications

The presented study quantitatively examines two research mindset attributes: open mindedness and being critical. Exploratory factor analysis revealed a two-factor solution in line with the research mindset's two constructs considered for the study. Kruskal Wallis test and multiple linear regression analysis were performed to explore differences in the scores of 'open mindedness' and 'being critical' based on gender and major of engineering doctoral students.

Analyses performed yielded intriguing results. There was no observed sufficient evidence to claim that there exist gender differences in 'open mindedness' and 'being critical' scores. Some differences in the scores of 'being critical' based on major were observed. Interestingly, the highest difference in the scores were obtained with students pursuing a Ph.D. in engineering education. This observation raises interesting questions and warrants further exploration.

The findings from our study hold significant implications for the cultivation of a research-oriented mindset among early career researchers, particularly within the realm of doctoral education. STEM doctoral programs in the U.S. are designed to promote critical thinking, persistence, teamwork, and communication (Golde, 2005; Golde & Dore, 2001; Leshner & Scherer, 2018), yet there is paucity of literature exploring gender and disciplinary or major differences in the critical thinking mindset of doctoral students. Numerous studies have shown that a critical thinking mindset can improve critical thinking (Abrami et al., 2008; Tiruneh et al., 2014). There are numerous studies suggesting no gender differences in critical thinking of undergraduate engineering students (Özyurt, 2015, Sola et al., 2017). This could be a possible reason for the lack of gender differences among engineering doctoral students' 'being critical' scores in this study.

The scores of Ph.D. students in the engineering education domain being lower than those in other engineering disciplines is worth noting. This could be because engineering education as a discipline connects more closely to the social sciences than other engineering disciplines. As argued by Brookfield, critical thinking is influenced by various traditions and assumptions, which essentially represent different epistemological positions. Disciplines may hold alternative views on the nature or meaning of critical thinking (Brookfield, 2012). This context sheds light on potential factors contributing to the observed differences and underscores the importance of understanding the nuances within different academic majors or disciplines.

Open-mindedness has been identified as a crucial attribute for success in doctoral education (Albertyn, 2022; Boud & Lee, 2005; Ortwein, 2015). There are minimal explorations of how doctoral education integrates open-mindedness into its curriculum and pedagogy nor investigations on whether gender or disciplines effect open-mindedness.

VI. Limitations and Future Scope

The most important limitation of this study is the small sample size. There is a lack of sufficient representation of all demographic variables considered in the study. Many levels within a few demographic variables were merged to form simple categorical variables. This affected how the interpretations could be made from the regression models built. Also, the adjusted r-squared value is close to zero, indicating that the regression models explain a minimal variance. In the case of EFA results, the variance in the inter-item correlations for both factors is greater than 0.01. The inter-item correlations for a few items are lesser than 0.30. These findings suggest that there may be differences in critical thinking skills among engineering disciplines, but further research is needed to confirm these results and explore potential factors that may explain these differences.

It is interesting to note that some significant differences in the 'being critical' scores were observed between a few disciplines. Conducting a focused study on these particular disciplines would be beneficial to understand how critical thinking varies among different engineering disciplines. In particular, a study to understand if differences exist between social science and engineering researchers could provide valuable insights to the community.

The current study focused on the attributes of the research mindset found by (Kavale & Carberry, 2023). The items in the survey instrument were created based on the codes generated in this qualitative study. The opinions on 'open mindedness' and 'being critical' are limited to this study and the biases of the participants. A deeper understanding of how 'open mindedness' and 'being critical' are available within doctoral education is needed.

Finally, the current study focused on generalizing only two attributes, 'open mindedness' and 'being critical' of the research mindset. Further studies are needed to explore the remaining four attributes of the research mindset to provide a fuller understanding of research mindset.

References

Abrami, P. C., Bernard, R. M., Borokhovski, E., Wade, A., Surkes, M. A., Tamim, R., & Zhang, D. (2008). *Instructional Interventions Affecting Critical Thinking Skills and Dispositions: A Stage 1 Meta-Analysis.* Review of Educational Research, 78(4), 1102–1134. https://doi.org/10.3102/0034654308326084

Albertyn, R. (2022). *Making a case for Doctoral Intelligence: Conceptualisation and insights for researcher development.* Innovations in Education and Teaching International, 59(6), 623–633. https://doi.org/10.1080/14703297.2021.1899033

Aron, A., Coups, E. J., & Aron, E. (2013). *Statistics for psychology* (6th ed). Pearson.

Ahern, A., Dominguez, C., McNally, C., O'Sullivan, J. J., & Pedrosa, D. (2019). *A literature review of critical thinking in engineering education.* Studies in Higher Education, 44(5), 816–828. https://doi.org/10.1080/03075079.2019.1586325

Boud, D., & Lee, A. (2005). *'Peer learning' as pedagogic discourse for research education.* Studies in Higher Education, 30(5), 501–516. https://doi.org/10.1080/03075070500249138

Bramhall, M. D., Gray, L., Corker, C., Garnett, K., & Hill, R. (2012). *Analysis of Critical Thinking Skills in an International, Cross-Institutional Group of Engineering Master's Students.* Industry and Higher Education, 26(4), 323–327. https://doi.org/10.5367/ihe.2012.0109

Branch, S. E., Woodcock, A., & Graziano, W. G. (2015). *Person Orientation and Encouragement: Predicting Interest in Engineering Research: Predicting Interest in Engineering Research.* Journal of Engineering Education, 104(2), 119–138. https://doi.org/10.1002/jee.20068

Brookfield, S. (2012). *Teaching for critical thinking: Tools and techniques to help students question their assumptions* (1st ed). Jossey-Bass.

Brooks, R., Brooks, S., & Goldstein, S. (2012). *The Power of Mindsets: Nurturing Engagement, Motivation, and Resilience in Students.* In S. L. Christenson, A. L. Reschly, & C. Wylie (Eds.), Handbook of Research on Student Engagement (pp. 541–562). Springer US. https://doi.org/10.1007/978-1-4614-2018-7_26

Brunhaver, S. R., Bekki, J. M., Carberry, A. R., London, J. S., & McKenna, A. F. (2018). *Development of the Engineering Student Entrepreneurial Mindset Assessment (ESEMA).* Advances in Engineering Education, 7(1). https://eric.ed.gov/?id=EJ1199672

Caratozzolo, P., Alvarez-Delgado, A., & Hosseini, S. (2019). *Strengthening critical thinking in engineering students.* International Journal on Interactive Design and Manufacturing (IJIDeM), 13(3), 995–1012. https://doi.org/10.1007/s12008-019-00559-6

Cattell, R. B. (1966). *The Scree Test For The Number Of Factors.* Multivariate Behavioral Research, 1(2), 245–276. https://doi.org/10.1207/s15327906mbr0102_10

Clark, J., & Johnstone, J. (2018). *Exploring the Research Mindset and Information-Seeking Behaviors of Undergraduate Music Students.* College & Research Libraries, 79(4), 499–516. https://doi.org/10.5860/crl.79.4.499

Clarkson, D. B., & Jennrich, R. I. (1988). *Quartic rotation criteria and algorithms.* Psychometrika, 53(2), 251–259. https://doi.org/10.1007/BF02294136

Cronbach, L. J. (1951). *Coefficient alpha and the internal structure of tests.* Psychometrika, 16(3), 297–334. https://doi.org/10.1007/BF02310555

Dougherty, D. (2013). *Design, make, play: Growing the next generation of STEM innovators* (M. Honey & D. Kanter, Eds.). Routledge.

Douglas, E. P. (2012). *Defining and Measuring Critical Thinking in Engineering.* Procedia - Social and Behavioral Sciences, 56, 153–159. https://doi.org/10.1016/j.sbspro.2012.09.642

Dweck, C. S. (2006). *Mindset: The new psychology of success* (1st ed). Random House.

Dweck, C. S. (2011). *Implicit Theories. In Handbook of Theories of Social Psychology* (pp. 43–61). Sage. https://www.torrossa.com/en/resources/an/4912667

Eilers, K., Peters, C., & Leimeister, J. M. (2022). *Why the agile mindset matters.* Technological Forecasting and Social Change, 179, 121650. https://doi.org/10.1016/j.techfore.2022.121650

Ericsson, K. A., & Simon, H. A. (1993). *Protocol Analysis: Verbal Reports as Data.* The MIT Press. https://doi.org/10.7551/mitpress/5657.001.0001

Fabrigar, L. R., Wegener, D. T., MacCallum, R. C., & Strahan, E. J. (1999). *Evaluating the use of exploratory factor analysis in psychological research.* Psychological Methods, 4(3), 272–299. https://doi.org/10.1037/1082-989X.4.3.272

Facione, P. A., Gittens, C. A., & Facione, N. C. (2016). *Cultivating A Critical Thinking Mindset.* Academia. Edu. Weekly Digest, 28. http://www.measuredreasons.com/

Fang, Z., Chang, B., & Dang, J. (2022). *Growth mindset matters: Influences of socioeconomic status on Chinese secondary vocational students' learning engagement.* Journal of Pacific Rim Psychology, 16, 183449092211419. https://doi.org/10.1177/18344909221141984

French II, R. P. (2016). *The fuzziness of mindsets: Divergent conceptualizations and characterizations of mindset theory and praxis.* International Journal of Organizational Analysis, 24(4), 673–691. https://doi.org/10.1108/IJOA-09-2014-0797

Golde, C. M. (2005). *The Role of the Department and Discipline in Doctoral Student Attrition: Lessons from Four Departments.* The

Journal of Higher Education, 76(6), 669–700. https://doi.org/10.1080/00221546.2005.11772304

Golde, C. M., & Dore, T. M. (2001). *At Cross Purposes: What the Experiences of Today's Doctoral Students Reveal about Doctoral Education.* Wisconsin Univ., Madison.

Gollwitzer, P. M., & Bayer, U. C. (1999). *Deliberative versus implemental mindsets in the control of action.* In Dual-process theories in social psychology (pp. 403–422). Guilford Pr.

Gupta, A. K., & Govindarajan, V. (2002). *Cultivating a global mindset.* Academy of Management Perspectives, 16(1), 116–126. https://doi.org/10.5465/ame.2002.6640211

Horn, J. L., & Engstrom, R. (1979). *Cattell's Scree Test in Relation to Bartlett's Chi-Square Test and Other Observations on the Number of Factors Problem.* Multivariate Behavioral Research, 14(3), 283–300. https://doi.org/10.1207/s15327906mbr1403_1

Kavale, S. M., & Carberry, A. R. (2023, June). *Attributes of Research Mindset for Early Career Engineering Researchers.* 2023 ASEE Annual Conference & Exposition, Baltimore, Maryland. https://peer.asee.org/42344

Kutner, M. H., Nachtsheim, C. J., Neter, J., & Li, W. (2005). *Applied linear statistical models* (5th ed). McGraw-Hill Irwin.

Kveven, A., Searle, J., & Araniva, R. (2014). *Creating a Research Mindset. In Tapping the potential of all: Undergraduate research at community colleges* (p. 51). Council on Undergraduate Research.

Lee Chuen, K., Chee Siaw, S., & Loh, L. (2019). *Young Researcher Programme: An Inquiry-based Learning to Cultivate Innovation and Research Mindset.* 730–740.

Leshner, A., & Scherer, L. (Eds.). (2018). *Graduate STEM Education for the 21st Century* (p. 25038). National Academies Press. https://doi.org/10.17226/25038

Maier, A., Škec, S., Kim, H., Kokkolaras, M., Oehmen, J., Fadel, G., Salustri, F., & Van der Loos, M. (2017). *Towards a DT mindset tool evaluation: Factors identification from theory and practice.* Design Processes, Design Organisation and Management, 2, 367–376.

McCoach, D. B., Gable, R. K., & Madura, J. P. (2013). *Instrument development in the affective domain: School and corporate applications* (3rd ed). Springer.

McEachern, K. P., & Horton, J. L. (2016). *Developing a Research Identity: Promoting a Research Mindset Among Faculty and Students.* The Educational Forum, 80(4), 444–456. https://doi.org/10.1080/00131725.2016.1206159

McGartland Rubio, D. (2005). *Content Validity. In Encyclopedia of Social Measurement* (pp. 495–498). Elsevier. https://doi.org/10.1016/B0-12-369398-5/00397-2

Naumann, C. (2017). *Entrepreneurial Mindset: A Synthetic Literature Review.* Entrepreneurial Business and Economics Review, 5(3), 149–172.

Ortwein, M. J. (2015). *The Regulation of Understanding through Intellectual Virtue: Some Implications for Doctoral Education.* Journal of Thought, 49(1–2), 71. https://doi.org/10.2307/jthought.49.1-2.71

Özyurt, Ö. (2015). *Examining the Critical Thinking Dispositions and the Problem Solving Skills of Computer Engineering Students.* EURASIA Journal of Mathematics, Science and Technology Education, 11(2). https://doi.org/10.12973/eurasia.2015.1342a

Prasad, S., & Bhat, R. S. (2021). *India Industry-University Collaboration—A Novel Approach Combining Technology, Innovation, and Entrepreneurship.* 2021 IEEE Global Engineering Education Conference (EDUCON), 373–380. https://doi.org/10.1109/EDUCON46332.2021.9454090

Sola, E., Hoekstra, R., Fiore, S., & McCauley, P. (2017). *An Investigation of the State of Creativity and Critical Thinking in Engineering Undergraduates.* Creative Education, 08(09), 1495–1522. https://doi.org/10.4236/ce.2017.89105

Someren, M. W. van, Barnard, Y. F., & Sandberg, J. A. (1994). *The think aloud method: A practical guide to modelling cognitive processes.* Academic Press.

Slocum-Gori, S. L., & Zumbo, B. D. (2011). *Assessing the Unidimensionality of Psychological Scales: Using Multiple Criteria from Factor Analysis.* Social Indicators Research, 102(3), 443–461. https://doi.org/10.1007/s11205-010-9682-8

Theodorsson-Norheim, E. (1986). *Kruskal-Wallis test: BASIC computer program to perform nonparametric one-way analysis of variance and multiple comparisons on ranks of several independent samples.* Computer Methods and Programs in Biomedicine, 23(1), 57–62. https://doi.org/10.1016/0169-2607(86)90081-7

Thurbon, E. (2016). *Developmental Mindset: The Revival of Financial Activism in South Korea.* Cornell University Press. https://doi.org/10.7591/9781501704178

Tiruneh, D. T., Verburgh, A., & Elen, J. (2014). *Effectiveness of Critical Thinking Instruction in Higher Education: A Systematic Review of Intervention Studies.* Higher Education Studies, 4(1), p1. https://doi.org/10.5539/hes.v4n1p1

Vargha, A., & Delaney, H. D. (1998*). The Kruskal-Wallis Test and Stochastic Homogeneity.* Journal of Educational and Behavioral Statistics, 23(2), 170–192. https://doi.org/10.3102/10769986023002170

Velicer, W. F. (1976). *Determining the number of components from the matrix of partial correlations.* Psychometrika, 41(3), 321–327. https://doi.org/10.1007/BF02293557

Zingoni, M., & Corey, C. M. (2017). *How Mindset Matters: The Direct and Indirect Effects of Employees' Mindsets on Job Performance.* Journal of Personnel Psychology, 16(1), 36–45. https://doi.org/10.1027/1866-5888/a000171

Copyright statement

Expanding Horizons: The Transformative Influence of Interdisciplinary Education on Engineering Students

Sakethreddy Narsareddygari, Madhu Pathlavath, Nishanthreddy Samala , Santosh Madeva Naik
Hyderabad Institute of Technology and Management
sakethreddy2903@gmail.com

Abstract

Context

This paper aims to fill the knowledge gap by highlighting the specific advantages of Interdisciplinary engineering education. This approach improves engineering student abilities to communicate, collaborate, and solve problems by incorporating ideas and techniques from various fields. Interdisciplinary education involves crucial abilities like teamwork and good thinking power, which are vital for engineering projects, in addition to improving problem-solving abilities.

Purpose or Goal

Through this research we would like to give a clear insight on how, multidisciplinary education improves engineering student job opportunities by providing them with a diverse skill set. As a result, they have greater capabilities to handle a wider range of problems and significantly improve society through their job. Through their various skills and expertise, they have the capacity to improve society and explore a wider range of career possibilities

Methods

For engineering students, interdisciplinary education approaches include integrated education systems, team-based projects, chances for interdisciplinary research, and cross-disciplinary courses and workshops. Engineering students gain greatly from these methodologies because they broaden their perspectives, improve their ability to communicate and collaborate, encourage innovation and adaptability, address complex problems, provide a variety of career options, and encourage ethical and societal considerations. Engineering students gain a complete skill set that equips them for the multifaceted challenges of the modern world by integrating interdisciplinary approaches into their curriculum.

Outcomes

Future through this research we will examine particular strategies and approaches that maximise the advantages of this pedagogical approach in order to progress the subject of interdisciplinary education in engineering. Educators may continue to improve the learning process and outcomes for engineering students by developing a greater grasp of how to successfully utilise interdisciplinary learning.

Conclusion

In conclusion, interdisciplinary education has several benefits for engineering students. They become more well-rounded professionals with better problem-solving skills, creative thinking, and stronger interpersonal and collaboration skills as a result. Interdisciplinary approaches are included in engineering curricula to better prepare students for complex problems in the real world. This method of instruction provides individuals with the knowledge and abilities they need to succeed in their vocations and have a beneficial influence on society.

Keywords—Cross-disciplinary courses, multifaceted challenges.

I. INTRODUCTION

The state of engineering education has seen an important transition from an interdisciplinary perspective as a result of the constantly changing global environment. This change is motivated by the realisation that conventional, siloed educational models could not effectively prepare engineering students to handle the complexity of contemporary situations. The necessity for education that spans these barriers is critical as the distinctions between engineering and several other fields become more hazy. This essay explores the enormous impact interdisciplinary education has on engineering students, demonstrating how this strategy broadens their perspectives and develops their skills.

Long-standing associations exist between technical aptitude, subject-specific knowledge, and engineering education. The requirements of today's interconnected and complex world still call for a broader knowledge base that includes efficient communication, teamwork ability, and creative thinking. A knowledge gap caused by the difference between technical competence and the development of all aspects of one's skills has led to the investigation of interdisciplinary engineering education as a potential remedy. This pedagogical change aims to improve students' capacity to engage with many ideas, methodologies, and viewpoints from a range of fields in addition to helping them solve challenging engineering challenges.

This paper's main objective is to clarify the particular advantages of a multifaceted engineering education. This methodology enables engineering students to manage the complexity of contemporary situations with improved problem-solving skills by merging concepts and approaches from other fields. This shift is being driven by the incorporation of crucial abilities like cooperation and strong analytical thinking, both of which are essential for the efficient completion of engineering

https://doi.org/10.52202/073963-0005

projects. Additionally, interdisciplinary education's diverse character gives students a comprehensive skill set and promotes flexibility and creative thinking that go beyond traditional disciplinary boundaries.

The value of engineering graduates in the workplace goes beyond their technical proficiency. It is necessary to have the capacity for effective cross-disciplinary communication, fluid teamwork with experts from various backgrounds, and creative problem-solving. This essay aims to investigate if multidisciplinary education gives students a competitive edge, advancing them towards a wider range of employment choices and enabling them to have a greater impact on society. Engineering students can better prepare themselves to tackle a wider range of problems by embracing interdisciplinary approaches and making a significant contribution to society.

This study explores the approaches used to integrate interdisciplinary education into engineering courses in the parts that follow. Students are exposed to the transformative impact of multidisciplinary learning through integrated educational systems, teamwork assignments, chances for interdisciplinary research, cross-disciplinary courses, and workshops. This study tries to identify tactics that maximise the advantages of interdisciplinary education by looking at the concrete results of such an approach, furthering the conversation about this pedagogical revolution.

In the end, a new approach to engineering education is now necessary due to the interaction between engineering and several other disciplines. Interdisciplinary techniques open the door to a new generation of engineers—those who are not only technically proficient but also have the insight to tackle complex problems and have a positive impact on society. Interdisciplinary education integrates engineering curricula with the needs of the modern world by producing well-rounded professionals with better collaboration abilities, problem-solving skills, and creative thinking. Through the prism of this study, we set out on a journey to understand the true meaning of multidisciplinary education and its enormous ramifications for engineering students and society at large.

(Costa et al., 2019)The impact of an interdisciplinary project on students' skill development is the main topic of this paper's discussion of an engineering school project. The project aims to cognitively challenge students by removing disciplinary borders and fusing knowledge to address challenging issues. After the project was finished, two focus groups (n = 16) were used in a qualitative study to better understand how students felt about the project, the skills they thought they had learned from it, and its significance in general. The findings show that students understood the importance of the project for their learning and skill improvement. They were also able to pinpoint the precise skills that the project aimed to target and develop.

(Van Den Beemt et al., 2020)In this abstract, interdisciplinary engineering education (IEE) research from 2005 to 2016 is reviewed. IEE seeks to give engineering students the abilities to combine knowledge from several fields in order to address societal concerns. 99 empirical studies are analysed in the review, which highlights the difficulties in emphasising collaborative effort in pedagogy (teaching) and creating explicit learning objectives. It draws attention to the necessity of strong pedagogy and immersive team-based learning experiences to nurture interdisciplinary abilities (teaching), as well as the lack of knowledge of resources impeding the development of IEE programmes (support).

(Gero, 2017)In order to introduce undergraduate students to interdisciplinary education, the "Science and Engineering Education: Interdisciplinary Aspects" course was developed and is now being taught. Each student in the class is expected to instruct their other classmates in an interdisciplinary lesson. Participants included sixteen advanced-stage students, and attitudes towards transdisciplinary science and engineering teaching and learning were assessed using qualitative methodologies. Despite the difficulties of teaching interdisciplinary courses, the results indicate a marked rise in the proportion of students who would be willing to do so in the future as the course went on.

(Spelt et al., 2017)Specifically in the context of a master's course on food quality management, this abstract covers a study that sought to prepare scientific and engineering students for interdisciplinary interaction. The study uses Illeris' learning theory to analyse 615 student experiences in order to understand the cognitive, emotional, and social aspects of learning. The results show that students often (214 times), cognitively (194 times), and socially (207 times) communicated their experiences. The emotional, cognitive, and social obstacles of applying discipline knowledge to complicated problems, as well as interacting with peers to uncover shared thoughts and experiences, were among the key interdisciplinary learning experiences. It's noteworthy that students valued the cognitive dimension more highly than the emotional and social dimensions.

(Zeidmane & Cernajeva, 2011)In order to improve the competencies of aspiring engineers and their market competitiveness, this abstract emphasises the importance of using an interdisciplinary approach in engineering education. It emphasises the value of general competencies, such as computer literacy, fluency in a professional language, and suggestions from educational psychology. A unified e-learning environment is thought essential for improving information literacy abilities, while the integration of CLIL (Content and Language Integrated Learning) methodologies is advised to increase foreign language proficiency. The abstract also highlights the importance of academic staff in curriculum design and promotes a harmony between theoretical understanding, real-world application, and the use of contemporary technologies. The Latvia University of Agriculture (LUA) and Riga Technical University (RTU) are two universities that use multidisciplinary methods as examples.

II. METHODOLOGY

A. Research Design

Type of Study: The mixed-methods technique used in this study combines the collecting of quantitative survey data with qualitative analysis. This strategy enables a thorough analysis of the influence of interdisciplinary education on engineering education.

Research Methodology: In order to investigate the relationships between multidisciplinary education and various outcomes within the framework of engineering education, this study employs an explanatory research technique.

B. Research Questions and Hypotheses:

Research Questions: This study addresses the following research questions:

1) How familiar are you with the concept of interdisciplinary education?
2) Have you participated in any interdisciplinary courses during your engineering education?
3) In your opinion, how does interdisciplinary education contribute to a well-rounded skillset?
4) How comfortable are you with collaborating with students from different disciplines?
5) Do you believe interdisciplinary education should be integrated into all engineering curricula?
6) How have interdisciplinary courses impacted your problem-solving approaches?
7) To what extent do interdisciplinary courses contribute to your overall academic experience?
8) In your opinion, should interdisciplinary education be integrated into professional development progsrams for engineers?
9) Would you recommend interdisciplinary education to future engineering students?
10) What challenges, if any, have you encountered while engaging in interdisciplinary coursework?

Hypotheses: Based on the literature that is currently available, we hypothesise that students who take part in multidisciplinary courses will perform better academically and have better problem-solving skills than those who do not.

C. Data Collection:

Data Sources: Through the use of Google Forms, data was gathered through a structured online survey. The selection of an online survey platform made it possible to gather data from a variety of engineering students effectively.

Sampling: Participants from diverse engineering courses were chosen using the random stratification technique to ensure representation across fields.

Participants: 50 engineering students from varied backgrounds, including undergraduate and graduate students, who represented different engineering specialties, received the survey.

D. Data Analysis:

Quantitative Analysis: To get a general sense of survey participants' perceptions, survey results were analysed using descriptive statistics, including means and standard deviations.

Qualitative Analysis: Following a comprehensive coding procedure, thematic content analysis was applied to open-ended survey replies. Using this method, we were able to systematically identify patterns and themes relating to student perception.

Data Validation: Inter-rater reliability tests were carried out for the qualitative data coding process in order to confirm the validity and reliability of our findings.

E. Interdisciplinary Education Metrics:

Multiple criteria, such as GPA improvement, course ratings, and self-reported skill development scores, were used to evaluate the impact of interdisciplinary education. This multifaceted strategy offers a comprehensive understanding of the effect.

F. Control Variables:

Participants' prior academic standing, course load, and demographics served as control factors. To account for the possibility of confounding factors and isolate the effect of interdisciplinary education, these variables were added.

G. Ethical Considerations:

All participants gave their informed consent after being informed of the study's objectives and the voluntariness of their participation. According to ethical research norms, which protected confidentiality and anonymity, data was gathered and preserved.

H. Data Presentation:

Quantitative data will be displayed using charts, while qualitative data will be displayed using thematic summaries. The results will be more understandable thanks to the verbal and graphic display.

I. Limitations:

Limitations include the possibility of response bias brought on by self-reporting and the fact that the study's scope was

https://doi.org/10.52202/073963-0005

restricted to one institution. There may be limitations on generalizability to different engineering education situations.

J. Data Interpretation:

The context of the study questions and hypotheses will be used to interpret data. To give a detailed knowledge of the influence of multidisciplinary education on engineering education, insights from both quantitative and qualitative data will be gathered.

K. Conclusion and Validation:

The methodology is created to be in line with the goals of the study, guaranteeing a thorough and robust approach to evaluating the influence of multidisciplinary education on engineering education. The study's validity is boosted by the use of mixed techniques, meticulous variable control, and ethical considerations.

The survey procedure includes examining how teachers and students perceive or have used interdisciplinary techniques, as well as determining the effects these methodologies may have on students' capacity for learning, critical-thinking abilities, problem-solving abilities, and other abilities. Basic information on the students, such as their name, the school they attend, their experience, and their area of specialization, were collected in the survey questionnaire.

We have asked the students to express their opinions through our questions, which include: Do they believe that interdisciplinary education can foster critical thinking skills among engineering students? How do interdisciplinary education methodologies contribute to increasing student participation? Do they agree that interdisciplinary techniques can prepare engineering students for the changing requirements of the industry? Interdisciplinary education approaches aim to give students a more thorough knowledge of scientific and technology ideas. All of these inquiries assisted us in analyzing their response.

III. RESULTS AND DISCUSSIONS

Let's examine the results of the analysis of each and every section of the questionnaire.

How familiar are you with the concept of interdisciplinary education?

This analysis provides a breakdown of respondents' familiarity with the concept of interdisciplinary education. The majority of respondents indicated that they are either "Somewhat familiar" (48%) or "Very familiar" (32%) with the concept. Only a smaller percentage mentioned being "Not very familiar" (12%), and "Not at all familiar" (8%).

These findings suggest that a significant portion of the respondents have at least some level of familiarity with interdisciplinary education, which may influence their perspectives on its impact on engineering education.

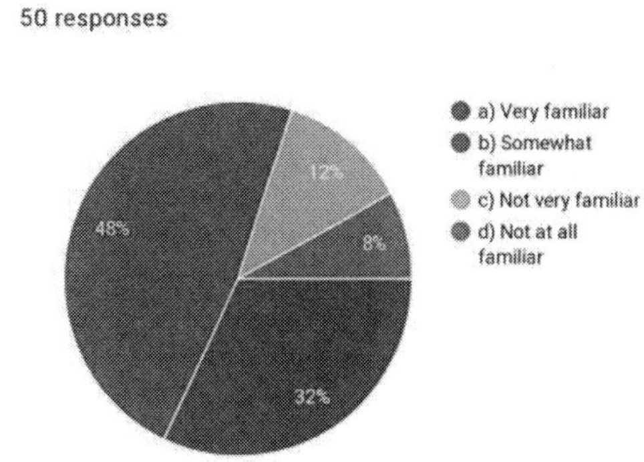

Fig. 1. Pie-chart describing familiarity of interdisciplinary learning.

Fig. 1 And Fig. 2 Describes about the familiarity about the interdisciplinary learning and to know better results initially we have asked the participants whether they are a part of any course or not.

Have you participated in any interdisciplinary courses during your engineering education?

This analysis presents the distribution of respondents' experiences with interdisciplinary courses during their engineering education. The majority of respondents have participated in interdisciplinary courses, with 38% indicating that they have done so frequently and 36% occasionally. A smaller portion (26%) reported that they have never participated in interdisciplinary courses during their engineering education.

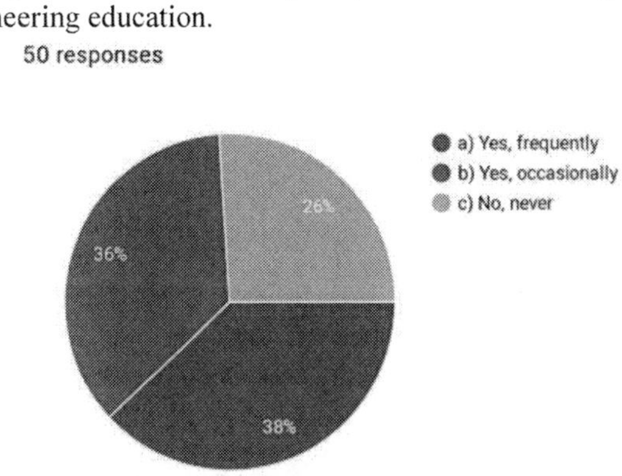

Fig. 2. Pie-chart describing about participating of any interdisciplinary education course.

In your opinion, how does interdisciplinary education contribute to a well-rounded skillset?

Interdisciplinary education is seen as a valuable contributor to a well-rounded skillset, with several key points of impact:

Broadened Perspective: A significant portion of respondents (32%) mentioned that interdisciplinary education provides a broader perspective. This suggests that exposure to multiple disciplines helps students develop a more comprehensive understanding of complex issues, which is crucial in engineering, where solutions often require considering various factors.

https://doi.org/10.52202/073963-0005

Enhanced Problem-Solving: An equally substantial percentage of respondents (32%) stated that interdisciplinary education enhances problem-solving abilities. This aligns with the expectation that tackling real-world engineering challenges often demands creative and multifaceted problem-solving approaches.

Fostered Creativity and Innovation: While a smaller percentage (14%) pointed out that interdisciplinary education fosters creativity and innovation, this aspect is still significant. Engineering students need to think creatively and innovate to address evolving technological and societal needs.

Improved Communication Skills: Another 14% of respondents mentioned that interdisciplinary education improves communication skills. This is particularly relevant since effective communication is crucial in interdisciplinary collaborations, allowing engineers to convey complex ideas to non-technical stakeholders.

No Noticeable Impact: A minority (8%) expressed that they did not notice any impact on their well-rounded skillset. This perspective highlights the need to ensure that interdisciplinary courses are designed effectively to achieve the desired outcomes.

Overall, these responses demonstrate that interdisciplinary education is generally seen as a valuable asset for engineering students, contributing to a well-rounded skillset by fostering a broader perspective, enhancing problem-solving abilities, nurturing creativity, and improving communication skills. These skills are essential for engineers to excel in their roles and adapt to the evolving demands of their field.

How comfortable are you with collaborating with students from different disciplines?

This analysis illustrates the distribution of respondents' comfort levels with collaborating with students from different disciplines. The majority of respondents either feel "Very comfortable" (34%) or "Somewhat comfortable" (30%) with interdisciplinary collaboration. A smaller percentage of respondents expressed "Neutral" feelings (20%). A minority reported feeling "Somewhat uncomfortable" (6%), while a similar percentage indicated being "Very uncomfortable" (10%) with interdisciplinary collaboration. Fig.3 describes the How at ease they are working with students from various academic fields.

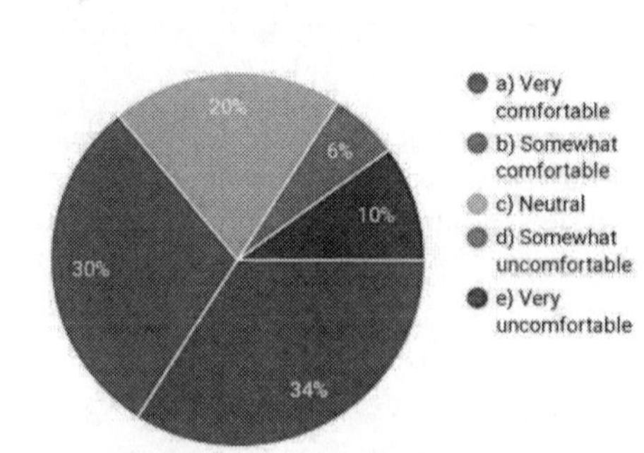

Fig. 3. Pie-chart describing about collaborating with students from different disciplines.

Do you believe interdisciplinary education should be integrated into all engineering curricula?

This analysis presents the distribution of respondents' opinions on whether interdisciplinary education should be integrated into all engineering curricula. A significant portion, 38%, believed that it should be integrated "Yes, definitely," while an additional 38% thought it should be integrated "Yes, to some extent." A smaller percentage of respondents (10%) felt that it should be optional, and 14% believed it is not necessary to integrate interdisciplinary education into all engineering curricula. Fig. 4 describes about the findings that interdisciplinary education should be part of the engineering curriculum or not.

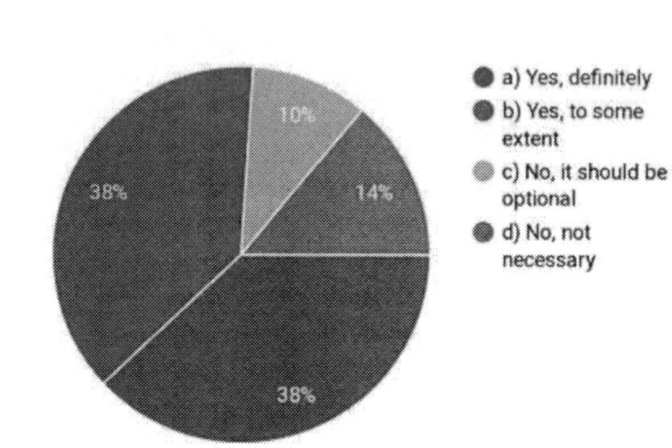

Fig. 4. Pie-chart describing about interdisciplinary education should be integrated into all engineering curricula.

How have interdisciplinary courses impacted your problem-solving approaches?

This analysis demonstrates the distribution of respondents' perceptions of how interdisciplinary courses have impacted their problem-solving approaches. The most common response was that interdisciplinary courses have helped identify multiple solutions to complex problems (42%). A significant portion of respondents also indicated that these courses encouraged thinking beyond traditional boundaries (36%). A smaller percentage reported no noticeable impact (10%), while 12% felt that interdisciplinary courses had hindered their problem-

https://doi.org/10.52202/073963-0005

solving abilities. Fig. 5 describes about the findings of interdisciplinary education impacted on problem solving approaches.

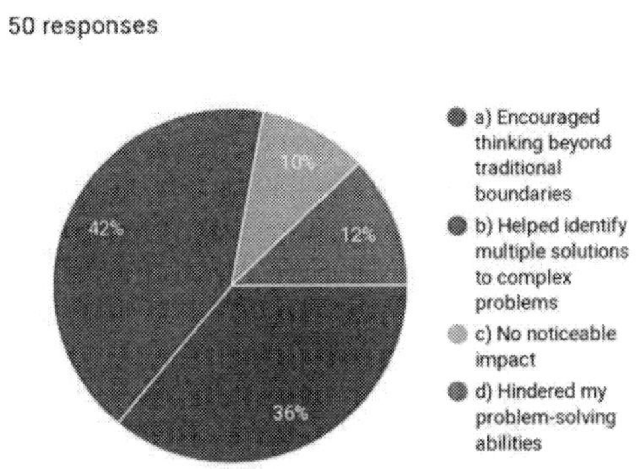

Fig. 5. Pie-chart describing about impact of problem-solving approaches

To what extent do interdisciplinary courses contribute to your overall academic experience?

This analysis presents the distribution of respondents' perceptions of the extent to which interdisciplinary courses contribute to their overall academic experience. The most common response was that interdisciplinary courses are considered a "Valuable addition to my education" (36%). A significant portion of respondents also considered them an "Essential part of my education" (34%). A smaller percentage expressed "Neutral impact" (12%), "Minor impact" (6%), or "Negligible impact" (12%) on their overall academic experience. Fig. 6 describes that about their experiences that interdisciplinary course contribute to their academics.

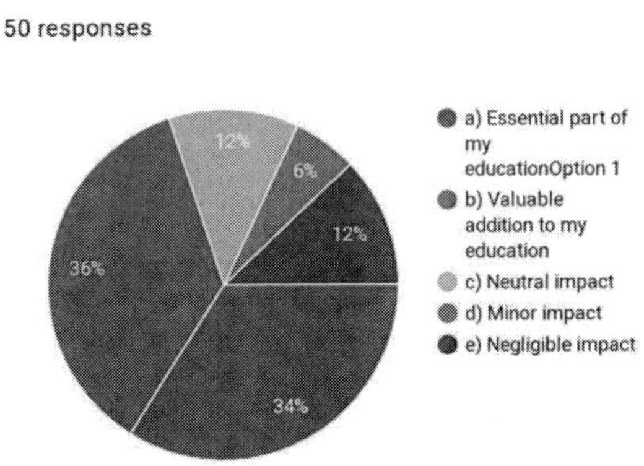

Fig. 6. Pie-chart describing about contribution of interdisciplinary course to overall academic experience.

In your opinion, should interdisciplinary education be integrated into professional development programs for engineers?

This analysis illustrates the distribution of respondents' opinions regarding the integration of interdisciplinary education into professional development programs for engineers. The majority, 54%, believed that interdisciplinary education should be integrated "Yes, definitely," while 34% thought it should be integrated "Yes, to some extent." A smaller percentage (12%) expressed that it is not necessary to integrate interdisciplinary education into professional development programs for engineers. Fig. 7 describes that interdisciplinary education be integrated into professional development programs or not.

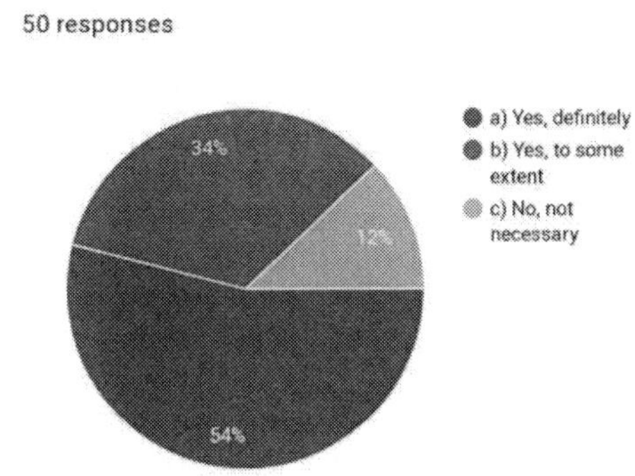

Fig. 7. Pie-chart describing about interdisciplinary education be integrated into professional development programs for engineers.

Would you recommend interdisciplinary education to future engineering students?

This analysis presents the distribution of respondents' recommendations regarding interdisciplinary education to future engineering students. The most common response was "Strongly recommend" (36%), followed by "Recommend" (36%). A smaller percentage expressed a "Neutral" stance (12%), while 6% "Do not recommend" and 10% "Strongly do not recommend" interdisciplinary education to future engineering student. Fig. 8 describes that they recommend interdisciplinary courses to future students or not.

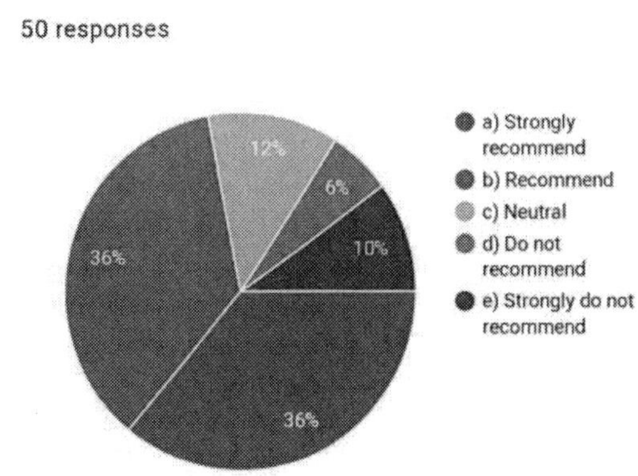

Fig. 8. Pie-chart describing about recommending interdisciplinary education to future engineering students.

What challenges, if any, have you encountered while engaging in interdisciplinary coursework?

1). Communication Issues: Many students mentioned difficulties in communicating effectively with peers from different disciplines, including understanding technical jargon and terminology.

2). Differences in Academic Backgrounds: Respondents highlighted varying levels of knowledge and skills among interdisciplinary team members, making it challenging to align academic backgrounds.

3). Integration of Knowledge: Some students struggled with integrating knowledge from multiple fields, finding it challenging to bridge the gap between different subject areas.

4). Time Management: Time management emerged as a concern, especially when juggling coursework from different disciplines simultaneously.

5). Group Work Challenges: Collaborative projects were often cited as a source of challenges, with some students noting issues with teamwork, leadership, and coordination.

6). Assessment and Grading: Concerns about how interdisciplinary coursework was assessed and graded were mentioned, with students seeking clarity in evaluation criteria.

7). Lack of Clear Structure: A few respondents pointed out a lack of clear structure or guidance in interdisciplinary courses, which made it difficult to navigate the curriculum.

Interdisciplinary education plays a pivotal role in shaping the educational experiences of engineering students, as evidenced by the survey responses. First, the data reveals that a significant portion of the respondents have some familiarity with interdisciplinary education (32% "Very familiar" and 48% "Somewhat familiar"). This suggests that interdisciplinary education is not an entirely foreign concept to engineering students, and there is an existing level of awareness. Moreover, the majority of respondents have participated in interdisciplinary courses during their engineering education (38% "Yes, frequently" and 36% "Yes, occasionally"). This demonstrates that interdisciplinary education is already an integral part of their academic journey. Respondents strongly believe in the benefits of interdisciplinary education. A substantial proportion recommends its integration into all engineering curricula (38% "Yes, definitely" and 38% "Yes, to some extent"). This reflects a consensus among students that interdisciplinary education can enhance their skillsets and contribute to a more holistic academic experience.

Furthermore, when asked about the impact of interdisciplinary courses on their problem-solving approaches, respondents largely felt that these courses have been beneficial. The majority mentioned that interdisciplinary courses encouraged thinking beyond traditional boundaries (36%) and helped identify multiple solutions to complex problems (42%). This highlights the role of interdisciplinary education in fostering critical thinking and creativity. In terms of comfort levels with collaborating across disciplines, the data suggests that a significant portion of students (34% "Very comfortable" and 30% "Somewhat comfortable") are open to and comfortable with such collaboration. This indicates that interdisciplinary coursework may contribute to developing important collaborative skills among engineering students.

Overall, the findings from the survey indicate that interdisciplinary education has a positive impact on engineering education. It enhances problem-solving abilities, encourages creative thinking, and prepares students for collaborative work in diverse environments. As engineering fields increasingly intersect with other disciplines, these skills and experiences become increasingly valuable for the next generation of engineers. Therefore, integrating interdisciplinary education into engineering curricula and professional development programs can further enhance the well-roundedness and adaptability of future engineers.

IV. CONCLUSIONS

Our survey, encompassing 50 engineering students from diverse backgrounds, aimed to explore the impact of interdisciplinary coursework on skill development. While our findings align with previous studies highlighting the benefits of interdisciplinary education in broadening perspectives and enhancing problem-solving skills (Costa et al., 2019; Van Den Beemt et al., 2020), our research delves deeper into the practical implications for engineering education. By uncovering students' perceptions of interdisciplinary coursework and their comfort levels in cross-disciplinary collaborations, our study provides empirical evidence that supplements existing theoretical frameworks. Moreover, while existing literature underscores the importance of interdisciplinary approaches, our results showcase nuanced insights into specific challenges faced by students, shedding light on potential areas for pedagogical enhancement. Incorporating demographic data, such as academic year, engineering specializations, and age range of the surveyed students, would further enrich the depth of our analysis, enhancing the contextual understanding of our findings

REFERENCES

Costa, A. R., Ferreira, M., Barata, A., Viterbo, C., Rodrigues, J. S., & Magalhães, J. (2019). Impact of interdisciplinary learning on the development of engineering students' skills. *European Journal of Engineering Education, 44*(4), 589–601. https://doi.org/10.1080/03043797.2018.1523135

Gero, A. (2017). Students' attitudes towards interdisciplinary education: A course on interdisciplinary aspects of science and engineering education. *European Journal of Engineering Education, 42*(3), 260–270. https://doi.org/10.1080/03043797.2016.1158789

Spelt, E. J. H., Luning, P. A., Van Boekel, M. A. J. S., & Mulder, M. (2017). A multidimensional approach to examine student interdisciplinary learning in science and engineering in higher education. *European Journal of Engineering Education, 42*(6), 761–774. https://doi.org/10.1080/03043797.2016.1224228

Van Den Beemt, A., MacLeod, M., Van Der Veen, J., Van De Ven, A., Van Baalen, S., Klaassen, R., & Boon, M. (2020). Interdisciplinary engineering education: A review of vision, teaching, and support. *Journal of Engineering Education, 109*(3), 508–555. https://doi.org/10.1002/jee.20347

Zeidmane, A., & Cernajeva, S. (2011). Interdisciplinary approach in engineering education. *2011 IEEE Global Engineering Education Conference (EDUCON)*, 1096–1101. https://doi.org/10.1109/EDUCON.2011.5773284

Student Use of Anchors and Metacognitive Strategies in Reflection

Anu Singh, Heidi A. Diefes-Dux
University of Nebraska-Lincoln
Corresponding Author Email: heidi.diefes-dux@unl.edu

Abstract

Context

Self-regulation, a skillset involving taking charge of one's own learning processes, is crucial for workplace success. Learners develop self-regulation skills through reflection where they recognize weaknesses and strengths by employing metacognitive strategies: planning, monitoring, and evaluating. Use of anchors assists learners' engagement in reflection.

Purpose or Goal

The purpose of this work was to gain insight into students' use of anchors when reflecting on their learning. The two research questions: (1) To what extent do students link their self-evaluation and learning objective (LO) self-ratings to their reflections? and (2) What dimensions and level of metacognitive strategies do students use in their self-evaluation of and reflections on weekly problem-solving assignments?

Methods

Data were upper-division engineering students' anchors (self-evaluations, LO self-ratings) and reflection responses for one assignment. Self-evaluations and reflections were analyzed for the presence of references to LOs. The number of students who linked the anchors to their reflection were tabulated. Additionally, a revised *a priori* coding scheme was applied to students' written work to determine type and level of metacognitive strategies employed.

Outcomes

Few students linked both anchors to their reflections. Students employed low to medium levels of the metacognitive strategies in their self-evaluations and reflections, even when they linked their anchors and reflections. The evaluating strategy dominated in the self-evaluations, while planning and monitoring dominated in the reflections.

Conclusion

Students have limited understanding of the use of anchors to guide their reflection responses. Students overall level of engagement in the metacognitive strategies indicates a need for formal instruction on reflection.

Keywords— Learning Objectives, Metacognition, Reflection

I. INTRODUCTION

Self-regulation is one of the critical skills required for workplace success in the 21st century (Rios et al., 2020). In the workplace, employees are expected to respond to changes that emerge due to global societal, economic, and technological transformations (Hager, 2004). To keep oneself prepared for changing situations, individuals must be able to regulate their learning by identifying their learning needs and monitoring their learning progress (Lord et al., 2009). ABET, the engineering program accreditation mechanism used by many institutions worldwide, emphasizes the need for engineering students to develop this skill with its Student Outcome 7: "an ability to acquire new knowledge as needed, using appropriate learning strategies" (ABET, 2023).

For a student to be a self-regulated learner, they must develop an understanding and awareness of their learning processes (or metacognition) and use that knowledge to control their learning processes (Colthorpe et al., 2019). Metacognitive skills can be developed in students by engaging them in activities that promote development of three metacognitive strategies: Planning, Monitoring, and Evaluating (Fridman et al, 2020). Reflection is one such technique that assists in shifting students' thinking from self-centeredness to self-awareness (Siewiorek et al., 2010); it provides opportunities for students to learn from their experience using their cognitive and metacognitive skills (Wegner et al., 2015). Hence, reflection takes students a step closer to being self-regulated learners.

However, there is evidence that in engineering classrooms, students need to improve their ability to reflect. Students' reflections show a lack of awareness of their performance and task knowledge, indicating their low metacognitive engagement (Seppanen, 2023). More precisely, students' engagement in all three metacognitive strategies (i.e., Planning, Monitoring, and Evaluating) are limited to low to medium levels while responding to weekly reflection prompts (Singh & Diefes-Dux, 2022). Reflection is a complex, rigorous, intellectual, emotional, and time-consuming process (Rodgers, 2002), but students' ability to reflect can be developed by providing multiple opportunities to reflect using anchors throughout a course.

https://doi.org/10.52202/073963-0006

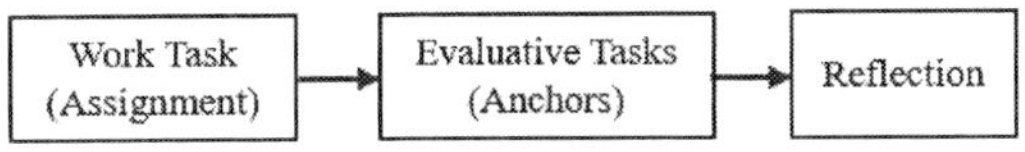

Fig. 1. Position of anchors in the sequence of activities

An anchor is a reference point that focuses the reflection activity. An anchor may be a work task providing a concrete experience on which to reflect. Anchors may also be formal self-evaluation tasks done between the work task and the reflection on the completed work task (Fig. 1). For instance, as in this study, the instructor used self-evaluation of the work task and learning objective (LO) self-ratings as anchors for reflection. During self-evaluation of the work task, the learner evaluates their work against a given standard (Tillema, 2010) and identifies what exactly they need to work on. During LO self-ratings, the learner rates their ability with an LO on a scale, which gives them opportunity to evaluate their proficiency with that LO. Overall, the use of anchors allows students to take a step back and identify specific knowledge, skill, and abilities that need improvement. Hence, the use of anchors set the stage for students to engage in deep reflection.

Studies have tended to only analyze either students' responses to self-evaluation (e.g., Baisley et al., 2022), LO self-ratings (e.g., Opanuga & Diefes-Dux, 2023), or reflection responses (e.g., Fong et al., 2023) in isolation. Separate analyses of anchors and reflections might not provide sufficient insights into students' learning challenges, metacognitive engagement, and self-regulation ability (Opanuga & Diefes-Dux, 2023; Singh & Diefes-Dux, 2023). Hence linking students' responses to the anchors (self-evaluating and LO self-ratings) and their reflections could assist in a better understanding of students' self-regulation ability.

The purpose of this quantitative-based qualitative study was to investigate the extent to which students link their work on anchoring activities to their reflections on their learning processes and to identify the dimension and level of metacognitive strategies used by engineering students during self-evaluation and reflection. Knowledge of students' propensity for linking the anchors to their reflection responses and their use of metacognitive strategies could help instructors design better instruction around the reflection activities.

II. THEORETICAL FRAMEWORK

Metacognition and self-regulated learning are theories used to guide the reflection intervention and analysis of students' work. Each theory is briefly described below.

Metacognition is commonly referred to as "thinking about thinking." Strong metacognitive skills typically result in better predicating, monitoring, and reflecting ability (Vogel-Walcutt, & Fiore, 2010). According to Flavell (1979), the two components of metacognition are knowledge of cognition and regulation of cognition. Knowledge of cognition refers to knowledge of one's own cognitive process and knowledge of strategies required to effectively perform the task, while regulation of cognition refers to strategies implemented to control one's cognitive processes: Planning, Monitoring, and Evaluating. Both components of metacognition are essential and interact with each other while performing a task (Schraw & Moshman, 1995). Overall, metacognition is important in self-regulation as "it enables individuals to monitor their current knowledge and skill levels, plan and allocate limited learning resources with optimal efficiency, and evaluate their learning state" (Schraw et al., 2006, p. 116). The present work focused on the regulation of the cognition component because its three elements (i.e., Planning, Monitoring, Evaluating) are crucial for self-regulated learning (Kittel et al., 2021).

Self-regulated learning (SRL) is a process wherein learners take responsibility for their own learning and metacognitively, motivationally, and behaviorally engage themselves in the pursuit of pre-determined goals (Zimmerman, 1989; 2002). The present study used Zimmermann's (2000) model of SRL, which considers SRL a cyclic process in three phases: Forethought (refers to phase before starting of the task and involves goal setting and strategic *planning*), Performance (refers to phase during the task where learner engage in *monitoring* their cognitive process, includes self-observation) and Self-reflection (refers to phase after completion of the task, where learners decide on the quality and impact of their performance or choices, includes *evaluating* and observing of oneself). The three phases of the SRL cycle indicate involvement in the three regulation of cognition elements (Planning, Monitoring, Evaluating, respectively). Anchoring (e.g., self-evaluation and LO self-rating) activities provide a means for students to engage in self-observation in a structured manner. Hence, integrating anchors and reflection activities can provide opportunities for students to use all three metacognitive strategies and engage deeply in an SRL cycle.

Reflection can be considered to be a self-regulation activity (Sandars, 2009) that supports the development of students' higher-order thinking and deep learning of skills (Wegner et al., 2015). In the learning context, reflection assists students in combining new learning with existing knowledge and skills (Mann et al., 2009) and prepares them for the workplace, where they must manage their learning according to task requirements (Schön, 1983). While reflection can provide opportunities for the learners to engage in all three metacognitive strategies, students do not automatically engage in deep metacognition, but they can be taught (Wedelin & Adawi, 2014) by providing suitable opportunities throughout a course.

III. LITERATURE REVIEW

Instructors can use a variety of activities to engage students metacognitively in a course (Lin, 2001). However, the present review will only focus on studies that investigated the use of self-evaluation, LO self-ratings, or self-reflection with an aim

https://doi.org/10.52202/073963-0006

of preparing students to become self-regulated learners. A few such studies are discussed below.

El-Maaddawy (2017) studied the impact of self-evaluation on students' grades. The author had students self-evaluate their work after receiving minimal feedback and a tentative grade on their submitted work. The self-evaluation activity, which was completed before revision of their work for a final grade, included identifying possible sources of errors and suggesting corrections. To set the standard for self-evaluation, the instructor discussed and provided model responses from previous assignments, including examples of excellent, good, and poor work. Analysis of students' homework assignments and in-class work using the above self-assessment paradigm improved students' grades throughout the semester, and students' perceptions collected through a survey showed that students agreed that the self-assessment technique improved their learning and developed self-regulation skills.

Ugulino and Ferreira (2021) studied the impact of students' self-ratings in combination with mentor feedback on course pass rates. They asked students to self-rate their proficiency on list of challenges provided by instructors for that week's topic covered in the classroom. Students rated their proficiency on the topics using a rubric consisting of three levels of proficiency (Entry, Medium, and Target). The results showed that the students' self-assessments, followed by mentors' feedback on submitted self-assessed work, resulted in an increase in the number of students who passed the course, indicating improvement in students' awareness of their learning. Opanuga and Diefes-Dux (2023) analyzed students' LO self-ratings on weekly assignments in isolation and suggested that the LO self-ratings be analyzed side by side with students' reflective responses to achieve a more in-depth understanding of students learning challenges.

Studies described above only analyzed students' self-evaluation responses and LO self-ratings in association with mentor feedback. However, these studies did not include reflection.

Reflection activities can be used in a course to achieve different objectives: metacognition, competency, and personal growth and change (Reflection Activities, n.d.). A few studies have implemented a guided reflection exercise called Exam Analysis and Reflection (EAR) in a mechanical engineering course (Benson & Zhu, 2015), an electrical circuit course (Claussen & Dave, 2017), and a microelectronic course (Clark & Dickerson, 2018) to investigate the effectiveness of reflection on students' performance and learning. The results of Benson and Zhu (2015) and Claussen and Dave (2017) emphasized the need for a more thorough integration of the reflection activity in the course, whereas Clark and Dickerson (2018) concluded that the effectiveness of reflection is sensitive to exam problem type.

The above studies focused on students' content learning and looked for depth in reflection responses. These studies did not examine students' use of metacognitive strategies during reflection. With the objective of gaining insight into students' metacognitive engagement and improvement in students learning, Diefes-Dux and colleagues (Stratman & Diefes-Dux, 2022; Singh & Diefes-Dux, 2022) analyzed students weekly reflection response using an *a prior* coding scheme based on Ku and Ho's (2010) reflection-in-action rubric. Stratman and Diefes-Dux (2022) examined the effect of differently worded reflection prompts on the level and metacognitive regulation strategy present in students' reflections. Results showed that students employed metacognitive strategies according to the reflection prompt. When the reflection prompt focused on using instructional team feedback to improve performance, students used Planning, Action, and Evaluating strategies. Whereas when the reflection prompt focused on one's proficiency with the LOs, reflections predominantly yielded use of the Monitoring strategy. Singh and Diefes-Dux (2022) identified the three metacognitive regulation strategies employed by upper-division engineering students in their reflections. The result showed that students predominantly employed low to medium Planning and Monitoring strategy, and a limited number of students were engaged in low to medium level Evaluation. In a follow-on study to better understand students' engagement in all three metacognitive strategies, Singh and Diefes-Dux (2023) analyzed both students self-evaluating comments and their reflection responses using an expanded coding scheme with four levels for each metacognitive strategy. Results of the study showed that pairing of self-evaluation and reflection activities provided opportunities for students to engage in the complete set of metacognitive strategies, though still at low to medium levels.

Overall, the studies described above underscore the effectiveness of using self-evaluation and LO self-ratings on students' learning and self-regulation ability. However, none of the above studies analyzed the link students make between the anchors and their reflection wherein a student would identify an error or a lack of proficiency with an LO and then reflect in depth on that finding. As a result, examining students' responses to the anchor activities and the extent to which they link those activities to their reflection will provide insight into students' ability to employ the anchors as they engage in metacognition.

IV. RESEARCH QUESTIONS

The study aims to address following research questions:

1. To what extent do students link their self-evaluation and LO self-ratings to their reflections?
2. What dimensions and level of metacognitive strategies do students use in their self-evaluation and reflections on weekly problem-solving assignments?

V.METHODS

This is a quantitative-based qualitative study (Chi, 1997). Specifically, students' self-evaluations and reflections are qualitatively coded for metacognitive strategy and level and presence of references to relevant LOs. The coded results are then treated as quantitative data.

A. Setting and participants

The study was set in a junior level process engineering course at a Midwest R1 U.S. university in Spring 2021 (N= 28). The course was required for some students and an elective for other students depending on each student's major and degree program. The course duration was shortened from 16 to 14 weeks and the delivery mode was synchronous via Zoom due to COVID-19 pandemic. Course instructional materials (e.g., videos, readings, list of learning objectives (LO), assignments, standards solution key, self-evaluation template file, and reflection prompts) were shared with students through Canvas, the learning management system. The study used convenience sampling, as this was a course in which reflection activities were being implemented.

B. Intervention

The course was divided into four modules: 1) Conservation of Mass, 2) Fluid Flow (Pipes, Fittings, and pumps for Newtonian and Non-Newtonian fluids), 3) Fan Selection, and 4) Thermal Preservation. Each unit of the course consisted of a minimum of three assignments called Trainings (TR). Each training consisted of parts A and B. Part A involved solving a computational problem set in an authentic context using Excel. After submission of part A, the instructor released a solution key. Part B consisted of two steps, self-evaluation and reflection as explained below.

1) Self-Evaluation (B.1): Students were asked to compare their solutions to the key and annotate their Excel work with comments on their errors or things they learned or needed to work on. To further assist students, the following prompt was provided.

When your method or answer is incorrect or either could be improved, you need to track down where the issues are and comment on what you figured out.

TABLE I
LO RATING SCALE (OPANUGA & DIEFES-DUX, 2023)

Scale	Text Options Provided to Students
5	I can do this on my own without referring to resources
4	I can do this on my own if I refer to some resources
3	I need more practice with this
2	I need someone to help me understand and do this
1	I am not sure what this means (I am very lost)

2) Reflection (B.2): After submitting their annotated Excel sheet, the reflection activity became available to students. In this activity, the students self-rated their abilities with the course learning objectives and responded to an open-ended reflection prompt.

LO Self-Ratings: Students were asked to rate their abilities with the training relevant LOs. The LO self-rating assignments were administered through Canvas-graded surveys. For each training-relevant LO, students were required to select one of the five text phrases that best described their proficiency level with the LO (Table I). The scale of 1 to 5 was for research purposes only and was not shown to students.

Open Ended Reflection Prompt: Students were then asked to respond to three open ended reflection prompts. The first prompt focused on students' plans to improve their learning; this one was analyzed in this study. The prompt asked students to reflect on the LOs using the corresponding proficiency indicators (Table II).

TABLE II
SAMPLE LEARNING OBJECTIVES AND PROFICIENCY INDICATORS FOR TR 3.3

Learning Objective		Proficiency Indicators
PS 01.00	**Employ a robust problem-solving process that clearly documents engineering work (PS 01.00-01.08)**	
PS 01.01	Write a clear problem description that contains some context and an indicator of what the goal of solving the problem is	• Sufficient context is provided to understand the nature of the problem • The goal indicates the result(s) that are being sought
FF 02.00	**Use the law of conservation of mass to find stream mass flow rates and compositions**	
FF 02.06	Perform material balances when measures of throughput, other than mass flow rates, are given	• Write material balances in terms of average velocity • Write material balances in terms of volumetric flow rate • Convert between mass flowrate and volumetric flow rate • Convert between mass flowrate and velocity • Identify whether the problem is solvable (degree-of-freedom analysis) • Select, with rationale, the independent equations needed to solve the problem • Complete problem using standard problem solving process
FF 02.08	Determine the operating point for a single fan or multiple fans given the system characteristic curve and the manufacturer's fan curve	• Overlay a system characteristic curve on a manufacturer's fan curve (single or multiple) • Determine the operating static pressure and volumetric flow rate
FF 03.00	**Characterize fluid flow**	
FF 03.01	Compute the Reynolds number for Newtonian fluids flowing in pipes	• Correctly use the Reynolds number formula to obtain a dimensionless number • Perform computations in SI or English units
FF 03.02	Classify fluid flow using the Reynolds number for Newtonian fluids flowing in pipes	• Classify fluid flow as laminar, turbulent, or transitional
FF 03.03	Determine the system characteristic curve for a fan used in a grain drying process	• Employ the six step process described in TR 3.2.3

https://doi.org/10.52202/073963-0006

Proficiency indicators were developed by the instructor to guide students about the aspects that constitute successful demonstration of LO. The first open-ended prompt read as follows:

For those learning objectives that you are not able to do on your own, what do you plan to do to improve your abilities? Refer to specific learning objectives and indicators of proficiency and be specific about your planned actions.

If there is nothing which you feel you need to improve upon, practice describing your newly acquired or strengthened skills (as if to a future employer or superior). What is the skill? How do you see that skill being useful in your work as an engineer?

When looking at the various tasks, the training serves as the experience on which the student reflects. The self-evaluation of work serves as the start of the reflection as students identify errors with the potential of connecting their successes and difficulties to the LOs. The self-evaluation also serves as an anchor for the open-ended reflection prompt. The LO self-ratings also serve as an anchor for the open-ended reflection prompt. When responding to the reflection prompt, the student optimally draws on what they learned about their learning from the experience and anchors.

C. Data collection

Students' self-evaluation of their computational work, their LO self-rating and their responses to first open-ended reflection prompt were collected from the Fan Selection (FA) unit. This module consisted of three trainings (TR 3.1-3.3). The data from TR 3.3 were used in the present work. For TR 3.3, students rated themselves on eight LOs; a few of them are shown in Table II. The data from TR 3.3 were used in the present work. For TR 3.3, students rated themselves on eight LOs; a few of them are shown in Table II. Students had access to the proficiency descriptions shown in Table II through the course list of LOs posted on Canvas.

TABLE III
BINARY ASSIGNMENT OF LOS ADDRESSED IN SELF_EVALUATION (ERROR) COMMENTS AND REFLECTIONS AND LO SELF_RATING FOR ONE SAMPLE STUDENT

Type	PS 01.00	PS 02.02	PS 03.01	FA 02.06	FA 02.08	FA 03.01	FA 03.02	FA 03.03
Error	1	1	0	0	1	0	1	1
Reflection	1	0	0	0	0	0	0	0
LO Rating	3	3	4	3	2	3	3	3

D. Data analysis

Students' self-evaluation comments (from B.1) on their computational work were submitted in a pre-defined Excel format. These comments were extracted and placed in a single Excel file for coding. Students' self-ratings of their proficiency with the LOs and responses to the open-ended reflection prompt were downloaded from Canvas and saved in Excel file. Data collected from students' self-evaluation comments, self-rating of LOs, and reflection responses were then analyzed in two steps to answer each research question. Twenty-five (n=25) of the 28 students enrolled in the course completed all three tasks.

TABLE IV
METACOGNITIVE STRATEGIES CODING SCHEME (ERTMER & NEWBY, 1996; KU & HO, 2010; SINGH & DIEFES-DUX, 2023)

Dimension	Description
Evaluating (E): Student's comments represent an assessment of their thoughts or performance influenced by outside factors (grades, feedback). Student identifies a problem/solution related to a task or goal (Ku & Ho, 2010).	
Low (EL)	Identifies a problem without any indication of trying to solve the problem (Ku & Ho, 2010). Comments identifying a solution but not the problem it helped solve. Acknowledgement of difference between students work and solution key by referencing to specifics of problem.
Medium (EM)	Identifies a solution(action) that was taken
High (EH)	Identifies a problem and a solution, and how the solution changed their thinking or something they can now do because they found a solution (Ku & Ho, 2010)
Very High (EVH)	Provides an assessment of the action(s) taken or describes obstacles overcome (Ertmer & Newby,1996)
Monitoring (M): Student's comments relate to task comprehension as a form of self-reflection (not influenced by outside factors). Response indicates an understanding/lack of understandingor known/unknown information (Ku & Ho, 2010); related primarily to course content.	
Low (ML)	Indicates an awareness of level of understanding, with no reference to a general topic or learning objective
Medium (MM)	Describes evidence or experience or things tried with topic or learning objective
High (MH)	Indicates an awareness of level of understanding with reference to specifics on theproficiency list for a learning objective
Very High (MVH)	Describes evidence or experience with reference to specifics (e.g., details concerning a learning objective)
Planning (P): Student comments on preparation for one's continued/improved learning or future task execution; related to course content learning or learning strategy (Ku & Ho, 2010)	
Low (PL)	Indicates an awareness of the need for planning (Ku & Ho, 2010)
Medium (PM)	Specifies an action a student plans to take and/or a clear goal (performance) they hope to achieve with indication of evidence of achievement
High (PH)	Specifies an action a student plans to take and/or a clear goal (learning) they hope to achieve with indication of evidence of achievement
Very High (PVH)	Given specific action(s) and clear goal, acknowledges potential obstacles or provides an explanation for choices being made to move forward (Ertmer & Newby, 1996)

https://doi.org/10.52202/073963-0006

1) Linking anchors to reflection

Students' self-evaluation (Error) comments and reflection responses were analyzed to determine whether students referred to the TR 3.3 related LOs, the primary anchor of concern in this study. The process of identifying these LOs within students' self-evaluation and reflection responses involved mapping the terms students used in their comments with proficiency indicators associated with each LO. Based on the presence of a reference to an LOs the response was assigned a 1 (present) or 0 (not present). For example, Table III shows the reference of the LOs addressed by a single student in their self-evaluation (error) comments and reflection response. The students mentioned LOs PS 01.00 and PS 02.02 (technical plotting) as well as FA 02.08, 03.02, and 03.03 (determine the system characteristics curve for a fan used in a grain drying process) in their self-evaluation. The students mentioned only PS 01.00 in their reflection. For the LO self-rating, the student rated their proficiency for each LO using the scale shown in Table I. The text options were converted to a scale of 1 to 5 (Table I).

Based on the information presented in Table III, three categories were created to track the references students made to the LOs in the self-evaluation and reflection. The first category, "Error+Reflection," indicates that students addressed an LO in their self-evaluation (Error) and their reflection, regardless of their self-rating of the LO. The next two categories take into consideration only LOs the students self-rated below 3 (Table I), which indicates a need for improvement with the LO. The second category, "LO<3+Reflection," indicates that a particular LO self-rating was below 3 and that LO was referenced in the reflection but not in the self-evaluation (Error) comments. The third category, "Error+LO<3+Reflection," indicates and that LO was rated below 3 and was referenced in both the self-evaluation (Error) comments and the reflection. Counts of comments in each category were made.

2) Self-Evaluation and Reflection Response

Students' self-evaluation comments and reflection responses for TR 3.3 were qualitatively analyzed in a deductive manner using a revised *a priori* coding scheme based on Ertmer and Newby (1996) and Ku & Ho (2010) with revisions by Stratman and Diefes-Dux (2022) and Singh and Diefes-Dux (2023) (Table IV). During analysis of students' self-evaluations and reflection responses, the texts were coded for the highest level of metacognitive strategy employed by students.

To ensure reliability of the developed coding scheme, two coders, one with experience in coding a dataset collected in the process engineering course and another coder with experience with a dataset collected in a first-year engineering course, coded ten training samples from the first-year engineering course dataset. After coding, both coders compared their coding results and calculated the similarity percentage; that is similarity achieved by coders on identification of dimension and level of metacognitive strategies. During the first round of coding, 60% of similarity rate was achieved. Coders agreed on the metacognitive strategy dimension, but differences emerged on assignment of the levels for a dimension. The difference in coding of levels was due to one coder's limited familiarity with the first-year context. Discussion and clarification on differences resulted in a similarity percentage of 80%.

VI. RESULTS

Results are presented to address each of the research questions separately.

A. Links to LOs

For each LO for TR 3.3, the frequency count of instances for "Error+Reflection," "LO<3+Reflection," and "Error+LO<3+Reflection" are shown in Fig. 2. Each category indicates the links students made between their work on the anchor activities and their reflection for TR 3.3.

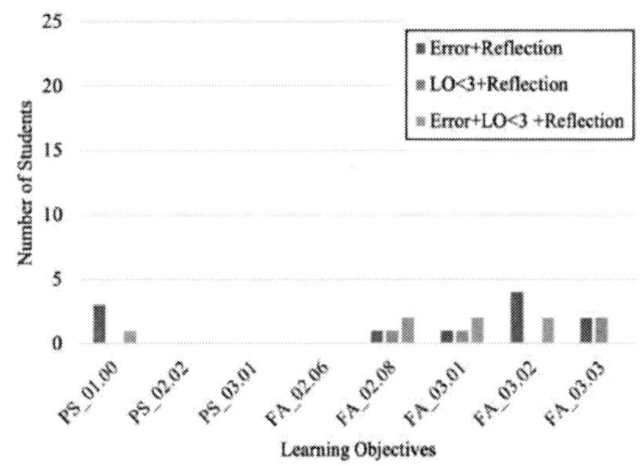

Fig. 2. Number of comments linking anchors and reflection for each TR 3.3 relevant LO.

Overall, only seven of the 25 students that completed the three parts of the assignment linked what they found in the anchor activities to their reflections. Three LOs (PS 02.02, PS 03.01, and FA 02.06) were neither commented on in the self-Among all three categories, the "Error+Reflection" category had the highest frequency counts (PS 01.00 and FA 03.02). This anchor-reflection link means students mentioned the LO in their error comments, rated themselves high (=3 or >3) on the LO, but reflected on the LO in their reflection response.

Few "LO<3+Reflection" and "Error+LO<3+Reflection" category anchor-reflection links were made for the Fan Selection LOs.

B. Metacognitive strategies

To address the second research question, the distribution of metacognitive strategies and highest-level of each metacognitive strategy employed by students in their self-evaluation comments and reflection responses are shown in Fig. 3 and 4.

https://doi.org/10.52202/073963-0006

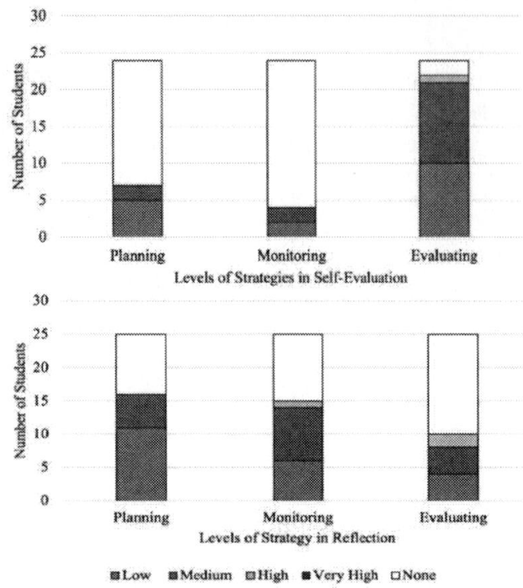

Fig. 3. Highest level of metacognitive strategies in self-evaluations and reflections.

Overall, among all three metacognitive strategies, students predominately used the Evaluating strategy during self-evaluation, whereas they used the Planning and Monitoring strategies during reflection. In addition, Fig. 3 shows the distribution of the levels of the metacognitive strategies employed by students in Self-Evaluation and Reflection.

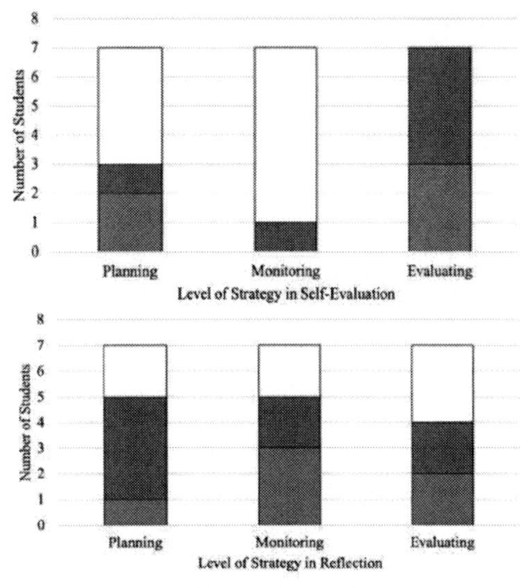

Fig. 4. Highest level of metacognitive strategies used in comments from Error+LO<3+Reflection category (n=7).

Overall, student engagement was mainly limited to the Low to Medium levels for all three metacognitive strategies, with one or two students engaging at the High level of the Monitoring and Evaluating strategies.

Fig. 4 shows the level of metacognitive engagement of students who commented on LOs in their self-evaluation, rated their proficiency low on those LOs, and mentioned them in their reflection ("Error+LO<3+Reflection"). While only seven students linked the anchors and reflection, they did not necessarily achieve higher levels for each metacognitive strategy. The one exception is Planning. These students had more medium level comments in their reflections than the class as a whole.

VII. DISCUSSION

With the aim of preparing students to be self–regulated learners, two anchors (i.e., self-evaluation and LO self-rating), were integrated with reflection into an engineering course. The study investigated (1) the extent to which students linked the anchors to their reflection responses and (2) the level of metacognitive strategies used by students during self-evaluation and reflection. Each research question is discussed below.

Regarding the first research question, results showed that only a few students linked both anchors to reflection, which means that these students mentioned the LOs that they needed to improve upon in their error comments while completing the self-evaluation, they then self-rated these LOs low, and finally reflecting on those LOs in their reflection response. The percentage of students with LO self-ratings at 3 or above for the eight LOs ranged from 52% to 96%. One of the reasons for high self-rating ratings of LOs on the scale could be students' low ability to evaluate their skills (Andaya et al., 2017) due to a lack of understanding of the what the skill should entail, which could have resulted in differences in their performance and their perception of those LOs. Also, students might have rated themselves high on the LO self-rating scale because completion of the LO self-rating activity contributed minimally to their course grade. As a result, students may not have thought through the activity and just completed the task. Or students perceived a risk to admitting their low ability with the LOs.

For the second research question, in the self-evaluation activity, one of the reasons for the predominance of Evaluating comments in the self-evaluation could be the nature of the assignment. Students compared their solution to the standard solution key provided by the instructor. However, the prompt provided for self-evaluation activity asked students to comment on things they missed, learned, and needed to work on. The prompt was intended to encourage engagement in the other metacognitive strategies. Perhaps students' lack of engagement in all three metacognitive strategies and their low level of engagement indicates students' lack of understanding of what they should do in response to the given instructions in the assignment. In academic settings, failure to follow instructions can hinder general learning, development of desired proficiency, and indicates low self-regulation ability in students (Dunham et al., 2020). It may not be completely an issue of the

https://doi.org/10.52202/073963-0006

ability to follow instructions as much as knowing what it means to sufficiently follow the instructions.

In the reflections, students predominantly employed low to medium levels of the Planning and Monitoring strategies rather than the Evaluating strategy. The levels of metacognitive strategies seen here were similar to those observed in the first two units of the course (Singh & Diefes-Dux, 2022, 2023). One of the reasons for the planning and monitoring emphasis in their work could be the first reflection prompt provided to students, which focused on discussing their learning proficiency with the LOs and strategies to improve on those LOs as needed. The prompt does not explicitly hint at a need for further evaluation. The instructor provided a single reflection prompt with the belief that upper-division students would be able to self-prompt themselves into making more meaning of their learning. However, this assumption proved false, as there is little evidence that students engaged in such self-prompting. Hence, this underscores the need for instruction on reflection and detailed feedback to direct students to improve their reflection abilities.

The second reason for the planning and monitoring emphasis could be that students may have felt they had completed their evaluation of their work during self-evaluation task. Students' limited use of the three metacognitive strategies aligns with the findings of Lew and Schmidt (2011) who described self-reflection as a complex process; students are poor at it, and instructors' guidance and supervision are needed to improve students' reflection abilities.

The few students who linked the anchors to their reflection employed low to medium levels of the three metacognitive. Studies have indicated that learners' self-evaluation skills influence their metacognitive engagement (Nisly et al., 2020; Steuber et al., 2017). Therefore, poor self-evaluation skills may be one of the reasons that students use low or medium level metacognitive strategies. To assist students in self-evaluation, external standards (solution key) were provided. However, offering external standards does not ensure that students will be able to think critically (Rawson & Dunlosky, 2007). A lack of critical thinking is demonstrated through low metacognitive engagement wherein students commonly describe what occurred but lacked evidence (Dewey, 1931) and depth of information. That is, students' engagement is limited to mere identification of their problems and not engagement in metacognition. Therefore, there is need to educate engineering students about the purpose of reflection and reflection writing (Csavina et al., 2016) to elevate the level of use of the metacognitive strategies.

The second reason for low metacognitive engagement could be the task value, which influences students' use of metacognitive strategies and the effort they expend on a given task (Buehl & Alexander, 2001). When students perceive a task as high value, they are motivated to use metacognitive skills (Bae & Kwon, 2021). This suggests that students may not have considered the anchor activities to be high-value tasks, highlighting their limited understanding of the importance of anchors in reflection.

Overall, metacognitive skills are difficult to develop over a short time or course (Nisly et al., 2020) but can be taught (Wedelin & Adawi, 2014) over an extended time. To ease the process of developing students' metacognitive strategies in a limited time, instructors can provide multiple opportunities in a course for students' metacognitive engagement and reflection writing (Jaiswal et al., 2021). Furthermore, instructors can improve students' level of use of metacognitive strategies by providing them sample responses for both desired and poor work for all dimensions and levels of metacognitive strategies (Zarestky et al., 2022).

VIII. IMPLICATIONS

This work has implications for both researchers and instructors. For researchers, the revised coding scheme allows for identification of both the metacognitive strategies and their levels of employment by students. Further, the detailed list of LOs provided a means for identifying whether or not students related their self-evaluations and LO self-ratings to their reflections. Without the LOs list, the relationships would have been more difficult to track.

For instructors, based on the lack of students' linking of the anchors to their reflection, instruction is needed at the start of the course that highlights the importance of the anchors and how anchors can be used effectively to improve engagement in reflection. Instructors should also provide reflection prompts for each of the three metacognitive strategies to engage students in all three dimensions of metacognition. Further, to improve students' level of metacognitive engagement, instructors can provide sample responses for each metacognitive dimension and level to highlight the differences among them. Finally, providing detailed feedback on students' reflection response can help students to work on points where their responses are insufficiently deep.

IX. CONCLUSION

This work focused on preparing students enrolled in a junior-level process engineering course as self-regulated learners. Students were provided with anchors with the aim of providing a means to sort out their learning difficulties so they could engage effectively in reflection. It was shown that students' ability to link the anchors to their reflections was limited and students employed the metacognitive strategies at only low to medium levels. Students' metacognitive engagement during self-evaluation and reflection were separately examined. Results showed that students mainly used low to medium levels of Evaluating in the self-evaluating activity, while the use of low to medium level of Planning and Monitoring dominated in their reflections. Overall, students' use of the three metacognitive strategies was at the superficial level.

 https://doi.org/10.52202/073963-0006

Acknowledgment

This work was made possible by a grant from the National Science Foundation (NSF 2235227). Any opinions, findings, and conclusions or recommendations expressed in this material are those of the authors and do not necessarily reflect the views of the National Science Foundation. The authors wish to also thank the UNL Nebraska SCIENCE Program of Excellence for its support for this research during the data collection phase.

References

Accreditation Board for Engineering and Technology. "Criteria for accrediting engineering programs, 2022-2023". https://www.abet.org/accreditation/accreditation- criteria/criteria-for-accrediting-engineering-programs-2022-2023/ (accessed Sept. 4, 2023).

Andaya, G., Hrabak, V. D., Reyes, S. T., Diaz, R. E., & McDonald, K. K. (2017). Examining the effectiveness of a postexam review activity to promote self-regulation in introductory biology students. *Journal of College Science Teaching*, *46*(4).

Bae, H., & Kwon, K. (2021). Developing metacognitive skills through class activities: what makes students use metacognitive skills? *Educational Studies*, *47*(4), 456–471. https://doi.org/10.1080/03055698.2019.1707068

Baisley, A., Hjelmstad, K., & Chatziefstratiou, E. (2022). *The accuracy of self-assessment in engineering mechanics*. Paper presented at the ASEE Annual Conference and Exposition, Minneapolis, MN. https://peer.asee.org/40837

Benson, D., & Zhu, H. (2015). *Student reflection, self-assessment, and categorization of errors on exam questions as a tool to guide self-repair and profile student strengths and weaknesses in a course*. Paper presented at the ASEE Annual Conference and Exposition, Seattle, WA. https://peer.asee.org/24763

Buehl, M. M., & Alexander, P. A. (2001). Beliefs about academic knowledge. *Educational Psychology Review*, *13*, 385-418.

Chi, M. T. H. (1997). Quantifying qualitative analyses of verbal data: A practical guide. *The Journal of the Learning Sciences*, *6*(3), 271–315. https://doi.org/10.1207/s15327809jls0603_1

Clark, R. M., & Dickerson, S. J. (2018). *The use of SPICE simulation to promote reflection and metacognition in a microelectronics course*. Paper presented at the ASEE Annual Conference and Exposition, Salt Lake City, UT.

Claussen, S., & Dave, V. (2017). *Reflection and metacognition in an introductory circuits course*. Paper presented at the ASEE Annual Conference and Exposition, Columbus, OH.

Colthorpe, K., Ogiji, J., Ainscough, L., Zimbardi, K., & Anderson, S. (2019). Effect of metacognitive prompts on undergraduate pharmacy students' self-regulated learning behavior. *American Journal of Pharmaceutical Education*, *83*(4), 6646. https://doi.org/10.5688/ajpe6646. PMID: 31223149

Csavina, K. R., Nethken, C. R., & Carberry, A. R. (2016). *Assessing student understanding of reflection in engineering education*. 2016 ASEE Annual Conference & Exposition, New Orleans, LA. https://doi.org/10.18260/p.26306

Dewey, J. (1931). *The way out of educational confusion*. Harvard University Press.

Dunham, S., Lee, E., & Persky, A. M. (2020). The psychology of following instructions and its implications. *American Journal of Pharmaceutical Education*, *84*(8), ajpe7779. https://doi.org/10.5688/ajpe7779

El-Maaddawy, T. (2017). *Enhancing learning of engineering students through self-assessment*. Paper presented at the IEEE Global Engineering Education Conference (EDUCON), Athens, Greece.

Ertmer, P. A., & Newby, T. J. (1996). The expert learner: Strategic, self-regulated, and reflective. *Instructional Science*, *24*(1), 1-24.

Flavell, J. H. (1979). Metacognition and cognitive monitoring: A new area of cognitive–developmental inquiry. *American Psychologist*, *34*(10), 906.

Fong, E., Yeter, I. H., Venkatesh, S., Kim, M. S., & Liu, J. (2023). *Studying the development of design thinking of undergraduate engineering students in Singapore: Qualitative reflection analysis*. Paper presented at the ASEE Annual Conference and Exposition, Baltimore, MD. https://peer.asee.org/44350

Fridman, R., Eden, S., & Spektor-Levy, O. (2020). Nascent inquiry, metacognitive, and self-regulation capabilities among preschoolers during scientific exploration. *Frontiers in Psychology*, *11*, 1790.

Hager, P. (2004). Lifelong learning in the workplace? Challenges and issues. *Journal of Workplace Learning*, *16*(1/2), 22–32. https://doi.org/10.1108/13665620410521486

Jaiswal, A., Lyon, J. A., Zhang, Y., & Magana, A. J. (2021). Supporting student reflective practices through modelling-based learning assignments. *European Journal of Engineering Education*, *46*(6), 987-1006.

Kittel, A. F. D, Kunz, R. A. C, & Seufert, T. (2021). Self-regulation in informal workplace learning: Influence of organizational learning culture and job characteristics. *Frontiers in Psychology*, *12*. https://doi.org/10.3389/fpsyg.2021.643748

Ku, K. Y., & Ho, I. T. (2010). Metacognitive strategies that enhance critical thinking. *Metacognition and Learning*, *5*, 251-267.

Lew, M. D., & Schmidt, H. G. (2011). Self-reflection and academic performance: Is there a relationship? *Advances in Health Sciences Education*, *16*, 529-545.

Lin, X. (2001). Designing metacognitive activities. *Educational Technology Research and Development*, *49*(2), 23–40. https://doi.org/10.1007/BF02504926

Lord, R. G., Diefendorff, J. M., Schmidt, A. M., & Hall, R. J. (2010). Self-regulation at work. *Annual review of psychology*, *61*, 543-568.

Mann, K., Gordon, J., & MacLeod, A. (2009). Reflection and reflective practice in health professions education: A systematic review. *Advances in Health Sciences Education*, *14*, 595-621.

Nisly, S. A., Sebaaly, J., Fillius, A. G., Haltom, W. R., & Dinkins, M. M. (2020). Changes in pharmacy students' metacognition through self-evaluation during advanced pharmacy practice experiences. *American Journal of Pharmaceutical Education*, *84*(1), 7489. https://doi.org/10.5688/ajpe7489

Opanuga, T., & Diefes-Dux, H. A. (2023). *Weekly self-rating of proficiency with course learning objectives: Gaining insight into undergraduate students' perception of their learning*. Paper presented at the ASEE Annual Conference and Exposition, Baltimore, MD.

Rawson, K. A., & Dunlosky, J. (2007). Improving students' self-evaluation of learning for key concepts in textbook materials. *European Journal of Cognitive Psychology*, *19*(4–5), 559–579. https://doi.org/10.1080/09541440701326022

Reflection activities. (n.d.). Retrieved September 4, 2023, from https://www.clemson.edu/otei/documents/Reflection%20Activities%20r.pdf

Rios, J. A., Ling, G., Pugh, R., Becker, D., & Bacall, A. (2020). Identifying critical 21st-century skills for workplace success: A content analysis of job advertisements. *Educational Researcher*, *49*(2), 80-89.

Rodgers, C. (2002). Defining reflection: Another look at John Dewey and reflective thinking. *Teachers College Record, 104*(4), 842-866.

Sandars, J. (2009). The use of reflection in medical education: AMEE Guide No. 44. *Medical Teacher*, *31*(8), 685-695.

Schön, D. A. (1983). From technical rationality to reflection-in-action. In *The Reflective Practitioner: How Professionals Think in Action* (pp. 21-69). New York: Basic Books.

Schraw, G., Crippen, K. J., & Hartley, K. (2006). Promoting self-regulation in science education: Metacognition as part of a broader perspective on learning. *Research in Science Education*, *36*(1), 111-139. https://doi.org/10.1007/s11165-005-3917-8

Schraw, G., & Moshman, D. (1995). Metacognitive theories. *Educational Psychology Review*, *7*, 351-371.

Seppanen, M. (2023). The quality of argumentation and metacognitive reflection in engineering co-design. *European Journal of Engineering Education*, *48*(1), 75-90. https://doi.org/10.1080/03043797.2022.2054314

Siewiorek, N., Shuman, L., Besterfield-Sacre, M., & Santelli, K. (2010). *Engineering, reflection and life long learning*. Paper presented at the ASEE Annual Conference and Exposition, Louisville, KY. https://peer.asee.org/16615

Singh, A., &. Diefes-Dux, H. A. (2022). *Students' metacognitive strategies revealed through reflections on their learning of process engineering concepts and skills*. Paper presented at the ASEE Annual Conference and Exposition, Minnesota, MN.

Singh, A., &. Diefes-Dux, H. A. (2023). Pairing self-evaluation activities with self-reflection to engage students deeply in multiple metacognition strategies. Paper presented at the ASEE Annual Conference and Exposition, Baltimore, MD.

Steuber, T. D., Janzen, K. M., Walton, A. M., & Nisly, S. A. (2017). Assessment of learner metacognition in a professional pharmacy elective course. *American Journal of Pharmaceutical Education*, *81*(10), 6034–28. https://doi.org/10.5688/ajpe6034

Stratman, E., & Diefes-Dux, H. A. (2022). *Impact of differently worded reflection prompts on engineering students' metacognitive strategies*. Paper presented at ASEE Annual Conference and Exposition, Minnesota, MN.

Tillema, H. (2010). Formative assessment in teacher education and teacher professional development. In P. Peterson, E. Baker, & B. McGaw (Eds.), *International encyclopaedia of education*. Elsevier Science.

Ugulino, W., & Pires, L. F. (2021). *The use of metacognition to develop self-regulated learning skills in students of a computer programming course*. Paper presented at the SEFI Annual Conference, Berlin, Germany.

Vogel-Walcutt, J., & Fiore, S. (2010). *Insights from empirical metacognitive research*. Paper presented at the 2010 Spring Simulation Multiconference. https://doi.org/10.1145/1878537.1878572

Wallin, P., & Adawi, T. (2018). The reflective diary as a method for the formative assessment of self-regulated learning. *European Journal of Engineering Education*, *43*(4), 507-521.

Wedelin, D., Adawi, T., Jahan, T., & Andersson, S. (2013). *Teaching and learning mathematical modelling and problem solving: A case study*. 1st International Conference of the Portuguese Society for Engineering Education (CISPEE), Porto, Portugal.

Wegner, J., Turcic, S. M., & Hohner, G. (2015). *Learning from experiences: Examining self-reflection in engineering design courses*. Paper presented at the ASEE Annual Conference and Exposition, Seattle, WA. https://peer.asee.org/24403

Zarestky, J., Bigler, M., Brazile, M., Lopes, T., & Bangerth, W. (2022). Reflective writing supports metacognition and self-regulation in graduate computational science and engineering. *Computers and Education Open*, *3*, 100085.

Zimmerman, B. J. (1989). A social cognitive view of self-regulated academic learning. *Journal of Educational Psychology*, *81*(3), 329-339. https://doi.org/10.1037/0022-0663.81.3.329

Zimmerman, B. J. (2002). Becoming a self-regulated learner: An overview. *Theory into practice*, *41*(2), 64-70. https://doi.org/10.1207/s15430421tip4102_2

Zimmerman, B. J. (2000). Attaining self-regulation: A social cognitive perspective. In *Handbook of self-regulation* (pp. 13-39). Academic press. https://doi.org/10.1016/B978-012109890-2/50031-7

Influential Factors in the Career Decision-Making of Gen Z in Engineering Education

Shradha Binani , Nayini Varshitha Reddy, Aashritha Kandra, Thakur Aryan Singh and Mihir Patel
Hyderabad Institute of Technology and Management
shradhabinani88@gmail.com

Abstract

Context

Generation Z individuals have a diverse and socially conscious cohort that values inclusivity, equality, ambition, and adaptability with a strong focus on career success. While numerous studies highlight the accomplishments of Gen Z across various domains such as technology, entrepreneurship, research and innovation. Very little research is done focusing on the career decision-making process of Gen Z.

Purpose or Goal

The purpose of the study is to exclusively addresses the gap on how parameters like passion and personal interest, gender bias, family and friends influence, career growth opportunities, and role models/mentors influence the career decisions of GEN Z, aiming to offer valuable insights into the complex dynamics of career decision-making and empower individuals to make informed and fulfilling choices for their careers.

Methods

Based on these five themes, a survey instrument was developed and administered to engineering students at an autonomous college in South India in the fall of 2023. A total of 260 responses were collected, and after data cleaning, 253 participants were included in the testing dataset. Exploratory factor analysis was performed to ascertain the survey instrument's factor structure, resulting in the confirmation of the hypothesized five factors.

Outcomes

The analysis revealed five factors as hypothesized with a minimum and maximum loading of 0.35 and 0.61. The internal consistency reliability index Cronbach's α ranged from 0.67 to 0.92, representing a strong consistency.

Conclusion

This survey instrument can be used in any education institutions to comprehensively capture the elements/factors that exert an influence on the career-decision making process of Generation Z. The insights gathered from this survey can serve as valuable input for shaping policies and interventions focused on mitigating gender biases and stereotypes, improving career guidance, and fostering environments that facilitate both personal and professional development.

Keywords—Career, Decision-Making, Family and Friends Influence, Passion, Gender Bias, Generation Z

I. INTRODUCTION

IN today's society, a career is not just a means of making a living; it's a lifelong journey of professional advancement within a chosen occupation can use it to prepare your manuscript. Students are often encouraged to envision careers in fields like education, sports, medicine, research, or politics, driven solely by their own preferences. (Hodkinson, P. et al.,1997)

However, the reality is more complex, as societal influences play a significant role. Past knowledge and societal norms can shape students' career choices, leading them to align their paths with prevailing trends rather than pursuing their genuine interests. The challenge lies in adapting to evolving corporate cultures, where students often struggle to update their approaches. (Duffy, R. D.et al., 2009)

Amid the ever-evolving landscape of higher education and workforce dynamics, a crucial aspect understands the driving forces behind career decisions. This quantitative research focuses on the influential factors guiding career choices among Generation Z (Gen Z) students within the context of engineering education. (Schwieger et al., 2018)

Born between the mid-1990s and early 2010s, Gen Z is characterized by their familiarity with advanced technology and a rapidly changing global landscape. As these digital natives embark on a journey in higher education, especially in fields like engineering, their career decisions are molded by a complex interplay of factors that distinguish them from earlier generations (Seemiller, C. et al., 2017)

The realm of engineering, a cornerstone of global innovation and progress, has witnessed the influx of Generation Z (Gen Z) students, who navigate a complex interplay of personal, societal, and educational influences when shaping their career choices. (Boutellier, R. et al., 2008)

This empirical research endeavors to unravel the intricate web of these multifaceted influences through a quantitative lens, shedding light on the decision-making patterns of this generation within the landscape of engineering education (Wendell, K. B. et al., 2017).

This dynamic landscape of career choices can lead to uncertainty and hinder the development of expertise in specific

https://doi.org/10.52202/073963-0007

sectors, potentially impeding their future success. Thus, the intricate dance between Generation Z's tech-savvy lifestyle and their career decisions presents a multifaceted panorama that merits thorough exploration and understanding (Haibo et al., 2018)

Against this backdrop, the primary objectives of this research are threefold: first, to examine the extent to which passion and personal interest impact Gen Z's career choices in the field of engineering; second, to elucidate the influence of gender bias on the decision-making process, particularly regarding the underrepresentation of women in engineering; and third, to explore the roles of family, friends, career growth opportunities, and role models/mentors (Binani, S. et al., 2023) in shaping the career trajectories of Gen Z in the engineering domain.

Understanding the influential factors in Gen Z's career decision-making process within the realm of engineering education has broader implications for educational institutions, policymakers, and industry stakeholders. (Aryani, F. et al., 2020) Insights gained from this research can inform targeted interventions aimed at promoting diversity, inclusivity, and informed career choices within the engineering institutions.

II. LITERATURE

The study's supporting literature addresses the concept of generations, defined as individuals sharing a common chronological, social, and historical context, and the corresponding generational theory that suggests similar characteristics and behaviors among individuals born within the same generation (Twenge, J. M. et al., 2010).

This notion has been widely explored in the context of evolving career preferences across different generational cohorts, such as baby boomers, generation X, and millennials. However, the focus now shifts to Generation Z (Gen Z), a cohort born between the mid-1990s and early 2010s, characterized by distinct experiences, values, and technological familiarity. As Gen Z enters higher education, particularly in engineering, understanding the factors shaping their career decisions becomes pivotal (Törőcsik, M. et al., 2014)

Generation Z, born between 1995 and 2010, emerged during a period of rapid information dissemination and technological accessibility (Ebadi, S. et al., 2021). The intrinsic connection to internet technology and smartphones is emblematic of their daily lives, with a substantial proportion consistently accessing online platforms (Szymkowiak, A. et al., 2021). An essential trait of Gen Z is their inclination to prioritize comfort and flexibility in career choices, displaying a tendency to eschew rigid routines and commitments. This characterization influences their selection criteria and their propensity to explore a diverse range of careers (Barhate, B. et al., 2022).

The Generation Z cohort holds a predominant presence in the digital workforce and exhibits both strengths and weaknesses when it comes to career selection. Research conducted by Paina & Irini (Racolţa-Paina et al., 2021) indicates their inclination toward opting for virtual roles, favoring positions that offer flexibility without stringent work routines or long- term commitments. Gen Z's adeptness in virtual communication, proficiency in utilizing diverse tools, video editing, content creation, and mastery of emerging job roles in the era of Industry 4.0 position them favorably to meet the demands of evolving job markets (Binani, S.et al., 2022).

Nevertheless, their propensity to gravitate towards roles lacking routine, characterized by low commitment and freelance arrangements, contributes to a frequent job- switching trend, often deviating from their inherent potential (Wheatley, A. C. et al., 2019). This predisposition engenders uncertainty in career selection, hindering the establishment of expertise in specific domains and impeding the development of a coherent career identity necessary for future success (Haibo et al., 2018).

Furthermore, the onset of the industrialization era 4.0 has substantially transformed perceptions about life, work, and career choices. Generation Z, predominantly composed of digital natives, is highly influenced by rapid information technology development, manifesting in their adoption of the internet, social media, and smartphones as integral components of their lifestyle (Ozkan, M et al., 2015). This tech- savvy orientation significantly informs their career selection, shaping their career preferences and trajectories (Gabrielova, K. et al., 2021).

Whereas, in the context of Asian society, a career choice extends beyond individual responsibility, encompassing a familial dimension that significantly impacts the sociocultural fabric (Gentina et al., 2020). The selection of a career path is intertwined with family obligations, and its repercussions reverberate within the family's societal context. Opting for a career aligned with personal aspirations and receiving parental endorsement yields comprehensive psychological and material backing. This robust support framework contributes to the cultivation of elevated career self-efficacy. Conversely, instances where a chosen career does not garner parental approval necessitate heightened support, particularly in the psychological realm. This circumstance leads to a tendency towards insecurity regarding the chosen career and the manifestation of diminished career self-efficacy. Concurring with the findings of (Kantamneni N. et al., 2018) an array of factors, including parental, peer, influencer, and social media support, intricately influence career self-efficacy through the channels of guidance, assistance, inspiration, and role modeling.

This literature review emphasizes the distinctive nature of Generation Z and the factors that impact their career decision-making, particularly in the field of engineering education. As this study delves into the empirical examination of these influential factors, it seeks to contribute to a more profound understanding of Gen Z's career choices and inform strategies

https://doi.org/10.52202/073963-0007

to enhance their decision-making processes within the engineering profession.

III. METHODS

After careful consideration of multiple factors influencing methodological selection, a quantitative approach has been adopted for this study. A comprehensive set of 25 items was initially employed to gather responses, aimed at assessing various constructs across five dimensions: Personal Interest and Passion (comprising 5 items), Gender Bias and discrimination/stereotypes (comprising 5 items), Influence of Family and Friends (comprising 5 items), Career Growth Opportunities (comprising 5 items), and Role Model/Mentor Influence (comprising 5 items). The administration of the instrument took place electronically during the spring of 2023, targeting engineering students within an autonomous institute situated in South India. The data collection employed the SPSS software, with subsequent analysis conducted through its utilization. The internal consistency of the five dimensions, as indicated by Cronbach's alpha coefficients, exhibited a range between 0.67 and 0.92, signifying a commendable level of coherence among the items. (Binani, S. et al., 2023).

The design of the survey instrument drew inspiration from an extensive literature review ((Binani, S. et al., 2022) shaped by the recognition of an impending landscape wherein the workforce is projected to be predominantly of Generation Z employees in the forthcoming years. The data collection phase occurred during the autumn of 2023. This survey was meticulously crafted to gauge the manner in which specific parameters such as personal passion, gender bias, familial and social influences, opportunities for career advancement, and the impact of role models and mentors collectively shape the career choices of individuals belonging to Generation Z. Additionally, the survey encompasses distinct demographic inquiries pertaining to students' backgrounds, encompassing aspects such as gender identity, field of engineering specialization, and educational board affiliation. The author meticulously developed a set of 25 interconnected items and questions, intricately designed to assess the intricate interplay of diverse factors influencing the career decision-making processes of Generation Z individuals.

Table-1 presents an in-depth understanding of the item formulation process for each scale, effectively clarifying the fundamental essence of the variables and showcasing exemplar items. Survey participants were guided to appraise each variable using a Likert scale spanning from 5 (strongly agree) to 1 (strongly disagree). Through the incorporation of this evaluative approach, the authors adeptly conducted a meticulous analysis, delving into the noteworthy impact of diverse factors on the career decision-making journey of Generation Z.

TABLE I
OUTLINE OF FIVE CONSTRUCTS

Construct	Definition of Construct	Example Items
Passion and Personal Interest	Passion and personal interest denote an individual's strong emotional inclination and intrinsic attraction towards a specific subjector pursuit.	I believe that my personal interests and passions will change over time, so it is not essential to pursue a career that aligns with them • I strongly agree that my personal interest align well with my career aspirations in the engineering field.
Gender Bias	Gender bias refers to the systematic and unequal treatment of individuals based on their gender, resulting in disparities and inequities.	• Gender plays a significant role in career decisions • I strongly agree that gender bias exists in the field of engineering.
Influence of Family and Friends	The impact exerted by family and friends on an individual's choices and decisions.	• I moderately agree that the opinion of my family and friends has an impact on my choice to pursue a career in engineering. • I strongly agree that I have the comfort to go against the opinions of my family and friends to pursue a career in Engineering.
Career Growth Opportunities	Career growth opportunities encompass pathways and prospects for professional advancement, skill development, and upwardmobility within one's chosen field.	• It is important to continue learning and gaining new skillsthroughout one's educational career. • I believe that there are amplecareer growth opportunities available to me in my field of work
Role Model/ Mentor Influence	Role model/mentor influence refers to the effect of exemplary individuals or guides on shaping the attitudes, behaviors, and decisionsof individuals.	• I feel that the influence of my role models or mentors onmy career decisions is positive and empowering . • I moderately agree that my career decisions have been influenced by role models or mentors in my life.

A. *Construct 1: Passion And Personal Interest*

This factor explores the extent to which personal passions align with chosen career paths, shaping decisions and aspirations. Through an examination of this construct, the study seeks to unveil the intricate interplay between inherent interests and the vocational trajectories pursued by young engineers.

B. Construct 2: gender bias and discrimination / stereotypes

This factor scrutinizes the impact of societal stereotypes, biases, and expectations on career pathways, shedding light on the multifaceted dynamics that shape the decision-making process. Through a meticulous examination of gender-related influences, this construct aims to contribute to a comprehensive understanding of the intricate interplay between gender bias and career aspirations within the Gen Z cohort.

C. Construct 3: influence of family and friends

This factor delves into the impact of familial and social networks on the career choices of Generation Z engineering students. Through a meticulous analysis, the research uncovers how interactions within this sphere shape and guide their career trajectories.

D. Construct 4: career growth opportunities

This factor investigates how Generation Z individuals perceive the prospects of skill development, upward mobility, and future job roles. It assesses their attitudes towards aligning their career choices with opportunities for growth and progress in the ever-evolving landscape of engineering.

E. Construct 5: role model/mentor influence

This factor explores how guidance from accomplished individuals shapes attitudes, decisions, and aspirations, ultimately influencing career trajectories. Investigating the role model/mentor influence construct provides insight into the dynamic interplay between personal inspiration and external guidance in the career decision-making process of young engineers.

IV. DATA ANALYSIS AND RESEARCH FINDINGS

Following the distribution of the survey questionnaire, responses were collected, categorized, and reviewed prior to conducting relevant descriptive statistical analyses using the SPSS software. Face validity was established through the engagement of three volunteers to review the questionnaire, furnishing feedback on language and terminology; the volunteers affirmed the adequacy of phrasing and word choices within the survey instrument. Out of a total of 260 responses, 253 were retained for analysis after data cleaning.

Data points from respondents who completed less than 50% of the survey were excluded, and respondents answering "yes" to all questions were omitted from analysis. Missing data was addressed using the group mean substitution method. The survey, with a duration of approximately 8 minutes, utilized a five-point Likert scale encompassing the options: strongly disagree, disagree, neutral, agree, and strongly agree. A reminder was issued to students after three days to encourage survey completion if they had not done so earlier.

TABLE II
PARTICAPANTS DEMOGRAPHIC INFORMATION

#	Category	N	%
	Total	253	100
1	Gender		
	Male	144	56.9
	Female	109	43.1
2	Engineering discipline		
	Computer science and engineering	89	35.2
	Data Science (DS)	26	10.3
	Cybersecurity (CS)	48	19
	Internet of things (IoT)	48	19
	Artificial intelligence and Machine learning (AI & ML)	42	16.6
3	Future Plans		
	Job	149	58.9
	Higher studies	90	35.6
	Other	14	6.5
4	Geographic Location		
	Urban	225	88.9
	Rural	28	11.1

The demographic data of participants, including gender identity, engineering discipline, future plans and Geographic

Location, are presented in Table 2. Among the 253 students who responded, 56.9% were male and 43.1% were female. Course distribution included various college disciplines (35.2% - CSE, 16.6% - AI & ML, 10.3% - DS, 19% - IOT, 19% - CSC). The predominant secondary education board among respondents was identified as having an 81.9% representation, highlighting demographic variance.

Table 3 presents the comprehensive descriptive statistics pertaining to all survey items employed in the study. An exploratory factor analysis methodology was undertaken in this research endeavor. To assess the appropriateness of items for factor analysis, Bartlett's test for sphericity was applied, with a significance level set at p=0.00. Additionally, the Kaiser–Meyer–Olkin (KMO) measure of sampling adequacy, with a threshold of KMO 0.84, was employed to evaluate the variance captured by the extracted factors (Kittur, J. et.al 2020)

Guidance for factor analysis was derived from multiple sources including parallel analysis, scree plots, and Kaiser's criterion. While parallel analysis of 5 factors and the scree plot / kaiser indicated the presence of seven factors. In alignment with the hypothesized number of factors, the decision was made to proceed with five factors. Given correlations exceeding 0.33 among the factors, the Promax rotation method was adopted (Kittur, J. et.al 2020)

The conclusive factor loadings for the five identified factors are detailed in Table 4. Within Table 3,it is observed that four factors (items 2, 3, 6, 7, 9, 13, 14, 15, 20, 22, 23, 24, 25) demonstrated cross-loading on more than one factor > 0.3 (

McNabb, D. E. 2020). Consequently, these 13 items were excluded from the analysis, resulting in a final compilation of four factors encompassing a total of 12 items. The factor loadings for factor 1 ranged from 0.32 to 0.50, while factor 2 exhibited loadings between 0.38 and 0.43. Similarly, factor 3 displayed loadings of 0.41 to 0.69, factor 4 ranged from 0.35 to 0.61, and factor 5 demonstrated loadings spanning to 0.38 ranging from 0.67 to 0.92, underscoring the strong reliability of the identified factors. (Kittur, J. et.al 2020) reliability coefficient Cronbach's α, showcased robust values The evaluation of internal consistency, as measured by the participants.

Table III. Descriptive Statistics of Five Constructs

#	**Measure**	Mean	SD
	Personal Interest and Passion		
1	Do you think your personal interests are in line with your career aspirations in the engineering field?	3.8	0.8
2	I believe that having a career aligned with my personal interests and passions will lead to greater job satisfaction.	4.1	0.7
3	When making career decisions, I prioritize my personal interests and passions over external factorssuch as salary or job availability.	3.7	0.9
4	I am comfortable deviating from societal or culturalexpectations in order to achieve a career aligned with my personal interests and passions.	3.6	0.9
5	I believe that my personal interests and passions will change over time, so it is not essential to pursue a career that aligns with them.	3.4	1.0
	Gender Bias and Discrimination/Stereotypes		
6	Do you believe that gender bias exists in the fieldof engineering?	3.0	1.1
7	Gender plays a significant role in career decisions	2.9	1.1
8	Do you think education and awareness programs can help overcome gender bias and stereotypes in engineering?	3.7	0.9
9	I feel that my gender has impacted my careeropportunities and advancements.	2.9	1.1
10	I feel that my educators have helped me navigategender-based barriers in my career.	3.3	0.8
	Role Models and Mentors Influence		
11	Have your career decisions been influenced by rolemodels or mentors in your life?	3.6	0.8
12	I feel that my role models or mentors have helped me develop a strong sense of purpose and direction in my career.	3.6	0.9
13	I feel comfortable discussing career-related issueswith my role models or mentors.	3.5	0.9
14	I feel that the influence of my role models or mentors on my career decisions is positive andempowering.	3.6	0.8
15	My role models or mentors have provided me with emotional support during challenging times in my career	3.4	0.9
	Influence of Family and Friends		
16	Does the opinion of your family and friends have an impact on your choice to pursue a career in engineering?	3.7	0.9
17	Do family and friends encourage you to pursue acareer that aligns with my interests and passion?	3.7	0.9
18	Do you have the comfort to go against the opinions of your family and friends to pursue a career in engineering.	3.5	0.9
19	I feel comfortable discussing career-related issueswith my family and friends.	3.7	0.9
20	Have you included your family and friends in yourcareer decision-making process?	3.7	0.8
	Career Growth Opportunities		
21	I believe that a higher education degree isnecessary for a successful career in engineering	3.7	1.0
22	It is important to continue learning and gaining newskills throughout one's educational career	4.1	0.8
23	I feel that my current institute values and supportsthe career growth and development of its students	3.6	0.8
24	I feel that my institute provides the necessary training and development programs to help me advance in my career	3.5	0.9
25	I believe that there are ample career growthopportunities available to me in my field of work	3.8	0.8

Table IV. Survey Instrument Ultimate Factor Loadings

#	Measure	F1	F2	F3	F4
	Personal Interestand Passion				
1	Do you think your personal interestsare in line withyourcareer aspirations in the engineering field?	0.501			
2	I am comfortable deviating fromsocietal or cultural expectations in order to achieve a career aligned withmy personalinterests and passions.	0.423			
3	I believe that my personal interestsand passions will change over time, so it is not essential to pursue a career that aligns with them.	0.329			
	Gender Bias and Discrimination/ Stereotypes				
4	Do you think education and awareness programs can help overcome gender biasand stereotypes in engineering?		0.382		
5	I feel that my educators havehelped me to navigate gender-based barriers in my career.		0.43		
	Role Models and Mentors Influence				
6	Have your career decisionsbeen influenced by rolemodels or mentors in yourlife?			0.412	
7	I feel that my role models or mentors have helped me to develop a strong sense of purpose and direction in my career.			0.696	
	Influence of Family and Friends				
8	Does the opinions of your family and friends have an impact on your choice to pursue a career inengineering?				0.358
9	Do family and friends have encouraged you to pursue a career that aligns with my interests and passion?				0.61
10	Do you have the comfort to go against the opinions of your family and friends to pursue a career in engineering.				0.429

V. CONCLUSIONS

The objective of formulating and creating a survey instrument was to comprehensively capture the elements/factors that exert an influence on the career-decision making process of Generation Z. Through an exploratory factor analysis, a total of five distinct factors were identified: personal passion and interests, the impact of gender bias, the role of family and peer influence, opportunities for career advancement, and the significance of role models and mentors in shaping career choices for Generation Z individuals. Subsequent to the collection of evidence for face validity, the factors displayed loadings ranging from 0.38 to 0.61, with Cronbach's α values spanning from 0.67 to 0.92. This survey instrument holds the potential to be adopted within various educational contexts, serving as a valuable tool to comprehend the influential determinants shaping Generation Z's decision-making process regarding careers, particularly within the domain of engineering education. Furthermore, the ethical implications of utilizing such a tool must be carefully considered, ensuring that participants' privacy and confidentiality are upheld throughout the data collection and analysis phases (Binani, S. et al., 2022). The broader implications of this research extend to educational institutions, policymakers, and industry stakeholders. The insights derived from this study offer guidance for targeted interventions aimed at fostering diversity, inclusivity, and well-informed career selections within the engineering profession

References

CHodkinson, P., & Sparkes, A. C. (1997). Careership: a sociological theory of career decision making. British journal of sociology of education, 18(1), 29-44.

Duffy, R. D., & Dik, B. J. (2009). Beyond the self: External influences in the career development process. The career development quarterly, 58(1), 29-43.

Schwieger, D., & Ladwig, C. (2018). Reaching and retaining the next generation: Adapting to the expectations of Gen Z in the classroom. Information SystemsEducation Journal, 16(3), 45.

Seemiller, C., & Grace, M. (2017). Generation Z: Educating and engaging the next generation of students. About campus, 22(3), 21-26.

Boutellier, R., Gassmann, O., & Von Zedtwitz, M. (2008). Managing global innovation: uncovering the secrets of future competitiveness. Springer Science & Business Media.

Wendell, K. B., Wright, C. G., & Paugh, P. (2017). Reflective decision-making in elementary students' engineering design. Journal of Engineering Education, 106(3), 356- 397.

Haibo, Y., Xiaoyu, G., Xiaoming, Z., & Zhijin, H. (2018). Career adaptability with or without career identity: how career adaptability leads to organizational success and individual career success?. Journal of Career Assessment, 26(4), 717-731.

Binani, S., Singh, T. A., & Shoeb, S. (2023). Freshmen Protégé Perspectives on Mentoring in Engineering Education-A Qualitative

https://doi.org/10.52202/073963-0007

Study. Journal of Engineering Education Transformations, 36(Special Issue 2).

Aryani, F., & Umar, N. (2020). Factors affecting z generation on selecting majors in the university: An Indonesian case. Journal of Social Studies Education Research, 11(3), 109-133.

Twenge, J. M. (2010). A review of the empirical evidence on generational differences in work attitudes. Journal of Business and Psychology, 25, 201-210.

Törőcsik, M., Szűcs, K., & Kehl, D. (2014). How generations think: research on generation z. Acta universitatis Sapientiae, communicatio, 1(1), 23-45.

Ebadi, S., Wang, T. T., Levine, H., Keesling, A., Semeghini, G., Omran, A., ... & Lukin, M. D. (2021). Quantum phases of matter on a 256-atom programmable quantum simulator. Nature, 595(7866), 227-232.

Barhate, B., & Dirani, K. M. (2022). Career aspirations of generation Z: a systematic literature review. European Journal of Training and Development, 46(1/2), 139-157.

Racolţa-Paina, N. D., & Irini, R. D. (2021). Generation Z in the workplace through the lenses of human resource professionals–A qualitative study. Calitatea, 22(183), 78-85.

Binani, S. (2022). Ethical Preparedness, Self-Efficacy and Challenges in Engineering Education–A Quantitative Study. Journal of Engineering Education Transformations, 35(Special Issue 1).

Wheatley, A. C., & Hibbler-Britt, L. M. (2019). The long tail of Generation Z and the future of a freelance economy. In Advances in the technology of managing people: Contemporary issues in business (pp. 83-92). Emerald Publishing Limited.

Haibo, Y., Xiaoyu, G., Xiaoming, Z., & Zhijin, H. (2018). Career adaptability with or without career identity: how career adaptability leads to organizational success and individual career success?. Journal of Career Assessment, 26(4), 717-731.

Ozkan, M., & Solmaz, B. (2015). The changing face of the employees–generation Z and their perceptions of work (a study applied to university students). Procedia Economics and Finance, 26, 476-483.

Gentina, E. (2020). Generation Z in Asia: a research agenda. The new Generation Z in Asia: Dynamics, differences, digitalisation, 3-19.

Gabrielova, K., & Buchko, A. A. (2021). Here comes Generation Z: Millennials as managers. Business Horizons, 64(4), 489-499.

Kantamneni, N., McCain, M. R., Shada, N., Hellwege, M. A., & Tate, J. (2018). Contextual factors in the career development of prospective first-generation college students: An application of social cognitive career theory. Journal of Career Assessment, 26(1), 183-196.)

Binani, S., Shoeb, S., & Singh, T. A. (2023). Development of Survey Instrument to Evaluate Freshmen Protégé Understanding on Mentoring in Engineering Education. Journal of Engineering Education Transformations, 36(Special Issue 2

Binani, S. (2022). Development of Survey Instrument to Assess Freshmen Engineering Students Understanding on Engineering Ethics. Journal of Engineering Education Transformations, 35(Special Issue 1).

Kittur, J., & Brunhaver, S. R. (2020, June). Developing an Instrument to Measure Engineering Education Research Self-Efficacy. In 2020 ASEE Virtual Annual Conference Content Access

Kittur, J., & Brunhaver, S. R. (2020, June). Developing an Instrument to Measure Engineering Education Research Self-Efficacy. In 2020 ASEE Virtual Annual Conference Content Access.).

McNabb, D. E. (2020). Research methods for political science: quantitative, qualitative and mixed method approaches. Routledge.

Kittur, J., Coley, B. C., & Kellam, N. N. (2020, June). Understanding how Novice Indian Faculty Engage in Engineering Education Research. In 2020 ASEE Virtual Annual Conference Content Access).

Binani, S. (2022). Freshmen Engineering Students' Perspectives on Engineering Ethics–A Qualitative Study. Journal of Engineering Education Transformations, 35(Special Issue 1)

Szymkowiak, A., Melović, B., Dabić, M., Jeganathan, K., & Kundi, G. S. (2021). Information technology and Gen Z: The role of teachers, the internet, and technology in the education of young people. Technology in Society, 65, 101565.

https://doi.org/10.52202/073963-0007

Review of the education policy to reduce the gender gap in engineering education in Ireland

Sandra I. Cruz Moreno; Shannon Chance
Technological University Dublin, Ireland
Corresponding Author Email: Shannon.Chance@TUDublin.ie

Abstract

Context

This paper investigates European and national Irish education policy geared toward improving representation of women in engineering at the tertiary level of education in the context of the persistent underrepresentation of women in engineering and other Engineering, Manufacturing and Construction (EMC) fields.

Purpose or Goal

Our concern for general equity prompted the research questions: To what degree do current strategies at the European level aim to reduce the gender gap in engineering courses, and what has been enacted to counteract gender imbalance at the third level in Ireland?

Methods

This paper provides macro-level analysis by mapping out the public policies aimed at enhancing women's access to ECM courses in Ireland. Employing scoping review procedures, the study collects and critiques existing literature, seeking to establish connections and assess alignment between gender-related goals and policy levers.

Outcomes

We located and assessed eleven strategies and/or policies at the European and Irish levels.

Conclusion

Several current Irish policies are geared toward addressing the gender gap. Yet their gender-related effectiveness is hindered in two primary ways. First, the policies lack concrete actions to promote women's participation in STEM fields and only a subset of them acknowledge the gender gap as a significant issue requiring resolution. Second, although many of them identify the demand for engineers in the labour market, they do not explicitly address gender-related aspects of the existing gaps. As a result, the pursuit of this objective may inadvertently exacerbate the gender disparity rather than alleviate it by attracting men but not women.

Keywords— Gender gap; STEM education policy; women in engineering.

I. Introduction

GENDER equality in engineering education goes beyond achieving a proportional enrolment of male and female students. To attain a meaningful representation of women in engineering education, women's active involvement in academia/research, and in leadership positions in higher education, is required (David, 2015). Nevertheless, the distribution of women students joining engineering programs at tertiary level serves as a preliminary indicator, providing some initial understanding of women's underrepresentation in this discipline.

Across all member countries of the Organization for Economic Cooperation and Development (OECD), in 2019 men represented 61% of the new entrants in the fields of Engineering, Manufacturing and Construction (EMC) (OECD, 2021). In the European Union, the second most common field of education in 2021 was EMC, which accounted for 15.8 % of all tertiary education students. However, almost three quarters (73.2 %) of all students in EMC were male (EUROSTAT, 2021). A similar distribution by sex is observed in Ireland where, despite the increase of women's enrolment in EMC programs in higher education over the last seven years, in 2021 only 23% of the graduates from these fields were women (Higher Education Authority, 2021).

Our concern for general equity lead to the research questions for this study: *To what degree do current strategies at the European level aim to reduce the gender gap in engineering courses, and what has been enacted to counteract gender imbalance at the third level in Ireland?*

II. Literature review

A. Relevance of gender in STEM from a policy perspective

Policy makers sense that supporting women in engineering is necessary:

(1) to achieve gender equality (Clavero & Galligan, 2021; Rosa & Clavero, 2022) under the premise that a wider access to a variety of degrees for a diversity of students (in terms of gender, socioeconomic and ethnic background) will have long

https://doi.org/10.52202/073963-0008

term benefits for both the students and their societal groups (Torotcoi et al., 2020). From this approach, gender equality is an intrinsic right to be addressed.

(2) to tackle the shortage of engineers by attracting people with new profiles into the workforce (Beede et al., 2011; Moloney & Ahern, 2022). In Ireland, a study on skills policy reported that 39% of the vacancies that were 'difficult to fill' in 2021 were associated with science, engineering and technology and were due to shortages of workers (OECD, 2023).

(3) to ensure better results from engineered solutions, by increasing diversity of viewpoints expressed in the design process. Research shows that having a diverse workforce with regard to race, ethnicity and gender (Hersh, 2000; Tannenbaum et al., 2019) leads to increased innovation, enhanced team effectiveness and organisational performance (Drew & Roughneen, 2004).

It appears that engineering, along with engineering education, is at the centre of questions of economic growth, innovative development, and social justice.

B. *Enhanced access of women in STEM*

Although across Europe a variety of policy frameworks have been developed to support female students undertaking engineering and other STEM (science, technology, engineering and mathematics) courses in higher education, little research has been published regarding the effectiveness of such 'access policies' in achieving increased participation by women in engineering higher education. Even where increases in women's STEM participation have been documented, such increase could be a result of context rather than of specific policy instruments. It is worth noting that a study by the European Commission (2008) reported higher proportion of women in STEM sectors but indicated these sectors were also characterized by lower budgets, reduced salaries or experiencing a male 'brain drain'. This implies that women may be filling positions that are not sufficiently attractive to men, perpetuating that certain fields of science are perceived as more masculine than others, and the higher paying, more illustrious STEM fields are still male dominated. More recent research is consistent with those findings and further indicates that the gender disparity in STEM is worsened by a tendency for female STEM graduates to exit the field during the early stages of their careers (Delaney & Devereux, 2022).

C. *Gender and education in Ireland*

The low proportions of females pursuing STEM careers in Ireland may result from gender stereotypes as well as the historical education systems and structural constrains that still prevail in Ireland, such as single-sex schools where subjects originally provided where different for boys and girls (Kelly et al., 2019; Kiernan et al., 2023). A gender-related disparity in educational offerings persists today, where the Leaving Certificate constitutes the final exam of the Irish secondary school system and is used as a university matriculation examination (on par with the SAT in the United States, for example) (DFHERIS & HEA, 2022, p. 17). Unfortunately, students in all-girls schools frequently lack access to as wide a range of STEM-related coursework as students in all-boys schools are offered in preparation for the Leaving Cert (often including technical graphics, physics, chemistry, and woodworking), and this represents a major barrier in exposing female students to Engineering and Technology subjects and helping them prepare to enter engineering and some other technical fields at third level (Kiernan et al., 2023).

III. Methods

This paper reports macro-level analyses: following a scoping of the literature on gender dimensions of STEM higher education, the lead author mapped currently existing public policies aimed at enhancing women's access to ECM courses in Ireland. Employing standard scoping review procedures (Arksey & O'Malley, 2005), the authors sought to establish connections and ensure coherence between the research problem statement (concerning women in engineering within the tertiary education sector) and policy instruments in Ireland (that were implemented to address this issue).

For the identification and characterization of policy related to the research question, we looked for European or Irish National policy frameworks in a variety of formats. For the purpose of this paper, we agreed to include policies plans, strategies, and policies reviews established to increase women recruitment, persistence and/or graduation rates in Engineering or STEM fields or to reduce gender gap in Engineering or STEM higher education programs. We selected the websites of relevant government institutions, European and international organisations and navigated their homepages to access to their documents. As the search strategy aimed to map the present policy context, any policy document prior to 2010 was excluded.

The authors located and assessed eleven strategies and/or policies at the European and Irish levels. Documents identified and analysed at the EU level included (1) the Gender Equality Strategy 2020-2025, the European Skills Agenda, and (2) the European Strategy for Universities. National policies considered in this study include (3) Ireland's National Skills Strategy 2025, (4) the National Strategy for Women and Girls 2017-2020, (5) the National STEM Education Policy Statement 2017-20126 and (6) its Implementation Plan, (7) the National Strategy for Higher Education 2030, (8) National Review of Gender Equality in Irish Higher Education Institutions, (9) the Gender Action Plan 2018-2020, (10) the National Development Plan 2021-2030, and (11) the National Access Plan for Higher Education 2022-2028.

IV. Findings

A. Commitments to gender equality in STEM at HE

1) Policy instruments in the EU

Despite the existence of an extensive policy framework within the European Union (EU), there remains a persistent gender imbalance between women and men across various sectors, occupations, and fields of study within STEM disciplines. Among the European initiatives to tackle this issue, the Gender Equality Strategy 2020-2025 (European Commission, 2020b) acknowledges, firstly, that women remain underrepresented in higher paid professions, and secondly, that the choice of educational and professional paths is constrained by stereotypical gender norms. The Gender Equality Strategy adopts a dual approach by promoting targeted measures, such as an EU-Wide communication campaign combatting gender stereotypes, and strengthened 'gender mainstreaming', a term that means incorporating a gender perspective consistently throughout every phase of policy formulation (United Nations Economic and Social Council, 1997) .

The European Skills Agenda, another current policy instrument, emphasizes the urgent need to increase the number of graduates from STEM tertiary education to foster innovation across Europe. Regarding the gender disparity, however, the initiative only suggests raising attractiveness of STEM studies and careers to involve girls and women "by encouraging a cross-disciplinary and innovative teaching-and-learning approach in schools, vocational education and training and higher education" (European Commission, 2020a, p. 13). Gender is not a central focus in this Agenda; it appears as an aside.

Finally, the European Strategy for Universities, introduced in 2022, prioritises increasing the gender balance of both students and academic staff as well as the overall reservoir of skills and competences in STEM within the region (European Commission, 2022). With this initiative, the European Commission committed to developing a European framework for diversity and inclusion and providing in the future a roadmap to address the underrepresentation of women in STEM fields through, which would include a manifesto on gender-inclusive STE(A)M education (whereby art is included). The current strategy also encourages universities across the region to implement institutional change such as developing and adopting gender equality plans.

Although these strategies function as guidelines for member states to articulate their national policies, serving as non-binding tools to address gender disparities in higher education within STEM fields, they could result in superficial efforts that fail to address the underlying issues because their compliance depends on the willingness of stakeholders to implement the suggested policies.

2) Ireland's national strategies

In general, the higher education policy landscape in Ireland aligns with the European framework. Ireland's National Skills Strategy 2025 states that careers in STEM need to be promoted, particularly to women. Main measures comprise raising public awareness of STEM, increasing the level of uptake of STEM at second level and supporting retention of students in STEM disciplines (Department of Education and Skills, 2016).

Regarding the gender equality strategy, it has not been updated in Ireland. The National Strategy for Women and Girls 2017-2020 is the last initiative published and it was extended up to 2021. Key outputs considered to reduce gender imbalance in STEM were: (1) a National STEM Education Policy Statement 2017-2026; (2) a literature review on the critical barriers to girls' participation in STEM and on the effective interventions for addressing gender balance in primary and post-primary education settings; and (3) guidelines for promotion of STEM careers to young people and parents. The first two products have been published. They consider a variety of stakeholders and long-term partnerships among students, families, schools, industry and society to help ensure the outcomes put forth within this policy framework, but they emphasize early childhood and early adolescence (Department of Education, 2017a, 2020, 2022) with little direct implication for higher education. Perhaps more importantly, this policy instrument recognizes a need to seek structural changes (instead of focusing solely on changing girls' attitudes and beliefs). Recently, a Gender Balance in STEM Education Advisory Group was created to support the enaction of this plan (Department of Education, 2022).

Regarding policy for the tertiary level, the National Strategy for Higher Education 2030 (Department of Education and Skills, 2011), issued in 2011, expressed concern over the low students demand on STEM courses in higher education "when the importance of these disciplines for enterprise strategy is growing" (Department of Education and Skills, 2011, p. 36). The gender issue was not acknowledged in this document and inequality was understood as a condition of student's socioeconomic background only. In consequence, no action was proffered to address the gender gap in higher education.

The National Review of Gender Equality in Irish Higher Education Institutions served as turning point for the advancement of policy interventions to tackle gender inequality (Dunne et al., 2022). In that document, the Expert Group suggested that every Higher Education Institution (HEI) should develop a Gender Action Plan, and strive to earn an Athena SWAN award, which eventually would become a requirement for research funding. As a result, a Gender Action Plan 2018-2020 was established to promote gender equality in HEI, addressing staff issues mainly, such as recruitment procedures, leadership, governance and management, and gender awareness (Higher Education Authority, 2018). Although this Plan is an important step in tackling gender inequality, it does not specifically target STEM fields or involve direct actions to increase the number of women students in those disciplines.

B. Reshaping higher education landscape for STEM

At the current time, the Irish government is transforming its

https://doi.org/10.52202/073963-0008

landscape of higher education by establishing a network of Technological Universities (TUs). These TUs are being formed by merging and expanding upon existing Institutes of Technology. The TUs are expected to align their programs with the specific needs of their respective region's sectors (OECD, 2022).

According to the National Development Plan 2021-2030 (Goverment of Ireland, 2021), the development of technological universities involves expanding their infrastructure through the Higher Education PPP programme to address increasing enrolments with a specific emphasis on areas of key skills needs, particularly in the STEM fields. Regarding the gender issues in HE: the Plan does not mention gender specifically for the sector, rather it acknowledges that gender equality is a national goal in alignment to the United Nations Sustainable Development Goals (SDG) and that each Minister is given the specific responsibility in implementing the SDG targets related to their sectors and ministerial functions.

As mentioned above, the Department of Education published the National STEM Education Policy Statement 2017-2026, accompanied with the STEM Education Implementation Plan 2017-2019 (Department of Education, 2017a, 2017b). Both instruments mainly focus on primary and secondary levels of education, although that influence is seen to extend to the areas of further and higher education. From the latest document, two annual indicators for success stand out. The documents ambitiously propose an (1) "Increased uptake of Leaving Certificate Chemistry, Physics, Technology and Engineering by 20%" and (2) "Increased uptake by females of STEM subjects by 40%" (Department of Education, 2017b, p. 4). However, a recent report of the progress in the plan implementation mentions that for the first indicator, "the increase is small", while for the second indicator, "the gender disparity in taking STEM subjects for examination is also clear" (Department of Education, 2023, p. 21). This suggests that the specific actions undertaken may require a longer period to yield the desired outcomes, additional steps need to be taken to achieve stated goals, or that the effectiveness indicators of these initiatives may need to be revised.

Regarding policies for higher education, the new National Access Plan for Higher Education 2022-2028 identifies three groups of students who are thus now formally understood as being underrepresented in higher education, and therefore prioritized. These three groups are persons who: (1) are socioeconomically disadvantaged; (2) are members of Irish Traveller and Roma communities, or (3) have disabilities including intellectual disabilities (DFHERIS & HEA, 2022, p. 22). These groups are targeted to measure the performance of the Plan, but the Plan itself does not specifically address specifically engineering or STEM education.

C. *Gender policies constrained to HEI initiatives*

Most education policies at both regional and national levels advocate a multidisciplinary or holistic approach within STEM curricula, aiming to raise the appeal of these disciplines and to reduce traditional stereotype barriers that affect women's engagement. However, there remains a lack of practical support or instruments for implementing these policies. Often universities are encouraged to implement institutional changes by fostering inclusive learning environments, setting access targets, or developing gender equality plans (European Commission, 2022; European Commission. Directorate General for Research and Innovation., 2023).

Universities in Ireland are mandated to establish equality policies, which encompass gender balance, as a key component. These regulations are framed by the Universities Act 1997 and the Technological Universities Act of 2018.

One of the HEI's initiatives to encourage girls into science, engineering and technology is focused on boosting mentorship through a program named 'Equality in Science and Technology by Engaged Educational Mentoring' (or ESTeEM) which aims to expand women students' depth of understanding regarding their career paths while providing a platform to network with fellow students (Devereux et al., 2022).

D. *Gender gap in engineering education in Ireland*

Claiming that Gender Equality Plans or the Athena SWAN awards have effectively achieved results in HEIs poses challenges, as changes in women's participation in STEM education are also result of broader social, political and economic factors that extend beyond specific policies (Drew, 2022). Nevertheless, according to HEA data (2021), there has been a gradual increase in the enrolment of women in EMC programs in Irish universities, rising from 16.6% in the academic year 2016/2017 to 21.2% in 2021/2022. It is worth noting that Muster Technological University (MTU) exhibited the most substantial growth in the proportion of female new entrants in EMC during this period. MTU's proportion of entrants into EMC who were women was 23.2% up from its previous level of 14.9%. In contrast, Technological University Dublin (TU Dublin) maintained the percentage of female new entrants at around 15%. University College Cork (UCC) and University College Dublin (UCD) have a more consistently increasing trend in the inclusion of women in EMC, with proportions of 40% and 36.8% in 2021/2022, respectively (see Table I). Complex factors may be at play here; for instance, women choosing to study in Dublin have may choices of multiple institutions for entering engineering, and TU Dublin is not the most prestigious of them (thus, highly prepared women are likely to choose UCD over TU Dublin). The emergence of TUs throughout the more rural areas of Ireland, e.g., Munster, may attract women who want to study engineering and live at, or closer to, home.

Another factor influencing student decisions could be that TU Dublin is larger in size than MTU and UCC. The size of the HEI appears to play a role in the representation of women in EMC studies: as the total number of students in the field increases, the proportion of women students tends to decrease.

https://doi.org/10.52202/073963-0008

Table I. Women as percentage of total new entrants into Engineering, Manufacturing and Construction, by year

Institute/Academic Year	**2016/2017**		**2017/2018**		**2018/2019**		**2019/2020**		**2020/2021**		**2021/2022**	
	(1)	(2)	(1)	(2)	(1)	(2)	(1)	(2)	(1)	(2)	(1)	(2)
Technological University Dublin	15.5%	4,950	15.4%	5,000	15.5%	4,905	16.1%	4,935	16.9%	4,945	15.8%	4,965
Munster Technological University	14.9%	2,650	16.1%	2,695	16.3%	2,795	18.7%	3,135	19.5%	3,055	23.2%	2,995
Atlantic Technological University	8.6%	2,160	10.3%	2,330	11.0%	2,315	12.2%	2,345	12.7%	2,565	14.2%	2,745
Technological University of the Shannon	8.5%	1,760	9.0%	1,830	10.5%	1,945	11.3%	2,125	12.0%	2,250	12.7%	2,210
University of Limerick	18.1%	1,710	22.7%	1,695	22.3%	1,705	23.3%	1,805	26.1%	1,860	26.4%	1,935
South East Technological University	11.8%	1,690	15.0%	1,605	16.2%	1,795	15.6%	1,830	14.8%	1,830	14.8%	1,860
University College Dublin	29.6%	1,505	32.7%	1,530	33.1%	1,540	36.1%	1,605	36.4%	1,595	36.8%	1,740
University of Galway	21.7%	1,130	24.3%	1,175	22.1%	1,130	24.2%	1,115	24.8%	1,190	25.2%	1,230
Trinity College Dublin	25.1%	875	26.9%	910	27.2%	975	27.2%	1,010	27.6%	1,070	29.9%	1,155
University College Cork	34.1%	865	34.6%	910	36.8%	925	38.8%	980	41.4%	990	40.0%	1,075
Dundalk IT	9.2%	595	10.8%	600	10.7%	560	12.8%	625	18.0%	750	18.5%	785
Dublin City University	16.5%	665	18.4%	570	20.8%	650	20.7%	605	20.4%	735	17.9%	670
Maynooth University	20.0%	275	19.4%	310	19.1%	340	16.9%	325	17.7%	310	18.8%	320
St. Angela's College	100.0%	40	90.0%	50	91.7%	60	100.0%	55	100.0%	65	100.0%	50
Grand Total	**16.6%**	**20,870**	**18.2%**	**21,210**	**18.6%**	**21,640**	**19.7%**	**22,490**	**20.5%**	**23,200**	**21.2%**	**23,740**

Notes:
(1) Women as percentage of total new entrants
(2) Total new entrants into Engineering Manufacturing and Construction (EMC)
Higher Education Institutions are ranked in descending order of female new entrants into Engineering, Manufacturing and Construction in 2021/2022
Data source: Higher Education Authority. (2021). *HEA Statistics*. https://hea.ie/statistics/data-for-download-and-visualisations/key-facts-figures/

Figure 1. Percentage of women students vs total undergraduate students in Engineering, Manufacturing and Construction, 2021/2022

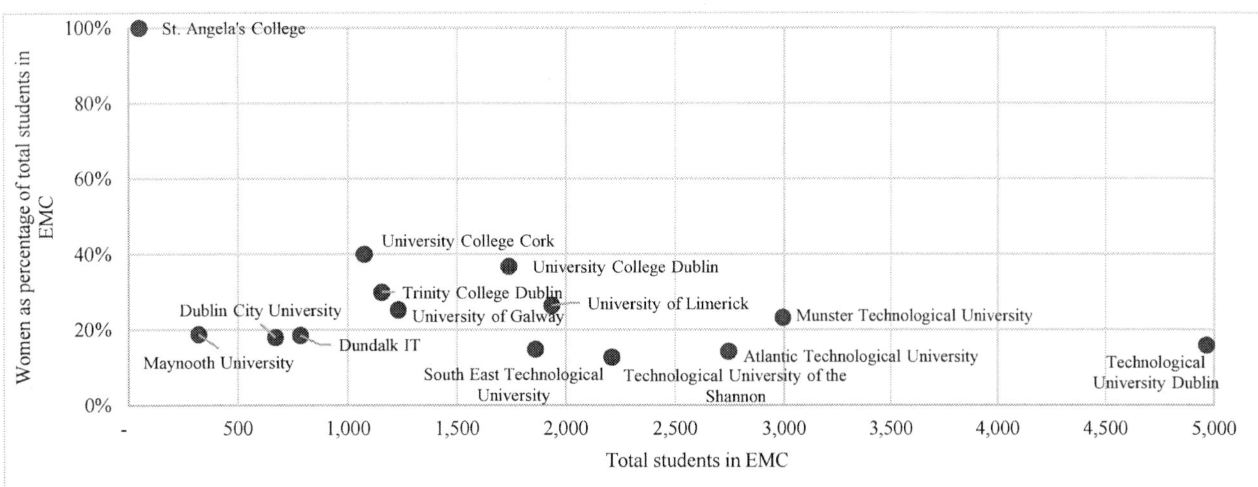

Data source: Higher Education Authority. (2021). *HEA Statistics*. https://hea.ie/statistics/data-for-download-and-visualisations/key-facts-figures/

https://doi.org/10.52202/073963-0008

Such is the case of the Technological University Dublin, the Atlantic Technological University (ATU), and the Technological University of Shannon (TUS), which each had more than 2000 newcomers in EMC in academic year 2021/2022, but women comprised less than 20%. In contrast, universities with a smaller number of students in EMC, such as UCD and UCC, have a more balance representation of women students (see Figure 1).

V. Conclusions

Our study focused on mapping the policy framework to promote the reduction of the gender gap in engineering education in Ireland at the tertiary level. As engineering is usually grouped with other disciplines, we conducted the policy analysis considering initiatives for STEM fields of education, which include engineering and other EMC fields.

European and Irish policy frameworks that promote STEM disciplines in higher education often advocate for gender equality not primarily for social justice reasons, but because of its potential to enhance economic competitiveness in the region and in the country (i.e., neoliberal economic reasons).

In Ireland, some key national strategies for higher education do not explicitly acknowledge the gender gap as a priority. Furthermore, despite the Athena SWAN charter becoming a key pillar of Ireland's strategy for gender equality in higher education (Drew, 2022; Dunne et al., 2022), universities are simply encouraged to enact the institutional changes, but the national strategies tend to prioritize primary and secondary levels of education. Accurately assessing the impact of various institutional gender policies to enhance women's participation in STEM education is problematic due to the measurement of explanatory factors (such as students' personal and family characteristics, educational systems and programmes, labour market, etc.), the context in which the institutions operate, and the availability of data. However, some Irish universities have successfully increased the share of women in EMC in the last six academic years. Further research is needed to explore their institutional policies, then identify and systematize good practices and lessons learnt – to serve as guidelines for others to accelerate women's access in engineering at higher education level in Ireland.

Ultimately, the gender gap in engineering goes beyond engineering higher education classrooms; the problematic gap extends into education leadership and engineering practice as well. This involves more research (1) on institutional initiatives to increase participation of women in decision-making of the tertiary level of engineering education; and (2) on the persistence of female engineering graduates to work in the industry sector.

References

Arksey, H., & O'Malley, L. (2005). Scoping studies: Towards a methodological framework. *International Journal of Social Research Methodology*, *8*(1), 19–32. https://doi.org/10.1080/1364557032000119616

Beede, D. N., Julian, T. A., Langdon, D., McKittrick, G., Khan, B., & Doms, M. E. (2011). Women in STEM: A Gender Gap to Innovation. *SSRN Electronic Journal*. https://doi.org/10.2139/ssrn.1964782

Clavero, S., & Galligan, Y. (2021). Delivering gender justice in academia through gender equality plans? Normative and practical challenges. *Gender, Work & Organization*, *28*(3), 1115–1132. https://doi.org/10.1111/gwao.12658

David, M. (2015). Women and Gender Equality in Higher Education? *Education Sciences*, *5*(1), 10–25. https://doi.org/10.3390/educsci5010010

Delaney, J. M., & Devereux, P. J. (2022). Gender Differences in STEM Persistence after Graduation. *Economica*, *89*(356), 862–883. https://doi.org/10.1111/ecca.12437

Department of Education. (2017a). *STEM education Policy statement 2017-2026*.

Department of Education. (2017b). *STEM implementation plan 2017-2019*.

Department of Education. (2020). *Review of Literature to Identify a Set of Effective Interventions for Addressing Gender Balance in STEM in Early Years, Primary and Post-Primary Education Settings*.

Department of Education. (2022). *Recommendations on Gender Balance in STEM Education*.

Department of Education. (2023). *STEM Education Implementation Plan. Phase 1 Enhancing Progress Report*. https://www.gov.ie/en/publication/055810-education-statistics/#latest-statistical-reports

Department of Education and Skills. (2011). *National Strategy for Higher Education 2030*. https://hea.ie/assets/uploads/2017/06/National-Strategy-for-Higher-Education-2030.pdf

Department of Education and Skills. (2016). *Ireland's National Skills Strategy 2025*. https://assets.gov.ie/137349/3b66360a-64f4-45db-881f-eb326950051e.pdf

Devereux, M., Heffernan, E., McKeever, S., Dunne, J., Shoemaker, L., & O'Leary, C. (2022). Reflections on Selected Gender Equality in STEM Initiatives in an Irish University. In F. J. García-Peñalvo, A. García-Holgado, A. Dominguez, & J. Pascual (Eds.), *Women in STEM in Higher Education* (pp. 69–83). Springer Nature Singapore. https://doi.org/10.1007/978-981-19-1552-9_4

DFHERIS, & HEA. (2022). *National Access Plan. A Strategic Action Plan for Equity of Access, Participation and Success in Higher Education 2022–2028*. Department of Further and Higher Education, Research, Innovation and Science. https://assets.gov.ie/233024/9771358b-667a-4172-867f-371d7e314f46.pdf

Drew, E. (2022). Navigating unChartered waters: Anchoring Athena SWAN into Irish HEIs. *Journal of Gender Studies*, *31*(1), 23–35. https://doi.org/10.1080/09589236.2021.1923463

Drew, E., & Roughneen, C. (2004). *Danger! Men at work: A study of the under-representation of women in third level engineering*. Department of Statistics, Trinity College Dublin.

Dunne, J., O'Reilly, A., O'Donoghue, A., & Kinahan, M. (2022). A Review of Irish National Strategy for Gender Equality in Higher Education 2010–2021. In F. J. García-Peñalvo, A. García-Holgado, A. Dominguez, & J. Pascual (Eds.), *Women in STEM in Higher Education* (pp. 21–49). Springer Nature Singapore. https://doi.org/10.1007/978-981-19-1552-9_2

European Commission. (2008). *Benchmarking Policy Measures for Gender Equality in Science*. Publications Office. https://data.europa.eu/doi/10.2777/55449

European Commission. (2020a). *European Skills Agenda*. https://ec.europa.eu/social/BlobServlet?docId=22832&langId=en

European Commission. (2020b). *Gender equality strategy 2020-2025*.

European Commission. (2022). *Communication on a European Strategy for Universities*. https://education.ec.europa.eu/document/commission-communication-on-a-european-strategy-for-universities

European Commission. Directorate General for Research and Innovation. (2023). *EU support to strengthen gender equality in STEM*. Publications Office. https://data.europa.eu/doi/10.2777/719121

EUROSTAT. (2021). *Tertiary education statistics*. https://ec.europa.eu/eurostat/statistics-explained/index.php?title=Tertiary_education_statistics

Goverment of Ireland. (2021). *National Development Plan 2021-2030*. https://www.gov.ie/en/publication/774e2-national-development-plan-2021-2030/

Hersh, M. (2000). The changing position of women in engineering worldwide. *IEEE Transactions on Engineering Management*, *47*(3), 345–359. https://doi.org/10.1109/17.865903

Higher Education Authority. (2018). *Gender Action Plan 2018-2020. Accelerating Gender Equality in Irish Higher Education Institutions*. HEA. https://hea.ie/assets/uploads/2018/11/Gender-Equality-Taskforce-Action-Plan-2018-2020.pdf

Higher Education Authority. (2021). *HEA Statistics*. Higher Education Authority. https://hea.ie/statistics/data-for-download-and-visualisations/key-facts-figures/

Kelly, R., McGarr, O., Lehane, L., & Erduran, S. (2019). STEM and gender at university: Focusing on Irish undergraduate female students' perceptions. *Journal of Applied Research in Higher Education*, *11*(4), 770–787. https://doi.org/10.1108/JARHE-07-2018-0127

Kiernan, L., Walsh, M., & White, E. (2023). Gender in Technology, Engineering and Design: Factors which influence low STEM subject uptake among females at third level. *International Journal of Technology and Design Education*, *33*(2), 497–520. https://doi.org/10.1007/s10798-022-09738-1

Moloney, G., & Ahern, A. (2022). Exploring the lower rates of entry into undergraduate engineering among female students through the application of the reasoned action approach. *European Journal of Engineering Education*, *47*(3), 483–500. https://doi.org/10.1080/03043797.2021.2025342

OECD. (2021). *Education at a Glance 2021: OECD Indicators*. OECD. https://doi.org/10.1787/b35a14e5-en

OECD. (2022). *A review of technological university academic career paths, contracts and organisation in Ireland* (OECD Education Policy Perspectives 64; OECD Education Policy Perspectives, Vol. 64). https://doi.org/10.1787/2b7ee217-en

OECD. (2023). *OECD Skills Strategy Ireland: Assessment and Recommendations*. OECD. https://doi.org/10.1787/d7b8b40b-en

Rosa, R., & Clavero, S. (2022). Gender equality in higher education and research. *Journal of Gender Studies*, *31*(1), 1–7. https://doi.org/10.1080/09589236.2022.2007446

Tannenbaum, C., Ellis, R. P., Eyssel, F., Zou, J., & Schiebinger, L. (2019). Sex and gender analysis improves science and engineering. *Nature*, *575*(7781), 137–146. https://doi.org/10.1038/s41586-019-1657-6

Torotcoi, S., Gologan, D., & Kurysheva, A. (2020). What Works for Underrepresented Groups? Identifying Effective Measures for Enhancing Access to Higher Education in Europe. In A. Curaj, L. Deca, & R. Pricopie (Eds.), *European Higher Education Area: Challenges for a New Decade* (pp. 177–196). Springer International Publishing. https://doi.org/10.1007/978-3-030-56316-5_13

United Nations Economic and Social Council. (1997). *Definition of Gender Mainstreaming*. International Labour Organization. https://www.ilo.org/public/english/bureau/gender/newsite2002/about/defin.htm

Copyright statement

Spatial Skills and Design Problem Scoping Behaviors in Undergraduate Engineering Students

Gibin Raju[a] and Sheryl Sorby[a]
University of Cincinnati [a]
Corresponding Author Email: rajugm@mail.uc.edu

Abstract

Context

Engineering design skills are essential for engineering students to succeed in their careers. Engineering design is a skill that is in high demand in the current job market and should be prioritized in education.

Purpose

While design has been acknowledged as a cognitive skill in research, there exists limited literature addressing the cognitive foundations of design thinking. Hence, engineering educators must understand the engineering design process, as well as the different ways students approach design problem-solving and the potential reason behind these differences. To understand how people solve design problems, we need to consider how their minds work and the strategies they use. Spatial ability stands out as a cognitive factor that is crucial for designers and holds significance in well-established theories and models of intelligence. However, to date, research exploring the impact of spatial ability on design thinking and its influence on problem-scoping behaviors remains limited. This paper examines how engineering students' spatial skills influence how they define the scope of open-ended design problems. The central research question that guides this paper is "How do design problem-scoping behaviors differ for engineering students based on their spatial scores?".

Methods

The researchers used a mixed methods research approach to answer their research question, collecting qualitative and quantitative data in two phases. One hundred twenty-seven undergraduate engineering students completed four tests that measure spatial reasoning skills in the quantitative phase and 101 students returned to finish the three design tasks in the second phase. This paper will examine the performance of students with low spatial and high spatial skills on one of the completed design tasks.

Outcomes

From the study, it was clear that spatial skills have an impact on the design-scoping behaviors of the undergraduate engineering students. It was inferred that high spatial skill visualizers emphasized the technical details of the design problem whereas low spatial skill visualizers emphasized the context of the design problem during their problem-scoping behavior. A Mann-Whitney test revealed there was a statistically significant difference in detail- and context-focused segments between the high and low spatial visualizer groups.

Conclusion

This research study confirms that a relationship exists between spatial and design skills. The study also found that undergraduate engineering students with different levels of spatial skills had different approaches to scoping design problems.

Keywords— *Spatial visualization skills, engineering design, design skills, problem-scoping behaviors, undergraduate engineering students*

I. INTRODUCTION

DESIGN is an important attribute of professional engineering practice. It is an important part of engineering education curriculum and a competency skill that is essential for student success in their chosen field. In our everyday lives, we see the benefits of engineering design, but we also experience the catastrophic consequences of engineers failing to consider the long-term effects of their design projects. As engineers it is important for us to develop the solutions of any design problem by taking into account of factors such as societal, cultural, and environmental. As engineering operates within real-world contexts, possessing the capacity to contemplate extensive ramifications, spanning technical, social, economic, political, cultural, and environmental facets, stands out as a crucial element in achieving success as an engineer (Cross, 1995; Nelson & Stoltermann, 2003; Cross, 2006).

Several reports, research studies, and accreditation criteria for engineering programs have indicated the need for consideration of non-technical contexts in the future of engineering practice (ABET Engineering Accreditation Commission, 2021; National Academy of Engineering, 2004; Lau, 2004). For instance, the Accreditation Board for Engineering and Technology (ABET) has included design as one of the outcomes of engineering programs.

Specifically, ABET says that undergraduates must attain:

https://doi.org/10.52202/073963-0009

"an ability to apply engineering design to produce solutions that meet specified needs with consideration of public health, safety, and welfare, as well as global, cultural, social, environmental, and economic factors."

ABET-accredited programs prepare graduates to be creative and innovative problem-solvers so that they can work with incomplete information, apply imagination to generate novel and unexpected solutions, as well as use drawings and other visual representations to communicate their ideas effectively (ABET Engineering Accreditation Commission, 2021). Engineering design is a cognitively demanding process that requires engineers to think about all the different parts of a system and how they work together (Lammi, 2013).

Research studies have identified design as a high-level cognitive skill that involves the production of successful iterative internal and external representations of an artifact to analyze and improve the design (Aurigemma et al., 2013; Kim & Maher, 2008; Lazar, 2018; Dorst, 2011). Still, the cognitive basis of the design thinking process is a relatively understudied area of research. While there is a growing body of research literature on the topic, there is still a lack of consensus on the specific cognitive processes involved in design thinking. Research studies have broken down the design process into "steps" such as defining the problem, researching solutions, coming up with ideas, building a prototype, choosing the best solution, implementing it, reframing the solution, learning from the experience, and so on (Ambrose & Harris, 2009; IDEO Education, 2012; Brown, 2009; Kueh & Thom, 2018). However, we still need to conduct additional research to fully understand the cognitive basis of design thinking. Spatial visualization skill is one of the key cognitive elements that is necessary for a designer (Williams & Sutton, 2011; Suh & Cho, 2020).

A. Spatial Skills and Engineering Design

Spatial skills are very important for engineering students, and there is more and more research showing that improving these skills can lead to significant benefits and help engineers to function more effectively in their respective fields of work (Serdar & deVries, 2015; Sorby & Baartmans, 2000; Duffy, et al., 2020). Many research studies have shown that students with strong spatial ability are more likely to be successful in STEM (Sorby et al., 2014; Sorby et al., 2018; Wai et al., 2009; Uttal et al., 2013). Spatial ability also helps individuals improve their capacity to imagine representations and mentally manipulate and transform these representations in different ways (Xue et al., 2017; Pylyshyn, 2003; Shepard & Metzler, 1971). Research studies have established spatial ability's crucial role in supporting and enhancing cognitive functions such as advanced thinking, abstract reasoning, and creative processes (Sorby et al., 2013; Ishikawa & Newcombe, 2021). These abilities are considered fundamental for navigating and interacting with our surrounding environment.

Engineers are known for their problem-solving skills. Research studies have shown that spatial skills are closely related to the ability to solve problems in mathematics and chemical engineering (Duffy, 2017; Loney, et al., 2019). There is a large body of research that shows the importance of spatial ability in engineering graphics. Engineers rely on their spatial visualization skills to effectively convey their design concepts (Sorby et al., 2013) and design projects of individuals with high spatial skills tended to show strengths in better design approach (Suh & Cho, 2020). Despite the importance of both spatial thiking and design thinking in engineering, there is still relatively little research on how the two relate to each other (Sutton & Williams, 2007; Sutton & Williams, 2010). Thus, this study aims to investigate the relationship between spatial skills and the engineering design scoping behaviors of undergraduate engineering students.

II. METHODOLOGY

In the present study, a sequential mixed methods research methodology was used to answer the central research question. This methodology consisted of two distinct data-collection and analysis strands as shown in Fig. 1. Firstly, the quantitative phase involved the collection and analysis of numeric data. Following this, the qualitative strand was implemented, involving the collection and analysis of textual data in a consecutive manner (Creswell & Clark, 2017). Sequential mixed methods research design methodology aims to purposefully select participants for the qualitative phase based on the quantitative data, rather than using random sampling. By doing so, we can leverage qualitative contextual data to enhance the interpretation of the findings (Subedi, 2016). We then put all of the data together and look at it closely to better understand the scientific findings and how they relate to our research questions (Creswell & Clark, 2017).

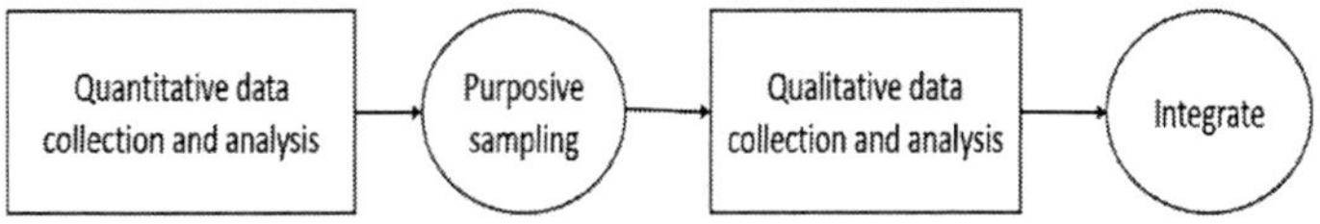

Fig. 1. Procedural diagram of the Mixed Methods Design –Sequential

https://doi.org/10.52202/073963-0009

This research aims to examine the relationship between the spatial skills and engineering design problem-scoping behaviors of undergraduate engineering students. This work is informed by answering the mixed methods research question: "How do design problem scoping behaviors differ for engineering students based on their spatial scores?".

A. Study Setting

The study was conducted at a public in the College of Engineering at the University of Cincinnati. Engineering students in the first and final years of their programs were recruited through emails and flyers that were posted around the college. In the initial phase, participants took four well established spatial ability tests online via Qualtrics while being proctored by the researchers. In the second phase, individual participants came back to complete three design tasks, while thinking aloud about their thoughts and processes. The Institutional Review Board (IRB) at the University of Cincinnati approved this study.

B. Quantitative Strand - Data Collection

A total of 127 undergraduate engineering students participated in the quantitative phase of the study. They took four well-established spatial ability tests online, proctored by a research assistant. The tests were the Paper Folding Test (PFT) (Ekstrom & Harman, 1976), the Mental Cutting Test (MCT) (College Entrance Examination Board, 1939), the Spatial Orientation Test (SOT) (Kozhevnikov & Hegarty, 2001) and the Mental Rotation Test (MRT) (Shepard & Metzler, 1971). A verbal analogy test was also included to control for general intelligence. Once the tests were graded, a principal component analysis was conducted to separate students into high and low spatial visualizers (data from medium-level visualizers were not included in this analysis).

C. Qualitative Strand - Data Collection

Thirty-one participants (15 high and 16 low spatial) were purposively chosen to participate in the phase 2 concurrent verbal protocol phase (Atman & Bursic, 2013). In this phase of the research, each participant was given three design problems to solve. For this study, the emphasis will be solely on one of these three problems, which involves listing factors for designing a retaining wall system. The problem statement for the third design task was:

> *"Over a typical summer the Midwest experiences massive flooding of the Mississippi River. What factors would you take into account in designing a retaining wall system for the Mississippi?".*

The qualitative phase of the study was conducted individually for each of the participants in a neutral and restful environment within the college premises. The room was chosen for its lack of distractions, which helped to ensure that the participants were able to focus on the task at hand. All of the participant's sessions were video- and audio-recorded, with the participants' permission, to facilitate the analysis of the data.

As each student completed the Midwest flood listing design task, their zoom session was recorded. The recording of each student was then analyzed using the following steps:

a) transcription – the students verbal protocol was transcribed from the video recording.
b) segmentation – the transcribed verbal text was divided into units that could be coded using a pre-defined two-dimensional coding scheme (Atman et al., 2008)
c) coding – the coding scheme shown in Table I (Raju et al., 2022, adapted from Atman et al. (2008)), was used to code each segment for physical location and frame of reference.

To ensure consistency in coding, two coders coded each part of the lists generated by each participant individually. The coders then compared their coding to make sure that they agreed with at least 90% of the codes assigned for each participant. After resolving any disagreements, the coders calculated their interrater reliability, which was a Cohen's Kappa of 0.965. This high value means that the two coders agreed very strongly on how they assessed the participants' design problem scoping behaviors.

D. Coding

In previous research studies, researchers used a two-dimensional coding scheme to describe how broadly participants scoped design problems (Adams et al., 2003; Bogusch et al., 2000; Rhone et al., 2001; Rhone et al., 2003; Raju et al., 2022). In this study, we use the same coding scheme where each of the responses was coded for frame of reference and physical location of the design problem. Researchers used physical location codes to record the physical area of focus that the participant focused on. There were four codes: *wall*, *water*, *bank*, and *surroundings*. The codes were ordered to show how participants' focus moved from the details of the wall to the context of the problem. The *wall* and *water* represent parts of the problem that are close to the retaining wall. These are considered detail issues because they are typical of bounded engineering problems that focus on core engineering science issues.

https://doi.org/10.52202/073963-0009

The frame of reference codes represents how participants thought about the design problem on a broader scale. They are divided into four categories: *technical*, *logistical, natural,* and *social*. These categories of codes are also arranged to show how participants' thinking moved from the *details* of the problem to the *context* of the problem. *Technical* and *logistical* factors are about the details of the problem, while *natural* and *social* factors are about the context of the problem. Table I shows a summary of the two-dimensional coding system and the four codes.

TABLE I
CODING DIMENSIONS AND ITS DESCRIPTION (Adams et al.,2003)

Physical Location	Description
Wall	The wall itself, what affects it, other options for having a wall, where to put it.
Water	River's length, aquatic fauna, flood (but not effects on flood on other locations), pressure problems (without mention of the wall).
Bank	Earth immediately adjacent to river, earth below the river (riverbed), wall's interface, river's edge, river's width.
Surroundings	Everything far from water, residential units, items along water, particular impacts of the flood to bank.
Frame of Reference	**Description**
Technical	Engineering or technical terminology such as design problems, choices about construction of the wall
Logistical	Expenses, financing, process of construction, maintainability issues, resources needed.
Natural	Water's level (volume), destruction, effects of flood, geography, animals, flora, climate, and climate projections.
Social	People, people's safety, views, cities, living areas, policies

III. RESULTS

A. Quantitative Phase

In the quantitative phase, spatial tests were graded in Excel after importing the data from Qualtrics by the research assistant. There were 127 undergraduate engineering students (42 Female and 85 Male) who participated in the study. Internal consistency reliability for each of the four spatial tests was calculated. The KR-20 score was found to be above 0.80 for each of the spatial tests expect SOT (KR20=0.65), which is generally considered to represent a reasonable level of internal consistency reliability (El-Uri & Malas, 2013). Considering the transition of paper pencil test to online, it was expected to have some impact. We performed principal component analysis to group the research participants into low, and high groups (Jolliffe & Cadima, 2016). We used the first principal component to divide the participants into three groups: those with low and medium spatial skills and those with high spatial skills. We only focused on the high and low spatial groups in this study. Table I shows the summary of the spatial scores of high and low spatial visualizers who participated in this phase. The average score and standard deviation for each spatial group was determined with a maximum score of 81.

TABLE II
SPATIAL SCORES – AVERAGE AND STANDARD DEVIATION

Spatial Scores	
Low Spatial Visualizer (n=16)	
Avg. Score	23.63
Std. Dev	5.39
High Spatial Visualizer (n=15)	
Avg. Score	61.2
Std. Dev	5.80

B. Qualitative Phase

As shown in Figure 2 (a), undergraduate engineering students generated an average of 12.22 coded segments. Looking at the detail- and context- focused segments independently, we found that participants focused more on detail-focused, or technical aspects, segments. The Mann-Whitney test revealed there was a statistically significant difference between the detail and context-focused segments overall. It was also observed that, on average, all four of the detail-focused nodes were covered and 10 out of 12 context-focused nodes were covered by the participants.

C. Integrating the data

From the purposive sampling, the data from 16 low spatial visualizers (6 Female and 10 Male) and 15 high spatial visualizers (2 Female and 13 Male) was included in this analysis. Following the coding scheme for design problem scoping behavior, we also studied how the time taken to solve this design problem varied between high and low spatial visualizers.

To investigate and characterize the breadth of design problem scoping behaviors among high and low spatial visualizer groups, we averaged the coded segments for the physical location and frame of reference and plotted them in a two-dimensional coding space. Figures 3 and 4 provide a detailed comparison of the coded responses from high and low spatial visualizers, showing what kind of factors were discussed while completing the design task. Each figure presents the average number of segments inside a circular disc by code pair for high and low spatial visualizer group. The circular disc size was proportional to the number of average numbers of coded segments at that node.

 https://doi.org/10.52202/073963-0009

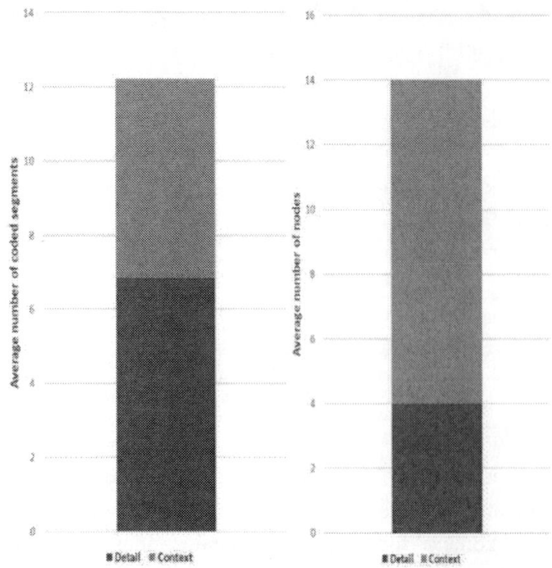

Fig. 2. (a) Average code pair segments by spatial visualizer groups. The bar division shows the average of detail- and context-focused segments. (b) Mean nodes covered by spatial visualizer groups for comparison. The bar division shows the average of detail and context-based nodes covered.

Upon plotting, it was inferred from the figure that high spatial visualizers focused more on the details of the Midwest flood listing problem as compared to the context of the problem when compared to low spatial visualizer group. This implies that the high spatial group focused more on the core engineering design problem because they generated more detail-focused segments. It was also very clear that the segments were not spread evenly across the coding space. Both high and low spatial visualizers tended to discuss more factors that were related to the wall and water compared to the bank and surroundings.

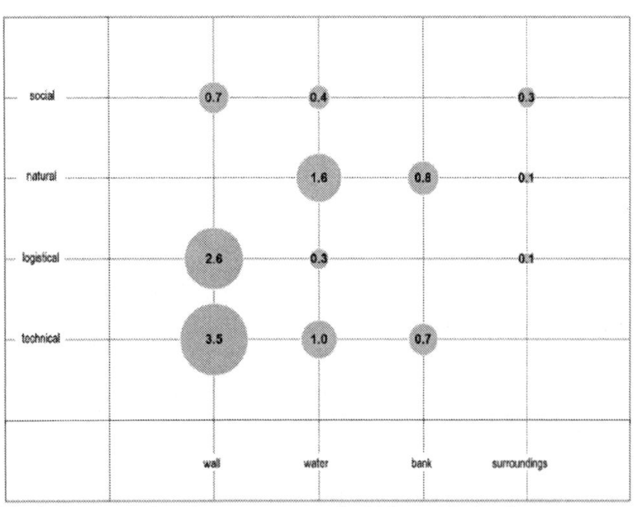

Fig. 3. Average code pair segments count for high spatial visualizers

The discussion of the wall factors focused on the technical details like wall dimensions and logistical considerations like cost and timeline of the project. The discussion of the water incorporated topics like flooding and wildlife. Contrasting these two figures (Fig.2 and Fig. 3.), it is clear from the averaged segment code values at each node that high spatial visualizers focused more on detail-oriented codes (WALL, technical and logistical) compared to low visualizers.

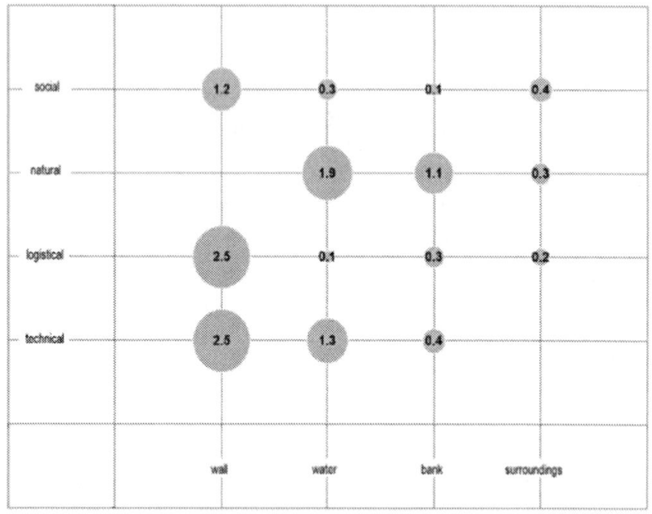

Fig. 4. Average code pair segments count for low spatial visualizers

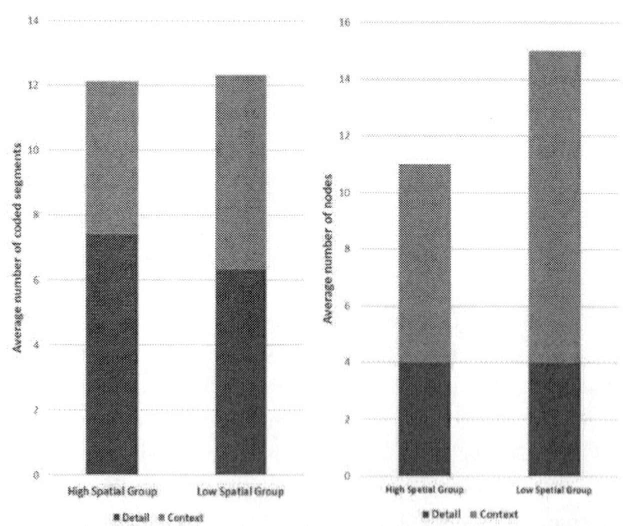

Fig. 5. (left) Average code pair segments by spatial visualizer groups. The bar division shows the average of detail- and context-focused segments. (right) Mean nodes covered by spatial visualizer groups for comparison. The bar division shows the average of detail and context-based nodes covered.

As shown in Figure 5 (left), high spatial visualizers generated an average of 12.13 coded segments and low spatial visualizers contained an average of 12.31 coded response segments. Looking at the detail- and context- focused segments

https://doi.org/10.52202/073963-0009

independently, we found that high spatial visualizer group focused more on detail-focused segments. A Mann-Whitney test revealed there was a statistically significant difference in detail-focused segments between the two groups. Also, it was inferred from the plot that the low spatial visualizer group focused more on context-focused segments as compared to the high spatial visualizer group. The Mann-Whitney test revealed that there was a statistically significant difference in context-focused segments between these groups ($p<0.05$).

As shown in Figure 5 (right), high and low spatial visualizers covered all nodes in the detail nodes. Meanwhile, low spatial visualizers had more nodes covered in the context nodes. This signifies that low spatial visualizer considered more factors that were away from the core issue of the problem. Also, it was found that low spatial visualizers took one minute more time on average to complete the problem when compared to high spatial visualizers.

Table II shows the results broken down by level of spatial skills. While looking at the high spatial group, it is inferred that they focus more on the technical issues which are related to the typical engineering problem. Whereas, the low spatial group focused more on the context issues, focusing on interactions between the design and the broader system such as social, environmental, and urban impacts.

TABLE II
INTEGRATING THE RESULTS

Level Spatia l Skills	**Avg. Scores**	**Average number of coded responses in the two-dimensional coding space**
High	61.2	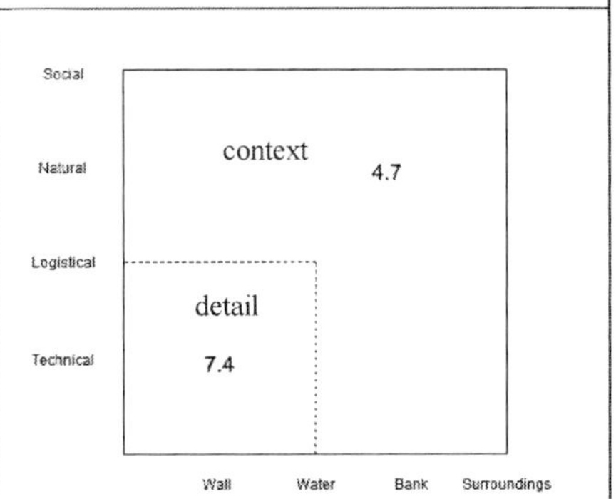
Low	23.6	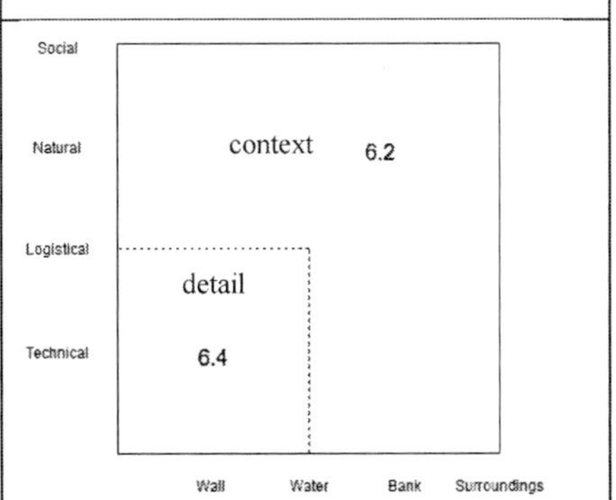

IV. DISCUSSION AND CONCLUSION

There is a strong consensus among researchers that spatial skills are important for success in engineering, but there has been limited research on the connection between engineering design behaviors and spatial skills. As described earlier in this paper, we sought to understand the relationship between spatial ability on the design problem scoping behaviors of undergraduate engineering students.

We found that the high spatial visualizers focused more on the core technical engineering and low spatial visualizer group generated more context-based segments. This indicates that high spatial visualizer emphasizes more on the technical issues of the phenomenon and low spatial visualizers focus on issues that are interactions from the proposed solution and broader system. One limitation is the fact that participant's year of study and gender were not considered during the analysis. So, future analysis is necessary to understand the impact of spatial skills based on their expertise level, gender and their impacts on design scoping behaviors. Currently, we are analyzing the data from a second year of data collection, which is anticipated to partially address the limitations of the ongoing study.

The Midwest flood listing task could serve as a valuable tool to understand the breadth of design problem-scoping. This research has helped us to understand how spatial visualization skills are related to engineering design skills. This understanding could be used to improve educational approaches to developing design capability in engineering education programs by helping the educators develop assessments and interventions to support design education.

ACKNOWLEDGMENT

This work was made possible by a grant from the National Science Foundation (DUE #2020785). Any opinions, findings, and conclusions, or recommendations expressed in this material are those of the authors and do not necessarily reflect the views of the National Science Foundation.

REFERENCES

ABET Engineering Accreditation Commission. (2021). *2022-2023 Criteria for Accrediting Engineering Programs.* Baltimore, MD: ABET.

Adams, R., Turns, J., & Atman, C. J. (2003). What Could Design Learning Look Like? *Design Thinking Research Symposium VI.* Sydney, Australia.

Ambrose, G., & Harris, P. (2009). *Basic design: Design thinking.* . NY: Fairchild Books.

Atman, C. J., & Bursic, K. M. (2013). Verbal Protocol Analysis as a Method to Document Engineering Student Design Processes. *Journal of Engineering Education , 87*(2), 121-132.

Atman, C. J., Cardella, M. E., Turns, J., & Adams, R. (2005). Comparing freshman and senior engineering design processes: an in-depth follow-up study. *Design Studies, 26*(4), 325-357.

Atman, C. J., Yasuhara, K., Adams, R. S., Barker, T. J., Turns, J., & Rhone, E. (2008). Breadth in Problem Scoping: A Comparison of Freshman and Senior Engineering Students. *International Journal of Engineering Education, 96*(4), 234-245.

Aurigemma, J., Chandrasekharan, S., Nersessian, N. J., & Newstetter, W. (2013). Turning Experiments into Objects: The Cognitive Processes Involved in the Design of a Lab-on-a-Chip Device. *Journal of Engineering Education, 102*(1), 117-140.

Bogusch, L. L., Turns, J., & Atman, C. J. (2000). Engineering Design Factors: How Broadly Do Students Define Problems? *American Society of Engineering Education Annual Conference & Exposition* (pp. 7-12). St. Louis, Missouri: American Society of Engineering Education.

Brown, T. (2009). Change by design. *Journal of product innovation management* , 381-383.

College Entrance Examination Board. (1939). *CEEB special aptitude test in spatial relations.* New York, NY.

Creswell, J. W., & Clark, V. L. (2017). *Designing and Conducting Mixed Methods Research.* Los Angeles: SAGE Publications, Inc.

Cross, N. (1995). Discovering Design Ability. In R. Buchanan, & V. Margolis, *Discovering Design: Explorations in Design Studies* (pp. 105-120). Chicago, IL: University of Chicago Press.

Cross, N. (2006). *Designerly Ways of Knowing.* London, UK: Springer-Verlag.

Dorst, K. (2011). The core of 'design thinking' and its application. *Design Studies*, 521-532.

Duffy, G. (2017). *Spatial Thinking in the Engineering Curriculum: an Investigation of the Relationship Between Problem Solving and Spatial Skills Among Engineering Students.* Dublin: TU Dublin.

Duffy, G., Sorby, S., & Bowe, B. (2020). An investigation of the role of spatial ability in representing and solving word problems among engineering students. *Journal of Engineering Education*, 424-442.

Ekstrom, R. B., & Harman, H. H. (1976). *Manual for kit of factor-referenced cognitive tests.* Princeton, NJ: Educational testing service.

IDEO Education. (2012). *Design thinking for educators.* IDEO.

Jolliffe, I. T., & Cadima, J. (2016). Principal component analysis: a review and recent developments. *Philosophical Transactions: Mathematical, Physical and Engineering Sciences*, 1-16.

Kim, M. J., & Maher, M. L. (2008). The impact of tangible user interfaces on spatial cognition during collaborative design. *Design Studies, 29*(3), 222-253.

Kozhevnikov, M., & Hegarty, M. (2001). A dissociation between object manipulation spatial ability and spatial orientation ability. *Memory & Cognition*, 745-756.

Kueh, C., & Thom, R. (2018). Visualising empathy: A framework to teach user-based innovation in design. In S. Griffith, K. Carruthers, & M. Bliemel, *Visual tools for developing student capacity for cross-disciplinary collaboration, innovation and entrepreneurship.* Common Ground Publishing.

Lammi, M. (2013). Engineering Design Thinking. *Journal of Technology Education*, 55-77.

Lau, A. S. (2004). Life-centered design - A paradigm for engineering in the 21st century. *American Society for Engineering Education Annual Conference* (pp. 9099-9108). Salt Lake City, Utah: American Society for Engineering Education .

Lazar, L. (2018). The Cognitive Neuroscience of Design Creativity. *Journal of Experimental Neuroscience, 12*, 1-6.

Loney, N. W., Duffy, G., & Sorby, S. A. (2019). The Relationship Between Spatial Skills and Solving Problems in Chemical Engineering. *126th ASEE Annual Conference and Exposition* (pp. 1-13). Tampa, Florida: ASEE.

National Academy of Engineering. (2004). *The Engineer of 2020: Visions of Engineering in the New Century.* Washington, D.C.: National Academy Press.

Nelson, H. G., & Stoltermann, E. (2003). *The Design Way: Intentional Change in an Unpredictable World : Foundations and Fundamentals of Design Competence.* Englewood Cliffs, NJ: Educational Technology Publications.

Pylyshyn, Z. W. (2003). *Seeing and Visualizing It's Not What You Think.* Cambridge, MA: MIT Press.

Raju, G., Sorby, S., & Reid, C. (2022). INVESTIGATING THE RELATIONSHIP BETWEEN SPATIAL SKILLS AND ENGINEERING DESIGN. *ICERI2022 Proceedings* (pp. 421-426). Seville, Spain: IATED.

Rhone, E., Adams, R., Turns, J., & Atman, C. J. (2003). Assessing IE Students' Preparedness for Practice:Do They Think Broadly? *Annual Industrial Engineering Research Conference.* Portland, OR.

https://doi.org/10.52202/073963-0009

Rhone, E., J, T., Atman, C., Adams, R., Chen, Y., & Bogusch, L. (2001). *Analysis of Senior Follow-up Data: The Midwest Floods Problem-Addressing Redundancies in Coding in CELT Technical Report #01-05.* Seattle: Center for Engineering Learning and Teaching, University of Washington.

Serdar, T., & deVries, R. H. (2015). Enhancing Spatial Visualization Skills in Engineering Drawing Course. *122nd ASEE Annual Conference & Exposition* (pp. 1-12). Seattle, WA: ASEE.

Shepard, R. N., & Metzler, J. (1971). Mental Rotation of Three-Dimensional Objects. *Science*, 701-703.

Sorby, S. A., & Baartmans, B. J. (2000). The Development and Assessment of a Course for Enhancing the 3-D Spatial Visualization Skills of First Year Engineering Students. *Journal of Engineering Education*, 301-307.

Sorby, S., Casey, B., Veurink, N., & Dulaney, A. (2013). The role of spatial training in improving spatial and calculus performance in engineering students. *Learning and Individual Differences*, 20-29.

Sorby, S., Nevin, E., Mageean, E., Sheridan, S., & Behan, A. (2014). Initial investigation into spatial skills as predictors of success in first-year STEM programmes. *SEFI 2014 42nd Annual Conference European Society for Engineering Education.* Birmingham, UK.

Sorby, S., Veurin, N., & Streiner, S. (2018). Does spatial skills instruction improve STEM outcomes? The answer is 'yes'. *Learning and Individual Differences*, 209-222.

Subedi, D. (2016). Explanatory Sequential Mixed Method Design as the Third Research Community of Knowledge Claim. *American Journal of Educational Research , 4*(7), 570-577.

Suh, J., & Cho, J. Y. (2020). Linking spatial ability, spatial strategies, and spatial creativity: A step to clarify the fuzzy relationship between spatial ability and creativity. *Thinking Skills and Creativity*, 1-17.

Sutton, K., & Williams, A. (2007). Spatial Cognition and its Implications for Design. *International Association of Societies of Design Research.* Hong Kong, China.

Sutton, K., & Williams, A. (2010). Implications of Spatial Abilities on Design Thinking. *Design and Complexity - DRS International Conference 2010.* London, UK.

Uttal, D. H., Meadow, N. G., Tipton, E., Hand, L. L., Alden, A. R., Warren, C., & Newcombe, N. S. (2013). The malleability of spatial skills: a meta-analysis of training studies. *Psychol Bull.*, 352-402.

Wai, J., Lubinski, D., & Benbow, C. P. (2009). Spatial ability for STEM domains: Aligning over 50 years of cumulative psychological knowledge solidifies its importance. *Journal of Educational Psychology*, 817-835.

Williams, A., & Sutton, K. (2011). Spatial Ability and its Influence on the Design Process. *Design Principles and Practices*, 141-152.

Xue, J., Li, C., Quan, C., Lu, Y., Yue, J., & C. Zhang. (2017). Uncovering the cognitive processes underlying mental rotation: an eye-movement study. *Scientific Reports*, 1-12.

Copyright statement

https://doi.org/10.52202/073963-0009

Transitioning from Engineering Practice to Engineering Education Research: Lessons Learned from Four Researchers

Srinivas Dustker[a]; Nusaybah Abu-Mulaweh[b]; Paul Leidig[c]; William Oakes[a]
Purdue University[a], The Johns Hopkins University[b], Thornton Tomasetti[c]
Corresponding Author Email: sdustker@purdue.edu

Abstract

Context
The field of engineering education research is still emerging and draws people from many career paths and backgrounds. This paper focuses on the stories of four researchers who came from engineering practice.

Purpose or Goal
This paper explores the stories of four researchers who transitioned from engineering practice to engineering education research. The research question is: What was the transitional experience from engineering practice to engineering education research, and how have their experiences been impacted by their background in the industry?

Methods
First-person accounts are presented as stories that span the early years of engineering education to the present from a large mid-western public university in the United States. The approach is modelled after Adams et al. (2007) and includes a summary of themes of similarities and differences analyzed by a fifth author.

Outcomes
In 2007, Adams et al, challenged the community to share stories and this paper continues that approach and gives glimpses into differences from early years of engineering education to the present. This paper prompts scholarly discussion, sharing of stories and lessons that can be learned as we seek to create a diverse research community.

Conclusion
We conclude by highlighting the importance of stories as humans are social beings who live a storied life. We provide the readers with different perspectives of transitional experiences to engineering education research from engineering practice that includes opportunities and challenges including the language, methods, and culture of engineering education research and how this community is different from practice.

Keywords— Transitioning, engineering education research, engineering practice

I. INTRODUCTION

ENGINEERING education has become a recognized field of study globally, albeit it is still emerging in several regions (Borrego & Streveler, 2014). A notable challenge in this burgeoning area is the preponderance of researchers primarily grounded in engineering disciplines rather than educational theory, thus encountering the arduous task of acquiring fresh expertise and viewpoints conducive to pedagogical research (Beddoes, 2014; Borrego & Bernhard, 2011)

The path of transitioning from an engineering academic to an education researcher is seldom documented, creating a void of representative narratives in the existing literature. A handful of resources offer a glimpse into such transitions, including 'Balancing Acts: The Scholarship of Teaching and Learning in Academic Careers'(Huber, 2004), where one engineer amongst four faculty members delineates their journey towards a scholarship that accommodates teaching and learning perspectives. Additional insightful narratives are encapsulated in 'Academic Pathfinders: Knowledge Creation and Feminist Scholarship'(Gumport, 2002), and the reflective accounts in 'Composing a life' underscore the experiences of women forging their academic paths (Bateson, 2001).

Despite the sparse literature on such transitions, there is a vibrant curiosity and eagerness to learn from the journeys of their peers, fostering a community grounded in shared experiences and stories(Adams et al., 2007). Recognizing this, we aspire to share our personal narratives of navigating from being engineering faculty to becoming engineering education researchers. Our aim is to stimulate scholarly discussions and embolden others to narrate their trajectories.

As we unfold our narratives in this paper, we will delve into the pivotal role stories have in sculpting the landscape of engineering education. Following an introduction to theoretical frameworks concerning narrative's role, we will transition to discuss the journeys to engineering-education researchers, embodying finding one's voice and articulating their story with authenticity. Emergent themes from the experiences offer valuable insights and guidance for others on similar paths. We endeavor to underline the potential of narratives in not only sharing individual stories but also in cultivating a rich repository of shared wisdom and experiences.

II. LITERATURE REVIEW

A. Narrative Research

Narrative inquiry takes various forms, one prominent one being storytelling, a method deeply rooted in human history for

https://doi.org/10.52202/073963-0010

conveying and discussing ideas, and as a vital research methodology, drawing upon established foundations laid out in the works of Jerome Bruner (J. S. Bruner, 2003, 2009) and Kieran Egan (Egan, 1993, 1999). This technique hinges on gathering firsthand accounts from individuals, offering a stage for voices that have previously been overshadowed or silenced, bringing to light a spectrum of experiences, including those of the marginalized.

In the sphere of engineering education research, one recognizes the pivotal role narratives play in unraveling culturally and socially contextual knowledge, fostering discourse, and building a shared "common ground" (Bromme, 2000). Moreover, it paves the way for nurturing a collaborative community of practice, a concept reiterated by Lave & Wenger, (1991) and fosters an environment conducive to interdisciplinary knowledge sharing and construction (Derry et al., 2020), drawing upon Bruner's seminal framework.

1) *The Foundations of Narrative Research in Education*

Narrative research, deeply rooted in qualitative and interpretative traditions, has been increasingly acknowledged as an indispensable tool in the field of education. It serves a dual role as both a phenomenon under study and a methodological approach, thereby presenting a rich yet complex landscape for educational research, encompassing various studies such as case and biographical studies(Connelly & Clandinin, 1990; Creswell & Poth, 2016; Goodson, 2014). Teachers are seen as innate storytellers, shedding light on the significance of narratives in the educational context (Connelly & Clandinin, 1990).

2) *Bruner's Perspective on Narratives*

Jerome Bruner has emphasized the intricate structure of narratives and their role in shaping realities through symbolic systems and cultural products(J. Bruner, 1991). He contended that narratives foster a deeper understanding and create vibrant frameworks for analyzing learning processes in educational settings, thus acting as fertile grounds for pedagogical development. It is crucial to leverage this understanding in engineering education research for a nuanced exploration of learning landscapes. Bruner's further insights into the narrative construction of reality underscore the pivotal role of cultural products, such as language, in facilitating sense-making processes. These narratives unravel complex learning processes by depicting characters navigating unprecedented scenarios, guided by discernible beliefs and values(J. Bruner, 1991).

This paper seeks to harness the power of storytelling to facilitate a deeper reflective practice, aiming to contribute to the collective understanding and knowledge base of emerging engineering education researchers through the rich tapestry of our personal narratives. By sharing our journeys, the hurdles we faced, our origins, and envisioned paths, we aim to foster a vibrant community grounded in shared experiences and wisdom. We aspire to delineate the emerging field of engineering education, guiding newcomers in their trajectories and enhancing the discipline's profile through a rich tapestry of multifaceted narratives.

III. Methodology

A. *Research Design*

The primary methodological approach adopted in this research was qualitative, seeking to draw out rich, detailed narratives from participants via structured interviews. The study was rooted in an interpretative paradigm, recognizing that the knowledge garnered would be constructed through the dynamic interaction between the researchers and the participants.

B. *Participants*

Four participants were engaged who all made the transition at the same large Midwestern University in the U.S. but came from diverse backgrounds and in different eras.

C. *Data Collection*

1) *Interview Protocol*

Participants were sent an email set of prompts, which allowed the participants to respond at their convenience within a stipulated timeframe. The email format also had the benefit of automatically documenting the responses, facilitating a transparent and straightforward data collection process. The prompts were developed to guide the participants and were crafted to facilitate deep exploration into the research questions. The participants were encouraged to freely articulate their thoughts, experiences, and reflections. The final protocol included the following questions:

1. Please describe your background before you entered the field of engineering education or started to be involved in engineering education research.
2. What got you into engineering education research?
3. What motivated you to transition from practice to engineering education research? What interests you in engineering education research?
4. Why research in engineering education and not in a technical engineering discipline?
5. How did you find the engineering education community vs. technical engineering community?
6. After your transition from practice to engineering education research, what was the transition like?
7. Please describe your current work in engineering education research.
8. Please describe the direction in which your current work is headed.
9. Are there things that engineering education can do more? Talk about lessons learnt or limitations the field has as of now that needs to be addressed.

D. Data Analysis

In the post data collection phase, our focus pivoted to analyzing the gathered data to unearth patterns and derive insights from the narratives of the participants. This involved a structured approach, where we meticulously went through the process of familiarizing ourselves with the data through repeated readings of the responses, a methodical exercise that set the stage for a detailed thematic analysis.

A three-tiered coding strategy – open, axial, and selective coding was applied through which we were able to cultivate an understanding of the narratives, leading to a structured, yet nuanced interpretation of the data at hand. By progressively building upon each stage of the coding process, we aimed to present a well-rounded analysis, grounded in theory and detailed observation, that unveils the intricate patterns and central themes vividly portrayed in the participants' responses.

IV. Our Stories

A. Empathy in Engineering: Nusaybah's Transition from Software Development to Engineering Education.

My background is in Electrical and Computer Engineering. After graduating with a bachelor's in computer engineering, I worked as a software developer for two years. In industry I realized that although I gained knowledge and experience, I was not satisfied. I wanted more out of my career. I went on to pursue my master's in electrical and computer engineering with the goal of transferring into academia. While I was working in industry I was also teaching a coding course, and I enjoyed that more than my actual full-time job. Fast forward to today - I am in my current position in EPICS as an instructor for 7 teams (subdivisions) and the coordinator for senior design.

While teaching, I was really interested in how students learn. How you can teach the same material, but one student will get it, and another didn't. Or how from semester to semester, you had to adapt your teaching. Because, what working one semester, may not work the same the next. So, learning, the process of learning, the dissemination of knowledge really intrigued me. Especially with my background in ECE and looking back to how I learned. I struggled in the field, but that was the status quo – you had to struggle in engineering. You had to earn it. That was the mentality. I want to change that.

I went from practice to education because the job was more satisfying. I love teaching. I love touching the lives of the next future engineer and having an impact on their learning and outlook. Over time, my interests changed. I became more interested in the learning process than technical aspects of ECE. The engineering education community is more like a family. They were more welcoming and understanding. Kind and inclusive of all. Never saw that in my experience in ECE. I think the transition came easy to me. Like I stated before, the shift was due to my interests, so everything flowed naturally in the direction of engineering education research. I think the only shock to my system was the lack of equations. Coming from the engineering technical background, and not needing my fancy calculator or a script to run over night or debugging code hours on end to find you missed a semi-colon...that was the biggest difference. No numbers, no code, no equations. And at first, I missed that life of ECE. There was too much I didn't know and didn't understand in engineering education. Having epistemology, methodology, frameworks, etc. thrown at me felt like I was in a foreign world. But like I said before, the community was beyond understanding, kind, and welcoming. So, I think after the first year of courses, my equations were left behind and really didn't mind any more.

My work in engineering education research has been focused on community-engagement in engineering. My current work focuses on empathy in engineering. Specifically how community engaged learning can help engineering students develop their empathic skills. Empathy is very important for engineering design, yet most don't think of 'empathy' when you talk about engineering. Research shows that by empathizing with stakeholders, engineers design more innovative solutions that focus on actual needs, discover new product applications, and avoid future mistakes before wasting money and resources. I want the field to shift to be more understanding of others. For them personally, for their work as an engineer, and for the world as a whole. I think my lessons learned really focus on me personally – I just wish I had exposure to this field early on. I think I would have found this home and begun this journey a lot earlier if I knew what engineering education research was and the possibilities.

B. The Story of Researcher B: From [Aerospace] Engineering to Educator and Innovator

I worked for a little more than five years in a [aerospace] company in aerodynamics and math modeling simulation. My degrees were in mechanical engineering, from a university that kept mechanical and aerospace completely separate. But I figure that my degrees proved that I could learn, and they wanted to teach me how they build their products, which is a different philosophy from other competitor companies. I learned engineering design and management philosophies. I participated in the flight test and then the process of making sense of the data that ultimately improved the flight simulator for pilot training.

I felt like I was drowning in uncertainty for 2 years. I learned aerospace-related skills on my own time and money to be useful in my job and to reduce my uncertainty, which my supervisor fully supported. And I was a teaching assistant briefly in my master's degree. I enjoyed teaching, but I was clear that I didn't know enough about it and that my degrees still left me unconfident in my own engineering abilities and knowledge. I wanted to bring practical experience to the classroom. And I would be fully vested in my 401(k)-matching money after 5 years of employment, so after 5 years, I started looking for PhD programs. The threshold for teaching at a university is usually a PhD. So I wanted a plan of study that could address how to learn and teach engineering, even if it was a secondary objective.

I will admit that I stumbled into my eventual PhD program. I didn't pursue a technical engineering discipline because I didn't know how to resolve master's research in heat transfer and nanotechnology with my work experience. I felt that I had equal proficiency in both. Also, I wanted to move somewhere else in the country, but did not have a definite choice yet. But I used the school's website to look at PhD programs offered. I landed on the Engineering Education page and eventually set up a face to face visit, where I met my advisor, who strongly advocated that I join the program. I had an epiphany that our PhD program would not intentionally make us better teachers. But I have always known that PhD programs must have research, so I stayed in the program.

When the university hired a new person for the chancellor/president role, that's when fundamental differences between pure academia and private industry workers became clearer. Academic leadership changes look like leadership changes in private industry to me, but now I know that not everyone holds this view. Even more differences between academics in engineering education research and in technical engineering became noticeable when I took a teaching role in engineering. I didn't understand the value of some social science courses in Engineering Education until 3 years after, so these weren't enjoyable while I was enrolled. However, I took an adjunct teaching position at a teaching focused university. When I started talking with these colleagues about designing classes and assignments and grading schemes, it was clear to me then that we had different philosophies about who should be an engineering student, because we deliberately researched engineering students' attributes and attributes of institutions and systems that influence students' recruiting and enrollment.

And now that I am on tenure track, I have mixed feelings about obtaining tenure. The decision-making structure in academia is exactly upside down compared to private industry. In private industry, top level leaders are supposed to use data to make decisions and it flows down to the lower levels. In academia, decisions are made by Roberts Rule of parliament voting at the department level and is supposed to flow up to higher levels of leadership. Since I started in private industry, I learned its decision-making structure and I prefer it over academic decision-making structure.

I have thought much about how my academic salary accounts for my industry experience, and vice versa. My industry experience to my department was a "nice to have" that did not get counted as credit on the 6-year path to tenure. But my colleagues who taught as adjunct in that institution or as tenure track at other institutions did receive credit. I have to think about money to take care of my family and my retirement. I do feel that my salary has lagged behind others in my age group with engineering bachelor's degrees.

But I also love teaching students, and I take some pleasure in research. I also still love tinkering, inventing, writing, and engineering. My institution is teaching focused. But we're experiencing a drop in enrollment and over 75% of our budget is from tuition.So we are strongly encouraged to bring money in. I tried a couple of avenues but so far, my proposals have been rejected.

By happenstance, a fellow classmate is now working at a government institution and just started up a research competition ([research competition]) where the undergraduate student winners would be offered a summer internship. This alleviates the need for me to generate my own research questions. This summer, I am working with 1 student, an incoming sophomore who loves using - a prescribed engineering design process that I learned in my PhD studies and has said that she has learned more with me than all her other classes, and she's so grateful.

Based on my pleasant experience with my student in the [research competition], I think I can offer this undergraduate research class annually. We have broad topics published by [research competition], and I help the students narrow the project to an actionable plan, according to their interest. It is applied research (technology readiness level 3 to 6 on a scale of 9), not fundamental research. I can write papers that are classified as "scholarship of teaching and learning" but I might not need external funding to conduct this type of research. For example, I just used the grading data from my own classes to write an engineering education research paper.

I took the qualitative research avenue. It really worked for my dissertation questions. But I believe that our technical only lengineering colleagues assume it's all social science research in Engineering Education. It works both ways; technical engineers who teach can conduct technical research and engineering education research, and engineering education researchers can conduct technical research, too, But engineering education research is its own discipline. Some engineering professors assume "I teach, I research, I am an engineer. Therefore, I am an engineering education researcher". They see each word in the title as separate and unrelated activities. But we in the discipline know that we lacked a unique word in English to name the discipline. Maybe if we were "engi"-gogy instead of pedagogy, then perhaps others on the outside would not claim to be in our discipline. That by itself may cause them to ask Engineering Education researchers what exactly we do, and that's a great conversation starter.

C. Paul's Journey: From Structural Engineering to Pioneering in Engineering Education Research

As a first-generation college graduate, I received my Bachelor of Science in Architectural Engineering from the Milwaukee School of Engineering and Master of Science in Civil Engineering from the University of Illinois at Urbana-Champaign. In both cases, I specialized in structures. Following this, I gained six years of industry experience in structural engineering consulting, becoming licensed as a Professional Engineer in the state of Colorado. Through this experience, I had the opportunity to work on transit hubs, high-rise towers, stadiums, university buildings, embassies, and high-end corporate facilities. For much of this time, I specialized in complex construction erection engineering, serving as the full-time staff engineer for this scope on an award-winning project that used the largest number of simultaneous stand jacks ever in the northern hemisphere, for example. Throughout my academic and professional experience, I had actively participated in and led community-engaged engineering and design projects for approximately a dozen years with a number of organizations before starting my PhD.

I have had a long-standing interest in topics related to education and, in particular, experiential learning. My curiosity in this space was sparked first when I attended an experimental new local public middle school which had a strong emphasis on project-based learning and multilevel education. Here, I participated in my first experience with what I would now know to call a community-engaged design project. Through the remainder of my education, I maintained involvement with this interest through serving in tutoring and teaching roles of various kinds, including teaching English to international students while studying abroad and taking on the role of co-instructor for a course in the Learning in Community (LinC) program at University of Illinois Urbana-Champaign (UIUC) while a graduate student there.

Throughout my time at university and in the workforce, I also had the opportunity to participate heavily in the organization Engineers Without Border USA (EWB-USA), which has a dual mission around projects that empower communities and equip leaders. Through my experiences with this and other similar organizations as well as within the industry setting, I found myself more and more drawn to the topics surrounding how the engineering and design project stakeholders are prepared, supported, and coordinated in the pursuit of objectives, beyond the specific technical details required for a given scope of work. I also observed what I perceived to be wide variations in the approaches taken and outcomes achieved by various groups over time. These experiences and observations drove me to want to learn more, create new actionable knowledge, and these cement high-impact positive practices in this spaces of experiential learning, civil engineering, community engagement, design, and engineers in professional practice. As I would come to find out, this basket of topics would fit nicely in the world of engineering education research.

The primary event which precipitated my decision to actually leave working in industry and begin my PhD in engineering education was finding a PhD advisor who I thought would be a good fit for me, in terms of research interests, industry background, funding, logistical support, and affiliated program and institution. For much of my time working in professional practice, I had been slowly exploring the possibilities of pursuing a PhD, through reaching out to faculty, having conversations with graduates, and the like. Many of these probes helped clarify my search path going forward and finding an opportunity that made sense was critical to deciding to pursue engineering education research. I enjoyed being a structural engineering practitioner, but I thought there was a good chance I might be able to find greater wellbeing and fulfillment working in engineering education. The day-to-day nature of the work appealed to me as well as the long-term goals I would be able to focus on. Ultimately, encouragement from those close to me and specific expressions of interest from my to-be advisor are what put me over the edge to take the risk to explore a new career path.

Throughout my career, I had an interest in widening my perspective and interacting with broader aspects of design projects. This was heavily influenced by the nature of my work as a leader in community-engaged engineering projects and organizations as a volunteer, where I had an opportunity to focus on people, projects, and process management. This was a significant factor in my move from working on new building design to construction engineering in my professional work as well, and I see pursuing a specialty in engineering education as a continuation down this path. I was also very interested in increasing my ability to address different types of challenges. Becoming more specialized in my existing technical area was not in line with this. Instead, I sought to widen my horizon and increase the potential leverage of my efforts by working to improve the human elements of engineering and design endeavors, doing so by supporting the empowerment of others who could have a much greater impact over may fields compared to my direct technical work on specific items.

Both of these communities [Engineering Education and Technical engineering] are expectantly large, and I can only speak to the small portions with which I have interacted. That said, the communities are naturally impacted by the environments and incentives that influence them. The largest cultural difference I have noticed between the two groups, as I have experienced them, is that the technical engineering community tends to be much more narrowly focused on their area of expertise while the engineering education community is generally open to casting a wider net and seeking out broader perspectives.

Like one might experience when traveling in a different country from one's home, a change in environment can help illuminate those things we take for granted as well as inform us about which aspects of our experience might be more universal and which are not. My transition brought to the forefront the diversity in ontological and epistemological lens. The timescale of most tasks I have encountered in academia are more extended than those I was accustomed to in engineering consulting, with lower concentrated intensity. Due to this and other environmental factors, the ways in which people prefer to go about collaborating on tasks can be quite different. I would say my experience in both teaching and research has aligned in some ways with the pace and procedures of my previous engineering work, but that they do so in different ways. Much of what is done in the two spaces is very similar, just with different specific content knowledge and contexts. The professional skills required and design approaches to problems are all much the same in my experience. This makes sense when we think about the idea of the T-shaped professional. Most of the items at the top of the T are common between the spaces and can be transferred reasonably easily, I think.

Transitioning from being a highly valued member of a professional team to being a new person in an academic program was a significant adjustment in terms of how people viewed and treated me. My previous experiences were often not valued by those in the engineering education spaces and I needed to prove myself in this new arena. Working with people mostly approximately my age or older in professional practice and then joining a program in which many of my classmates were up to a decade younger than me was a meaningful cultural adjustment as well. Finally, the reading and studying styles that worked well in my previous engineering coursework were no longer appropriate for the classes in engineering education; accepting this and learning new methods took significant time and energy. I found making connections with others who also had previous industry experience expectantly helpful; establishing ways to make these connections easier to find and build would be a positive development, I think. Also, broadly speaking, the whole system of how universities compensate graduate students for their work is also something that should be reviewed; I think changes here could help those coming from industry but also go far beyond this. After getting used to the new environment, I find the work much less stressful while still interesting and engaging. Academic life also affords the opportunity to meet many new people and explore fascinating ideas to an extent far beyond my experience in engineering practice. The flexibility in how I decide to manage my schedule is also a benefit in my view.

My research interests remain in the areas of experiential learning, civil engineering, community engagement, design, and engineers in professional practice. I often prefer to take a mixed methods approach to research questions and may be described as an action researcher, given my great interest in connecting scholarship with practice. My specific work at this time covers a number of areas centered around community-engaged learning. This includes creating a new Model for Project-Based Community Engagement, writing about various case studies, and beginning work on an alumni study of former EWB-USA student members.

The EWB-USA alumni study is intended to serve as my dissertation topic over the following two years. In addition to this, I am interested in exploring opportunities to investigate the teaching, coaching, and learning of design as well as continue to publish on practical aspects of conducting experiential learning in the design space, such as looking at assessment methods. [As far as it pertains to the lessons or limitations] Focusing on and valuing more highly the transition of scholarship to practice. Many of the largest challenges in engineering education appear to be matters of execution. Build better understandings of the differences between specialties within the very large and ill-defined space of engineering, targeting scholarship to the individual disciplines as appropriate. Connect and engage scholarship and student learning more with professional practice and the broader community.

D. Bill's Journey: From Aviation Design Engineer to Pioneer in Engineering Education Research

I graduated with a master's in mechanical engineering and entered a career as a design engineer in aviation. I loved the work and the industry and was selected to be a corporate recruiter which got me back to campus. The recruiter role opened doors to speaking to engineering classes and working on how to transition students into their professional careers. I found these parts of my work very interesting and saw the gaps in education that I thought I could help address.

To be a professor I needed a PhD and began with the goal of doing a traditional faculty path with technical research, teaching and service. While I was in the PhD program at Purdue University, I met Prof. Jim Jones who was doing innovative work in education and was a leader in active learning, assessing the impact and publishing on the work. He invited me to be a part of a group of graduate students who would become the first ASEE student chapter. ASEE was not initially enthused and we had to advocate and were successful. That changed my thinking about a traditional Mechanical Engineering career. An opportunity opened to join what was called Freshman Engineering, at first as a visiting assistant professor and then as an assistant professor. Those positions were focused on teaching, advising and service. As a former head advised me, those positions had a viable path to tenure but likely not to a full professor and the pay and respect from other faculty would not be high. That meant that I would spend a career as an underappreciated (by peers) associate professor, underpaid in engineering and have the opportunity to impact thousands of first-year students. I thought that was exciting.

The idea of doing educational research was not on my agenda at first. However, Professors Goranka Bjedov and George Bodner would change that. Goranka was an associate professor in our department and brilliant. She was the leader of a large multi-college grant between engineering, science and math and I was included as a co-PI. She was so brilliant that she left our department and went to work with one of the largest tech companies on the west coast. This left me as a lead on the grant. Prof. George Bodner was a distinguished professor in Chemistry Education and also on the leadership team for the grant. He saw the opportunity with the grant funds to do significant qualitative work with first-year students and faculty. He took me under his wing, and we worked with three of his graduate students. I got a personalized introduction to qualitative educational methods from one of the country's best. I started to get hooked. At the same time, I was added to a multi-campus committee as part of a very large retention grant. Our committee was responsible for first-year seminars modeled after the successes at the University of South Carolina. As a naïve assistant professor, when we formed the committee, it needed a chair. I said, sure, I'd be willing to be chair and was nominated and confirmed at that first meeting. Three of the full professors who were on the committee pulled me aside at our next meeting and said that if I was going to do this as an assistant professor, I needed to get some papers out of it and to ask for a graduate student to help with assessment. It was granted and we hired one of the best students I have ever worked with, Brian French, now a distinguished professor in educational psychology. That work got me into psychometrics, and we created an academic motivation instrument and measured the impact of the first-year seminars. While learning these skills, I was taking a leadership role in the emerging EPICS Program especially in the area of expanding EPICS to other institutions. EPICS offered opportunities for more research as well as means to apply what we were learning. The work we did in early scholarship helped lay the foundation to create our school of Engineering Education.

When I made the transition, there were no formal engineering education departments or schools. The engineering education community was a collection of colleagues who were mostly isolated. I got involved in ASEE and the ERM division and with the Frontiers in Education (FIE) Conference. These were the places where the community came together. In those early years, there were many people who were researchers and innovators in education. The ERM division is named for Educational Research and Educational Methods and there were both in the community. I gravitated to the M – methods people. I think this was because I saw how the current research showed conclusively that many innovations did work and could address many of the needs we had. Early in my career I became involved in service-learning and the dissemination of the pedagogy through faculty development. I saw the research as a means to validate these approaches. That early community was very, very supportive and collaborative.

While my identity was not primarily as a researcher, my CV was one of the examples used to convince our administration to create the first department of engineering education. In those early years, we talked a lot about what we would be and how we would measure success. I advocated that we use a broad view of scholarship and be different than the traditional disciplines. The counter argument was that if we were different, we would not get academic respect and credibility. Engineering Education Research has established a global identity and credibility but have become mostly restricted to traditional views of scholarship measures in journal papers.

For me, the transition was to teach and engage with communities and other faculty. When I made that career change, the research followed. As I described earlier, I was blessed with early opportunities to gain experience qualitative and quantitative research methods that directly impacted what we were doing. I have found engineering education research with direct connection to what we are teaching and how we are engaging invaluable to improving and refining our approaches.

The current work is focused on the areas of community-engaged learning in how students learn, how we impact communities and how we assist other faculty implementing the pedagogy in their own institutions. We mostly use qualitative approaches, but we continue to use mixed methods too with significant quantitative components. Our field has focused on establishing credibility with other education, social science, and science researchers. As we mature as a field we have opportunities to connect with the engineering fields addressing important challenges to increase learning, engagement, persistence, and diversity in engineering. We also have opportunities to move findings into practice. We can reach out and re-establish connections with colleagues in the traditional engineering disciplines to work together to integrate research findings into the classrooms of the future.

V. Discussion

In analyzing our narratives, it became apparent that despite the individualities in our stories, there are several recurring themes, or points of convergence, that emerged prominently. These encompass the unexpected discovery of our passions, the non-linear pathways undertaken, varying degrees of support and resistance, the forging of multifaceted identities, and a relentless spirit of perseverance (Beddoes, 2014; Borrego & Bernhard, 2011).

All of us were propelled by persistent inquiries that eventually morphed into our core passions. The terminology used in narrating our journeys often mirrored the fortuitous and somewhat unplanned nature of our explorations into the realm of engineering education research. Our stories echo a sense of non-linearity, lacking a predetermined roadmap, instead exhibiting comfort with a winding, undefined trajectory. There was a pronounced element of instinctiveness and spontaneity in our journeys, evoking an imagery of intuitively crafting a unique bouquet while ambling through a meadow. It is clear that our paths were characterized by purposeful intent, coalescing naturally through our individual actions and narrative constructions, echoing an appreciation for diversified perspectives gleaned through cross-disciplinary engagements (Adams et al., 2007).

A ubiquitous sentiment across our narratives was the necessity of navigating through a myriad of support systems. Interactions with diverse individuals and communities emerged as a foundational scaffold shaping our careers, albeit accompanied by instances of disagreement, disregard, and even hostility towards our endeavors. This landscape birthed a dynamic identity, sometimes taking on the role of a pioneering engineering education researcher, at other instances leveraging a cross-disciplinary approach, and occasionally reverting to our original engineering roots. This narrative reflects a rich tapestry of identities interwoven as engineers, educators, and engineering educators, underscoring the importance of harmonizing these varied identities and fluidly transitioning among them in varied contexts(Borrego & Streveler, 2014).

Reflecting retrospectively on our experiences reveals an inherent thread of tenacity running through our stories. Unpacking this further, we noticed that we constantly embraced either a "learner's stance" or a "researcher's stance", fostering a readiness to step into uncharted territories and relinquish our comfort zones. This approach has not only honed our observational and synthesizing skills to a level unanticipated but also rejuvenated our commitment to lifelong learning, invigorating our professional identities through enriched experiences and learnings.

VI. Lessons

As we reflect on our journeys, we distill several pieces of advice that might steer the paths of emerging engineering education researchers. While these suggestions are rooted in personal experiences, they encompass universal values and strategies that could be beneficial for anyone stepping into this arena. Below are the distilled pieces of advice:

A. Cherish and Chase Your Dreams

Delving deep into the wells of our aspirations, we find the invigorating spirit of dreams, a vital aspect that could be transformative, especially for budding engineering education researchers. The mantra "Because dreams need doing," has echoed powerfully, resonating deeply with the youthful hearts and energetic minds embarking on a path of discovery (Baranowski & Delorey, 2007). Encouraging a culture that cherishes dreams could be a potent driver, propelling individuals to break through boundaries and foster innovation. It nurtures a vibrant dynamism where dreams are not mere figments of imagination but catalysts for real-world change. Through the pursuit of dreams, one can cultivate resilience and ingenuity, both of which are cornerstone qualities in the field of engineering.

B. Cultivate Community Connections

In a world replete with opportunities for global collaboration, fostering community connections stands as a pillar in the progressive growth of any individual in the educational sphere. Building networks beyond one's immediate surroundings can open up rich avenues for learning and mutual growth. It is not just about expanding your social network but creating a synergistic ecosystem where diverse perspectives coalesce, offering a rich tapestry of insights and experiences. This proactive approach could be a springboard for international collaborations, further elevating the scope and impact of engineering education research.

C. Engage Deeply with Your Field

The landscape of engineering education research is vast and constantly evolving. A deep and immersive engagement with this dynamic field requires a receptive mindset, one willing to adapt and grow with the evolving paradigms. While the journey may initially appear daunting, with continuous effort and a spirit of inquiry, one can traverse this landscape proficiently. Engaging deeply fosters a symbiotic relationship with the field, allowing one to draw from a well of knowledge while also contributing significantly to its expansion.

D. Embrace the Learner's Perspective

To nurture a rich and fruitful educational ecosystem, adopting a learner's perspective is indispensable. It encourages a state of perpetual curiosity, where the desire to explore and learn is not confined to the students but is a fundamental principle guiding the educator's approach. This perspective beckons one to venture into unknown realms with an open heart, fostering a conducive environment for exploration and discovery where finding solutions becomes a collaborative and enriching journey rather than a solitary task.

E. Shape Your Own Career Path

Embarking on a self-directed journey of career development heralds a path of self-discovery and purposeful growth. It encourages individuals to be vigilant, seizing opportunities that resonate with their personal and professional aspirations. This path is characterized by a dual approach, where one seeks to foster internal growth through self-improvement and external fulfillment by enhancing the learning outcomes in students. By carving out a personalized career trajectory, one stands to create a fulfilling journey marked by milestones of personal achievement and broader educational impact.

F. Adopt a Researcher's Perspective

Adopting a researcher's perspective infuses one's educational journey with a critical lens, transforming persistent curiosities into research-driven queries with real-world implications. Envision your educational spaces – be it a class or

https://doi.org/10.52202/073963-0010

an entire campus – as fertile grounds for research, constantly offering questions begging for deeper exploration. This perspective nurtures a culture of inquisitive thinking, fostering a rich dialog between one's experiences and the evolving questions that shape the educational landscape.

It is imperative to note that these pieces of advice function as heuristics, implying that while they are grounded in reason, their efficacy is not absolute. However, drawing from Billy Koen's articulation of the engineering method (Koen, 2003)– utilizing heuristics to optimize outcomes in complex, relatively undefined scenarios within available resources – these suggestions represent our best current strategies. We aspire that our contemplative exercise aids in enriching the existing corpus of knowledge in this domain.

VII. Conclusion

In the discourse surrounding various professions, Robin Adams noted a conspicuous absence, articulated by a PhD student, of "engineering lore" in comparison to the rich narratives enveloping artists, writers, and individuals in other professions (Adams et al., 2007). This observation extends to the relatively uncharted territory of engineering education researcher lore, where the landscape of personal and professional narratives remains significantly unexplored.

In this exposition, we ventured to fill this gap to a certain extent, weaving tales from our individual journeys with the hope that they echo with others in our community. We envision these stories as potential linchpins, binding us together through shared experiences and familiar struggles, offering a glimpse into the rich tapestry that encapsulates the essence of being an engineering education researcher. We believe that these narratives can stand as a testament to our professional engagement in the field, illustrating the varied pathways and rich experiences that bring one into this sphere.

Stories, beyond being mere recounting of events, serve as discursive instruments, fostering exploration, sharing, and reflection. They are platforms that foster communal understanding, allowing for the cultivation of common grounds regarding the identity and evolution of engineering education researchers. This tapestry of tales serves as a nurturing ground where more stories can take root, facilitating transformative shifts and providing insight into the largely untread paths of this profession.

Through the lens of storytelling, the obscured becomes apparent, unveiling the underlying intentions in our deeds and shedding light on invaluable life learnings. It metamorphoses into a pedagogical tool, a mirror for reflective practice, and a wellspring for research inquiry. A well-articulated story has the power to reach out and touch others, sparking recognition, understanding, and empathy.

As we delve into crafting our narratives, it is essential to be mindful of the core elements that make a story resonate deeply with its audience. Drawing on resources from Stephen Denning (Denning, 2004, 2005), it is emphasized that a compelling narrative encapsulates essential details about the situations and the personas involved. It weaves a coherent narrative thread, encompassing plots with their respective resolutions, thereby immersing the reader into the lived experiences of the storyteller.

We extend an invitation to you to reflect and embark on the journey of narrating your story, utilizing this rich array of tools and insights to carve out a narrative that is uniquely yours yet resonates universally, contributing to the vibrant mosaic of engineering education research lore.

References

Adams, R., Fleming, L., & Smith, K. (2007). Becoming an engineering education researcher: Intersections, extensions, and lessons learned among three researchers' stories. *Proceedings of the International Conference on Research in Engineering Education (ICREE), 10.*

Baranowski, M., & Delorey, J. (2007). Because dreams need doing: New messages for enhancing public understanding of engineering. *Global Strategy Group-NAE.*

Bateson, M. C. (2001). *Composing a life*. Grove Press.

Beddoes, K. (2014). Methodology discourses as boundary work in the construction of engineering education. *Social Studies of Science*, *44*(2), 293–312.

Borrego, M., & Bernhard, J. (2011). The emergence of engineering education research as an internationally connected field of inquiry. *Journal of Engineering Education*, *100*(1), 14–47.

Borrego, M., & Streveler, R. A. (2014). Preparing engineering educators for engineering education research. *Cambridge Handbook of Engineering Education Research*, 457–475.

Bromme, R. (2000). Beyond one's own perspective: The psychology of cognitive interdisciplinarity. *Practicing Interdisciplinarity*, 115–133.

Bruner, J. (1991). The narrative construction of reality. *Critical Inquiry*, *18*(1), 1–21.

Bruner, J. S. (2003). *Making stories: Law, literature, life*. Harvard University Press.

Bruner, J. S. (2009). *Actual minds, possible worlds*. Harvard university press.

Connelly, F. M., & Clandinin, D. J. (1990). Stories of experience and narrative inquiry. *Educational Researcher*, *19*(5), 2–14.

Creswell, J. W., & Poth, C. N. (2016). *Qualitative inquiry and research design: Choosing among five approaches.* Sage publications.

Denning, S. (2004). *Squirrel Inc.: A fable of leadership through storytelling*. John Wiley & Sons.

https://doi.org/10.52202/073963-0010

Denning, S. (2005). *The leader's guide to storytelling: Mastering the art and discipline of business narrative*. John Wiley & Sons.

Derry, S. J., Gance, S., Gance, L. L., & Schlager, M. (2020). Toward assessment of knowledge-building practices in technology-mediated work group interactions. In *Computers as cognitive tools* (pp. 29–68). Routledge.

Egan, K. (1993). Narrative and learning: A voyage of implications. *Linguistics and Education*, *5*(2), 119–126.

Egan, K. (1999). *Children's minds: Talking rabbits and clockwork oranges.* Teacher's College, Columbia University.

Goodson, I. F. (2014). Storying the self: Life politics and the study of the teacher's life and work. In *Curriculum* (pp. 3–20). Routledge.

Gumport, P. J. (2002). *Academic pathfinders: Knowledge creation and feminist scholarship*. Bloomsbury Publishing USA.

Huber, M. T. (2004). *Balancing acts: The scholarship of teaching and learning in academic careers*. Routledge.

Koen, B. V. (2003). *Discussion of the method: Conducting the engineer's approach to problem solving*. Oxford University Press.

Lave, J., & Wenger, E. (1991). *Situated learning: Legitimate peripheral participation*. Cambridge university press.

 https://doi.org/10.52202/073963-0010

Bringing life: an inclusive national conversation to develop integrated engineering curricula in South Africa

Teresa Hattingh[a] and Helen M. Inglis[b]
University of Johannesburg, South Africa[a], University of Pretoria, South Africa[b]
Corresponding Author Email: teresah@uj.ac.za

Abstract

Context and Purpose

This paper describes a national engineering curriculum renewal initiative designed to meaningfully integrate technical and professional competencies, to prepare graduates for the world of work and the challenges faced by society. The paper presents a descriptive case study to identify the underlying critical success factors of the project.

Approach

Using a social constructivist perspective of curriculum design, we adopt Kotter's model as a theoretical lens for the analysis. The case study draws on the personal reflections of the authors, two members of the project team.

Outcomes

The project is described in detail, and the importance and relevance of key phases and steps in the process are highlighted. The crucial roles of broad stakeholder engagement, structured interventions to provoke thinking differently, and sharing of best practices are discussed. Several challenges are identified specifically in relation to stakeholders entering and leaving the process at different points. The paper further shows the additional benefits that can arise through a national initiative of curriculum change.

Implications for curriculum renewal projects

Our reflections reveal the differences and similarities between curriculum renewal initiatives at a national and institutional level. A national project, as described in this paper, presents many opportunities, and yet there are complexities that need to be understood and managed throughout the process. We end this paper with insights gained regarding these complexities and how they can be mitigated.

Keywords—curriculum; integration; renewal; collaboration; national; community of practice

I. INTRODUCTION

ENGINEERING education in South Africa struggles with low throughput rates and overloaded academic staff. Engineering curricula are typically content heavy, with the first two years focused on mathematics and science fundamentals, and higher years centred around the technical, discipline-specific content. The development of professional competencies is often tacked on at the end of programmes to align with the expectations of accrediting bodies. Concurrently, universities are continuously questioned about the relevance of curricula and the preparedness of students to meet the needs of industry and society. Many of these challenges are not unique to South Africa, and globally institutions have responded by shifting to engineering curricula that are more holistic and integrated.

Inspired by such international initiatives, a project entitled *Bringing life to Engineering Curricula* (iecurricula.co.za) was formed to explore how South African engineering curricula could be transformed to address the experienced challenges by reimagining engineering curricula in an integrated way. This project is unique in that it aims to do this at a national level, involving all institutions that offer programmes in engineering. Changing curricula at a department or even faculty level can be a daunting task and therefore, the project identified the importance of clear change strategies from the outset. This paper describes the context, design and implementation of the project activities. The project is interpreted through the lens of change theory, and critical success factors and challenges that have emerged from the process are presented. The findings and implications of this paper provide insights into the principles that underpin curriculum design projects and specifically highlight the complexities at a national level where contexts, capacities and strategies differ. The paper further foregrounds the additional benefits that can be achieved by initiating national conversations around teaching, learning and curriculum in engineering education. The philosophy and thinking behind the change strategy design will be the focus of a separate paper.

Engineering curriculum renewal initiatives that cut across an entire faculty (Mitchell et al., 2019) are uncommon while those at a national level are even more rare. A national project was run in Australia in 2011 that looked at ways of re-imagining engineering design curricula. This national initiative brought together a range of stakeholders to workshop the possibilities of bringing together the three dimensions of an engineering graduate (technical, professional and personal competence) in relation to engineering design (Goldsmith et al., 2011). The

https://doi.org/10.52202/073963-0011

authors are however not aware of any other national project of this scale that aims to rethink entire engineering programme curricula.

II. Literature Review

A. The concept of an integrated curriculum

In the past few decades, there have been global calls for engineering curricula renewal from industry, professional institutions and government (Mitchell, 2021). Curriculum responsiveness is needed to adapt to a world that increasingly requires graduates who: exhibit strong professional skills without losing core technical competence, have an awareness of the societal context in which engineering activities take place and can transfer these competencies to the workplace. Many argue that this can only be successfully achieved with a curriculum that is coherent and interconnected (Graham, 2012). Ideas around a holistic or integrated curriculum have existed since the early 1990s (Shaeiwitz et al., 1994; Olds & Miller, 2004) and yet, many institutions around the globe still have programmes with traditional curricula that are often disconnected or heavily focused on the development of technical knowledge and skills.

While thinking that encourages the integration of communication, teamwork, creativity and hands-on experience into engineering curricula has also existed since the early 2000s (Tryggvason et al., 2001), this is often done with modules still arranged in engineering discipline streams and fundamental and complementary modules separated from engineering modules. Integration of engineering curricula also frequently only occurs in the higher years, sometimes through a capstone project with little integration in earlier years (Bailey et al., 2002).

There are however an increasing number of institutions around the world that are designing and implementing engineering curricula that shake up traditional approaches and consider integration at multiple levels. While many of these take place in smaller private institutions that are more agile (Mitchell et al., 2021), there are several examples of successful programmes at larger institutions where constraints and contextual complexities can play a more significant role in curriculum change. Some of these examples include programmes at The Massachusetts Institute of Technology (MIT) in the United States of America, Tecnologico de Monterrey in Mexico, and University College London (UCL) in the United Kingdom.

The MIT New Engineering Education Transformation (NEET) programme was started in 2016 and is a student-focused, project-centred curriculum that includes inter-disciplinary content and engagement and is designed to be relevant and to narrow the gap between theory and practice. (Crawley, 2018). Tecnologico de Monterrey uses challenge-based learning in their curriculum for Sustainable Development Engineering that is designed to be student-centred, encourage a real-world perspective, and concurrently develop technical and professional competencies while focusing on sustainable development (Caratozzolo, 2021).

UCL implemented an engineering curriculum that is based on a student-centred pedagogy that integrates discipline-specific content with professional skills using a backbone of problem-based learning experiences in 2011 (Mitchell et al., 2021). The curriculum is integrated in two key ways: firstly it brings together theoretical knowledge, practical skills and transferable skills or professional competencies including teamwork, communication and awareness of social impact and secondly, the curriculum adopts an integrated view of engineering that encourages multi-disciplinary approaches to creative problem solving and innovation. What makes this example particularly unique is that it spans eight departments across a faculty. These case studies are an inspiration to engineering educators and show that an integrated approach to engineering education is possible even with large student numbers and in resource-constrained environments.

B. Approaches to curriculum renewal

Many examples of curriculum renewal highlight specific considerations that can affect the overall success of the change process and the effectiveness of the redesigned curriculum. Most significantly, the importance of context is foregrounded by several authors (Mitchell et al., 2021; Case et al., 2015). Case et al. (2015) show that context should affect both the process followed to redesign the curriculum and the ultimate curriculum that is designed and warn against merely adopting approaches or models developed elsewhere (Case et al. 2015).

Many studies reflect on foundational elements that support the development and change of a new curriculum. Walkington et al. (2002) emphasise the importance of a broad perspective of the curriculum that requires decisions to be made in relation to content, teaching, assessment, teaching resources and facilities. They further advocate for a broad range of stakeholders to inform this decision-making. Many studies highlight the need for a focus on individual educators which includes professional development (Caratozzolo, 2021; Jamieson & Lohmann, 2012; Dai et al., 2022), facilitating collaboration and building educator agency (Jamieson and Lohmann, 2012) and the development of communities of practice (Wenger 2000). Furthermore, the design of competency assessment standards (Caratozzolo, 2021) and a holistic assessment mechanism for the programme (Bailey et al., 2002) and the change (Walkington et al., 2002) are needed.

Overwhelmingly, studies on curriculum renewal reflect on the importance of managing the change process. Mitchell et al. (2021) discuss how curriculum renewal involves systematic change at three levels: the individual level, the organisational level, and the level at which these two integrate and interact. Walkington et al. (2002) specifically adopted guiding principles that included that change is a journey, non-linear and uncertain.

Case studies also indicate that a curriculum cannot be changed in isolation from the organisational culture (Kolmos, Hadgraft & Holgaard, 2016). Mitchell et al. (2021) report that even ten years after implementation, the organisational culture is still changing in relation to the curriculum.

Underlying many of these elements is the role of people, individually and collectively, in driving success. And while strong leadership is required (Mitchell et al., 2021), sustainability needs engagement with a broad community where every person involved is a change agent (Walkington et al., 2002). The UCL case study discusses how their curriculum renewal was a response to staff who wanted to bring about change and introduce innovations but were either not senior enough or did not have enough leadership support to do this. They unpack how these people were a key part of the process as they became the change agents and formed the core team that led and drove the project (Mitchell et al., 2021).

III. Theoretical Framework and Methodology

This reflective case study views curriculum development as a collaborative process of social construction, adopting Dai et al.'s view that "curriculum development can be conceptualised as an evolving dialogue between stakeholders with different interests, beliefs, and commitments to education, where they collaboratively navigate, negotiate, and construct new meanings and practices" (Dai et al., 2022, page 25). Kotter's 8-step change model (Kotter, 1996) is adopted as a theoretical framework for analysing the project process. Kotter's model is believed to be particularly useful at encompassing the behavioural, cognitive and affective elements of change (Calgarie et al., 2015) and its use is well established in engineering education literature (Borrego & Henderson, 2014; Goncher et al., 2023). Kotter's model includes 8 steps that are designed to guide the change process:

1) **create a sense of urgency:** this includes identifying and involving key stakeholders and recognising opportunities and potential threats, and can be done through an analysis of the current state;

2) **build a guiding team:** this includes establishing a team of committed individuals who have sufficient power to initiate the change;

3) **get the vision right:** both the development of a vision and a strategy to achieve the vision are important;

4) **communicate the vision for buy-in:** this includes handling any concerns or issues as they arise;

5) **empower action:** this considers potential obstacles or barriers that exist or could arise to empower stakeholders involved in the change;

6) **create short-term wins:** this includes breaking the longer-term goal into short-term targets to maintain momentum;

7) **don't let up:** can be achieved by analysing success stories and identifying areas for further improvement; and

8) **make change stick:** often requires focus on the underlying structures and support to ensure that change is not superficial.

When using Kotter's model in higher education contexts, Calgary et al. (2015) noticed the importance of transparency in the process and adaptation of steps to contextual needs. Furthermore, they noticed that the model should not be seen as a linear journey and that cycling back through the steps may be inevitable. The model is therefore used to interpret the process taking this thinking into consideration.

This paper makes use of a descriptive case study methodological approach (Yin, 2014). The paper first describes the national curriculum project design and experiences. Thereafter, the findings and discussion present an interpretation and analysis of the project process in relation to the chosen theoretical framework, highlighting critical success factors and challenges experienced.

Multiple dimensions of the broader project are currently being analysed and documented for publication. This first paper focuses on the implementation of the first stages of the project. We are two active members of the core project team. The second author contributed from the proposal writing stage, and the first author joined the team in the early months of the project. This case study is written based on our perceptions of the project, supplemented by the notes and resources which have been generated over the life of the project. As we participate and roll out new stages, this paper has offered us the opportunity to look back critically and gaze forward as we think about how the interactions described in this paper can continue to have a growing impact on our universities and on others around the world. The perspectives in the paper are necessarily our own, and we cannot claim objectivity, although we have attempted to ground our discussion in the project resources.

IV. Description of the Case

A. Context of the Project

South Africa has 16 universities which are accredited by the Engineering Council of South Africa (ECSA) to educate students towards registration as Professional Engineers or Professional Engineering Technologists. Of these universities, six are research-intensive universities, graduating engineers, seven are Universities of Technology, graduating engineering technologists, and three are comprehensive universities, with programs for both engineers and engineering technologists.

ECSA is a member of the International Engineering Alliance (IEA) and a signatory of the Washington, Sydney and Dublin Accords, which allow professional recognition of graduates from other member countries. ECSA ensures compliance with these Accords by requiring universities to provide evidence that every graduate has demonstrated competence in eleven Graduate Attributes, spanning technical and professional competencies (ECSA, 2020a; ECSA, 2020b).

As already discussed, global engineering education trends

show an increased focus on developing professional competencies (Graham, 2018). A leading example of this is the University College London (UCL)'s Integrated Engineering Programme (IEP), which was introduced in 2014 and extends across 8 programmes in different departments. The IEP adopts a student-centred pedagogy that integrates existing discipline-specific content with the development of professional skills through a backbone of project-based learning experiences (Mitchell et al., 2021).

A Royal Academy of Engineering grant call in 2020 provided the catalyst for several South African universities and organisations to begin a collaboration with University College London (UCL). The aim of this collaboration was to develop a framework for implementing integrated and holistic curricula in South Africa, adapting the ideas to be feasible with financial as well as human resource constraints, to accommodate entering students who have received schooling with wide variations in quality, and to be applicable across our diverse institutional contexts. A core idea underpinning the project was that the re-imagined curricula should promote both staff and student well-being. The project objectives are:

1. To develop a framework within the South African context for implementing an integrated curriculum in engineering programs;
2. To identify areas where it is feasible to implement integrated curricular approaches as pilots within faculties, including strategies for approval; and
3. To develop a training program for staff to become expert facilitators of active integrated learning.

The initial project team included representatives from five South African universities, the South African Society of Engineering Education (SASEE), as well as UCL. Over the duration of the project, the team has expanded to include representatives from an additional four South African universities. The project has always aimed for inclusivity and open sharing of ideas between all institutions. Leveraging the cooperation and participation of multiple South African universities has enabled wide sharing of ideas, experiences, and best practices. Collaborating across institutions and academic departments has also allowed us to think broadly about the principles and practices of engineering education, separate from specialised disciplinary knowledge or current institutional structures.

B. Project Design and Implementation

From the project launch in early 2021, the project team met online every two weeks for two hours. These project team sessions helped us to articulate and evolve our understanding of the curricular needs of our universities, and to understand each other's contexts and perspectives. Through these meetings we co-created the project activities (listed in Table 1) which included a range of interactions, meetings, and workshops.

The first principle of the project was to intentionally structure activities to meaningfully include the voices of a wide range of stakeholders. Deans from all the South African universities were consulted at the beginning of the project with a presentation and open discussion at the Engineering Deans' Forum and agreed to support the project by promoting participation from their staff. Throughout the project, presentations at the Engineering Deans' Forum have kept the Deans updated on progress. The project represents voices from research-intensive universities, universities of technology and comprehensive universities, ensuring that the perspectives on curriculum are not biased towards one type of institution. It was vitally important to have the contribution of committed lecturers, who know what is happening in classrooms and who will be responsible for implementing and teaching new curricula. Lecturers were recruited via multiple channels and, through participation in workshops, made an important contribution to shaping the direction of the project. A strong relationship was formed between the project lead and ECSA representatives, with regular meetings to explore the regulatory constraints and opportunities. This has opened the possibility of influencing the framing of regulations. Industry partners were invited to participate in several workshops, sharing their expertise on graduate competencies and workplace training and mentoring. In some universities, students were also involved in a few workshops.

Before imagining the design of new curricula, it is important to understand the regulations, processes and constraints which govern change both at an institutional and a national level. One of the first major project activities was to conduct interviews with departmental undergraduate programme coordinators, to understand current practices around curriculum renewal. All programme coordinators were invited to be interviewed, and 28 programme coordinators from 14 universities participated in 16 online focus group discussions. The information from these interviews informed our understanding of the current state, explored the range of practices in different departments, and allowed the identification of potential barriers to curriculum renewal.

A series of online workshops focused on understanding what can be integrated into the curriculum, what an integrated curriculum could look like, and what it can offer to improve the education of engineering students in South Africa. A key concern of many participants was that it would not be feasible to implement a model which has been successful in UCL in our considerably more resource-constrained institutions. However, as mentioned in the literature (Graham, 2012), integrated curricula have been developed in response to the need to educate more professionally competent engineers for workplaces which are more multidisciplinary, multicultural, and sustainability-focused than ever before. This is equally true in South Africa, and these workshops have provided a lens for critically examining why we teach the way we do, and whether our practices remain relevant and appropriate. To counter the

https://doi.org/10.52202/073963-0011

TABLE I
PROJECT TIMELINE AND ACTIVITIES

Date	Activity
June 2021	Presentation to the national Deans' forum to launch the project and get buy-in for participation of staff
July to December 2021	Interviews with departmental undergraduate coordinators to understand current practices and constraints around curriculum renewal.
January to April 2022	Design and run an online workshop series to develop a shared framework for an integrated curriculum: 1. What is an integrated curriculum? And how does it bring life? 2. Integrating ECSA 3. Assessment strategies for an integrated curriculum 4. Sustainable integration: Collaborations to ensure sustainability of the curriculum change 5. Cultivating Life: how do we make our integrated curriculum ideas reality
July 2022	Release of showcase videos highlighting current cases of integration as examples of best practices in South African universities: 1. Integrating theory and practice 2. Integrating professional competencies 3. Integrating the workplace
July to November 2022	Online / hybrid workshop series of structured conversations to imagine change 1. Imagining the ideal classroom 2. Imagining an integrated first year 3. Unpacking graduate attributes: Teamwork 4. Joint workshop with industry / ECSA / academics

fear that integration would not be possible in our context, a series of videos were developed to showcase current examples of integrated practices in South Africa.

An important contribution of this project has been the development of a series of conversations which allowed us to imagine different ways of teaching and learning. These structured conversations were built around simple reflective questions which prompt conversation without triggering resistance. For example, thinking about their ideal classroom experience prompted lecturers to notice the gap between what they experience and what they hope for. Asking what experiences engineering students should have in their first year of study provided an opportunity for recognising the benefits of early integration. Similarly, identifying what our graduates will do in the workplace allowed a reality check on what is essential in the curriculum. For each Graduate Attribute, asking the question: "What distinguishes an expert from a novice?" (in, for example, teamwork) promotes an open and reflective conversation that can help lecturers to design activities and assessments which could promote the development of those attributes.

C. *Tracking the progress and success of the project*

The progress and success of the project was tracked through the achievement of activities, engagements and overall project objectives. More importantly, the success of the project has been monitored through stakeholder engagement and feedback. For each activity that is run, the number of people who participate is noted. This enables us to see if participants (and by extension, institutions) are returning for multiple sessions and if new people are joining. After each engagement a feedback survey is used to elicit participant feedback that can be used for project team reflections on the design of future engagements. The team reflected collectively after each engagement about what was effective, and what was learned.

V. ANALYSIS

In this section, we use the 8 steps of Kotter's change model as a theoretical lens to interpret and analyse our experience of this ongoing national conversation. Our analysis of key strengths and lessons learned provides insight into how to implement further initiatives in South Africa as well as other contexts.

1) **Creating a sense of urgency:**

The project began by identifying and engaging key stakeholders, starting with those in senior leadership positions to create an awareness of the project and to make space in institutions for individuals to participate. The UCL partner was drawn in to create excitement about the possibilities of an integrated curriculum. The focus group sessions held with programme coordinators at institutions enabled the mapping of the current state to understand the national landscape.

2) **Building a guiding team**:

The guiding team was created from enthusiastic educators who were passionate about the project and volunteered and committed their time to achieving the longer-term purpose. The team was small enough to ensure that strong connections between team members could be established while still providing room for different perspectives. The regular meetings of the team kept momentum and focus.

3) **Getting the vision right** and 4) **Communicating the vision for buy-in:**

The vision for the project was collectively developed. This process was not rushed, and no assumptions were made about individual or institutional needs and complexities. This also ensured that a broad range of stakeholders were involved in the process of developing the curriculum framework and that issues were collectively tackled as they arose.

5) **Empowering action:**

Many individuals who are active participants in the project are isolated but enthusiastic individuals within departments and institutions. Supporting these individuals specifically is therefore a project priority. The project actively seeks to build communities of practice and develop capacity to empower individuals and drive agency for change.

6) **Creating short-term wins:**

Although the project has had guiding timelines, the focus has not been on meeting deadlines but rather on ensuring that the shared understanding and development of individuals is a priority. Momentum has been maintained by regular and frequent engagements and communication and tracking of

https://doi.org/10.52202/073963-0011

participation and convergence of thinking. The project team values and appreciates that holistic change of engineering education curricula is a long-term endeavour and that breaking the project up into achievable parts can more easily maintain focus and energy.

7) **Not letting up:**

The project activities were designed to continuously reflect on gains made and opportunities for further improvement. The regular nature of activities kept momentum going. Through the showcase videos, specific success stories were shared to encourage and excite stakeholders about the possibilities for change. Regular and detailed updates were shared with Deans specifically drawing attention to the progress made and how this fits into the bigger picture.

8) **Making change stick:**

It is not easy to integrate engineering curricula. Anyone implementing curricular change in their department or university will require significant support. We felt that the project developed our capacity through building a robust community of practice. Educators were empowered to embark on sustainable long-term change by a combination of tested approaches to initiate conversations, along with connections with supportive colleagues at multiple institutions.

In our experience of this project we have identified two important levels at which change has happened. The first level was the national conversations working towards a shared understanding of an integrated approach to curriculum design for South Africa. The second level was the replication of elements of this process within departments at institutions, in which participants became facilitators. This experience is illustrated schematically in Figure 1. The process is shown as a labyrinth, to represent the complex and nonlinear nature of the work, in which it can feel like we are re-walking the same path, although each time it is different and we are changed in new ways. A labyrinth allows different people to engage with the process at different times and different speeds. The arrows suggest that the process will not reach a final conclusion, but that participants will become guides and will re-walk the journey with other people, as the process is re-enacted with wider groups and in different contexts. This highlights the importance of having a team, in which some facilitators may spend more time forging ahead with the early adopters, while others remain available to welcome and guide new participants through the process.

VI. Discussion

Our personal reflection on the process in relation to the chosen theoretical lens has enabled us to identify what we believe are critical success factors for this project.

Purposeful engagement with a broad range of stakeholders: Our purposeful engagement sought out a broad range of voices that could add to the richness of the curriculum renewal conversations. This intentional inclusion proactively built buy-in from key stakeholders at various levels and places within the eco-system. This contributed to the development of a robust shared understanding of the vision. The importance of

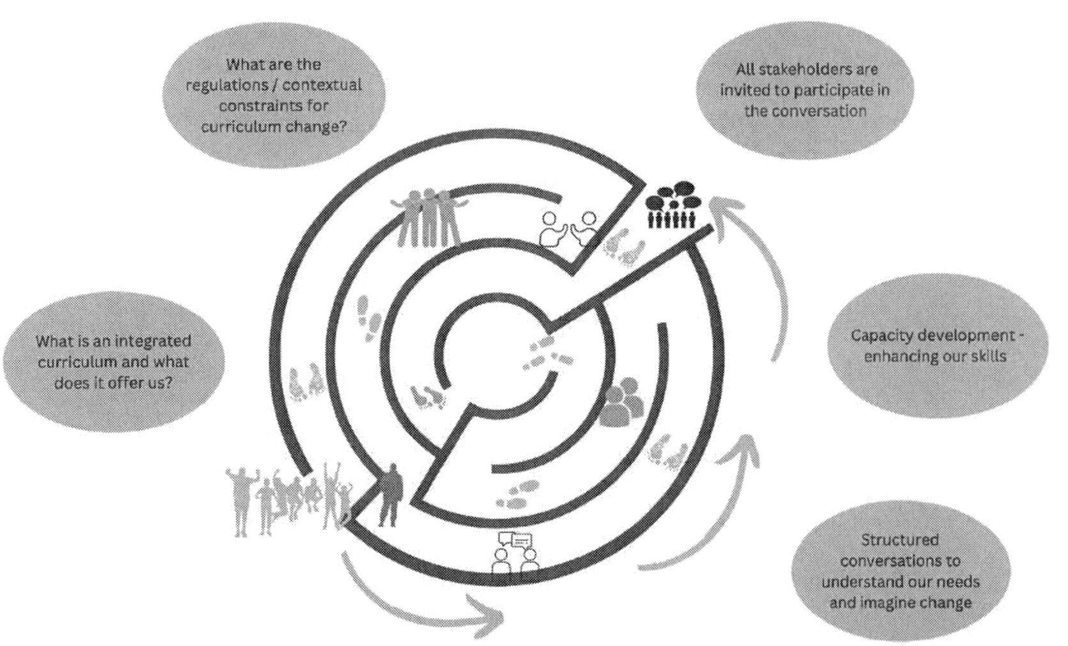

Fig. 1. A labyrinth as a metaphor for the nonlinear, interactive, and co-created curriculum change process.

https://doi.org/10.52202/073963-0011

stakeholder engagement emerged as a key critical success factor as predicted by literature (Walkington, 2002).

Facilitation of a national conversation: The national nature of this project meant that experiences, perceptions and needs from different contexts were incorporated into all activities. This provided different perspectives which enabled creativity and the generation of new and different ideas. It also facilitated engagements that were sensitive to context and did not make any assumptions about what would or would not work. These conversations privilege all voices including those that are often marginalised at higher education institutions, such as students and low-rank academic staff. Reducing power relationships, both between individuals and between institutions, created spaces which valued every voice. We believe that this sensitivity ensured that context was built into the experience from the outset (Case et al., 2015; Kolmos et al., 2016).

Critical and reflective conversations: The project centred around critical and reflective conversations that took place within the project team and were designed into all broader national stakeholder engagements. One of the guiding principles for the project was the importance of creating spaces for conversations. This enabled collective construction of ideas and understanding and the building of individual confidence and agency (Jamieson & Lohmann, 2012). This ensured that the project did not pre-determine or dictate what solutions should be, allowing solutions to be adapted to unique contexts and sharing the process to reach solutions to facilitate capacity building and sustainability. The conversations also enabled many participants to share their frustrations and constraints with the current state which built trust and openness.

Agile and evolving process: This success factor relates to the evolving and agile nature of the process. Acknowledging the non-linear process of change (Walkington, 2002), although the high-level nature and progression of activities was carefully designed by the project team from the outset, the detail remained fluid so that engagements could adapt to the collective needs of the stakeholders. The bi-weekly project team meetings allowed for a responsive approach, incorporating co-creation through the stakeholder feedback and observations of activity effectiveness. For each engagement, although the intended purpose was clear, no predetermined outcome was assumed. Space was created to allow stakeholder current needs to emerge.

Involving stakeholders in the process: The importance of taking each stakeholder on a journey of discovery and individual development emerged as one of the most valuable aspects of the project. We discovered that when new partners joined the journey, it was not possible to shortcut their experience of learning and sharing. This confirms the notion that curriculum renewal is as much about the process as it is about the outcome (Case et al., 2015). Throughout this process, not only is a curriculum designed but capacity is built, communities are formed, and confidence and agency are initiated. This also supports sustainability as the individuals who form part of the process become the change agents (Mitchell et al., 2021).

VII. Conclusions and Implications

This paper details an ongoing conversation which has progressed over three years, and which continues to evolve. The implementation has been flexible and responsive, leaving the outcomes of each engagement open. The project has solicited input from the ground up, embracing different opinions, developing capacity, creating a common understanding, and consciously privileging voices which are often marginalised by power structures. This co-creation of knowledge incorporated interviews with program coordinators to understand the drivers of and constraints on curriculum renewal, and online national workshops to capture and understand the experience of academics, industry, and the national accreditation body. Through this broad engagement, the project has created spaces and templates for conversations to overcome internal resistance to curriculum renewal.

This case study has implications for research and practice. The reflective analysis reveals the opportunities and complexities of developing a curriculum renewal strategy that is suitable across institutions at a national level while working towards a framework that is more universally applicable than those that are developed within a specific faculty or institution.

This project amplified the importance of context as discussed in the literature. Different contexts have different needs, opportunities, and constraints even within a single country. The different journeys of each participant and institution drove diverse narratives and experiences. National conversations illuminate these differences, challenging thinking and prompting further agility in the curriculum renewal process and outcome. Institutional activities will further need to incorporate the voice of the student to interrogate and address contextual nuance more deeply.

Curriculum review processes can often be focused on reaching an end product which can be implemented. This project emphasised that curriculum renewal is not only about the outcome – the new curriculum – but also about the individual and collective value experienced through the process of building trust and deepening shared understanding of what the curriculum is for. Implementing a curriculum or even a combination of curriculum ideas from elsewhere is not only potentially contextually inappropriate but misses the value of capacity development and the shifting perspectives and paradigms that can occur as individuals immerse themselves in thinking about how a curriculum enables and encourages different approaches to teaching and learning.

As the project unfolds, capacity development and communities of practice will become key areas to ensure sustainable engagement with curriculum renewal initiatives at institutional levels while retaining the value of national collaboration and support. This is critical as the initial excitement wears off and the hard work of change continues. Curriculum renewal, like any broad-reaching shift in an operational environment, must be seen as a long journey that cannot be rushed, and that requires intentional engagements and ongoing encouragement to be sustained.

ACKNOWLEDGMENT

This project was supported by a Royal Academy of Engineering Transforming Systems through Partnership grant, Grant Number RAE - TSP2021\100157.

We gratefully acknowledge the vibrant collaborations we have had with the project team: Lelanie Smith (project lead), John Mitchell (UCL partner), Reuben Govender, Zach Simpson, Eudes Tshitshonu, Chris von Klemperer, and Karin Wolff.

REFERENCES

Bailey, M., Floersheim, R. B., & Ressler, S. J. (2002). Course assessment plan: A tool for integrated curriculum management. *Journal of Engineering Education*, *91*(4), 425-434.

Borrego, M., & Henderson, C. (2014). Increasing the use of evidence-based teaching in STEM higher education: A comparison of eight change strategies. *Journal of Engineering Education*, 103(2), 220-252.

Calegari, M. F., Sibley, R. E., & Turner, M. E. (2015). A roadmap for using Kotter's organizational change model to build faculty engagement in accreditation. *Academy of Educational Leadership Journal*, *19*(3), 31.

Caratozzolo, P., & Membrillo-Hernández, J. (2021). Evaluation of challenge based learning experiences in engineering programs: The case of the Tecnologico de Monterrey, Mexico. In *Visions and Concepts for Education 4.0: Proceedings of the 9th International Conference on Interactive Collaborative and Blended Learning (ICBL2020)* (pp. 419-428). Springer International Publishing.

Case, J. M., Fraser, D. M., Kumar, A., & Itika, A. (2016). The significance of context for curriculum development in engineering education: a case study across three African countries. *European Journal of Engineering Education*, *41*(3), 279-292.

Crawley, E. F. (2018). Redesigning undergraduate engineering education at MIT–the New Engineering Education Transformation (NEET) initiative. In *2018 ASEE Annual Conference & Exposition*.

Dai, Y., Liu, A., Qin, J., Guo, Y., Jong, M. S. Y., Chai, C. S., & Lin, Z. (2023). Collaborative construction of artificial intelligence curriculum in primary schools. *Journal of Engineering Education*, *112*(1), 23-42.

ECSA (2020a). Engineering Council of South Africa. Qualification Standard for Bachelor of Engineering Technology (BEng Tech), NQF Level 7, Document E-02-PT, available at: https://www.ecsa.co.za/EcsaDocuments/sitepages/ecsa%20documents.aspx

ECSA (2020b). Engineering Council of South Africa. Qualification Standard for Bachelor of Science in Engineering (BSc(Eng))/ Bachelors of Engineering (BEng): NQF Level 8 E-02-PE, available at: https://www.ecsa.co.za/EcsaDocuments/sitepages/ecsa%20documents.aspx

Goldsmith, R., Reidsema, C., Campbell, D., Hadgraft, R., & Levy, D. (2011). Designing the future. *Australasian Journal of Engineering Education*, *17*(1), 1-9.

Goncher, A., Hingle, A., Johri, A., & Case, J. M. (2023). The role and use of theory in engineering education research. In *International handbook of engineering education research* (pp. 137-155). Routledge.

Graham, R. (2018). The global state of the art in engineering education. *Massachusetts Institute of Technology (MIT) Report, Massachusetts, USA*.

Kolmos, A., Hadgraft, R. G., & Holgaard, J. E. (2016). Response strategies for curriculum change in engineering. *International Journal of Technology and Design Education*, *26*, 391-411.

Kotter, J. P. (1996). Leading Change, Harvard Business School Press, Boston.

Mitchell, J., Nyamapfene, A., Roach, K., & Tilley, E. (2019). Philosophies and pedagogies that shape an integrated engineering programme. *Higher Education Pedagogies*, *4*(1), 180-196.

Mitchell, J. E., Nyamapfene, A., Roach, K., & Tilley, E. (2021). Faculty wide curriculum reform: the integrated engineering programme. *European Journal of Engineering Education*, *46*(1), 48-66.

Olds, B. M., & Miller, R. L. (2004). The effect of a first-year integrated engineering curriculum on graduation rates and student satisfaction: A longitudinal study. *Journal of Engineering Education*, *93*(1), 23-35.

Shaeiwitz, J. A., Whiting, W. B., Turton, R., & Bailie, R. C. (1994). The holistic curriculum. *Journal of Engineering Education*, *83*(4), 343-348.

Tryggvason, G., Thouless, M., Dutta, D., Ceccio, S. L., & Tilbury, D. M. (2001). The new mechanical engineering curriculum at the University of Michigan. *Journal of Engineering Education*, *90*(3), 437-444.

Walkington, J. (2002). A process for curriculum change in engineering education. *European Journal of Engineering Education*, *27*(2), 133-148.

Wenger, E. 2000. *Communities of Practice: Learning, Meaning and Identity*. Cambridge: Cambridge University Press.

Yin, R. K. (2014). *Case study research: Design and methods*. 5th edition. Thousand Oaks: Sage.

Review on empathy in engineering education: Conceptions, interventions, and challenges

Preethi Baligar[a], Radhika Amashi[a], Vijayalakshmi M[b], Juebei Chen[c], Aida Guerra[c], Xiangyun Du[c]
[a]Centre for Engineering Education Research, [b], School of Computer Science and Engineering, KLE Technological University, Hubballi, India
[c] UCPBL Center, Aalborg University, Aalborg, Denmark
Corresponding Author Email: preethi.b@kletech.ac.in

Abstract

Context

Despite an innate need for empathetic engineering, the dialogue on what constitutes empathy in engineering, its conceptualization, and the curriculum development to ingrain empathy as a transversal skill is still in its infancy.

Purpose or Goal

This study develops a broad conceptualization of empathy for engineering education and aggregate interventions to develop and measure it in engineering education. It specifically focuses on exploring the conceptions/frameworks/elements of empathy in engineering, curricular interventions for its development, and approaches and instruments used to measure it.

Methods

This study is a scoping review of Scopus and Web of Science databases for articles on empathy in engineering education published between 2013 and 2023.

Outcomes

This research illustrates that while the concept of empathy is not new, its inclusion, conceptualization, and measurements for engineering education are emerging. The conceptualization of empathy in engineering has taken various forms, with emerging frameworks highlighting its multi-dimensional nature. Eleven challenges come to light when examining the interventions for developing empathy at the module, course, and program levels.

Conclusion

Despite all that we know about the need for empathy in the engineering profession, its foray into engineering education is still in its infancy. The proposed scoping review has implications for both research and practice. It provides a synergistic view of literature on empathy development in engineering with critical findings for conceptions of empathy, recommendations for operationalizing the elements of empathy across progressive years of engineering study and disciplines leading to an array of curricular interventions with appropriate approaches for measuring it. Further research opens opportunities to develop instruments for measuring empathy from discipline-dependent and independent perspectives.

Keywords— challenges; engineering education; empathy; interventions

I. INTRODUCTION

Empathy has been a cornerstone on which social understanding, progress and sustenance have rested, as summarised by (Battarbee et al., 2014) in their insightful reflection, "We must intentionally seek opportunities to connect with people in meaningful ways ...". In this regard, the Organization for Economic Co-operation and Development research includes empathy as one of the critical elements for social progress (Marta, 2015). This need for meaningful engagement with people applies to general humanity and is also an innate need among professions that seek to build socio-technical artefacts for human consumption based on society's needs, experiences, and aspirations.

While the roots of empathy are often seen entrenched in professions like Law (Hoffman, 2011), Social work (Eriksson & Englander, 2017), Health care (Decety, 2020) and allied professions (Alzayed, 2019), its need in engineering are evidenced when we view engineering as not just restricted to "an application of Math and Science" but a vehicle for engaging with and eliciting perspectives of stakeholders , which are influenced by their culture, race, religion, location, economic and social status (Mohedas, Sienko, Daly & Cravens, 2020), among other dependent factors; and shape the objectives, constraints and functions (Dym et al., 2005) of the technical artefacts designed for societal consumption: eventually guiding the principles of "human-centred design"(Oehlberg et al., 2012), and "empathetic design" practices (Tang, 2018). Academies of engineering, accreditation bodies and professional ethics for engineers all mandate the development of solutions, processes, devices and components that cater to public health, welfare and safety (Shuman et al., 2005). This focus on "public" requires engineers to act empathetically (Battarbee et al., 2014), i.e., understand users' needs, develop user-centric solutions, and act altruistically throughout the design process.

Historically, mature disciplines like social welfare, medicine, nursing, and law have a well-articulated conceptualisation of empathy. However, a definitive description of what it means to be an empathetic engineer is evolving, as seen in an emerging

https://doi.org/10.52202/073963-0012

landscape of interpretive frameworks that identify elements of empathy (Hess et al., 2017; Sanz et al., 2023; Walther et al., 2020). While empathy is often seen as a trait, something an individual is born with (Kunyk & Olson, 2001), several studies have projected empathy as a teachable skill (Sanz et al., 2023; Walther et al., 2016), subsequently leading to several initiatives in engineering education, which have developed engaging curriculum, programs and courses designed to develop empathy (Bairaktarova, 2022; Yeaman et al., 2020). Despite the need and initiatives for fostering empathy in engineering education (Wilson & Mukhopadhyaya, 2022) have revealed that engineering students often face "cognitive dissonance" when their myopic perception of the engineering profession is juxtaposed with society, policy and education's thrust towards empathetic design. This is further compounded by the inhibitory relationship between students' degree of responsibility towards the public vis a vis their technical and analytical thought that engineering education develops (Levy, 2018; Hess et al., 2015; Shannon, Jones & Mina, 2019)

Despite all that we know about empathy, its need and initiatives to promote its development, scholars have revealed the need for a coherent framework (Guanes et al., 2022; Strobel et al., 2013; Surma-aho & Hölttä-Otto, 2022), a set of directed guidelines for its development (Alzayed et al., 2021) and a description of "direct, measurable impact of interventions on the empathy of engineers" (Wilson & Mukhopadhyaya, 2022, p.03). Considering the length and breadth of these needs, this study undertakes a scoping review of the literature to develop a broad conceptualisation of empathy for the engineering profession and aggregate interventions to develop and measure it in engineering education. It specifically focuses on exploring the conceptions/frameworks/elements of empathy in engineering, curricular interventions for its development, and approaches and instruments used to measure it through the following research questions:

1. How is empathy conceptualized in engineering?
2. What are the interventions for developing and measuring empathy in engineering education?
3. What kinds of challenges are experienced in designing learning contexts for developing empathy?

II. Methods

This study follows the scoping review methodological framework proposed by (Arksey & O'Malley, 2005), which outlines the following steps in conducting the scoping review: 1) identifying the research question, 2) identifying relevant studies, 3) study selection, 4) charting the data, 5) collating, summarising and reporting the results. With the identified research problem and questions, the following sections detail how the protocol was adopted.

A. Identifying relevant studies

Inclusion criteria

1. The article is written in English and is peer-reviewed
2. The article focuses on the engineering/technology education
3. The article focuses on the development of empathy among engineering students.
4. The article identifies elements of empathy/conceptualisation of empathy in engineering education and practice OR It applies an existing empathy framework for its course/interventions.

Exclusion criteria

1. The article focuses on non-engineering streams like arts, medicine, sports, nursing, early childhood and economy
2. Empathy in engineering practice towards customers
3. Empathy as a perspective while identifying problems in engineering

The search was conducted for the Scopus and Web of Science (WoS) databases by using the search strings: "Empath*" AND "Engineer*" OR "Technology*" published between the years 2013 and 2023. Both conference and journal articles were included. Of the 148 and 276 records from the Scopus and WoS databases, 95 and 132 were selected by screening the title, keywords and abstracts. After screening the full papers, 61 records were included for review. Of which 52 belonged to Scopus and 09 to WoS databases. The remaining records were excluded from the study due to the following reasons: duplicity; empathy between team members; teacher empathy; empathy for inclusion and diversity; entrepreneurial mindset learning spaces, promoting engagement of female students; historical and intercultural empathy, empathetic Technology, digital empathy, non-engineering education context, and detection of empathetic dialogue in digital conversations.

B. Tracking and Analysis

Based on the study's objective, initial themes were identified based on the research questions: conceptualization of empathy, interventions, engineering discipline, course name, research design, data collection instruments, participants, recommendations, challenges, and scope for the future. In order to validate the paper screening and data analysis, the first three authors independently screened the abstracts and the title for the first 10 records, followed by a discussion for convergence between them. The 4th author then screened 10% of the records. At the second level, the first three authors independently appraised the full papers based on the themes mentioned earlier. With an initial agreement of around 60%, the researchers converged through a discussion to reach a consensus of 89%. The 4th research then assessed the final coding of all records, leading to 61 records being included in this scoping review.

https://doi.org/10.52202/073963-0012

III. FINDINGS

A. *Demographics of selected records*

In order to explore publications on empathy development in engineering education, demographic information was classified into the following aspects: country, publication source, engineering discipline, research design and article type.

In terms of the 24 publication sources, the distribution of articles is as follows: Journal of Engineering Education(3), Engineering Studies(3), European Journal of Engineering Education(2), International Journal of Engineering Education(2), Journal of Mechanical Design(2), IEEE Transactions on Education(1), Journal of creative behaviour (1), Knowledge Management and E-Learning(1), Pedagogies(1), Social Sciences(1), Sustainability(1), TechTrends(1), Australasian Journal of Engineering Education(1), Design Journal(1), Design Studies(1), Education for Chemical Engineers(1) and Education Sciences(1). The 37 papers were presented at the following conferences: ASEE Annual Conference and Exposition, Conference Proceedings (21), IEEE International Professional Communication Conference(3), Proceedings - Frontiers in Education Conference, FIE (4), IEEE International Professional Communication Conference(2), among other conferences related to engineering education.

Among the 11 countries, 46 papers were from the USA, followed by the UK, Japan and Canada at two each. There was only 1 record from India.

For many records, the engineering discipline was not mentioned, as the intent of the records was to conceptualise empathy in engineering and not specific to a discipline. Further, the records revealed specific articles from biomedical engineering (2) and chemical engineering (2), each from Computer science and electrical engineering. Two studies also focused on developing empathy in the context of entrepreneurial mindset and design thinking. Six articles focused on developing empathy in the first year of engineering education.

Based on the scope of the records, three categories of papers were identified: conceptualisation of empathy (15), design of a course for the development of empathy and subsequent effect on it (35) and literature review (11). The records can be grouped into qualitative studies-36% (22), quantitative studies-23% (14), literature review-20% (12), mixed methods- 15% (9) and multi-methods-6.6% (4).

B. *Conceptualization of empathy in engineering*

From the earliest conceptualisations of empathy describing empathy as an ability to "think and feel oneself into the inner life of another person" (Kohut, 1959, p.82), several dimensions of empathy have been identified to reveal the complex and multi-dimensional facets of empathy in engineering education. While it appears that the cornerstone perspectives of empathy come from its cognitive, affective and behavioural aspects, which are widely cited in the literature, its interpretation for engineering education is still evolving. The literature review presents 08 conceptualisations of empathy, as indicated in Table I. A closer look at the conceptualisations reveals the means (communication, relationship, listening, resonance, connection, expansive empathy) vs ends (orientation, skills, professional way of being) dichotomy between the elements. Thus, this indicates that empathy is both a process and an outcome.

C. *Interventions for developing and measuring empathy in engineering education*

Table II showcases different courses and their interventions that promote the development of empathy among students. It captures the approach, empathy frameworks applied and measurement instruments.

1) *Course Levels and Duration*

The development of empathy as a skill is implemented at different levels of education, from course-level interventions to broader program-level experiences varying from a single university to multiple universities, which span weeks to longer, semester-length engagements. For example, the "Design for Sustainable Development" course extends over 20 weeks, allowing students to engage deeply with real-world case studies and iterative design processes (SCO09). In contrast, the "global innovation program" extends for a year, bridging academia and industry (SCO36).

2) *Intervention: Real-World Problems and Community Engagement*

The interventions involve students in real-world projects or challenges, such as waste management (SCO18), bridge design (SCO50), accessibility to older adults (SCO59), assistive technology for old age (SCO45, SCO06), drones for social good (SCO30) and sustainable development (SCO45, SCO42), which are designed in collaboration with nonprofit clients or industry. Notably, service learning and community engagement courses like "Technical Communication" (SCO41, SCO60), Design for Sustainable Development (SCO09), Introduction to Engineering Design (SCO58), Introduction to Engineering (SCO48) highlight the positive impact on local communities and society. Interventions in SCO58, SCO48, and SCO51 emphasise problem-based learning (Chen, Kolmos & Du (2021). Empathy often involves effective written communication and collaboration, evident in (SCO47, SCO49, SCO17). Interventions in (SCO26, SCO18) explicitly connect empathy with ethical behaviour. Empathy develops via reflective assignments, as in (SCO22, SCO16).

These interventions demonstrate several opportunities to develop empathy through hands-on experiences, problem-solving, collaboration, ethical considerations, and self-reflection. Ultimately, these efforts aim to produce engineers and professionals who excel in their technical abilities and understand and address the human aspects of their work, contributing to positive social impact.

https://doi.org/10.52202/073963-0012

TABLE I
CONCEPTUALISATIONS OF EMPATHY

ID	Conceptualisation of empathy
[SCO40]	Empathy in Design for Engineers exists at two levels: Self-Centered and user-Centered State 1 Self-Centred: Self-Awareness - understanding of the self and Other awareness - differentiating the self from others State 2: User-Centered: 1). Listening - the engineer is pulled into the client's world, exploring, absorbing, and experiencing without judgement. 2). Resonance - engineer shares emotional state with client related to client's needs 3) Connection - The engineer uses shared resonance to form a bond with the client, understanding emotions and needs 5) Detachment - The engineer steps back from interaction with the client ;switches modes from empathy to analytic to design for client's needs
[SCO47]	Conceptualizes empathy as a skill, a practice orientation, and a professional way of being
[SCO57]	Empathic engineering education framework that includes a set of four categories of learning theory and three categories of analytical skills
[SCO81]	Expansive empathy has been defined as the capacity to comprehend and provide inclusive design solutions that consider the intricate relationships between the engineering system and the requirements of various stakeholders, including those who are vulnerable, marginalized, and mainstreamed.
[SCO95]	A Skills, Professional, and Citizenship Model for Developing Empathy From Situational, Systemic, and Global Perspective/Framework for developing empathy in Computer science
[WEB010]	A framework to teach higher-order skills which includes empathy in the context of engineering and entrepreneurial skills: Knowledge, Persuasiveness, and Empathy (KPE)
[SCO02]	Empathy is a human trait, professional state, communication process, caring and a special relationship.
[WEB24]	Five core concepts that form the overall concept of empathy in design: understanding, action, research, orientation and mental processes

3) Pedagogy

The table II reveals a spectrum of pedagogical approaches, from active learning (SCO40, SCO50), training (SCO22), workshops (SCO59), collaborative teamwork (SCO47, SCO30), design thinking (SCO17) and project-based learning (SCO58, SCO36) to service learning (SCO41, SCO60, SCO49) and laboratory experiences (SCO18). For instance, the "Engineering Ethics" course employs a Sequential Interactivity, Reflection, and Application (SIRA) framework to facilitate ethical discussions through case studies (SCO26). Further, the "Exploring LegaCs" program (SCO85) incorporates narrative and storytelling as tools to develop empathy. Researchers and educators must consider these pedagogies to design interventions to foster empathy among students.

4) Data Collection Approaches

The most widely used data collection approaches for measuring empathy include reflections, interviews, classroom observations, and self-report surveys. This diversity highlights the complexity of measuring empathy, which requires quantitative and qualitative data following the mixed or multi-method approach for data analysis, especially within curricular constraints.

5) Frameworks for Empathy Measurement

Table II highlights several frameworks and scales used to measure empathy. These tools allow researchers to assess different dimensions of empathy, such as perspective-taking, fantasy, and empathetic distress among participants. The Davis's Interpersonal Reactivity Index (IRI) is a widely used psychological assessment tool designed to measure an individual's dispositional empathy (SCO58, SCO95, SCO85, SCO48, SCO36, SCO42). The IRI consists of four subscales or dimensions, each of which assesses a different aspect of empathy: Perspective-taking (PT), Fantasy (FS), Empathic Concern (EC) and Personal Distress (PD). Toronto Empathy Questionnaire is used to measure an individual's empathic tendencies (SCO41, SCO60). It is also measured as a state of empathy, i.e., State 1, Self-centred; this state involves self-awareness, where one understands oneself and develops awareness of others. State 2 is User-centred: this state shifts the focus to the user or client (SCO40). In (SCO18) the authors use the framework of Empathy, Care, and Ethics to understand and practise empathy. Goleman's Three Types of Empathy (SCO09) refers to three main types: cognitive empathy (understanding another person's perspective), emotional empathy (feeling the emotions of others), and empathic concern (caring about the well-being of others). (SCO22) measures empathy through Cognitive Empathy or Perspective-Taking using a prominent aspect of empathy studied in social psychology. The Engineering Professional Responsibility Assessment Tool (SCO30) assesses personal and social awareness, professional development, and professional connectedness. Zaki's Framework of Empathy (SCO85) considers aspects such as sharing, thinking about, and caring about others when understanding and measuring empathy. Baron-Cohen's Model outlines a skills, professional, and citizenship approach to developing empathy from situational, systemic, and global perspectives (SCO95). The Empathy Assessment Index (EAI) (SCO45) assesses empathy and comprises five affective and cognitive constructs: affective response, mentalising, self-other awareness, emotion regulation, and perspective-taking. Another views empathy as a multifaceted concept encompassing skill practice, orientation and a professional way of being (SCO47, SCO32, SCO49, SCO51). Measuring empathy is challenging due to its multifaceted and context-dependent nature. The non-linear progression of empathy development and its context specificity makes it difficult to assess straightforwardly.

https://doi.org/10.52202/073963-0012

TABLE II
INTERVENTIONS AND FRAMEWORK FOR EMPATHY MEASUREMENT

ID	Learning Context /Level/Approach	Data Collection	Framework for empathy measurement	Duration	Pedagogy	Intervention/Activity
SCO60 /41	Technical Comm [C][Mixed]	Surveys & Reflections	Toronto Empathy Questionnaire (Spreng et al., 2009)	6 weeks,	Service Learning	Students partnered with a nonprofit client
SCO18	Design Lab Course [C][Qual]	field notes, observations, interviews	Empathy, care and ethics. (Campbell, Yasuhara, & Wilson, 2015), (Campbell, Yasuhara, & Wilson, 2012)	six hours each week	Laboratory	waste management challenges in developing country
SCO22	Principles of Design[C][Quan]	Survey	NA	2-day training	Design-based	complex, open-ended problems affecting a fictitious world called "Planet Vayu,"
SCO58	introduction to engineering design [C][Mixed]	Survey, essay Reflection Interview	IRI scale. (Davis, 1983)	8 weeks	Project Based	Project to identify affected stakeholders and their needs
SCO47	Engineering and Society course [C][Qual]	Reflections	(Walther et al., 2017)	NA	Design-based	Team-based design challenges, readings/ discussions
SCO32	Engineering and Society course [C] [Qual]	Reflections	(Walther et al., 2017)	1 hour 15 mins per module	Service Learning	
SCO49	NA [C] [Qual]	semi-structured interviews	NA	10 hours	Community service project	Field visits, simulations of disability experiences
SCO51	Engineering and Society Course [C] [Qual]	Reflection prompt	(Walther et al., 2017)	NA	Project Based	Real-world problem
SCO42	Introductory Engineering Design [C][Quan]	NA	IRI scale. (Davis, 1983)	8 weeks	Active-learning reflections, role plays	project on United Nation's Sustainable Development Goal 3,
SCO09	Design for Sustainable Development [C] [Qual]	Observations, Reviews and Interviews	cognitive empathy, emotional empathy, and empathic concern. (Goleman & Senge, 2014)	20 weeks	Design-based	Design problems
SCO17	product design course [C][Multi]	Testimonies, Observation, Interviews or Focus Groups	NA	NA	group-based, open-ended design challenges	human-powered Washing machine to be used in developing countries.
SCO26	Engineering Ethics [C] [Qual]	semi-structured interviews	NA	NA	Case-based	Ex-Development and distribution of tissue-engineered heart valves
SCO16	Engineering and Society course [C] [Qual]	Skill Activity, Applied Activity, Reflective Homework Prompts	Walther et al., 2017)	1 hour 15 mins per module	Project Based	Real-world problem
SCO95	Programming and Programming Fundamentals [Multi]	Survey, Teacher observations and student perceptions	(Baron-Cohen, 2012)	NA	practical sessions, problem-solving, and lab practice.	Sudoku Programming
SCO45	Rehabilitation engineering course [Multi] [Mixed]	Survey, interviews	(Segal et al., 2017)	10 weeks	Design-based societal challenge	Projects on disabilities or recreation with local nonprofit organizations
SCO30	NA [Multi] [Mixed]	Survey and Open-ended design challenge	(Canney & Bielefeldt, 2016) (Davis, 1983)	One Sem	Design-based societal challenge	Teamwork, collaboration, conversations, workshops, group discussions

TABLE III(CONTD)
INTERVENTIONS AND FRAMEWORK FOR EMPATHY MEASUREMENT

ID	**Learning Context /Level/Approach**	**Data Collection**	**Framework for empathy measurement**	**Duration**	**Pedagogy**	**Intervention/Activity**
SCO36	Global Innovation Program [Multi] [Quan]	Survey	(Davis, 1983)	1 year		open-ended industry projects
SCO40	Electromagnetic Fields and Waves [Course] [Qual]	Reflections	Goleman's 5 key elements of EQ (Goleman, 2020)	One sem	Active Learning	NA
SCO85	Exploring Life Stories of Engineers [Multi] [Mixed]	Survey, Open-ended responses, Semi-structured Interviews	(Davis, 1983) (Zaki, 2019)	8 weeks	Story focused learning	Prompts and group discussions culminating in a "Story Slam."
SCO48	Introduction to Engineering [course] [Qual]	observation, interviews and peer feedback	(Davis, 1983)	One semester	Cooperative PBL	sustainable development-
SCO50	NA [Multi][Mixed]	daily observations, Interviews and surveys	NA	NA	think pair share, teamwork	electrical circuits, water access issues, bridge design, and biomedical engineering, boat-float challenge

TABLE III
CHALLENGES IN DESIGNING LEARNING CONTEXTS FOR DEVELOPING EMPATHY

Sl. No	**Category of challenge**	**Challenges of developing empathy in engineering education**
1.	Non-separation with technical content	Empathy is not a standalone concept [SCO02], pedagogical[SCO26], Integration of Empathy and Care into Engineering Education[SCO27], danger of disciplinary separation of content[SCO38]
2.	Development of real-world context	To be developed in the context of Societal and ethical responsibilities [SCO02], classroom interventions[SCO26], proximity, similarity, and familiarity biases, temporal, spatial[SCO32], micro-to-macro contexts of problems[SCO49]
3.	Empathy needs Multicultural perspectives.	Empathy in a multicultural setting and international setting [SCO06], Cultural Analysis[SCO18], international students[SCO26], contextualize empathy training within broader cultural norms[SCO31]
4.	Academic/curriculum-related	Optimal engagement via a standalone module, course or program [SCO09], against established learning practices and instructional norms[SCO16], Instructional Discomfort[SCO18], Time Constraints, Instructional Consistency[SCO30], Balancing Curriculum and Participation[SCO30], Short duration of courses[SCO45], Time in high-enrollment courses[SCO72], Single extra-curricular experience[SCO81], Balancing Curriculum and Participation[WEB103], Instructional Consistency[WEB103],
5.	Threats to measuring/assessment empathy/reliable measures	Prolonged nature of courses affected by real-time, non-academic events [SCO09]. It is challenging to measure the delta change in empathy [SCO09], Assessment of Empathy Progress[SCO30], data collection is primarily self-report questionnaires[SCO33], requires emotional understanding and reflecting[SCO36], Lack of sufficient Time to show a marked difference[SCO81], Assessment of Empathy Progress[WEB103],
6.	Multi-dimensionality of construct	Intricate and multifaceted relationships between empathy measures and innovative behaviours [SCO12], unintended Outcomes[SCO23]
7.	Discipline-independent measures of empathy	It is challenging to create a survey instrument that reliably assesses the influence on participation and gauges views of empathy across a range of engineering specialities [SCO14]
8.	Discipline-specific empathy interpretation and interpretation	Absence of training to nurture empathy in an engineering context[SCO21] Limited Focus on Empathy and Care Training in Engineering Education, Need for Engineering-Specific Training Methods[SCO27], the challenges of students' gradual transition to accepting a concept such as empathy as relevant to engineering[SCO38]
9.	Cognitive dissonance	Tension and reluctance among students caused by the difference between their expectations from engineering[SCO16], lack of alignment, technical overemphasis [SCO18], the disparity between the technical and empathic mindsets[SCO21], the need to overcome the perception of empathy as external to technical work[SCO21], attracting empathetic individuals to engineering, changing perceptions of engineering, developing empathy and care in engineers [SCO27], the role of epistemological differences for both students and instructors [SCO38], connection between empathy and engineering [WEB103]
10.	Dependent on prior conditioning	Inspired by prior experiences[SCO16], prior engagement with the community[SCO89]
11.	Psychological biases	Comfort Zone Bias, Narrow Empathic Horizon, Emotional Complexity [SCO18], Empathic Biases, Experience, Internalization, Emotional Regulation[SCO23]

https://doi.org/10.52202/073963-0012

D. *Challenges in designing learning contexts for developing empathy*

The challenges experienced in developing empathy in engineering education are listed in Table III. A nuanced and granular look at the challenges reveals that the challenges exist at different levels: Non-separation with technical content, development of real-world context, empathy needs multicultural perspectives, academic/curriculum-related, threats to measuring/assessment empathy/reliable measures, multi-dimensionality of construct, discipline-agnostic measures of empathy, cognitive dissonance, dependent on prior conditioning, psychological biases, discipline-specific empathy interpretation and interpretation. Among these challenges, those related to academics and curriculum form the most extensive set which focuses on how to design interventions and what level: module, course, or program level to ensure prolonged and effective engagement with the community. However, this prolonged engagement introduces threats on external validity of the measures and pose challenges for interpreting the success of the assessments. The second set of challenges comes from the epistemological assumptions of undergraduate engineering students, which portrays a conflict between what they believe about engineering, what they majorly learn in engineering and what they are expected to accomplish as professional engineers. This is further compounded by the analytical-heavy coursework, which eventually leads to a dip in their empathetic understanding.

IV. Discussion and Conclusion

This paper scrutinizes evidence regarding the conceptualization of empathy, interventions for developing it and challenges faced in engineering education based on a scoping review of literature from two databases (Web of Science and SCOPUS) published during 2013-2023. This research illustrates that while the concept of empathy is not new, its inclusions, conceptualization, and relevance for engineering education are emerging.. The conceptualization of empathy in engineering has taken various forms, with emerging frameworks highlighting its multi-dimensional nature. Empathy is viewed as a skill, a practice orientation, a professional way of being, (Walther et al., 2017) and a crucial component of engineering ethics. It involves cognitive, affective, and behavioural aspects (Goleman & Senge, 2014), each dimension playing a distinct role in engineering education and practice.

All the initiatives that seek to develop empathy portray that empathy cannot be developed by separating technical content from societal connections (Rivas & Husein, 2022). They seek to leverage technical knowledge for community engagement through service and community-based learning. These experiences expose students to the realities of the communities they serve, fostering empathy as they work to address real-world issues, which are commonly addressed using design-based pedagogy. Further, on examining the interventions, eleven challenges for developing empathy are seen at different levels: module, course, and program levels; faculty and student related; and discipline specific and discipline independent. These challenges present opportunities for further design of interventions.

While literature supports immersion in real-world problems for developing empathy, the openness and ill-structuredness of the problem pose a challenge in terms of time, effort and technical feasibility. This is interpreted by (Guanes et al., 2022) as the micro-meso-macro focus of problems that must be carefully arbitrated by the instructors much ahead of Time. In addition, instructors also need to develop learning contexts that are consonant with the students in terms of familiarity, similarity, proximity of the situation (Brewer et al., 2017.), and stakeholders, which, if avoided, may contribute to further cognitive dissonance.

Further, while this study sees empathy as a unitary entity, scholars often view it in conjunction with ethics, entrepreneurial thinking, design-based research, creative thinking, innovation, and care. Thus, this opens up avenues for instructional designers to design a holistic ramp of courses that focus on these transversal skills intertwined with technical competency.

Although engineering education can strive to develop effective and efficient interventions to develop empathy, empathy often depends on personal dispositions and characteristics, such as civic-mindedness (Lin et al., 2021), which refers to valuing community engagement and empathic interpersonal communication. Thus, this opens opportunities to explore how civic-mindedness can be developed even before students join engineering education. Specific studies also highlight the gender-dependent nature of empathy, with female students demonstrating higher levels of empathy than male counterparts (Christov-Moore et al., 2014).

Measuring empathy is challenging (Hall & Schwartz, 2022) due to its multifaceted and context-dependent nature. The non-linear progression of empathy development, context specificity, and its generic nature make it difficult to assess straightforwardly. The instruments and scales commonly used to measure empathy are dated and do not reflect engineering-specific and discipline-dependent nuances of engineering, which opens up thrust for further investigation.

Lastly, this study has several limitations: search is restricted to just the last decade, is confined to the in-person engagement and does not delve into the empathic responses in technology-enabled environments and does not focus on the assessment and evaluation structure of the courses.

References

Afroogh, S., Esmalian, A., Donaldson, J. P., & Mostafavi, A. (2021). Empathic design in engineering education and practice: an approach for achieving inclusive and effective community resilience. Sustainability, 13(7), 4060. **[SCO57]**

Alzayed, M. A. (2019, October). An exploration of the role of student empathy in engineering design education. In 2019 IEEE Frontiers in Education Conference (FIE) (pp. 1-2). IEEE. **[SCO42]**

Alzayed, M. A., McComb, C., Menold, J., Huff, J., & Miller, S. R. (2021). Are you feeling me? An exploration of empathy development in engineering design education. Journal of Mechanical Design, 143(11), 112301. **[SCO58]**

Arksey, H., & O'Malley, L. (2005). Scoping studies: Towards a methodological framework. International Journal of Social Research Methodology, 8(1), 19–32. https://doi.org/10.1080/1364557032000119616

B. Bush, S., Edelen, D., Roberts, T., Maiorca, C., Ivy, J. T., Cook, K. L., ... & Cox Jr, R. (2022). Humanistic STE (A) M instruction through empathy: leveraging design thinking to improve society. Pedagogies: An International Journal, 1-20. **[WEB003]**

Bairaktarova, D. (2022). Caring for the future: Empathy in engineering education to empower learning. Journal of Engineering Education, 111(3), 502-507. **[SCO79]**

Battarbee, K., Suri, J. F., & Howard, S. G. (2014). Empathy on the edge: Scaling and sustaining a human-centered approach in the evolving practice of design. IDEO. Http://Www. Ideo. Com/Images/Uploads/News/Pdfs/Empathy_on_the_Edge. Pdf.

Besser, D., Brown, K., Nowariak, A. H., Brass, T., Leininger, R. A., & Thomas, A. (2020, June). Empathy, engineering, and girls. In 2020 ASEE Virtual Annual Conference Content Access. **[SCO50]**

Beyond Surface Knowledge: An Exploration of How Empathic Design Techniques Enhances Engineers Understanding of Users' Needs. (2016). International Journal of Engineering Education. **[SCO17]**

Brewer, M. A., Sochacka, N. W., Walther, J., & Miller, S. E. (2017, July). How do students meaningfully interpret the role of empathy in engineering? A social phenomenological study. In Proceedings of the 2017 Research in Engineering Education Symposium, Bogota, Colombia (pp. 6-8). **[SCO32]**

Burns, H. D., & Lesseig, K. (2017, June). Infusing empathy into engineering design: Supporting under-represented student interest and sense of belongingness. In 2017 ASEE Annual Conference & Exposition. **[SCO31]**

Busu, T. N. Z. T. M., Mohd-Yusof, K., & Rahman, N. F. A. (2020, December). Empathy enhancement among engineering students through cooperative problem-based learning. In 2020 IEEE International Conference on Teaching, Assessment, and Learning for Engineering (TALE) (pp. 889-894). IEEE. **[SCO48]**

Campbell, R. C., Yasuhara, K., & Wilson, D. (2012, October). Care ethics in engineering education: Undergraduate student perceptions of responsibility. In 2012 Frontiers in Education Conference Proceedings (pp. 1-6). IEEE.

Canney, N. E., & Bielefeldt, A. R. (2016). Validity and reliability evidence of the engineering professional responsibility assessment tool. Journal of engineering education, 105(3), 452-477.

Chen, J., Kolmos, A., & Du, X. (2021). Forms of implementation and challenges of PBL in engineering education: a review of literature. European Journal of Engineering Education, 46(1), 90-115.

Christov-Moore, L., Simpson, E. A., Coudé, G., Grigaityte, K., Iacoboni, M., & Ferrari, P. F. (2014). Empathy: Gender effects in brain and behavior. Neuroscience & Biobehavioral Reviews, 46, 604–627.

Cooper, L. A., Johnston, A., Hubbard, E. H., & Self, B. P. (2020, June). Development of Empathy in a Rehabilitation Engineering Course. In 2020 ASEE Virtual Annual Conference Content Access. **[SCO45]**

Davis, M. H. (1983). Measuring individual differences in empathy: Evidence for a multi-dimensional approach. Journal of personality and social psychology, 44(1), 113.

De Lira, C. (2017, August). Improving the Learning Experiences of First-Year Computer Science Students with Empathetic IDEs. In Proceedings of the 2017 ACM Conference on International Computing Education Research (pp. 293-294). **[WEB84]**

Decety, J. (2020). Empathy in medicine: What it is, and how much we really need it. The American Journal of Medicine, 133(5), 561–566.

Dostál, D., Plháková, A., & Záškodná, T. (2017). Domain-specific creativity in relation to the level of empathy and systemizing. The Journal of Creative Behavior, 51(3), 225-239. **[WEB82]**

Dym, C. L., Agogino, A. M., Eris, O., Frey, D. D., & Leifer, L. J. (2005). Engineering design thinking, teaching, and learning. Journal of Engineering Education, 94(1), 103–120.

Eriksson, K., & Englander, M. (2017). Empathy in social work. Journal of Social Work Education, 53(4), 607–621.

Fila, N. D., & Hess, J. L. (2016, June). In their shoes: Student perspectives on the connection between empathy and engineering. In 2016 ASEE Annual Conference & Exposition. **[SCO21]**

Goleman, D. (2020). Emotional intelligence. Bloomsbury Publishing. (Goleman, 2020)

Goleman, D., & Senge, P. M. (2014). The triple focus: A new approach to education.(Goleman & Senge, 2014)

Gray, C., De Cresce El Debs, L., Exter, M., & Krause, T. (2016). Instructional Strategies for Incorporating Empathy in Transdisciplinary Technology Education. 2016 ASEE Annual Conference & Exposition Proceedings, 25746. https://doi.org/10.18260/p.25746 **[SCO18]**

Guanes, G., Wang, L., Delaine, D. A., & Dringenberg, E. (2022). Empathic approaches in engineering capstone design projects: student beliefs and reported behaviour. European Journal of Engineering Education, 47(3), 429-445. **[SCO72]**

Hall, J. A., & Schwartz, R. (2022). Empathy, an important but problematic concept.The Journal of Social Psychology,162(1), 1–6.

Hauke, E., Cresswell-Maynard, K., & Craddock, D. (2015). An interdisciplinary approach to embedding the global dimension into engineering design: empathy, engagement and creativity. In DS 82: Proceedings of the 17th International Conference on Engineering and Product Design Education (E&PDE15), Great Expectations: Design Teaching, Research & Enterprise, Loughborough, UK, 03-04.09. 2015 (pp. 087-092). **[SCO09]**

Hess, J. L., Beever, J., Strobel, J., & Brightman, A. O. (2017). Empathic perspective-taking and ethical decision-making in engineering ethics education. Philosophy and engineering: Exploring boundaries, expanding connections, 163-179. **[SCO33]**

Hess, J. L., Carrillo-Fernandez, A., Fila, N. D., & Schimpf, C. T. (2021, July). Exploring how empathy manifests with/for teammates in a junior-level biomedical engineering course. In 2021 ASEE Virtual Annual Conference Content Access. **[SCO52]**

Hess, J. L., Sanders, E., & Fila, N. D. (2022, January). Measuring and promoting empathic formation in a multidisciplinary engineering design course. In ASEE annual conference exposition proceedings. **[SCO80]**

Hess, J. L., Sprowl, J. E., Pan, R., Dyehouse, M., Morris, C. A. W., & Strobel, J. (2012, June). Empathy and caring as conceptualized inside and outside of engineering: Extensive literature review and faculty focus group analyses. In 2012 ASEE Annual Conference & Exposition (pp. 25-520). **[WEB111]**

Hess, J. L., Strobel, J., & Brightman, A. O. (2017). The development of empathic perspective-taking in an engineering ethics course. Journal of Engineering Education, 106(4), 534-563. **[SCO26]**

Hess, J. L., Strobel, J., & Pan, R. (2016). Voices from the workplace: Practitioners' perspectives on the role of empathy and care within engineering. Engineering Studies, 8(3), 212-242. **[SCO24]**

Hess, J. L., Strobel, J., Pan, R. (Celia), & Wachter Morris, C. A. (2017). Insights from industry: A quantitative analysis of engineers' perceptions of empathy and care within their practice. European Journal of Engineering Education, 42(6), 1128–1153. https://doi.org/10.1080/03043797.2016.1267717

Hess, J. L., Strobel, J., Pan, R., & Morris, C. A. W. (2014, June). Practicing engineers' perceptions of empathy and care: derived exploratory factor structure from a 37-item survey. In 2014 ASEE Annual Conference & Exposition (pp. 24-991). **[WEB103]**

Hess, J. L., Strobel, J., Pan, R., & Wachter Morris, C. A. (2017). Insights from industry: a quantitative analysis of engineers' perceptions of empathy and care within their practice. European Journal of Engineering Education, 42(6), 1128-1153. **[SCO27]**

Hess, J., Fila, N., Purzer, S., & Strobel, J. (2015). Exploring the Relationship between Empathy and Innovation among Engineering Students. 2015 ASEE Annual Conference and Exposition Proceedings, 26.740.1-26.740.12. https://doi.org/10.18260/p.24077 **[SCO12]**

Hoffman, M. L. (2011). Empathy, justice, and the law. Empathy: Philosophical and Psychological Perspectives, 230–254.

Hoople, G. D., & Choi-Fitzpatrick, A. (2017, June). Engineering empathy: a multidisciplinary approach combining engineering, peace studies, and drones. In 2017 ASEE Annual Conference & Exposition. **[SCO30]**

Jacobs, E., De Jongh Curry, A., Deaton, R., Astorne-Figari, C., & Strohmer, D. (2016). Empathy and Gender Inequity in Engineering Disciplines. 2016 ASEE Annual Conference & Exposition Proceedings, 26936. https://doi.org/10.18260/p.26936 **[SCO14]**

Jaycox, H., Hess, J. L., Zoltowski, C. B., & Brightman, A. O. (2014). Developing novel practices of somatic learning to enhance empathie perspective-taking for ethical reasoning and engineering design. 2014 IEEE Frontiers in Education Conference (FIE) Proceedings, 1–5. https://doi.org/10.1109/FIE.2014.7044278 **[SCO06]**

Johnson, N., Wang, L., Delaine, D., & Walther, J. (2022, August). Evidence-based Opportunities for the Development of Empathy in Engineering through Community-based Learning. In 2022 ASEE Annual Conference & Exposition. **[SCO73]**

Kohut, H. (1959). Introspection, empathy, and psychoanalysis an examination of the relationship between mode of observation and theory. Journal of the American Psychoanalytic Association, 7(3), 459–483.

Kunyk, D., & Olson, J. K. (2001). Clarification of conceptualizations of empathy. Journal of Advanced Nursing, 35(3), 317–325.

Levy, M. (2018, March). Educating for Empathy in Software Engineering Course. In REFSQ Workshops (pp. 1-9).

Levy, M. (2018, March). Educating for Empathy in Software Engineering Course. In REFSQ Workshops (pp. 1-9). [SCO38]

Lin, A., Hess, J. L., & Fernández, A. P. C. (2021, October). The Relationship Between Empathy and Civic-Mindedness in First-Year Engineering Students. In 2021 IEEE Frontiers in Education Conference (FIE) (pp. 1-9). IEEE. **[SCO63]**

Lingappa, A. K., Kamath, A., & Mathew, A. O. (2022). Engineers and Social Responsibility: Influence of Social Work Experience, Hope and Empathic Concern on Social Entrepreneurship Intentions among Graduate Students. Social Sciences, 11(10), 430. **[SCO86]**

Litvinov, A., Gardner, A., Pradhan, S., & Childers, J. (2022, October). The role of empathic experiences of entrepreneurial engineers within accelerators: a phenomenological study. In 2022 IEEE Frontiers in Education Conference (FIE) (pp. 1-7). IEEE. **[SCO76]**

Lunn, S., & Bell-Huff, C. (2022, August). What Story Do You Want to Tell? Developing Empathy in Engineering Students through an Extra-Curricular Narrative Sharing Experience. In 2022 ASEE Annual Conference & Exposition. **[SCO85]**

Marinelli, M., Male, S., Chapman, E., & Strobel, J. (2022). Engineers' perceptions of the importance of empathy and care: initial insights from engineers practicing in Australia. In Towards a new future in engineering education, new scenarios that european alliances of tech universities open up (pp. 2066-2071). Universitat Politècnica de Catalunya. **[SCO75]**

Marta, R. (2015). "Skills for Social Progress: The Power of Social and Emotional Skills.

Mashhadi, A. R., & Svihla, V. (2022, August). Expansive Empathy: Defining and Measuring a New Construct in Engineering Design. In 2022 ASEE Annual Conference & Exposition. **[SCO81]**

Mohedas, I., Sienko, K. H., Daly, S. R., & Cravens, G. L. (2020). Students' perceptions of the value of stakeholder engagement during engineering design. Journal of Engineering Education, 109(4), 760–779.

Neubauer, D., Paepcke-Hjeltness, V., Evans, P., Barnhart, B., & Finseth, T. (2017). Experiencing technology enabled empathy mapping. The Design Journal, 20(sup1), S4683-S4689. **[SCO25]**

Oehlberg, L., Leighton, I., Agogino, A., & Hartmann, B. (2012). Teaching human-centered design innovation across engineering, humanities and social sciences. International Journal of Engineering Education, 28(2), 484.

Pappa, C. I., Reinhold, S., Pittich, D., Vogel-Heuser, B., & Strobel, J. (2020, April). Engineers' job-related perceptions of empathy in Germany. In 2020 IEEE Global Engineering Education Conference (EDUCON) (pp. 741-745). IEEE. **[SCO43]**

Patterson, L. (2019, July). Engineering students' empathy development through service learning: Quantitative results from a technical communication course. In 2019 IEEE International Professional Communication Conference (ProComm) (pp. 194-200). IEEE. **[SCO41]**

Patterson, L. (2021, October). Engineering Students Empathy Development through Service Learning: Examining Individual Student Experiences in a Technical Communication Course. In 2021 IEEE International Professional Communication Conference (ProComm) (pp. 93-100). IEEE. **[SCO60]**

Rivas, D. F., & Husein, S. (2022). Empathy, persuasiveness and knowledge promote innovative engineering and entrepreneurial skills. Education for Chemical Engineers,40, 45-55. **[WEB010]**

Sanz, C., Coma-Roselló, T., Aguelo, A., Álvarez, P., & Baldassarri, S. (2023). Model and Methodology for Developing Empathy: An Experience in Computer Science Engineering. IEEE Transactions on Education. **[SCO95]**

Segal, E. A., Gerdes, K. E., Lietz, C. A., Wagaman, M. A., & Geiger, J. M. (2017). Assessing empathy. Columbia University Press.

Shannon, R. A., & Jones, S. K., & Mina, M. (2019, June), Designing a Multi-Cycle Approach to Empathetic Electrical Engineering

Courses Paper presented at 2019 ASEE Annual Conference & Exposition , Tampa, Florida. 10.18260/1-2--32612 [SCO40]

Shuman, L. J., Besterfield-Sacre, M., & McGourty, J. (2005). The ABET "professional skills"—Can they be taught? Can they be assessed? Journal of Engineering Education, 94(1), 41–55.

Sochacka, N. W., Youngblood, K. M., Walther, J., & Miller, S. E. (2020). A qualitative study of how mental models impact engineering students' engagement with empathic communication exercises. Australasian Journal of Engineering Education, 25(2), 121-132. **[SCO51]**

Spreng*, R. N., McKinnon*, M. C., Mar, R. A., & Levine, B. (2009). The Toronto Empathy Questionnaire: Scale development and initial validation of a factor-analytic solution to multiple empathy measures. Journal of personality assessment, 91(1), 62-71.:Toronto Empathy Questionnaire

Strobel, J., Hess, J., Pan, R., & Wachter Morris, C. A. (2013). Empathy and care within engineering: Qualitative perspectives from engineering faculty and practicing engineers. Engineering Studies, 5(2), 137–159. https://doi.org/10.1080/19378629.2013.814136

Surma-aho, A. O., Björklund, T. A., & Holtta-Otto, K. (2018, June). Assessing the development of empathy and innovation attitudes in a project-based engineering design course. In 2018 ASEE annual conference & exposition. [SCO36]

Surma-Aho, A., & Hölttä-Otto, K. (2022). Conceptualization and operationalization of empathy in design research. Design Studies, 78, 101075. **[WEB24]**

Tang, X. (2018, June). From'Empathic Design'to'Empathic Engineering': Toward a Genealogy of Empathy in Engineering Education. In 2018 ASEE Annual Conference & Exposition. [SCO37]

Wallisch, A., Briede-Westermeyer, J. C., & Luzardo-Briceno, M. (2021). Fostering user-empathy skills of engineering students by collaborative teaching. International Journal of Engineering Education, 37(1), 223-243. **[SCO59]**

Walther, J., Brewer, M. A., Sochacka, N. W., & Miller, S. E. (2020). Empathy and engineering formation. Journal of Engineering Education, 109(1), 11–33. https://doi.org/10.1002/jee.20301 **[SCO47]**

Walther, J., Miller, S. E., & Sochacka, N. W. (2017). A model of empathy in engineering as a core skill, practice orientation, and professional way of being. Journal of Engineering Education, 106(1), 123-148. **[SCO34]**

Walther, J., Miller, S., Sochacka, N., & Brewer, M. (2016). Fostering Empathy in an Undergraduate Mechanical Engineering Course. 2016 ASEE Annual Conference & Exposition Proceedings, 26944. https://doi.org/10.18260/p.26944 **[SCO16]**

Wilson, E., & Mukhopadhyaya, P. (2022). Role of empathy in engineering education and practice in North America. Education Sciences, 12(6), 420. **[SCO89]**

Wong, K., Norris, R. L., Siddique, Z., Altan, M. C., Baldwin, J., & Merchan-Merchan, W. (2016, August). Cognitive empathy in design course for a more inclusive mechanical engineering. In International Design Engineering Technical Conferences and Computers and Information in Engineering Conference (Vol.50138, p. V003T04A005). American Society of Mechanical Engineers. **[SCO22]**

Yeaman, A., Bairaktarova, D., & Reid, K. (2020). A Qualitative Study of Empathy in the Experiences of Students in a First-Year Engineering Service Learning Course. 2020 ASEE Virtual Annual Conference Content Access Proceedings, 34040. https://doi.org/10.18260/1-2--34040 **[SCO49]**

Zaki, J. (2019). The war for kindness: Building empathy in a fractured world. Crown

Copyright Statement

Investigating How Students Perceive Assessment in the Context of Engineering Education

Ousman, Badjie[a]; Yogesh, Velankar[b], and Md. Abdullah, Al Mamun[c].
School for Sustainable Futures, Amrita Vishwa Vidyapeetham, Amritapuri, Kollam, Kerala, India [ab],
Department of Technical and Vocational Education (TVE), Islamic University of Technology (IUT), Gazipur, Dhaka, Bangladesh [c]
Corresponding Author Email: yogeshvelankar@am.amrita.edu

Abstract

Context

Assessment in engineering education is crucial for evaluating student achievement and teaching effectiveness. Engineering universities should align programs with objectives, curriculum, instruction, and assessment strategies.

Purpose or Goal

Recognizing that students are fundamental in the teaching and learning process, their feedback on content delivery and assessment methods is invaluable for refining and improving the overall instructional process. Consequently, this study investigates students' perceptions of assessment practices in engineering education. This study also examines the significant disparity in the perception of undergraduate engineering students at different academic levels.

Methods

Through a quantitative survey, data was collected from 557 undergraduate engineering students at various academic year levels from six (6) programs across four engineering universities in Bangladesh. Descriptive statistics have been used to get insights into the demographics of the students. Cronbach's alpha was used to assess the reliability of the data. The ANOVA analysis investigates the significant differences in students' perception of assessment practices at different academic levels.

Outcomes

The results show high reliability of the survey instrument with a Cronbach's alpha value of 0.801. The ANOVA findings reveal that first-year students' perceptions of assessment practices differ significantly from third-year and fourth-year students.

Conclusion

We observed that there is a significant variance in the perception of first-year and third-year students on the alignment of assessment with planned learning. Similarly, a significant difference exists between the perception of first-year and fourth-year on the authenticity and transparency in assessment. However, the magnitude of the difference is small.

***Keywords*—** assessment; engineering education; engineering universities; student perception;

I. INTRODUCTION

EMPLOYING suitable assessment approaches aids in cultivating quality graduates for a country. In contrast, discrepancies in assessment methods across engineering institutions can lead to disparities in the caliber of graduates (Stehle & Peters-Burton, 2019). The significance of valid and consistent assessment practices is paramount for the success of formal education (Taber, 2018). To fulfill student anticipations and achieve labour market demands, engineering universities should align their program goals, curriculum, pedagogy, and evaluation strategies to the industry needs (Ali, 2018).

Engineers play a crucial role in shaping our technological, infrastructural, and economic landscapes (Pleasants, 2023). They drive technological advancements, infrastructure development, address global challenges and foster innovation (Raman et al., 2015; Stehle & Peters-Burton, 2019). They are essential in designing and maintaining systems and structures that underpin our daily lives, ensuring that society benefits from technological progress (Jónsson, 2023). Competent engineers possess a unique blend of technical skills and creative problem-solving, enabling innovation and entrepreneurship (Huang-Saad et al., 2018; Sneider, 2016). They contribute to economic growth by creating efficient products, reducing costs, and improving quality of life (McGowan & Bell, 2020). Their role is essential in creating a sustainable, prosperous, and innovative society, as the world becomes interconnected and forms a complex network within (Pleasants, 2023).

Therefore, quality education and training to produce competent engineers are essential for tackling modern challenges (Ali, 2018; McGowan & Bell, 2020; Pleasants, 2023). Providing the graduates with skills, knowledge, and mindset to navigate complex technological, societal, and environmental landscapes should be of importance to achieving a sustainable future. This includes specialization, problem-solving abilities, interdisciplinary collaboration, ethical practices, adaptability, global perspectives, entrepreneurial

https://doi.org/10.52202/073963-0013

innovation, quality assurance, safety, and job market competitiveness (Chen et al., 2021; Nedungadi et al., 2018; Stitt-Bergh et al., 2018). Continuous learning and investment in comprehensive training cultivate a skilled workforce capable of driving progress, solving complex challenges, and contributing positively to the world (Frey et al., 2017; 'Raman et al., 2013; Sneider, 2016). However, despite their enormous efforts, engineering universities continue to encounter difficulties in producing qualified engineers who can fulfill global expectations (Shahid et al., 2022). Effective curricula delivery and assessment methodologies and practices to ensure desired outcomes are one of the challenges faced by these universities (Denton, 1998; Raman et al., 2021; Shahid et al., 2022; Sneider, 2016). Therefore, this paper focuses on exploring the perception of undergraduate students on assessment practices in four (4) engineering universities in Bangladesh. The perception of students was explored in relation to the assessment's Alignment with Planned Learning (APL), Authenticity of the Assessment (AA), Student Consultation on Assessment (SCA), Transparency in Assessment (TA) and Diversity of Assessment (DA)(Koul et al., 2006; Mussawy, 2009; Trochim, 2007; Waldrip et al., 2008) as depicted in Table I.

Assessment in engineering education plays a vital role in shaping the quality, effectiveness, and relevance of engineering programs (Ali, 2018). It involves the systematic evaluation of students' knowledge, skills, and competencies to determine their understanding of engineering concepts and their capability to apply them in practical scenarios (Diwakar et al., 2023; Ghaicha, 2016). Assessment is a multifaceted process that goes beyond measuring rote memorization. It evaluates students' ability to apply knowledge, solve problems, think critically, innovate, collaborate, and uphold ethical standards (Gürdür Broo et al., 2022). Effective assessment enhances the quality of engineering education, produces capable graduates, and supports the continued advancement of the engineering profession (Diwakar et al., 2023). Recognizing that students are at the center of any instructional process, their feedback on content delivery and assessment methods is invaluable for refining and improving the overall instructional process (Jónsson, 2023). Consequently, in this study, we investigated students' perceptions of assessment practices in engineering education. Furthermore, we examined whether there is a significant disparity in the perception of assessment practices among undergraduate engineering students at different academic levels. Measurement scales on the student perception of assessment questionnaire (SPAQ) used in previous studies (Dhindsa et al., 2007; Koul et al., 2006; Mussawy, 2009; Waldrip et al., 2008) to elicit students' perception of assessment practices were used for the data collection. The following hypotheses have been tested to evaluate students' perception of assessment used in undergraduate engineering education programs.

1. Hypothesis 1 (H1): There is no significant difference in the perception of students about the assessment alignment with the planned learning between different academic years.
2. Hypothesis 2 (H2): There is no significant difference in the perception of students about the authenticity of the assessment between different year levels.
3. Hypothesis 3 (H3): There is no significant difference in the perception of students about student consultation on assessment between different year levels.
4. Hypothesis 4 (H4): There is no significant difference in the perception of students on the transparency of assessment between different year levels.
5. Hypothesis 5 (H5): There is no significant difference in the perception of students on the diversity of assessment between different year levels.

II. RELATED WORK

Engaging students in the assessment process is a fundamental shift in education that empowers learners to take an active role in their own educational journey. This approach recognizes students as partners in their learning and fosters a deeper understanding of the learning objectives (Cao & Tech, 2023) (Jogan, 2019). Involving students in assessment process can foster student empowerment, self-regulation, enhanced understanding, feedback, diverse perspectives, co-creation of assessment, individualization, accountability, transparency, improved communication, and preparation for the real world (Ozan, 2019). By actively participating in assessment, students gain a clearer understanding of learning objectives, align their efforts with desired outcomes, and develop metacognitive skills and overall education effectiveness (El-Maaddawy, 2017; Hattingh & Dison, 2020). This approach can foster a growth in mindset and commitment for continuous learning. Students can contribute to setting personalized goals and selecting assessment methods that align with their strengths, interests, and learning preferences (Jogan, 2019). The lack of involving students in assessment can lead to missed opportunities for engagement, authentic learning experiences, feedback, and personalized growth (Raaper, 2023). To foster a holistic learning environment, it is crucial to recognize students as active partners in assessment and to leverage their insights to enhance the overall learning journey (Jónsson, 2023).

Bangladesh's higher education sector faces resource shortages and struggles to improve education standards. Despite government efforts, many institutions struggle with high-quality instruction, poor student learning outcomes, and poor performance in real work environments (Chowdhury, 2016). Assessment plays a critical role in supporting or undermining students' education. Teachers should focus on standardized assessment criteria to better understand students and make informed judgments (Ghaicha, 2016).

Assessment is valid when it aligns with planned learning objectives and is continuous throughout the semester. Studies concerning the learning of engineering students through essay

 https://doi.org/10.52202/073963-0013

tests, particularly in the cognitive learning domain, recommend that educators in Bangladesh ought to concentrate on the application, analysis, evaluation, and creation aspects, rather than merely emphasizing recall when dealing with the subdomains within the cognitive domain. (Raihan et al., 2013). To ensure validity in assessment, it should at least address the cognitive, affective, and psychomotor domains of learning (Mohd-Yusof et al., 2015). An authentic assessment simulates a real work environment and evaluates students' ability to apply knowledge in real-world situations (Bhatia et al., 2023; El-Maaddawy, 2017; 'Fan et al., 2015). It aims to build minds capable of performing in social and economic environments. According to Ozan (2019), the implementation of authentic assessment led to a notable enhancement in academic accomplishment and a positive shift in the attitude of aspiring educators toward educational measurement.

Back in 2010, university students in New Zealand voiced apprehensions about their educational journey, primarily due to the prevalence of a grading-centered environment. This culture induced stress, absenteeism, and disrupted coordination (Harland et al., 2015). Educators observed a lack of communication and hesitancy in scaling down assessments. The consensus among academics was that the existing methods curtailed students' capabilities, and they advocated for more manageable assessments. Nonetheless, a single lecturer opined those recurrent assessments that readied students for the demanding real world, even though it constrained their autonomy to explore beyond the set curriculum.

The significance of impartiality in evaluating education outcomes in higher education was highlighted by Zlatkin-Troitschanskaia et al (2019), in which they advised avoiding superficial judgments and utilizing valid, reliable, and transparent assessment and evaluation approaches as diverse learning needs and styles require inclusive assessment approaches (Jónsson, 2023; Stitt-Bergh et al., 2018).

A study conducted by Koul et al (2006), developed and validated a five-scale measurement instrument called Students Perceptions of Assessment Questionnaire (SPAQ) and analyzed the relationship between these scales with students' attitudes toward science. The findings revealed that Congruence with Planned Learning, Authenticity, Transparency, and Diversity were all favorably connected with students' views toward science.

Dhindsa et al (2007), also conducted a study in Brunei where the results showed that students highly perceived Congruence with Planned Learning (CPL) and Transparency in Assessment (TA), Assessment Applied Learning (AAL), and Transparency in Assessment (TIA). However, students had a low perception of Student Consultations, which contradicts the findings of Mussawy (2009), which states that authenticity of the assessment (AA) has the highest perception. Students also had low perceptions of Students Consultation on Assessment (SCA) and Transparency in Assessment (TA).

Hence, taking into account the viewpoints expressed in existing literature regarding the utilization of SPAQ and the significance of gathering students' perceptions on assessment practices to guide instructional content, pedagogies, and assessment approaches, it is appropriate we undertook a study of this nature within the realm of engineering education in the context of Bangladesh.

III. METHODOLOGY

We adopted a quantitative methodological approach in our study. By conducting a survey, a substantial volume of data was collected from four (4) different engineering universities in Bangladesh, comprising both teachers and students. Nonetheless, the outcomes and insights presented in this study are drawn exclusively from the dataset involving students.

Bangladesh has many engineering universities, encompassing both public and private institutions. The investigation centers on four (4) different engineering universities (designated as U1, U2, U3, and U4). The selection of these universities was deliberate, taking into consideration their unique attributes and also the convenience of the data collection process. U1 was singled out due to its diverse student and faculty composition, being an international institution that welcomes learners from over twenty nations. U2 was chosen for its specialized approach to student training as a public university, with a primary focus on cultivating professional engineers across various disciplines. Its admission policy exclusively admits diploma engineers to pursue degree-level studies. U3's inclusion is attributed to its specialization in textile engineering programs, making its student body significant for this study, given Bangladesh's substantial textile industry. Lastly, U4 was designated to represent the domain of private engineering universities of Bangladesh. All these universities are situated in the Dhaka Division of Bangladesh. The study specifically targeted six engineering departments, namely Electrical and Electronic Engineering (EEE), Computer Science and Engineering (CSE), Mechanical Engineering (ME), Civil Engineering (CE), Textile Engineering (TE), and Industrial and Production Engineering (IPE). Only the undergraduate engineering students were selected from these universities via convenience sampling method.

A. Data Collection Instrument

We used SPAQ instrument, developed and validated by Koul et al (2006) and Waldrip et al (2008), to investigate how students perceive assessment practices in Bangladesh. Additionally, Dhindsa et al (2007) and Mussawy (2009) further substantiated the instrument's reliability and validity. With slight adjustments, we used the 22 items SPAQ instrument with a Likert scale ranging from 1 to 5 (1 representing "strongly disagree," 2 indicating "disagree," 3 representing "neutral," 4 signifying "agree," and 5 denoting "strongly agree").

B. Data collecting Procedure

Prior approval from the relevant authorities of the selected institutions has been taken before the data collection. Upon obtaining approval, participants were informed about their

TABLE I
SCALES OF MEASUREMENT

Scale	Description
Alignment with planned learning	The extent to which learning program e goals, objectives, and activities are aligned with assessment tasks.
Authenticity	The extent to which assessment tasks are relevant to the learner and also features real-life situations.
Student Consultation	The extent to which students are consulted and informed about the forms of assessment tasks.
Transparency	The extent to which assessment tasks are well-defined and clear to the learner.
Diversity	The extent to which all students have an equal chance at completing assessment tasks.
Alignment with planned learning	The extent to which assessment tasks align with the goals, objectives, and activities of the learning program.

voluntary participation and the option to withdraw from the study at any time. It was emphasized that the collected data would be utilized exclusively for academic purposes. A Google form was developed and its link was distributed to educators and student leaders, as well as shared on specific social media platforms utilized by students for educational purposes. The Google form remained accessible for a duration of two (2) months. A total of 557 respondents completed the questionnaire, and their provided answers were meticulously examined to derive conclusive findings.

C. *Analysis*

Various statistical techniques were utilized to contextualize and comprehend the collected data, as well as to discern the respondents' demographics. Cronbach's alpha reliability test was executed to gauge the internal consistency of the instrument items. Employing descriptive analysis, the percentage distribution of respondents was depicted concerning gender, academic year, and academic program. An assessment was made to identify the significant outliers and the normal distribution of the dataset to ensure its integrity. To confirm or refute the stipulated hypothesis, an Analysis of Variance (ANOVA) together with Leven's test was carried out, aiming to ascertain significant variations in the means and similarity of the variances respectively. The significant value (p) was considered to be $< .05$ throughout the analysis.

IV. RESULTS AND DISCUSSION

Reliability analysis was conducted on each of the 5 item variables or constructs (APL, AA, SCA, TA and DA) using Statistical Package for Social Sciences (SPSS) version 26 software. Cronbach's Alpha showed in Table II that the reliability of each construct in the questionnaire (SPAQ) exceeded the minimum standard value of $\alpha = .7$ indicating an acceptable level of internal consistency (Taber, 2018).

TABLE II
RELIABILITY STATISTICS

Construct	No of Item	Cronbach's alpha
APL	4	.775
AA	4	.755
SCA	4	.732
TA	6	.751
DA	4	.798
Overall Reliability	22	.874

A. *Descriptive Statistics*

There were 82.41% of the participants were male and 17.59% were female from the 557 participants. The students were from six engineering departments, with 17.41% affiliated with Computer Science and Engineering, 13.10% from Electrical and Electronic Engineering, 8.62% from Mechanical Engineering, 8.62% from Industrial and Production Engineering, 17.05% from Civil Engineering, and 35.19% from Textile Engineering. Furthermore, the student population was distributed across various academic years: 15.62% represented the first-year cohort, 15.79% were second-year students, and 38.78% and 29.80% were in their third and fourth-year students respectively.

B. *Assessment of Outliers and Normal Distribution*

As revealed, the scores within each measurement scale or construct demonstrate the absence of significant outliers (Hoaglin & Iglewicz, 1987). Moreover, the assumption of a data normality within each construct was satisfied, with skewness and kurtosis values ranging from -.735 to .422 and -.826 to .816 respectively, falling within acceptable limits (Demir, 2022). Ideally, both the skewness and kurtosis coefficients must be zero for normally distributed data. However, given that skewness and kurtosis values mostly deviate from zero, acceptable thresholds are defined for these values. Various studies suggest different ranges for these thresholds. However, most studies propose that the ranges should be less than ± 2 (Demir, 2022; Field, 2013; George & Mallery, 2010; Gravetter et al., 2020; Trochim, 2007).

C. *Test of Hypotheses*

1) *H1: Alignment with the Planned Learning (APL)*

One-way ANOVA analysis was conducted to compare students' perceptions of the assessment's APL across different academic year levels as shown in Table III. The assumption of equal variance was met, as evidenced by Levene's test result $F(3, 553) = .942$, $p > .05$. Notably, a statistically significant distinction in students' perceptions emerged across distinct year levels concerning the assessment's APL ($F(3, 553) = 3.163$, $p < .05$). The observed difference in means and effect size indicated a small effect size (partial eta squared = .017), suggesting that 1.7% of the variability in APL scores could be explained by year levels. Upon further analysis using the Tukey honestly significant difference (HSD) test for post hoc comparisons, we found that the mean score for the first-year

https://doi.org/10.52202/073963-0013

TABLE III
ONE-WAY ANOVA RESULTS OF H1: APL

Test of Homogeneity of Variance					One-Way ANOVA Results		
Year Level	Mean	Standard Deviation	Levene's Statistics	Sig(p)	F	Sig(p)	η^2p
First-year	3.8937	.59017	F (3, 553) = .942	.420	F (3, 553) = 3.163	.024	.017
Second year	3.7045	.65164					
Third year	3.6481	.67085					
Fourth-year	3.6943	.59313					

Group Difference				
Year Level	Mean Difference	Sig(p)	95% Confidence Interval	
			LB	UB
First – Third year	.24553*	.013	.0384	.4527

(M = 3.8937, SD = .59017) students significantly differed from that of the third-year students (M = 3.6481, SD = .67085), p < .05. However, no significant differences were observed between the second year (M = 3.7045, SD = .65164) and the fourth year (M = 3.6943, SD = .59313), nor between the second and fourth year, and first and third year.

2) H2: Authenticity of Assessment (AA)

One-way ANOVA was again conducted to compare students' perceptions of the AA across different academic year levels as shown in Table IV. The assumption of equal variance was met, as evidenced by Levene's test result F (3, 553) = 1.181, p > .05. Notably, there is a statistically significant variance in students' perceptions across distinct year levels concerning AA, (F (3, 553) = 3.642, p < .05). The observed difference in means and effect size indicated a small effect (partial eta squared = .019), suggesting that 1.9% of the variability in AA scores could be explained by year levels. Further analysis using the Tukey HSD test for post hoc comparisons found that the mean score for the first-year students (M = 3.6293, SD = .90128) significantly differed from that of the fourth-year students (M = 3.2892, SD = .83210), p < .05. However, no significant differences were observed between the second year (M = 3.3494, SD = .76896) and the third year (M = 3.4595, SD = .80791), nor between the second and third year, and first and fourth year.

3) H3: Student Consultation on Assessment (SCA)

Through a one-way ANOVA analysis, students' perceptions of SCA across different academic year levels were examined as shown in Table V. The assumption of equal variance was met, as evidenced by Levene's test result F (3, 553) = .736, p > .05. Notably, there was no statistically significant variance in students' perceptions across the different year levels concerning SCA, (F (3, 553) = 1.695, p > .05).

4) H4: Transparency of Assessment (AA)

A one-way ANOVA analysis was also conducted to compare students' perceptions of the TA across different academic year levels as shown in Table VI. The assumption of equal variance was met, as evidenced by Levene's test result F (3, 553) = 1.408, p > .05. Notably, a statistically significant difference in students' perceptions emerged across different year levels concerning TA, (F (3, 553) = 2.723, p < .05). The observed difference in means and effect size indicated a small effect (partial eta squared = .015), suggesting that 1.5% of the variability in TA scores could be explained by year levels. Upon further analysis using the Tukey HSD test for post hoc comparisons, we found that the mean score for the first-year students (M = 3.8276, SD = .79746) significantly differed from that of the fourth-year students (M = 3.5612, SD = .70718), p < .05. However, no significant differences were observed between the second year (M = 3.6307, SD = .72514) and the third year (M = 3.6883, SD = .70732), nor between the second and third year, and the first and fourth year.

5) H5: Diversity of Assessment (DA)

One-way ANOVA analysis was finally conducted to compare students' perceptions of DA across different academic year levels as shown in Table VII. The assumption of equal variance was met, as evidenced by Levene's test result F (3, 553) = .254, p > .05. Notably, there was no statistically significant variance in students' perceptions across distinct year levels concerning DA, (F (3, 553) = 2.134, p > .05).

Based on the ANOVA findings presented in Tables III, IV, and VI, it can be deduced that a statistically significant distinction prevails in the perceptions of undergraduate engineering students concerning assessment practices related to APL, AA, and TA. The associated p-values are .024, .013, and .043, respectively, all of which fall below the established threshold for statistical insignificance (p = .05). Consequently, hypotheses H1, H2, and H4 are rejected, as documented in sources like Field (2013), George & Mallery (2010), and

TABLE IV
ONE-WAY ANOVA RESULTS OF H2: AA

	Test of Homogeneity of Variance				One-Way ANOVA Results		
Year Level	Mean	Standard Deviation	Levene's Statistics	Sig	F	Sig	η^2p
First-year	3.6293	.90128	F (3, 553) = 1.181	.316	F (3, 553) = 3.642	.013	.019
Second year	3.3494	.76896					
Third year	3.4595	.80791					
Fourth-year	3.2892	.83210					

Group Difference				
Year Level	Mean Difference	Sig	95% Confidence Interval	
			LB	UB
First – Fourth year	.34015*	.010	.0590	.6213

TABLE V
ONE-WAY ANOVA RESULTS OF H3: SCA

	Test of Homogeneity of Variance				One-Way ANOVA Results	
Year Level	Mean	Standard Deviation	Levene's Statistics	Sig	F	Sig
First-year	3.7443	.87436	F (3, 553) = .736	.531	F (3, 553) = 1.695	.167
Second year	3.4915	.82128				
Third year	3.6157	.82118				
Fourth-year	3.5346	.85054				

TABLE VI
ONE-WAY ANOVA RESULTS OF H4: TA

	Test of Homogeneity of Variance				One-Way ANOVA Results		
Year Level	Mean	Standard Deviation	Levene's Statistics	Sig	F	Sig	η^2p
First-year	3.8276	.79746	F (3, 553) = 1.408	.240	F (3, 553) = 2.723	.043	.015
Second year	3.6307	.72514					
Third year	3.6883	.70732					
Fourth-year	3.5612	.70718					

Group Difference				
Year Level	Mean Difference	Sig	95% Confidence Interval	
			LB	UB
First–Fourth year	.26634*	.029	.0191	.5135

Gravetter et al. (2020). Nevertheless, there is no notable statistical difference observed with regard to student perception in SCA and DA, where the corresponding p-values stand at .167 and .095 respectively. Hence, hypotheses H3 and H5 are failed to reject (Field, 2013; George & Mallery, 2010; Gravetter et al., 2020).

Furthermore, we delved into group disparities to illustrate variations among different year levels. This was accomplished by conducting supplementary examinations utilizing the Tukey

https://doi.org/10.52202/073963-0013

TABLE VII
ONE-WAY ANOVA RESULTS OF DA

Test of Homogeneity of Variance					One-Way ANOVA Results	
Year Level	Mean	Standard Deviation	Levene's Statistics	Sig	F	Sig
First-year	3.1897	.91178	F (3, 553) = .254	.858	F (3, 553) = 2.134	.095
Second year	3.0313	.85921				
Third year	3.2778	.87260				
Fourth-year	3.1084	.85253				

HSD test for post hoc comparisons. The outcomes of this analysis revealed statistically significant mean differences between first and third-year students in their assessment perception score for APL (M = .24553*, p = .013). Similarly, a considerable divergence in perception score emerged between first and fourth-year students in both AA and TA, showcasing a mean difference of (M = .34015*, p = .010) and (M = .26634*, p = .029) correspondingly.

Notably, no additional assessments were undertaken for SCA and DA, as the perceptions of students within these constructs did not present any statistically significant differences as the p-values are .167 and .095 respectively. However, this does not validate the notion that students have reached to a consensus that the assessment practices take into account students' viewpoints and involve consultation between educators and students in assessment-related decisions. In fact, students perceived the DA as the lowest scale with an average mean score of 3.1746, which exhibited agreement in their viewpoint regarding less inclusivity and diversity of the assessment practices compared to other scales. They believe that assessment practices lack the inclusion of a broad array of approaches. Furthermore, despite the differences in their perception of APL, AA and TA, students still consider APL, TA and AA to be the highly observed scale with an average mean value of 3.7092, 3.6613 and 3.4179 respectively. However, the findings of this study confirmed the results of Mussawy (2009).

V. CONCLUSION

The SPAQ has been used in K-12 education settings and at the time of this study we have not come across any study that uses SPAQ in engineering higher education. Furthermore, we observed that there is a significant variance in the perception of first-year and third-year students on the alignment of assessment with planned learning. Similarly, a significant difference exists between the perception of first-year and fourth-year on the authenticity and transparency in assessment. However, the magnitude of the difference is small. Therefore, this study has depicted the significant importance of involving students in decision-making on assessment practices in engineering education as it will further enhance their learning capabilities. Additionally, assessment practices have to be inclusive and authentic. Further studies may explore online assessment practices in engineering education. The perception of teachers can also be explored and compared to that of the students for a better understanding of assessment practices in engineering education context.

ACKNOWLEDGMENT

The authors would like to thank the E4LIFE International Ph.D. Fellowship Program offered by Amrita Vishwa Vidyapeetham, Kerala, India. We extend our gratitude to the Amrita Live-in-Labs® academic program for providing all the necessary support. We also like to thank the Department of Technical and Vocational Education, Islamic University of Technology (IUT), Dhaka, Bangladesh for facilitating the conduct of this study.

REFERENCES

Ali, L. (2018). The Design of Curriculum, Assessment and Evaluation in Higher Education with Constructive Alignment. *Journal of Education and E-Learning Research*, *5*(1), 72–78. https://doi.org/10.20448/journal.509.2018.51.72.78

Bhatia, S., Elhadad, S., Deshmukh, A., Yellela, M. K., & Vangala, O. S. R. (2023). Hack the Problem: A Problem-Based Learning Approach for Ethical Hacking and Network Defense Curriculum. *SIGCSE 2023 - Proceedings of the 54th ACM Technical Symposium on Computer Science Education*, *2*, 1346. https://doi.org/10.1145/3545947.3576292

Cao, M. Y., & Tech, V. (2023). *A Comparative Literature Review: Comparing Approaches to Teamwork Assessment in Engineering Education in the US and China.*

Chen, C. H., Hung, H. T., & Yeh, H. C. (2021). Virtual reality in problem-based learning contexts: Effects on the problem-solving performance, vocabulary acquisition and motivation of English language learners. *Journal of Computer Assisted Learning*, *37*(3), 851–860. https://doi.org/10.1111/jcal.12528

Chowdhury, F. (2016). Employment of Active Learning at HEIs in Bangladesh to Improve Education Quality. *International Education Studies*, *9*(10), 47. https://doi.org/10.5539/ies.v9n10p47

Demir, S. (2022). Comparison of Normality Tests in Terms of Sample Sizes under Different Skewness and Kurtosis Coefficients. *International Journal of Assessment Tools in Education*, *9*(2), 397–409. https://doi.org/10.21449/ijate.1101295

Denton, D. D. (1998). Engineering education for the 21st century: Challenges and opportunities. *Journal of Engineering Education*, *87*(1), 19–22. https://doi.org/10.1002/j.2168-9830.1998.tb00317.x

Dhindsa, H. S., Omar, K., & Waldrip, B. (2007). Upper Secondary Bruneian Science Students' Perceptions of Assessment. *International Journal of Science Education*, *29*(10), 1261–1280.

Diwakar, S., Kolil, V. K., Francis, S. P., & Achuthan, K. (2023). Intrinsic and extrinsic motivation among students for laboratory courses - Assessing the impact of virtual laboratories. *Computers & Education*, *198*, 104758. https://doi.org/10.1016/j.compedu.2023.104758

El-Maaddawy, T. (2017). Innovative assessment paradigm to enhance student learning in engineering education. *European Journal of Engineering Education*, *42*(6), 1439–1454. https://doi.org/10.1080/03043797.2017.1304896

'Fan, X., 'Zhou, J., & 'Tong, M. (2015). Teaching Methods of Engineering Practice Course for Senior Undergraduate Students. *2015 IEEE International Conference on Teaching, Assessment, and Learning for Engineering (TALE)*, 111–114.

Field, Andy. (2013). *Discovering statistics using SPSS.* (3th ed.). Sage Publications.

Frey, M., Pokkiyarath, M., Mohan, R., Sai Shibu, N. B., Gracia, V. C., Mohan, V., & Siddhan. (2017). Elevating education of India's rural village girls through distance learning technology supported by sustainable electricity. *GHTC 2017 - IEEE Global Humanitarian Technology Conference, Proceedings*, *2017-January*, 1–8. https://doi.org/10.1109/GHTC.2017.8239300

George, D., & Mallery, P. (2010). *SPSS for Windows Step by Step: A Simple Guide and Reference, 17.0 Update* (10th, illustrated ed.). Allyn & Bacoon.

Ghaicha, A. (2016). Theoretical Framework for Educational Assessment: A Synoptic Review. *Journal of Education and Practice*, *7*(24), 2222–1735.

Gravetter, F. J., Wallnau, L. B., Forzano, L.-A. B., & Witnauer, J. E. (2020). *Essentials of statistics for the behavioral sciences.* (10th ed.). Cengage Learning.

Gürdür Broo, D., Kaynak, O., & Sait, S. M. (2022). Rethinking engineering education at the age of industry 5.0. *Journal of Industrial Information Integration*, *25*. https://doi.org/10.1016/j.jii.2021.100311

Harland, T., McLean, A., Wass, R., Miller, E., & Sim, K. N. (2015). An assessment arms race and its fallout: high-stakes grading and the case for slow scholarship. *Assessment and Evaluation in Higher Education*, *40*(4), 528–541. https://doi.org/10.1080/02602938.2014.931927

Hattingh, T., & Dison, L. (2020, November 16). Exploring student perspectives on assessment practices in an engineering context. *2020 IFEES World Engineering Education Forum - Global Engineering Deans Council, WEEF-GEDC 2020*. https://doi.org/10.1109/WEEF-GEDC49885.2020.9293656

Hoaglin, D. C., & Iglewicz, Boris. (1987). Fine-Tuning Some Resistant Rules for Outlier Labeling. *Journal of the American Statistical Association*, *82*(400), 1147–1149.

Huang-Saad, A. Y., Morton, C. S., & Libarkin, J. C. (2018). Entrepreneurship Assessment in Higher Education: A Research Review for Engineering Education Researchers. *Journal of Engineering Education*, *107*(2), 263–290. https://doi.org/10.1002/jee.20197

Jogan, S. N. (2019). "ASSESSMENT FOR LEARNING OR OF LEARNING: A PERCEPTION OF TRAINEES." In *Journal Homepage: International Journal of Research in Social Sciences* (Vol. 9, Issue 1). http://www.ijmra.us,http://www.ijmra.us,

Jónsson, Í. R. (2023). Student involvement in assessment and power relations: teacher's perspective. *Pedagogy, Culture and Society*. https://doi.org/10.1080/14681366.2023.2186930

Koul, R., Fisher, D., & Earnest, J. (2006). Using Student Perceptions in Development, Validation and Application of an Assessment Questionnaire. *Proceedings of the 2006 Conference of the Australian Association of Environmental Education* , 294–305.

McGowan, V. C., & Bell, P. (2020). Engineering Education as the Development of Critical Sociotechnical Literacy. *Science and Education*, *29*(4), 981–1005. https://doi.org/10.1007/s11191-020-00151-5

Mohd-Yusof, K., Ahmad Helmi, S., Aliah Phang, F., & Mohammad, S. (2015). Future Directions in Engineering Education: Educating Engineers of the 21st Century. In *ASEAN Journal of Engineering Education* (Vol. 2, Issue 1).

Mussawy, S. A. J. (2009). *Assessment Practices: Student's and Teachers' Perceptions of Classroom Assessment*. University of Massachusetts Amherst.

Nedungadi, P., Mulki, K., & Raman, R. (2018). Improving educational outcomes & reducing absenteeism at remote villages with mobile technology and WhatsAPP: Findings from rural India. *Education and Information Technologies*, *23*(1), 113–127. https://doi.org/10.1007/s10639-017-9588-z

Ozan, C. (2019). The effect of authentic assessment on academic achievement and attitude towards educational measurement and opinions of prospective teachers. *International Journal of Evaluation and Research in Education*, *8*(2), 299–312. https://doi.org/10.11591/ijere.v8i2.18564

Pleasants, J. (2023). Rethinking the Nature of Engineering: Attending to the Social Context of Engineering. *Science and Education*. https://doi.org/10.1007/s11191-023-00445-4

Raaper, R. (2023). Students as consumers? A counter perspective from student assessment as a disciplinary technology. In *Neoliberalism and Education* (1st Edition, pp. 1–16). Routledge.

Raihan, M. A., Shamim, R. H., Clement, C. K., & Lock, H. S. (2013). A STUDY ON ASSESSMENT & EVALUATION OF ENGINEERING STUDENTS' LEARNING BY ESSAY TEST BASED ON THE COGNITIVE DOMAIN OF BLOOM'S. *International Journal of Advances in Engineering & Technology, Mar*, *6*(1), 1–11.

'Raman, R., 'Achuthan, K., & 'Nedungadi, P. (2013). Virtual Labs in Engineering Education: Modeling Perceived Critical Mass of Potential Adopter Teachers. *Scaling up Learning for Sustained Impact: 8th European Conference on Technology Enhanced Learning, EC-TEL 2013*, 288–300.

Raman, R., Sairam, B., Veena, G., Vachharajani, H., & Nedungadi, P. (2021). Adoption of online proctored examinations by university students during COVID-19: Innovation diffusion study. *Education and Information Technologies*, *26*(6), 7339–7358. https://doi.org/10.1007/s10639-021-10581-5

Raman, R., Venkatasubramanian, S., Achuthan, K., & Nedungadi, P. (2015). Computer science (CS) education in Indian schools: Situation analysis using darmstadt model. *ACM Transactions on Computinig Education*, *15*(2). https://doi.org/10.1145/2716325

Shahid, M. A., Rahman, M., & Ahmed, F. (2022). Challenges in Engineering Education: A Review. *Proceedings of the 5th International Conference on Industrial & Mechanical Engineering and Operations Management,* 1–7.

Sneider, C. (2016). Connecting Science and Engineering Education Practices in Meaningful Ways. Contemporary Trends and Issues in Science Education. In L. A. Annetta & J. Minogue (Eds.), *Grand Challenges for Engineering Education* (1st ed., Vol. 44, pp. 19–35). Springer, Cham.

Stehle, S. M., & Peters-Burton, E. E. (2019). Developing student 21st Century skills in selected exemplary inclusive STEM high schools. *International Journal of STEM Education, 6*(1). https://doi.org/10.1186/s40594-019-0192-1

Stitt-Bergh, M., Kinzie, J., & Fulcher, K. (2018). *Refining an Approach to Assessment for Learning Improvement.*

Taber, K. S. (2018). The Use of Cronbach's Alpha When Developing and Reporting Research Instruments in Science Education. *Research in Science Education, 48*(6), 1273–1296. https://doi.org/10.1007/s11165-016-9602-2

Trochim, W. M. (2007). *The Research Methods Knowledge Base Systems Evaluation View project Systems Evaluation Protocol View project.* http://trochim.human.cornell.edu/kb/index.htm

Waldrip, B. G., Fisher, D. L., & Dorman, J. P. (2008). Students' perceptions of assessment process: Questionnaire development and validation. *In Proceedings of the 5th International Conference on Science, Mathematics and Technology Education*, 561–568.

Zlatkin-Troitschanskaia, O., Schlax, J., Jitomirski, J., Happ, R., Kühling-Thees, C., Brückner, S., & Pant, H. A. (2019). Ethics and Fairness in Assessing Learning Outcomes in Higher Education. *Higher Education Policy*, 537–556.

APPENDIX: SURVEY QUESTIONS

Alignment with planned learning

1. My assessment in engineering courses tests what I understand.
2. My assessment in engineering department tests what I memorize.
3. My assignments/tests are about what I have done in class.
4. I am assessed on what the teacher has taught me.

Authenticity of Assessment

5. I find engineering department assessment tasks are relevant to what I do outside of school.
6. Assessment in the engineering department tests my ability to apply what I know to real-life problems.
7. Assessment in the engineering department examines my ability to answer everyday questions.

Student Consultation on Assessment

8. I can show others that my learning has helped me do things.
9. In the engineering department, I am clear about the types of assessment being used.
10. I am aware of how my assessment will be marked.
11. My teacher does explain to me how each type of assessment is to be used.
12. I can have a say in how I will be assessed in the engineering department through the assessment system.

Transparency in Assessment

13. I am told in advance when I am being assessed.
14. I am told in advance on what I am being assessed.
15. I am clear about what my teacher wants in my assessment tasks.
16. I know how particular assessment tasks will be marked.
17. My relation with the teacher does not have any influence on my assessment scores.
18. I am always provided with the feedback by the teacher on my assessment.

Diversity in Assessment

19. I can complete the assessment tasks by the given time.
20. I am given a choice of assessment tasks.
21. I am given assessment tasks that suit my ability.
22. When I am confused about an assessment task, I am given another option to answer it.

https://doi.org/10.52202/073963-0013

Research in Engineering Education Symposium-2024
Jan 4-6, 2024 - Hubballi, India

Capturing Latent Learning Indicators Through Technology: New Paradigms for Measuring Student Learning

Kaushik Mallibhat; Uma Mudenagudi
School of Electronics and Communication Engineering
KLE Technological University, Hubballi, India
Kaushik@kletech.ac.in

Abstract

Context

Technology has been helpful in the field of education for the design, delivery, and assessment of courses. Though academicians quickly adopted the new technology for delivery, they still use the traditional written exams to assess student learning, even in professional courses, including medical, engineering, yoga, and music education systems.

Purpose

The paper focuses on the investigation of how recent technological advancements help capture the hidden and accurate learning indicators of student learning, what devices are found helpful by researchers towards capturing the latent learning indicators, what the trends are, and what are the publicly available datasets that can catalyze the research in the field of learning analytics.

Methods

The study was carried out by adopting the PRISMA template of Systematic Literature Review (SLR), and the four phases, including identification, screening, selection, and inclusion methods, were carried out towards investigating the research questions.

Outcomes

This paper helps academicians and researchers in the field of education and learning analytics to get an overview of the current trend and identify the research gaps towards integrating data from multiple sources and connecting the educational theories with the captured parameters.

Conclusion

A drift has been observed from unimodal data sources to multimodal sources, capturing the data from the perceived behavior of the student to the hidden cognitive and affective domain characteristics.

***Keywords*—**Learning analytics, Cognitive domain, Affective domain, Diagnostic analytics, Predictive analytics, Prescriptive analytics.

I. Motivation

STUDENT'S learning is often measured through formative and summative assessments, which are majorly in time-bound written exams, even in professional courses like medicine, engineering, fashion technology, yoga, and music education. Despite several pedagogical initiatives like Project- Based Learning (PBL) (Mallibhat et al., 2022), Problem-Based Learning (Wood, 2003), Activity-Based Learning (Sharma et al., 2018), and Blended Learning (Vijaylakshmi et al., 2021) methods listed in the literature, in the majority of the assessment occasions, students are assessed through written exams, which may not be the reflection of the actual competencies and skills acquired by students. The problem of assessing students' learning through written exams has drawbacks, including the ability of the student to comprehend the learnings within a given time, language constraints, and the nature of the written exams emphasizing rote learning (Condon & Kelly-Riley, 2004) The problem is much more significant in non-native English-speaking countries like India.

On the other hand, Technology Enabled Learning Environments (TELE) are rapidly expanding, especially in the post-pandemic age, to meet the needs of millennial and Gen-Z learners. These TELEs can record students' digital footprints, which aids teachers in assessing students learning.

The limitations of traditional assessment systems and the advantages of TELE motivated the authors to carry out the Systematic Literature Review (SLR) to understand the existing body of knowledge and identify the research gaps towards capturing learning indicators through technology. This study helped the first author crystallize the doctoral degree's research questions and objectives.

II. Introduction

Intelligent Computer Assisted Instruction (ICAI) was first used in 1960 to collect student log data to analyze the student's learning patterns. However, a new area of study known as "Learning Analytics" has evolved in recent years as a result of the confluence of "learning," "analytics," and "human-centered design." Learning analytics is a topic of research that deals with acquiring, measuring, analyzing, and reporting information on students, learning environments, and their surroundings, according to the Society for Learning Analytics Research

https://doi.org/10.52202/073963-0014

(SoLAR). The growth of technology has catalyzed learning analytics research and provides capabilities to capture data much beyond log data.

Learning analytics has categories, including descriptive, diagnostic, predictive, prescriptive, adaptive, and causal analytics.

Descriptive analytics, often known as dashboard analytics, is the most fundamental type of data analysis and focuses on identifying trends from historical and present data. This type of analytics investigates the answers to the question, "What happened?"

Diagnostic analytics[1] is used to analyze data to identify the factors behind trends and variable correlations. This progressive descriptive analytics and analysis step can be carried out manually, algorithmically, or through statistical tools. This type of analytics investigates answers to the question, "Why did this happen?"

The use of data to forecast upcoming trends and occurrences is known as predictive analytics. It projects prospective future situations using historical data to guide strategic decision-making. Regression analysis is one of the most widely used predictive analytical tools. It investigates the answer to the question, "What might happen in the future?"

The technique of analyzing data to decide on the best course of action is known as prescriptive analytics. Considering all relevant factors, this analysis produces recommendations for the subsequent stages. Prescriptive analytics is a valuable technique for making data-driven decisions.

Predictive and prescriptive analytics are integrated to make real-time adjustments in adaptive learning analytics, while predictive and diagnostic analytics are integrated to understand the cause-and-effect relationship in causal analytics methods.

TELEs offer a wide range of data capturing facilities (how to capture); however, 'what to capture,' 'when to capture,' 'why to capture,' and 'what information do they convey are grounded in the learning theories. This paper tries to bridge the gap by investigating the following research questions.

1. How technology-enabled data sources are capturing the latent learning indicators of the student?
2. What devices and software are the state of the art that enables the capture of latent learning indicators of the student?
3. What type of analytics is currently the state of the art?
4. What data sets are currently available that can help to build a machine learning/deep learning model for predictive and prescriptive analytics?

To investigate the research questions, the authors found that there needs to be a comprehensive literature review that can address the above research questions. This motivated the authors to perform the Systematic Literature Review (SLR).

The SLR process using the PRISMA method is described in Section II; Section III discusses the findings concerning each of the research questions, followed by inferences in Section IV.

[1] https://online.hbs.edu/blog/post/diagnostic-analytics

III. SYSTEMATIC LITERATURE REVIEW (SLR) PROCESS

The objective of SLR is to examine, summarize, and "reconcile the evidence to inform research policy and practice." (Petticrew &Roberts, 2006). Preferred Reporting Items for Systematic Reviews and Meta-Analyses (PRISMA) data flow method was adopted (Moher et al., 2009; Page et al., 2021), and the flow consists of four phases: identification, screening, selection, and inclusion. The entire flow diagram is shown in Figure 1.

A. Identification

The process of SLR began with the identification of keywords and databases. An exploratory way of searching was used on Google Scholar to identify the initial keywords. Databases, including IEEE, Science Direct, Google Scholar, ACM, and Scopus, were considered to find the relevant papers. In addition to this, to address the fourth research question, author had to search for data bases including Kaggle and papers with code and found additional papers.

The keywords used to find the appropriate papers are smart learning environments, learning management systems, and student learning. Advanced search options in the databases were used to select the year of publication. Duration from 2003 to 2023 was used to identify the papers. 219 papers were identified, including 115 conference papers and 104 journal articles. 18 papers were removed during the elimination of duplicates.

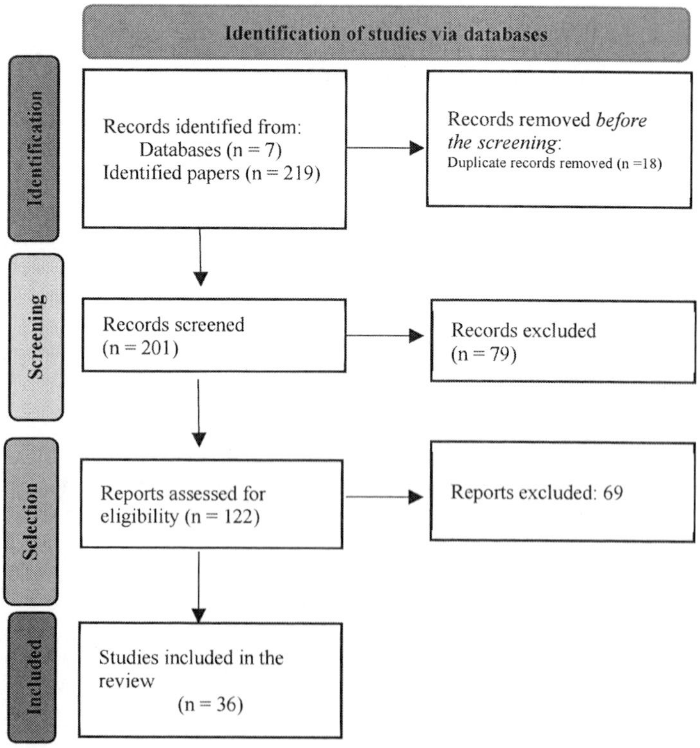

FIGURE 1 PRISMA FLOW DIAGRAM USED TO CARRY OUT SLR

https://doi.org/10.52202/073963-0014

B. Screening

Screening criteria -year (exclusion criteria- E1) was used to identify papers, while the other exclusion criteria included E2- Short papers/ papers with only abstracts. E3- Non-English papers, E4- Only conceptual works. Inclusion criteria I1- Papers from early access, I2- papers related to Educational Data Mining were also considered. I3- Primary source articles and secondary source articles were also considered.

C. Selection and Inclusion

A total of 122 articles were selected and passed for round 1 and tried categorizing the papers into three categories: strongly selected category, weakly selected category, and reject category. There were 19 papers in the strongly selected category and 34 papers in the weakly selected category. All the secondary search articles were categorized into weakly selected categories, while 69 papers were classified into the reject category. The papers where only qualitative analysis was carried out based on focus group discussions or surveys without technology were excluded from the study as it is beyond the scope of the paper. Finally, 36 papers were shortlisted for the following study.

IV. Literature Summary

SLR was carried on with the lenses of formulated research questions. The subsections summarize the literature about each of the research questions.

A. How technology-enabled data sources are capturing the latent learning indicators of the student?

Technology has enabled data capture both in in-classroom (formal learning setting) and online (in-formal learning settings) environments.

The data obtained from the in-classroom environment enables to capture the data related to student's behavioral activities (physical domain), including attendance (Bhattacharya et al., 2018), (Chango et al.,2021) posture (Henderson et al., 2020), body movements (Ashwin et al., 2023), yawning (sleepy) (Omidyeganeh et al., 2016) interaction with peers (Liu et.al., 2019) interaction with teachers (Liu et.al., 2019), detection of malpractices (Prathish, S., & Bijlani, K, 2016)

On the other hand, the data from the online settings enables the capture of student data beyond behavior characteristics and extends to the cognitive and affective domains. Figure 2 shows the tree diagram representing various data sources in formal and informal learning settings.

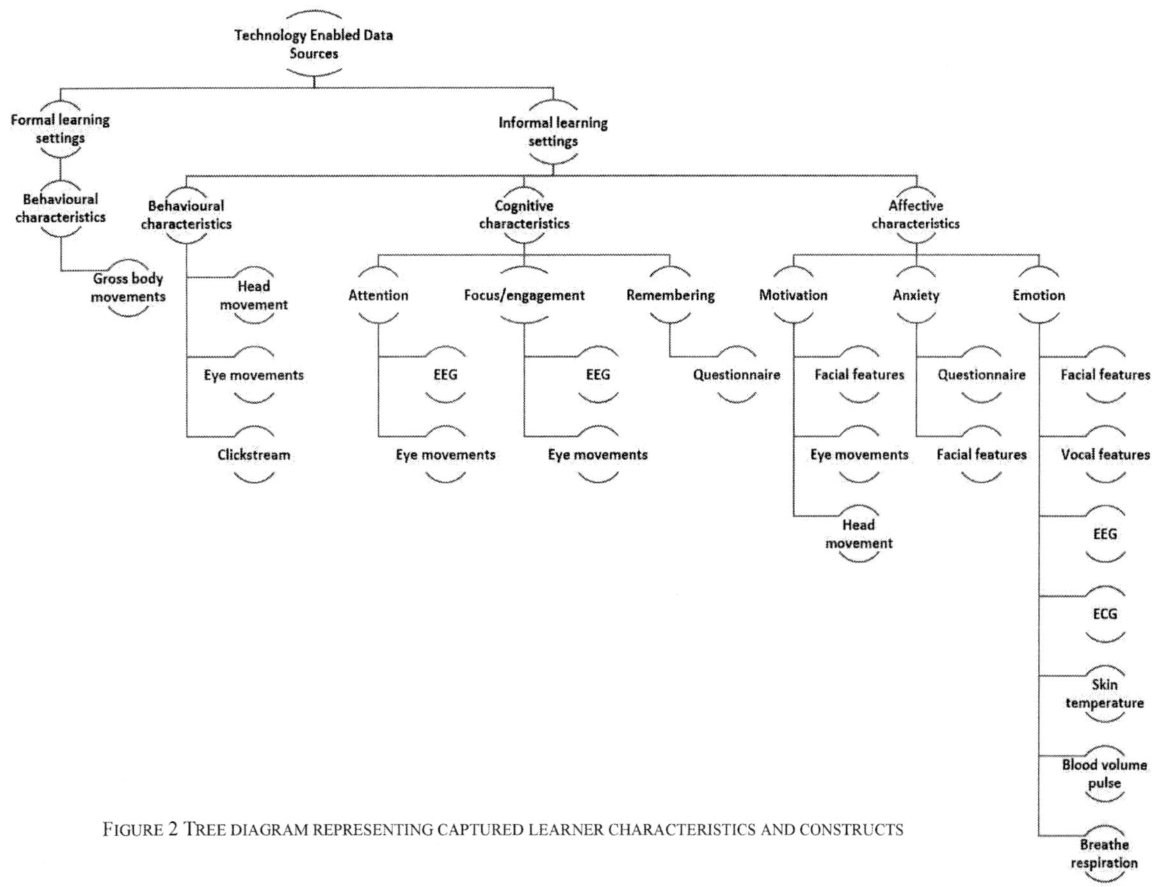

FIGURE 2 TREE DIAGRAM REPRESENTING CAPTURED LEARNER CHARACTERISTICS AND CONSTRUCTS

https://doi.org/10.52202/073963-0014

The branches of the tree diagram represent the characteristics captured, and at the next level of the tree, the measured construct is shown, and the leaf node represents the data source.

Each of the constructs is measured through a set of parameters, which serve as learning indicators. Table 1 shows the constructs' mapping, with measured parameters and the learning indicators.

The breadth of the literature has spread in different dimensions. Another dimension of literature is the data sources based on the space from which the data is captured (Mu, S et.al., 2020). The data spaces are classified into digital, physical, physiological, psychometric, and environmental spaces.

Physical space refers to the data space that captures the learner's behavioral characteristics, including gestures and gross body movements, captured through sensors and cameras. Digital space refers to the data space that captures the digital footprints of the learner during the learning process.

While the learner is engaged in learning, the learner exhibits several physiological changes that convey information about the amount of learning. This data space refers to physiological space.

Psychometric space refers to the data captured through surveys or questionnaires that can serve as feedback on learning from the learners. It can be captured using technology like Mentimeter[1], Google Forms, LMS-based survey forms, or pop-up questions while using the instructional material.

Environmental space refers to the data related to the environmental parameters like temperature and weather conditions affecting the learning.

The data from each of the mentioned spaces is a rich source of information about the learner and the learning. The data can exist in either time series data, including sensor data, video, textual data, or in the form of images click stream data.

Researchers have used multiple hardware software to capture various forms of data and used data from multiple sources to draw inferences about the learner and the learning.

B. What devices and software are the state of the art that enables the capture of latent learning indicators of the student?

Technological advancement has enabled us with multiple hardware and software that help capture the learning indicators. One of the challenges that a researcher faces is choosing the appropriate hardware and software for the data capture. Appropriate hardware and software can orient the researcher and reduce the efforts during data collection. Thus, a table summarizing the various tools and techniques used by researchers to capture learning indicators in formal and informal learning settings is shown in Table 1.

[1] https://www.mentimeter.com/
[2] https://www.mendeley.com/datasets
[3] https://www.cmu.edu/datalab/tools/datashop.html
[4] https://dataverse.harvard.edu/

C. What are the trends and types of analytics currently being used?

Following are the observations by the authors in the direction of investigating the trends and types of analytics currently being used. The summary is represented in Figure 3.

1. Predictive analysis techniques are becoming more popular than descriptive analysis techniques.
2. Methods for prescriptive, adaptive analytics are still in the development stage or not integrated with the existing systems.
3. The transition from unimodal to multimodal sources of information has also been noted as a trend.
4. The information is combined at the characteristics level (for example, behavioral characteristics with cognitive characteristics, cognitive characteristics with affective characteristics), construct level (for example, attention with gross body movements), and feature level (for example, blink rate with pupil diameter).

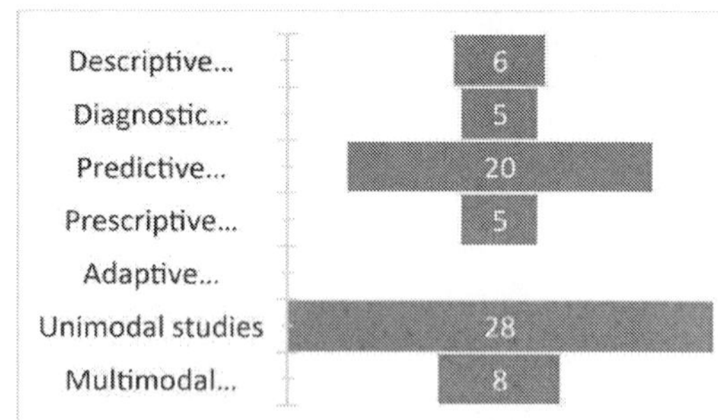

FIGURE 3 TYPES AND TRENDS OF ANALYTICS

D. What data sets are currently available that can help to build a machine learning/deep learning model for predictive and prescriptive analytics?

Among 36 papers included in the study, only 9 papers presented the data set description. As a result, the first author had to search for the datasets through papers and platforms like Kaggle, Papers with code towards addressing the fourth research question.

The authors found three types of datasets, namely.

1. Data sets are released by organizations/ research labs to facilitate the other researchers to carry out the research. It included Mendeley data repository[2], Carnegie Mellon University's DataShop and DataLab[3], Harvard dataverse[4]. These datasets are publicly available.
2. Data sets using crowd-sourced platforms and as a part of Educational Data Mining (EDM) conferences. These datasets are part of hackathons and made available on platforms, including Kaggle and Papers with code. These datasets are publicly available, along with associated research papers and code.
3. Data sets are released by individual researchers. The associated datasets may be made available to researchers upon request. Table II shows the summary of such available datasets.

https://doi.org/10.52202/073963-0014

TABLE I
SUMMARY SHOWING THE VARIOUS HARDWARE AND SOFTWARE USED TO CAPTURE DIFFERENT CONSTRUCTS

Context of the studies	Construct	Parameters (Operationalized through)	Hardware and software used	Indicator/feature	References
Learners' attention detection, Cognitive profiling of learners, Quantification of user engagement, Mind wandering study, Drowsiness detection	Attention	Eye movements	i-Trace, Tobii eye trackers with Tobii pro lab software, Webcams	Eye fixations, saccades, pupil position, gaze vector	Anwar et.al., 2021 Grandchamp et.al., 2014 Olsen et.al., 2022 Sharif et.al., 2019 Wang et.al., 2021 Wong et.al., 2023,
		EEG	Emotiv, NeuroSky MindWave Mobile BrainWave Starter Kit, Brain Vision, Ant Neuro, BrainVision actiCamp with BrainVision PyCorder, Open BCI, MATLAB with EEGLAB, BCILAB and ERPLAB	Frequency bands, Event-Related Potentials (ERP), power spectral density, coherence values	Baker et.al., 2010 Chakladar et.al., 2021 Hassan et.al., 2020 LaRocco et.al., 2020 Li, X et.al., 2011 Souza et.al., 2021 Toa et.al., 2021
		Eye movements and EEG	Wearable eye trackers NeuroSky MindWave Mobile BrainWave Starter Kit	Pupil movement, Frequency bands, Event-Related Potentials (ERP)	Khosravi et.al., 2022 Lai et.al., 2019
Determining the effectiveness of instructional resources, Student thought patterns and reading behaviors	Engagement, attention (focused and instantaneous), working memory, and visual perception	EEG	14-channel Emotiv EEG device	Statistical features, including minimum, maximum, mean from the time domain, and average power and power of alpha, beta, and gamma, were captured.	Benitez et.al., 2016 Masood et.al., 2017
	Engagement	EEG, Eye movements, heart rate variability, Galvanic Skin Resistance	Muse	alpha, beta, and gamma absolute band power, eye gaze coordinates, eye motion velocity, inter-beat interval (R-R), skin conductance value	Giannakos et.al., 2020 Hussain et al., 2011 Krigolson et.al., 2017
		Click patterns participation in discussions.	Moodle-based LMS, Vimeo, YouTube	Time spent on tasks; number of tasks/milestones completed	Botelho et.al., 2019 Brodny,2017, Joshi et.al., 2022, Yue et.al., 2019
		Eye movements(gaze)	Webcam and Webgazer	Area of Interest (AOI) and off-screen proportions	Yang et.al., 2021, Papoutsaki et.al., 2015
Malpractice detection	Engagement	Facial features, open/ closed eyes, head movement, object detection, hand signs	Webcam, microphone, OpenCV	Facial key points, hand key points	Hussain et al., 2011, Prathish, S., & Bijlani, K, 2016
Student engagement with the video content	Remembering	Questionnaire	H5P tool	Response from the learner, time taken to give a response, number of attempts taken to give the correct answer	Amashi et.al., 2021 Amashi et.al., 2023
Dance tutoring system, yoga studies, Student interaction with teachers and peers	Gross Body Movements, head movements, hand gestures	Posture	Kinect, RGB camera, Myo, Real sense	Power and wavelength of the reflected light	Ashwin et.al., 2023 Henderson et al., 2019
Emotional meter	Engagement	EEG, Eye movements	SMI eye-tracking glasses	Pupil diameter, blink rate	Mele et.al., 2012 Zheng et.al., 2018
Understanding the learner experience through multimodal data	Attention + Engagement	EEG, ECG, BVP,	Eye tracker, Empatica E4 wristband, 20-channel EEG machine, webcam	Blood volume pressure, heart rate, body temperature. Keystrokes, facial key points	Monkaresi et.al., 2016, Villarroel et.al., 2018

https://doi.org/10.52202/073963-0014

TABLE II

SUMMARY SHOWING THE VARIOUS AVAILABLE DATA SETS ALONG WITH CAPTURED CHARACTERISTICS, DATA SPACES, AND LINKS TO ACCESS THE DATASETS

Name of the dataset	Dataset description	Characteristics captured	Data Space	Link for the dataset
Stanford University's Social Network: MOOC User Action Dataset	The user activities on a well-known MOOC platform are represented by the MOOC user action dataset. The actions are shown as a directed temporal network. Edges reflect user actions on targets, whereas nodes represent users and course activities (targets). The actions contain timestamps and properties.	Behavioral (Clickstream)	Digital space	act-mooc.tar.gz
Carnegie Mellon University's DataShop and DataLab	Holds 358,000 student records from online courses, intelligent tutors, simulators, and educational games, totaling more than 705,000 hours of student data spread among 1466 datasets.	Behavioral (Clickstream)	Digital space	https://www.cmu.edu/datalab/tools/datashop.html
Open University Learning Analytics dataset	It includes information for seven courses on the Virtual Learning Environment (VLE), students, and their interactions with it.	Behavioral (Clickstream)	Digital space	https://analyse.kmi.open.ac.uk/open_dataset
NUS Multi-Sensor Presentation (NUSMSP) Dataset	It comprises four different categories of data: educational data, sensor data, EMA data, and pre-and post-survey replies.	Behavioral + Affective	Physiological, Digital, and Psychometric space	https://studentlife.cs.dartmouth.edu/dataset.html
EdNet	Santa, a multiplatform AI tutoring service with more than 780K users, has aggregated all student-system interactions into a dataset called EdNet.	Behavioral (Clickstream)	Digital space	https://github.com/riiid/ednet
MUTLA: A Large-Scale Dataset for Multimodal Teaching and Learning Analytics	In this dataset, students' time-synchronized multimodal data records from the Squirrel AI Learning System (SAIL) are used to answer tasks of increasing degrees of complexity. These records include learning logs, videos, and EEG brainwaves.	Behavioral + Cognitive + Affective	Digital space + Physiological + Psychometric space	https://paperswithcode.com/dataset/mutla
SEED dataset	This data set consists of 15 Chinese movie clips representing the positive, neutral, and negative emotions selected to serve as the stimuli. The other variants of the data set include SEED GER and SEED FRA.	Affective	Digital space + Physiological	https://bcmi.sjtu.edu.cn/home/seed/.

V. INFERENCES

The following study was initiated to investigate how technology has enabled the capture of latent learning indicators, what devices will be helpful for data capture, what the current trend and what data sets are available that enable the research in learning analytics. This paper provides an overview for the researchers to understand the essentials and directions of the learning analytics research area. It was observed that there is a shift in emphasis from capturing behavioral characteristics of a learner in the formal learning settings to capturing cognitive and affective characteristics in the informal learning settings. EEG capturing devices can help to capture both cognitive and affective characteristics. Further the inferences drawn from the study are described in two dimensions.

1. From the perspective of handling the data from multiple sources and their integration.
2. From the perspective of connecting the learning theories to the constructs and measured parameters.

A. Handling the data from multiple sources and its integration

Several studies have shown that the direction toward using multiple data sources to capture student learning and research is toward integrating data from multiple sources towards predictive, prescriptive, and adaptive analytics.

Due to the enormous amount of data from multiple sources, many challenges are encountered concerning data pre-processing, data quality, and data alignment at one level. At another level, making the right choice of appropriate data integration methods.

Fusing information from either different data spaces or different characteristics refers to 'Multimodal data fusion'; an emerging field in this direction is 'Multimodal Learning Analytics.'

Literature categorizes the data fusion methods into three broad categories: rule-based, classification-based, and estimation-based.

On the other hand, researchers have fused the data based on the application at the feature level (early fusion), parameter level (decision level), or hybrid level.

Researchers started with the simple concatenation of data using mathematical operators followed by other techniques, including similarity-based approaches, probability-based approaches, and ensemble-based approaches, and the recent approach is the use of attention-based mechanisms. The authors identified a need for the studies that involves investigation of statistical association between the multimodal data thus helping to model the joint distribution of data.

B. *Connecting the learning theories with the measured constructs and parameters*

The authors observed that several researchers have focused on only capturing one or more parameters and their integration towards describing the data, diagnosing the results, or predicting. However, the string connecting the measured parameters with the learning theories needs more substantial and further research.

VI. Conclusion

Though enough research is available for capturing the latent learning indicators, integrating the data from multiple sources, and the availability of suitable data sets, one of the bottlenecks for implementing on the various online platforms or LMS, is the use of external devices, which adds up to the cost. The behavioral characteristics can be easily captured due to the availability of ubiquitous cameras in the form of webcams and CCTVs, but to capture the cognitive and affective characteristics, there is a need for affordable devices and software for the effective utilization of the carried-out research work. The technology has enabled to capture the latent learning indicators including attention, cognitive load, cognitive stress, interest, excitement, engagement, frustration and boredom with the help of devices including emotiv EEG sensors. The drift from measuring the behavioral characteristics of the learners to cognitive and affective characteristics has enabled the research towards affordable devices that can capture the data with temporal resolution from milli seconds to hours and seamlessly get integrated with the existing learning management systems.

References

Amashi, R., Koppikar, U., & Vijayalakshmi, M. (2023). Investigating the Association Between Student Engagement With Video Content and Their Learnings. IEEE Transactions on Education.

Amashi, R., Vijayalakshmi, M., Kotturshettar, B. B., Kodancha, K., & Shettar, A. (2021, October). FAQ as a tool in Asynchronous Mode of Delivery in Blended Learning. In 2021 IEEE Frontiers in Education Conference (FIE) (pp. 1-5). IEEE.

Anwar, S. N. S. S., Abd Aziz, A., & Adil, S. H. (2021, November). Development of real-time eye tracking algorithm. In 2021 4th International Conference on Computing & Information Sciences (ICCIS) (pp. 1-6). IEEE.

Ashwin, T. S., Prakash, V., & Rajendran, R. (2023). A Systematic Review of Intelligent Tutoring Systems based on Gross Body Movement Detected using Computer Vision. Computers and Education: Artificial Intelligence, 100125.

Baker, R. S., D'Mello, S. K., Rodrigo, M. M. T., & Graesser, A. C. (2010). Better to be frustrated than bored: The incidence, persistence, and impact of learners' cognitive–affective states during interactions with three different computer-based learning environments. International Journal of Human-Computer Studies, 68(4), 223-241.

Benitez, D. S., Toscano, S., & Silva, A. (2016, April). On the use of the Emotiv EPOC neuroheadset as a low cost alternative for EEG signal acquisition. In 2016 IEEE Colombian Conference on Communications and Computing (COLCOM) (pp. 1-6). IEEE.

Bhattacharya, S., Nainala, G. S., Das, P., & Routray, A. (2018, July). Smart attendance monitoring system (SAMS): a face recognition based attendance system for classroom environment. In 2018 IEEE 18th international conference on advanced learning technologies (ICALT) (pp. 358-360). IEEE.

Botelho, A. F., Varatharaj, A., Patikorn, T., Doherty, D., Adjei, S. A., & Beck, J. E. (2019). Developing early detectors of student attrition and wheel spinning using deep learning. IEEE Transactions on Learning Technologies, 12(2), 158-170.

Chakladar, D. D., Dey, S., Roy, P. P., & Iwamura, M. (2021, January). EEG-based cognitive state assessment using deep ensemble model and filter bank common spatial pattern. In 2020 25th International Conference on Pattern Recognition (ICPR) (pp. 4107-4114). IEEE.

Chango, W., Cerezo, R., & Romero, C. (2021). Multi-source and multimodal data fusion for predicting academic performance in blended learning university courses. Computers & Electrical Engineering, 89, 106908.

Condon, W., & Kelly-Riley, D. (2004). Assessing and teaching what we value: The relationship between college-level writing and critical thinking abilities. Assessing Writing, 9(1), 56-75.

Giannakos, M. N., Sharma, K., Pappas, I. O., Kostakos, V., & Velloso, E. (2019). Multimodal data as a means to understand the learning experience. International Journal of Information Management, 48, 108-119.

Grandchamp, R., Braboszcz, C., & Delorme, A. (2014). Oculometric variations during mind wandering. Frontiers in psychology, 5, 31.

Hanley, T., & Cutts, L. (2013). What is a systematic review. Counselling Psychology Review, 28(4), 3-6.

Hassan, R., Hasan, S., Hasan, M. J., Jamader, M. R., Eisenberg, D., & Pias, T. (2020, May). Human attention recognition with machine learning from brain-EEG signals. In 2020 IEEE 2nd Eurasia Conference on Biomedical Engineering, Healthcare and Sustainability (ECBIOS) (pp. 16-19). IEEE.

Henderson, N., Rowe, J., Paquette, L., Baker, R. S., & Lester, J. (2020). Improving affect detection in game-based learning with multimodal data fusion. In Artificial Intelligence in Education: 21st International Conference, AIED 2020, Ifrane, Morocco, July 6–10, 2020, Proceedings, Part I 21 (pp. 228-239). Springer International Publishing.

Hussain, M. S., AlZoubi, O., Calvo, R. A., & D'Mello, S. K. (2011). Affect detection from multichannel physiology during learning sessions with AutoTutor. In Artificial Intelligence in Education: 15th International Conference, AIED 2011, Auckland, New Zealand, June 28–July 2011 15 (pp. 131-138). Springer Berlin Heidelberg.

Joshi, Y., Mallibhat, K., & Vijayalakshmi, M. (2022, November). Students' performance prediction using multimodal machine learning. In 2022 IEEE IFEES World Engineering Education Forum-Global Engineering Deans Council (WEEF-GEDC) (pp. 1-5). IEEE.

Joshi, Y., Mallibhat, K., & Vijayalakshmi, M. (2022, November). Systematic literature review: An investigation towards finding constructs for performance prediction of students in an online engineering course. In 2022 IEEE IFEES World Engineering Education Forum-Global Engineering Deans Council (WEEF-GEDC) (pp. 1-5). IEEE.

Khosravi, S., Khan, A. R., Zoha, A., & Ghannam, R. (2022, October). Employing a Wearable Eye-tracker to Observe Mind-wandering in Dynamic Stimuli. In 2022 29th IEEE International Conference on Electronics, Circuits and Systems (ICECS) (pp. 1-4). IEEE.

Krigolson, O. E., Williams, C. C., Norton, A., Hassall, C. D., & Colino, F. L. (2017). Choosing MUSE: Validation of a low-cost, portable EEG system for ERP research. Frontiers in neuroscience, 11, 109.

Lai, S., Liu, J., Niu, B., Tian, H., & Wu, F. (2019, December). Combining facial behavioral cues, eye movements and EEG-based attention to improve prediction of reading failure. In 2019 International Joint Conference on Information, Media and Engineering (IJCIME) (pp. 485-489). IEEE.

LaRocco, J., Le, M. D., & Paeng, D. G. (2020). A systemic review of available low-cost EEG headsets used for drowsiness detection. Frontiers in neuroinformatics, 42.

Li X., Hu, B., Dong, Q., Campbell, W., Moore, P., & Peng, H. (2011, October). EEG-based attention recognition. In 2011 6th International Conference on Pervasive Computing and Applications (pp. 196-201). IEEE.

Liu, R., Stamper, J., Davenport, J., Crossley, S., McNamara, D., Nzinga, K., & Sherin, B. (2019). Learning linkages: Integrating data streams of multiple modalities and timescales. Journal of Computer Assisted Learning, 35(1), 99-109.

Mallibhat, K., Chikkamath, S., Tabib, R., Nirmala, S. R., & Mudenagudi, U. (2022). Project-based learning in machine learning course: experience and

outcomes. In EDULEARN22 Proceedings (pp. 1545-1551). IATED.
Masood, N., & Farooq, H. (2017). Emotiv-based low-cost brain computer interfaces: A survey. In Advances in Neuroergonomics and Cognitive Engineering: Proceedings of the AHFE 2016 International Conference on Neuroergonomics and Cognitive Engineering, July 27-31, 2016, Walt Disney World®, Florida, USA (pp. 133-142). Springer International Publishing.
Mele, M. L., & Federici, S. (2012). Gaze and eye-tracking solutions for psychological research. Cognitive processing, 13, 261-265.
Moher, D., Liberati, A., Tetzlaff, J., Altman, D. G., & PRISMA Group*. (2009). Preferred reporting items for systematic reviews and meta-analyses: the PRISMA statement. Annals of internal medicine, 151(4), 264-269.
Monkaresi, H., Bosch, N., Calvo, R. A., & D'Mello, S. K. (2016). Automated detection of engagement using video-based estimation of facial expressions and heart rate. IEEE Transactions on Affective Computing, 8(1), 15-28.
Mu, S., Cui, M., & Huang, X. (2020). Multimodal data fusion in learning analytics: A systematic review. Sensors, 20(23), 6856.
Olsen, J. K., Ozgur, A. G., Sharma, K., & Johal, W. (2022). Leveraging eye tracking to understand children's attention during game-based, tangible robotics activities. International Journal of Child-Computer Interaction, 31, 100447.
Omidyeganeh, M., Shirmohammadi, S., Abtahi, S., Khurshid, A., Farhan, M., Scharcanski, J. & Martel, L. (2016). Yawning detection using embedded smart cameras. IEEE Transactions on Instrumentation and Measurement, 65(3), 570-582.
Page, M. J., McKenzie, J. E., Bossuyt, P. M., Boutron, I., Hoffmann, T. C., Mulrow, C. D., ... & Moher, D. (2021). The PRISMA 2020 statement: an updated guideline for reporting systematic reviews. International journal of surgery, 88, 105906.
Papoutsaki, A. (2015, April). Scalable webcam eye tracking by learning from user interactions. In Proceedings of the 33rd Annual ACM Conference Extended Abstracts on Human Factors in Computing Systems (pp. 219-222).
Prathish, S., & Bijlani, K. (2016, August). An intelligent system for online exam monitoring. In 2016 International Conference on Information Science (ICIS) (pp. 138-143). IEEE.
Sharif, B., Peterson, C., Guarnera, D., Bryant, C., Buchanan, Z., Zyrianov, V., & Maletic, J. (2019, May). Practical eye tracking with itrace. In 2019 IEEE/ACM 6th International Workshop on Eye Movements in Programming (EMIP) (pp. 41-42). IEEE.
Sharma, S., & Kumar, R. (2018). Activity based learning system in higher education: A student centred approach. International Journal of Social Science, 7(4), 469-474.
Souza, R. H. C. E., & Naves, E. L. M. (2021). Attention detection in virtual environments using EEG signals: a scoping review. frontiers in physiology, 12, 727840.
Toa, C. K., Sim, K. S., & Tan, S. C. (2021). Electroencephalogram-based attention level classification using convolution attention memory neural network. IEEE Access, 9, 58870-58881.
Vijaylakshmi, M., Baligar, P., Mallibhat, K., Kavale, S. M., Joshi, G., & Shettar, A. (2021, October). Transition from in-person learning to technology enhanced learning in engineering education: Faculty challenges. In 2021 IEEE Frontiers in Education Conference (FIE) (pp. 1-5). IEEE.
Villarroel, R., Villalobos, C., Merino, E., Barcelos, T., & Munoz, R. (2018, November). Developing a dashboard to support the analysis of multimodal educational data. In 2018 37th International Conference of the Chilean Computer Science Society (SCCC) (pp. 1-6). IEEE.
Wang, Y., Lu, S., & Harter, D. (2021). Towards Collaborative and Intelligent Learning Environments Based on Eye Tracking Data and Learning Analytics: A Survey. IEEE Access, 9, 137991-138002.
Wong, A. Y., Bryck, R. L., Baker, R. S., Hutt, S., & Mills, C. (2023, March). Using a Webcam Based Eye-tracker to Understand Students' Thought Patterns and Reading Behaviors in Neurodivergent Classrooms. In LAK23: 13th International Learning Analytics and Knowledge Conference (pp. 453-463).
Wood, D. F. (2003). Problem based learning. Bmj, 326(7384), 328-330.
Xu, T., Zhou, Y., Wang, Z., & Peng, Y. (2018). Learning emotions EEG-based recognition and brain activity: A survey study on BCI for intelligent tutoring system. Procedia computer science, 130, 376-382.
Yang, X., & Krajbich, I. (2021). Webcam-based online eye-tracking for behavioral research. Judgment and Decision making, 16(6), 1485-1505.
Yue, J., Tian, F., Chao, K. M., Shah, N., Li, L., Chen, Y., & Zheng, Q. (2019). Recognizing multidimensional engagement of e-learners based on multi-channel data in e-learning environment. IEEE Access, 7, 149554-149567.
Zheng, W. L., Liu, W., Lu, Y., Lu, B. L., & Cichocki, A. (2018). Emotionmeter: A multimodal framework for recognizing human emotions. IEEE transactions on cybernetics, 49(3), 1110-1122.

Copyright statement

https://doi.org/10.52202/073963-0014

Research in Engineering Education Symposium-2024
Jan 4-6, 2024 - Hubballi, India

Exploring Motivational Factors for Faculty Engagement in Service-Learning Courses: Perspectives from Indian Undergraduate Engineering

Surendra Bandi[a], Gopalkrishna Joshi [b], Ashok Shettar[c], Rohit Kandakatla[d]
Hyderabad Institute of Technology and Management, Hyderabad, Telangana, India [a], MIT Vishwaprayag University, Solapur, Maharastra, India[b],
KLE Technological University, Hubli, Karnataka, India[c], KG Reddy College of Engineering and Technology, Hyderabad, Telangana, India[d]
surendra.mca@hitam.org, ghjoshi@gmail.com, vc@kletech.ac.in, rohit.kandakatla@kgrcet.ac.in

Abstract

Context

Service-learning is a teaching and learning strategy which integrates academic learning with meaningful community engagement and helps in an increased sense of civic responsibility (Oakes, 2014). Most research on service-learning so far have focused on the impact of service-learning on student learning, community benefits and institutional development and there is little literature available on the role of faculty in service-learning (S. R. Bandi et al., 2023). The article will investigate the motivational factors for engineering faculty in India to engage in service-learning programs.

Purpose

There have been increasing calls on the wider adoption of service-learning in undergraduate engineering in India, as it helps engineering students achieve graduate attributes as recommended by international accreditation agencies (Kandakatla et al., 2023). However, for engineering institutions to adopt service-learning, it is critical for to get the faculty engaged as they are one of the key stakeholders who will lead the design and implementation of these programs The study attempts to answer the research question - "What are the factors that influence faculty engagement in service-learning?" with a goal to understand how engineering faculty be motivated to engage in service-learning programs in India.

Approach

Qualitative approach is chosen as the methodology for the study as we intend to investigate the experiences of engineering faculty which will reveal their motivational factors to engage in service-learning programs (Michael Quinn Patton, 2002). Data is collected from faculty at an engineering institution in South of India which has systematically integrated service-learning into their curriculum. Semi structured interviews were used to collect the data and we employed thematic analysis for data analysis to get an in-depth exploration of the faculty perspectives.

Outcomes

The pilot study revealed primary insights on the impact of service-learning courses on faculty learning in undergraduate engineering-education. Demographic background of faculty, peer support and institutional encouragement are the motivational factors for faculty members to be part of service-learning facilitation. The results will include a detailed thematic analysis and discussion on the motivational factors for faculty in service-learning.

Conclusions

India announced the National Education Policy (NEP) in 2020 and service-learning has been one of the most relevant pedagogies to achieve the vision laid out in the policy (S. Dustker et al., 2023). Findings from the study will enable engineering education leaders to gain insights on how build capacity to enhance faculty skills in service-learning, make available required for effective planning and implementation, and provide incentives to motivate and encourage engineering faculty to engage in service-learning.

Keywords— Service-learning; Faculty Engagement; Motivational Factors

I. INTRODUCTION

The adoption of service-learning practices in higher education has been steadily increasing across the globe (Clayton et al., 2023). Service-learning provides an approach for faculty to bridge the gap between theory and practice by integrating them in a meaningful way. Through service-learning, students are engaged in real-time problem-solving experiences aimed with a goal to develop solutions for the benefit of the partner communities. Service-learning offers students with several opportunities to experience empathy (Kandakatla et al., 2022), become socially responsible, and build professional skills such as communication, teamwork, and project management (Kandakatla et al., 2023). Involvement in service-learning experiences have been reported to also have a positively impact on student's academic achievement, self-

https://doi.org/10.52202/073963-0015

efficacy, and leadership (Astin et al., 2000). In the last few decades, there has been a growing acceptance of service-learning as a pedagogy among engineering institutions across the globe.

Same is the case in India, where there has been significant push to enable higher education institutions to engage with local communities. The government of India launched a national level program called Unnat Bharat Abhiyan UBA) that aims to bring transformational change in the rural development of India with the support of higher education institutions, while students get an opportunity to understand the society (Radhakrishnan et al., 2022). Community engagement and service has been an integral part of Indian education system. Students who are part of implementing UBA activities at higher education institutions in India immensely benefit by building their social, technical and leadership skills (S. Bandi & Naik, 2020). The National Education Policy (NEP) announced by India also highlights the importance of holistic development and multidisciplinary education. Service-learning addresses achieving objectives of holistic development of the learners and multidisciplinary approaches (S. Dustker et al., 2023).

The announcement of the UBA program in India has led to many institutions to launch community engagement programs and adopt service-learning as a pedagogy for students learning. However, effective implementation of service-learning requires tremendous level of preparations and efforts from the institutions especially the faculty. Faculty engaged in service-learning efforts are one of the most important stakeholders who could influence the success or failure of the initiatives (Bandi et al., 2023). However, faculty can find it discouraging to engage in service-learning as it has been known to significantly increase their responsibilities and workload (Camus et al., 2022). It is therefore essential that the faculty engaged in service-learning are highly motivated and committed to both the community impact and students' learning. This study aims to understand the underlying factors that motivated engineering faculty in India to engage in service-learning programs. A qualitative methodology was used to design and conduct the study at a private technological university in South India. Data was collected using semi-structured interviews that were conducted with faculty members who are part of teaching service-learning courses to undergraduate engineering students at the university. The results are presented through different themes that indicate how faculty could be motivated to adopt service-learning in their teaching practices. The article will enable faculty to explore if they belong to the service-learning initiative and help leadership of the higher education institutions to sustain motivation among the faculty for service-learning. The next sections of the paper will cover literature survey, methodology, results, discussion, and conclusion.

II. LITERATURE SURVEY

This section provides an overview of the literature on faculty engagement in service-learning.

A. *Background of Service-learning*

There are several definitions for service-learning in the literature. The following definition by Bringle and Hather is widely used - *"Service-learning is a credit-bearing, educational, experience in which students participate in an organized service activity that meets identified community needs and reflect on the service activity in such a way as to gain further understanding of course content, a broader appreciation of the discipline, and an enhanced sense of civic responsibility."* (Bringle & Hatcher, 1995). Service-learning has four components – academic learning, service, reciprocal partnerships (Delaine et al., 2023) and reflection (Oakes et al., 2002). Credit is given for the academic learning with the standards of the respective regulators. Learners take part in services that attempt applying their knowledge and skills in proposing solutions to the problems of the community. The long-lasting partnerships built among the partners (Students, Faculty, Institutions, Community etc.) benefit from each other. Student often reflect on the service-learning experiences to analyze and further their learning (Dustker et al., 2021). There have been different models of how institutions integrated service-learning into their curriculum. Some of the popular models include EPICS (Engineering Projects in Community Service) of Purdue University, SLICE (Service-learning integrated throughout College of Engineering) of University of Massachusetts-Lowell and HESE (Humanitarian Engineering and Social Entrepreneurship) of Penn State University, USA, which successfully implemented service-learning in their curriculum (Bielefeldt & Pearce, 2012).

B. *Faculty engagement in service-learning*

A study on the enablers and barriers to implementation of service-learning report that faculty are one of the key stakeholders the success of service-learning initiatives (Bandi S, 2021). Institutions take high level decisions on allowing integration of service-learning in curriculum and students are the recipients. Faculty are the bridge between the strategic planning of courses and delivery to the students. Faculty spend significant time with students and become very important in success or failure of the initiative. Faculty also plays an important role in making students work with community. The role of faculty in service-learning to improve connection with community is highlighted by Rona (Karasik, 2020). There is an appreciation to bring real time problems to the classroom and work collaboratively to achieve solutions. If the faculty are not motivated to integrate service-learning in academics, the outcome will be uncertain. Faculty agree to the pedagogy for better understanding and learning (Blakey et al., 2016).

https://doi.org/10.52202/073963-0015

C. Factors influencing faculty engagement in service-learning

A range of factors motivate faculty to incorporate service-learning in their teaching. Darby et al. explored how faculty are motivated to be involved in service-learning (Darby & Newman, 2014). The study reported that faculty realized benefits of the pedagogy and the value on students learning and development. A discourse analysis in this context elicited a teaching learning model, tool for personal identity, institutional mission, and community participation (O'Meara & Niehaus, 2009). Wide range of factors were revealed by the faculty in a study by Chris (Hammond, 1994). The factors include the experience of faculty in the past, unselfish desire to serve society, support from the institutions, civic values, societal values, and teaching. Choi et al described how students work closely with their classmates and faculty where the bonding gets stronger (Choi et al., 2023) which becomes a motivation for faculty. Following are the categories under which the faculty motivation happens for choosing service-learning, as per the literature explored.

D. Institutional and Administrative Support

Not all faculty undergo mandatory training before they take up teaching, particularly in higher education. But it is required to understand different learning styles of the students and plan the curriculum accordingly. Faculty needs a specific training to take up service-learning courses (Derreth et al., 2022). Few institutions established suitable mechanism to sustain community engagement in academic learning. Different models are designed to cater the needs of the undergraduate engineering students across the world. Suitable models have to be developed for effective outcomes of the service-learning pedagogy (Kiely, 2005). Training to the faculty is required to continuous learning for improvement of teaching. Grace Ngai presented the obstacles faced and the respective strategies to address the challenges in implementing service-learning (Ngai & Chan, 2019). There are challenges ranging from faculty background, expertise, financial benefits, time, patience, and commitment to delivering etc. Sufficient funding and time are required for faculty to properly plan and implement the community engagement initiatives for academic enhancement (Kelli, 2020). The challenges once mitigated or resolved, the same become motivators to practice the same in different contexts for learning. There are studies on how the challenges have been addressed and later became potential contributors. While there are challenges for practitioners of service-learning, there are suitable solutions to overcome (Garvin & Acosta Lewis, 2022).

E. Gaps & research question

Studies on service-learning so far focused on the benefits to the students, community and to some extent on the institutions. However, a critical gap exists regarding the faculty motivations for community engagement in service learning within Indian context. There is less literature available on the role of faculty and motivation, particularly in Indian context. According to Anderson et al practitioners should reimagine the service-learning interventions and role of faculty (Anderson et al., 2019). Though the sources available are limited, the literature survey provided useful information and insights. But it lacks a comprehensive analysis of factors driving faculty participation and the impact of faculty engagement in service-learning. For Indian context, with the increase in adoption of service-learning, it is important to understand how faculty get motivated to engage in service-learning. Hence this study on "Faculty motivations to engage themselves in community engagement for academic service learning of the students in the Indian context" is taken up to fill the gap.

III. METHODOLOGY

A. Research design

The aim of the study is to analyze faculty motivations in service-learning by considering a case in the Indian context. The research question is "What are the factors that influence faculty engagement in service-learning?". We took a qualitative approach to design the study as we could investigate the experiences of faculty who were engaged in service-learning programs. The experiences of faculty are multifaceted in this case. We recognize that the responses from the faculty are complex and diverse which necessitates an inquiry to capture depth of their perspectives. The open-ended nature of the faculty responses and outcomes of the study justifies the need to adopt qualitative methodology for the study (John W. Creswell & J. David Creswell, 2018).

B. Sample Strategy

The study is planned to conduct on the faculty of a private technological university in southern India. There are 20 faculty members from two campuses of university who taught the service-learning course. The criteria guiding participant selection are as follows:

a) Experience in Facilitating Service-Learning Courses: Participants were chosen based on their experience in facilitating service-learning courses. This criterion ensures that the insights gathered are grounded in the practical engagement of faculty members with service-learning pedagogy.

b) Diversity in Years of Experience: To capture a comprehensive range of perspectives, the sample includes faculty members with varying years of experience, ranging from lower to higher levels. This diversity aims to know potential differences and commonalities in motivations across different career stages.

c) Representation Across Engineering Disciplines: Recognizing the multidisciplinary nature of engineering academics, the sample includes faculty members from different engineering disciplines. This diversity enhances the study's richness by exploring how motivations for engaging in service-learning may vary across distinct engineering fields.

The university authorities granted permission to interview the identified 20 participants, and 19 faculty members were successfully interviewed for the study. The cooperation of the university and the high participation rate indicates the relevance and interest of faculty members in contributing to the understanding of motivational factors in service-learning within the context of this private technological university in southern India.

C. *Data collection*

Target participants of the study are the faculty who are facilitating service-learning courses. As there is a smaller number of institutions offering such courses, the number of faculty who have experience of teaching service-learning courses is less. Investigation of the research question required us to elicit rich qualitative data from the participants and semi-structured interviews were used for collecting data. An interview instrument was designed to systematically collect the required qualitative analysis. To establish the context of the participants' background, the interview questions on their demographics and educational backgrounds are added after they introduce themselves. The next question is to directly understand what made them to be in service-learning including their personal choices and enforcement from the institution. We also wanted to know if their previous service-learning experiences were influencing them. So a question was added on what was their involvement in the community engagement activities prior to taking up this assignment. The next questions were to understand the motivation from the stakeholders including colleagues, community and institution. A couple of pilot interviews were conducted to validate the instrument and there were few changes made in the instrument to avoid binary, leading and biased questions. The interviews are audio recorded, transcribed, and analyzed.

D. *Data Analysis*

The data transcripts are cleaned to remove filler words and repetitions. After exploring the transcripts multiple times, a codebook was created using the relevant quotes which are in line with the objective of the study. Themes evolved out of the codes with internal connections. Thematic analysis is done to answer the research question proposed. The interview transcripts were read several times to see the emerging patterns. Codes have been assigned to the specific quotes which highlight the motivational factors of faculty. Broad themes evolved out of the data excerpts that were coded during the analysis.

E. *Reliability and Validity of the Study*

The reliability of the semi-structured interview protocol was checked through cognitive pilot interviews conducted prior to the start of the data collection process. The codebook prepared during the data analysis was validated in collaboration with a fellow researcher to verify the initial round of the coding process. An approval from the Ethics board of the university is taken before proceeding to the participant interviews. The discussion section highlights the outcomes of the study and comparison of the findings with the existing body of literature on the motivational factors for faculty to accept teaching academic service-learning courses.

IV. RESULTS

The results of the study are presented using themes and sub themes. Every theme specifies a particular factor that motivated faculty members to choose service-leaning in their teaching. Quotes from the participants responses are shown in the results to exemplify the findings. Themes and subthemes emerged out of the data collected after familiarizing the data, generating initial codes and by combining the related codes. We will provide details results accompanied by the supporting evidence.

A. *Theme 1: Professional Growth Opportunities for Faculty with Service-Learning*

Several participants mentioned that their experience of teaching service-learning courses helped in their professional growth. The following subthemes support the analysis further.

1) *Sub Theme 1.1: Improvement in Teaching Skills*

Some faculty members did not agree to take up service-learning courses by their own choice while some are interested from the beginning but most of them started liking it as they progressed. The following two quotes exemplify the same.

"I thought it will be better opportunity to me to learn the teaching skills, so I agreed to it, and it worked for me." (Participant HP2)

"Initially it was not a choice, but as I mean as a part of what we called teaching learning methodology or process. I gained interest with relation to this course." (Participant H1)

One participant explained that the positive impact of the experience was beyond the service-learning courses to other courses as well.

"Service learning. I think it is very, very important and it has enriched my knowledge. Whatever theoretical knowledge I had, I have learned it practically now and I'm more empowered to teach any other subjects and implement service learning in other subjects. That has motivated me." (Participant B13)

These insights highlight the broader value of integrating service-learning which help in improving teaching methods and professional growth.

2) *Sub Theme 1.2: Student Mentoring*

In this sub theme, we delve into the experiences of the faculty members in service-learning significantly improved their ability to mentor students effectively. Through their narrations, we can gain further insights.

The following statement from the participant indicates the impact of service-learning on the faculty member's ability and comprehend and engage students at a deeper level. It says that the service-learning helped faculty to understand students' behavior and team dynamics.

"Certainly, these things will help us in understanding the behavior of the students of behavior of the teams." (Participant: H1)

Another participant reflected on her growth as mentor. The narrative reveals that the multifaced nature of faculty member's experience emphasizes on the improvements in managing students, stakeholder interactions, communication, and teaching methods. It shows a strong transformative effect on service-learning on faculty's mentoring capabilities.

"I learned how to manage the students, how to interact with the stakeholders, how to use different methods to teach the difficult concept. These are the things what I learned from. Even my communication how to engage all things. It helps me a lot." (Participant B8)

A participant described the reciprocal nature of mentoring in service-learning. This illustrates the two-way impact of mentoring by faculty members in service-learning. With this faculty understand the students better and help them better.

"When we start mentoring the students, we also learn many more things. We were also not aware of personalized learning aspects of the students. Once we started into being with questions, start mingling with students and we'll come to know." (Participant B12)

The participants' narratives conclude that there is a deeper understanding of the students behavior and team dynamics as a result of continuous interaction between faculty and students in the community engaged projects.

3) Sub Theme 1.3: Enhanced Personal Learning Through Service-Learning Courses for Faculty

A participant has seen an opportunity to learn how to see a real-world problem from the user and solver ends.

"When we go to the community, we will go into real field testing. We will understand the things and improve observation skills as well, and then how to communicate with the stakeholders and all and then how to define a problem. We will have both the perspective of faculty and community be mixed and we will define a problem properly and then we'll go to ideate." (Participant: HP2)

Another participant revealed that the course helped in improving his skills and research skills.

"It has helped in presentation skills and in designing the activities for the courses. So actually, we are thinking of bringing out all these ideas as papers this time, once papers and all, so may be at that time we can expect the professional growth." (Participant H3)

Another participant spoke about the way the service-learning experience was useful in dealing with students while addressing their contextual issues.

"The learning was happening from day one and every time I learned the accumulated knowledge helped me to pass on and deal with the other students. Who came later and prepared them better than what I used to do in the past. So, accumulation of knowledge and experience makes you stronger in in in trying to tell and teach the students to defend themselves and work in their situations." (Participant H2)

Another faculty member shared that there is improvement in the learning when he repeated the course for the next batches of the students.

"It has made me a better person personally and professionally. I mean they say that you teach service learning now in a better manner than compared to earlier. So, I've got a positive result when I take this course." (Participant B10)

Participants highlighted the valuable opportunity to approach real-world problems from both user and solver perspectives, refining their observation skills and communication with stakeholders.

4) Sub Theme 1.4: Research Opportunities

One participant described the increased interest and opportunity to extend research on Design Thinking and Social Innovation further.

"I'm able to do a lot of research related to DTSI. I've done one paper already and trying to communicate it now and. In other subjects also I'm able to get many research ideas. This will help me in my professional group also. Professionally, I'm able to meet people here and interact with the principal, interact with many courses' coordinator, and all. It brings a kind of visibility also on the campus, if you are involved in service learning." (Participant B13)

This subtheme focused on the participant's increased interest and expanded research opportunities in Design Thinking and Social Innovation through service-learning, demonstrating its positive impact not only on professional growth but also on campus visibility and networking.

5) Sub Theme 1.5: Institutional Support

The faculty members acknowledged their willingness and motivation to visit communities to facilitate service-learning course with the required logistic support by the institution. The same is presented here by a participant as

"All sorts of support from institution were given. Like for example if we wanted to go to community visit bus facilities was provided by the institution itself. We never made our own vehicles work or we never went by ourselves. It was all made easier for us by the institution." (Participant: B10)

The critical role of institutional support in sustaining faculty members' willingness and motivation to engage in community visits for service-learning courses is exemplified by the provision of facilities for these visits.

6) Sub Theme 1.6: Interdisciplinary Approach

Many participants reported that there are no boundaries between subjects and courses when dealing with the service-learning courses.

"Service-learning subject has taken more depth in terms of me having to acquire more knowledge than required. When I'm teaching any other subject, I have a boundary in which I gain the knowledge and I leave it there. But this is not the case with service-learning subject." (Participant H2)

"Problem or certain situation we would diversify in apart from our own discipline, we would tend to think in a multidisciplinary aspect. So, it could help even in professional growth." (Participant B9)

The participants emphasized the nature of service-learning courses, by removing the traditional subject boundaries and the encouragement of interdisciplinary thinking.

7) Sub Theme 1.7: Career Progression

The faculty reflected that there is some weightage in their appraisal process for those who are part of service-learning. Here is a quote in support.

"Ultimately, we do have an appraisal system where certain aspects in which the questions are related to whether faculty members are engaged in developing the newer courses, whether he or she is engaged in developing newer methodologies? Yes. Surely this service-learning experience helps in it." (Participant: H1)

One participant was specific in briefing the details of learning and how it relates to his career progression.

"It has helped me in my career also because I have learned many more things focusing on problems. Real world applications like community engagements and some experience. Like reflection and integrations, interdisciplinary approach, soft skill development, personal growth, motivation and engagement, and ethical consideration, and also can take challenges and complexities." (Participant B12)

Participants emphasized the impact of service-learning on career progression, noting its positive influence on the faculty appraisal process.

B. Theme2: Opportunity to contribute to society.

1) Subtheme 2.1: Motivation Through Community Engagement in Service-Learning

When we asked the participants what motivated them to accept service-learning courses, they mentioned that engaging with society helps both faculty and students to learn with ease.

"This course offers that flexibility, and the course main objective is to bring a kind of change and make students learn to engage with the society and all. Yes, definitely that was the main motivation factor." (Participant: B13)

Many other participants highlighted that the community engagement was never part of their regular courses. One example from the data is.

"This course is helping us to develop our experiential learnings and we could have a community engagement which was not there in the any of the courses. And here it also involves interdisciplinary approaches. Means which helping us to have a cultural competence, along with the ethical consideration with respect to the society and also the long-term commitment." (Participant: B14)

The participants expressed that the motivation to embrace service-learning courses comes from the opportunity they provide for both faculty and students to learn seamlessly by engaging with society.

2) Subtheme 2.2: Passion to Serve Community

Some of the faculty members have an intrinsic motivation to serve the community. In their words, it gives them satisfaction that they are part of social service.

"Initially it was by deputation. We were into DTSI and once we got into FDP, it was something interesting. What I got to know because we were actually getting into the social cause, knowing their problem, and trying to simplify the various aspects." (Participant: B9)

"Because to it is a societal cause to serve the society that is the main objective. That's what I feel. That's what will remain in the future. Something like that to whatever the service we provide or serve the society that will be the remaining long term or achievement you will have that satisfaction." (Participant B1)

One participant expressed his passion towards contributing to society and how it was made true through teaching a course.

"I had the passion from the beginning, like something I have to do for the society. So, I have to contribute something to society. So, in this regard, I came to know we have to take up one service learning. Courses like social innovation, so I thought like, why not join this group so that I can take this subject for the students and with the help of student group I can reach the society." (Participant H3)

Another participant shared his thoughts on the responsibilities of a citizen towards a society to show gratitude.

"Giving back to society is a big thing because I believe since I used to deal or teach system simulation, so nothing exists in isolation. Everything is interdependent on each other so as we grow in our lives. You can't say that I don't have to look at people around me or I have nothing to do with them. That kind of an attitude would not help me survive or grow on a longer run because we're all interdependent on each other." (Participant H2)

An interviewee realized the scope for faculty and students to create an impact on the society through the design thinking course.

"I found that here there is a great opportunity for us to make students go out there and whatever they have learned and implement that interact with the society and try to bring in a kind of change in the society." (Participant: B13)

As per a participant's viewpoint on the value of design thinking course is already known from an online course and it motivated further to accept the invitation to facilitate the similar course to the students. Following is the quote from his own words.

"Openly accepted the invitation to be a part of the service learning because I had done a NPTEL course on design thinking for social design thinking course and I found it very exciting, and I thought that I'll get to learn something in the service-learning course. And it was a kind of choice also." (Participant: B13)

The theme of 'Passion to Serve Community' outlines faculty members' intrinsic motivation to contribute to society, finding satisfaction in social service, and recognizing the societal impact achievable through service-learning courses.

3) Subtheme 2.3: Engineers Commitment to Real-World Impact

Few participants conveyed how they are motivated to use the service-learning courses reminding the role of engineers towards society.

https://doi.org/10.52202/073963-0015

"Whatever they have studied in engineering, it should be helpful for society. It means they have to help society. This course is easy to set context." (Participant B7)

"We live in society and society also has many problems. We as engineers have a greater role to play because we are the creators." (Participant H2)

One participant was narrating his motivation in terms of connections between concepts and real-world applications using DTSI course.

"Whenever the students will learn theoretical concepts in the classroom, students learning will get connected to the real-world problems and that also mainly societal problems. So, for that reason I felt I should be part of it." (Participant: B11)

When we asked participants about the challenge that they faced during facilitating the service-learning course, most of them mentioned that challenges and few of them explained the contexts on how the challenges were the source of motivation to work further.

"The challenges are part of our life and I'm always keen to take up challenges. And work over them. So that is quite a push motivation. You could call it as I get a push out of this when I get some challenges to be worked upon. (Participant H2)

"I can understand more problems where engineering can be applied. What problems to take, what to solve? How? What are the expectations of the sufferers?" (Participant B4)

The theme 'Engineers' Real-World Impact' is showcased as participants use service-learning to apply engineering knowledge for societal benefit, bridging theory and practice, with facilitation of challenges motivating continued engagement.

C. Theme3: Opportunity to experience Empathy.

The faculty members reflected upon their experiences with the community. They voiced their opinions on how they started being empathetic towards societal issues and learnt empathy in service-learning courses. One example is from the data is followed.

"When someone face the problems, I will try to understand why they are facing, how they're facing, and the biggest advantage of this service-learning course is that like we will get into empathize with all the people who are facing the problems." (Participant: HP2)

One participant opined that all the educated people need not really know the pain of society but the ones who observe the people who are in pain will learn empathy which is through service-learning courses.

"We will come to know lot of incidents what all is going on around us. I mean, in the community, how people are responding. So even though if they are highly qualified, some of them are not at all concerned about poor people and needy people. So, we need to learn many more things about society. That is possible with the DTSI course." (Participant: B12)

Few participants reported that they were not aware of certain bigger issues that the community around face and courses like "Design Thinking and Social Innovation" only can provide opportunities to understand those issues.

"If we look, there is one more picture apart from our regular life. That is what they are facing the social difficulties in their lives. There is no other way in traditional courses to explore this." (Participant: B5)

A similar opinion is shared by another participant on how he started taking benefit of learning empathy in his respective department after learning it from the service-learning course.

"I thank my colleagues. From the training that we had, I came to know about empathy and all. The basic standards, the rules and regulations, and the journey started. Now, in my department automation and robotics, I am using empathy in the design thinking." (Participant HP1)

The theme encapsulates the impact of service-learning courses on faculty members, giving a deeper understanding of societal issues and instilling empathy.

V. DISCUSSSION

The discussion brings a comprehensive analysis and synthesis of findings with the support of existing literature on the topic. Figure 1 shows the representation of the themes of discussion in a relative positioning of the factors that motivate faculty to adopt service-learning in their courses. The observations are related to the personal growth in faculty members, their professional growth, institutional support given to them followed by community development because of implementing service-learning in academic curriculum. Any change that starts at personal level leads to community development with an opportunity of academic enhancement for the students. Further discussion focuses on the analysis of the same.

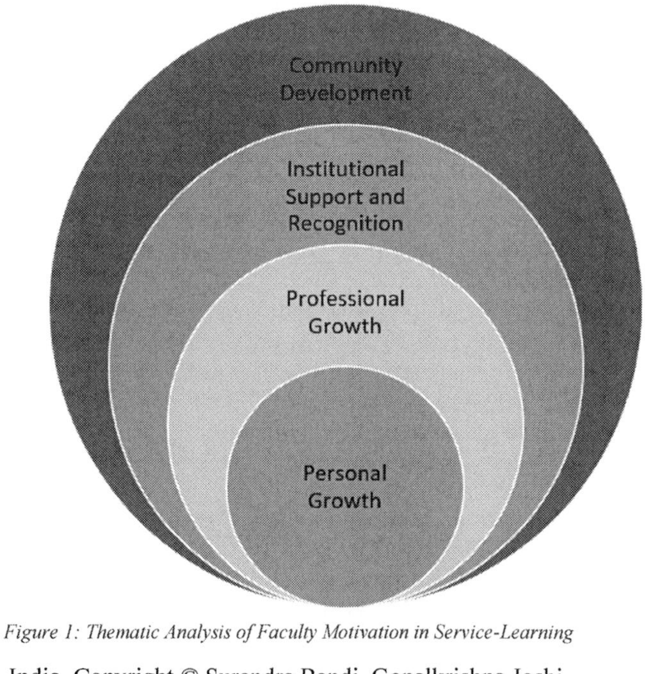

Figure 1: Thematic Analysis of Faculty Motivation in Service-Learning

https://doi.org/10.52202/073963-0015

A. *Personal Growth*

The participants expressed about their personal learning as a motivation factor for them during their participation in service-learning activities as part of the academic course. They discussed the factors like their enhanced skills and learning the concept of empathy. Every individual irrespective of being a faculty member or student has unique learning style and preferences. Faculty motivation is influenced by personal experiences and fulfillment (Holland, 2019). In our qualitative case study, responses from some participants justify that the motivations which come from the perspective of personal benefits stay longer. The benefits include satisfaction, growth, and fulfillment etc. Even the interdisciplinary approach is one opportunity for faculty to go beyond their comfort zones of their departments and learn concepts of other areas which help faculty understand overall problem.

B. *Professional Growth*

From the results section we can infer that service-learning experience contributed to the professional growth of faculty members. The improvement in the teaching quality, participating research and career opportunities can be considered as significant benefits for faculty.

According to Camus at all, service-learning experience of faculty helped in professional development (Camus et al., 2022). Different faculty members come together to offer community engagement courses during pandemic (Sylvan & Becker, 2022). Professional growth of the faculty in exploring interdisciplinary approach and leveraging institutional support is also worth noting. The following section elaborates on the same in detail.

C. *Institutional Support and Recognition*

Almost all the participants in the study expressed that the support provided by the institution to implement DTSI course was outstanding. Since the university is a private and has a visionary leadership with a motto of contributing to the society, the initiative was driven from the top. As Elisa et al reported, impact on faculty to use service-learning depends on type of institution as well (Elisa et al., 2002). Even type of engagement also influenced faculty members to get involved in community engagement (O'Meara, 2008). The faculty who involved in service-learning in this case study are the full time and they have reasonable connect with their local communities. As per Clayton et al., there is need for institutional initiatives to promote faculty taking up research on service-learning to understand potential benefits of the pedagogical practices involved with community engagement (Clayton et al., 2023). Institutional practices can bring positive change in adaptation of service-learning for a meaningful learning experience to the students (Leigh & Kenworthy, 2018). The outcome of this study is an example of Institutional support being a motivational factor for the faculty.

D. *Community Development*

Passion to serve community, meaningful community engagement, fulfilling the role of engineer and multi discipline approach with real-world problems are the motivating factors mentioned by several participants. An existing article by Cooper endorses the same outcome. It allows interventions for the community using interdisciplinary approaches (Cooper, 2014). Connecting with community as mentioned by Krebs (Krebs, 2006) is very significant in successful outcomes of service-learning courses. The impact which pedagogy creates on communities is one of the major drivers for faculty (Richard et al., 2022). Interactions and interventions with community help students, faculty, and institutions to improve learning environments (Karasik, 2020). Several participants in this study expressed the same. Example quote from their interviews are already presented in the results section.

E. *Implications to Practice*

It is observed through literature and the current study that there is an opportunity to improve success of service-learning outcomes by understanding the role of faculty members and support them. There may be challenges to the faculty in the journey. Grace Ngai presented the obstacles faced and the respective strategies to address the challenges in implementing service-learning in their study (Ngai & Chan, 2019). The challenges may be internal or external to the institution. There are challenges ranging from faculty background, expertise, financial benefits, time, patience, and commitment to delivering etc. Sufficient funding and time are required for faculty to properly plan and implement the community engagement initiatives for academic enhancement (Kelli, 2020). The challenges once mitigated or resolved, the same become motivators to practice the same in different contexts for learning. There are studies on how the challenges have been addressed and later became potential contributors. While there are challenges for practitioners of service-learning, there are suitable solutions to overcome (Garvin & Acosta Lewis, 2022).

Overall, the factors which motivate faculty engagement in service-learning courses to undergraduate engineering students in Indian context are discussed and compared with the existing literature. Policy makers at institution level have to analyze the factors that motivate faculty in community engagement and incorporate initiatives addressing their challenges to improve outcomes of service-learning courses. This is a case study of one private technological university from its two campuses. The interpretations are from a researcher's point of view and may require further member checks to get deeper insights from different dimensions of research interest.

VI. CONCLUSION

The qualitative study explored into the motivational factors driving faculty engagement in service-learning courses, offering valuable insights into the dynamics within the context of Indian undergraduate engineering education. By analyzing and synthesizing the broad themes in results and discussion sections, we understand the pivotal role that faculty members play as key stakeholders in the success of service-learning implementation. Recognizing the critical importance of understanding what motivates faculty to embrace these courses, our paper identified specific factors influencing their engagement. The implication of this study extends beyond the immediate findings to understand faculty motivation in service-learning further in various conditions. It is imperative to recognize various dimensions that may shape these motivations. Hence, we recommend future research on a comparative study considering the factors like gender, different programs and locations which influence faculty motivation in service-learning in Indian context for undergraduate engineering students.

REFERENCES

Anderson, K. L., Boyd, M., Ariemma Marin, K., & McNamara, K. (2019). Reimagining Service-Learning: Deepening the Impact of This High-Impact Practice. *Journal of Experiential Education*, *42*(3), 229–248. https://doi.org/10.1177/1053825919837735

Astin, A. W., Vogelgesang, L. J., Ikeda, E. K., & Yee, J. A. (2000). *How Service Learning Affects Students*. 7.

Bandi, S., & Naik, S. M. (2020). Unnat Bharat Abhiyan, A Government of India Initiative Complementing Service Learning in Engineering. *Journal of Engineering Education Transformations*, *34*(Special Issue). https://doi.org/10.16920/jeet/2021/v34i0/157234

Bandi, S. R., Joshi, G., Shettar, A., Technological, K., & Kandakatla, R. (2023, May). A Systematic Literature Review on Faculty Learning in Service-Learning. *IEEE EDUCON 2023 Proceedings*. IEEE EDUCON Global Education Conference, American University Kuwait, Kuwait.

Bandi, S., Dustker, S. M., Kandakatla, R., Oakes, W., & Kotabagi, S. (2021, November). Enablers and Barriers to Implementing Service-Learning in India–A Case-Study of Two Service-learning Models Integrated into Undergraduate Engineering Curriculum. In 2021 World Engineering Education Forum/Global Engineering Deans Council (WEEF/GEDC) (pp. 241-249). IEEE.

Bielefeldt, A., & Pearce, J. (2012). Service Learning in Engineering. *Convergence: Philosophies and Pedagogies for Developing the Next Generation of Humanitarian Engineers and Social Entrepreneurs*, 24–52.

Blakey, J. M., Theriot, S., Cazzell, M., & Sattler, M. (2016). Is Service-Learning Worth It?: A Mixed-Methods Study of Faculty's Service-Learning Experiences. *International Journal of Research on Service-Learning and Community Engagement*, *3*(1). https://doi.org/10.37333/001c.21572

Bringle, R. G., & Hatcher, J. A. (1995). A Service-Learning Curriculum for Faculty. *Michigan Journal of Community Service Learning*, 11.

Camus, R. M., Ngai, G., Kwan, K. P., & Chan, S. C. F. (2022). Transforming Teaching: Service-Learning's Impact on Faculty. *Journal of Higher Education Outreach and Engagement*, *26*(2), 47–64.

Choi, Y., Han, J., & Kim, H. (2023). Exploring key service-learning experiences that promote students' learning in higher education. *Asia Pacific Education Review*. https://doi.org/10.1007/s12564-023-09833-5

Clayton, P. H., Bringle, R. G., & Hatcher, J. A. (2023). *Research on Service Learning: Conceptual Frameworks and Assessments: Volume 2B: Communities, Institutions, and Partnerships*. Taylor & Francis.

Cooper, J. R. (2014). Ten Years in the Trenches: Faculty Perspectives on Sustaining Service-Learning. *Journal of Experiential Education*, *37*(4), 415–428. https://doi.org/10.1177/1053825913513721

Darby, A., & Newman, G. (2014). Exploring Faculty Members' Motivation and Persistence in Academic Service-Learning Pedagogy. *Journal of Higher Education Outreach and Engagement*, *18*, 91–119.

Delaine, D. A., Redick, S., Radhakrishnan, D., Shermadou, A., Smith, M. M., Kandakatla, R., Wang, L., Freitas, C., Dalton, C. L., Dostilio, L. D., & DeBoer, J. (2023). A systematic literature review of reciprocity in engineering service-learning/community engagement. *Journal of Engineering Education*, *n/a*(n/a). https://doi.org/10.1002/jee.20561

Derreth, R. T., Jones, V., & Levin, M. (2022). Measuring Cognitive and Social-Emotional Development in Faculty Preparing for Service-Learning Facilitation. *Journal of Higher Education Outreach and Engagement*, *26*(3), 5–20.

Dustker, S., Bandi, S., & Oakes, W. (2023). Scope of Service-Learning in Accomplishing Objectives of National Education Policy 2020, India—A Conceptual Analysis. *Journal of Engineering Education Transformations*, *36*(Special Issue 2). https://doi.org/10.16920/jeet/2023/v36is2/23054

Dustker, S. M., Reddy, B. S., Kandakatla, R., Joshi, G. H., & Oakes, W. C. (2021, July 26). *Role of Reflection in Service Learning-based Engineering Programs: A Cross-cultural Exploratory and Comparative Case Study in India and the USA*. 2021 ASEE Virtual Annual Conference Content Access. https://peer.asee.org/role-of-reflection-in-service-learning-based-engineering-programs-a-cross-cultural-exploratory-and-comparative-case-study-in-india-and-the-usa

Elisa, S. A., Golden, J., & Susan, R. J. (2002). *Factors That Motivate and Deter Faculty Use of Service-Learning*.

Garvin, M. C., & Acosta Lewis, E. (2022). Reasons Faculty Teach, or Do Not Teach, Service-Learning Courses in a Pandemic: The Role of Faculty Investment and Clues for the Future of Service-Learning. *Higher Learning Research Communications*, *12*(0). https://doi.org/10.18870/hlrc.v12i0.1311

Hammond, C. (1994). Integrating Service and Academic Study: Faculty Motivation and Satisfaction in Michigan Higher Education. *Michigan Journal of Community Service Learning*, *1*(1). http://hdl.handle.net/2027/spo.3239521.0001.102

Holland, B. A. (2019). Factors and Strategies that Influence Faculty Involvement in Public Service. In *Building the Field of Higher Education Engagement*. Routledge.

John W. Creswell & J. David Creswell. (2018). *Research Design Qualitative, Quantitative, and Mixed Methods Approaches* (5th ed.). SAGE Publications, Inc.

Kandakatla, R., Dustker, S. M., Bandi, S., & Oakes, W. (2023). Achieving Indian National Board of Accreditation Engineering Graduate Attributes through Project-Based Service-Learning: Conceptual Analysis. *International Journal for Service Learning in Engineering, Humanitarian Engineering and Social Entrepreneurship*, *18*(1), Article 1.

Kandakatla, R., Penta, S., B., L., Sinha, S., D., C., & K., U. S. (2022). Role of Empathy to Identify Unmet Needs in Indian Villages through Service Learning Program—A Case of the Unnat Bharat Abhiyan Program. *2022 IEEE IFEES World Engineering Education Forum - Global Engineering Deans Council (WEEF-GEDC)*, 1–5. https://doi.org/10.1109/WEEF-GEDC54384.2022.9996259

Karasik, R. J. (2020). Community Partners' Perspectives and the Faculty Role in Community-Based Learning. *Journal of Experiential Education*, *43*(2), 113–135. https://doi.org/10.1177/1053825919892994

Kelli, M. H. (2020). *A Phenomenological Study of Faculty Experience with Service-learning—ProQuest*. https://www.proquest.com/openview/39c9fac15825ddbe0ceae5e57580c481/1?pq-origsite=gscholar&cbl=18750&diss=y

Kiely, R. (2005). A Transformative Learning Model for Service-Learning: A Longitudinal Case Study. *Michigan Journal of Community Service Learning*, *12*(1), 5–22.

Krebs, M. M. (2006). *Service-Learning: Motivations for K-12 Teachers*.

Leigh, J. S., & Kenworthy, A. (2018). Exploring the "Three Ps" of Service-Learning: Practice, Partnering, and Pressures. In *Oxford Research Encyclopedia of Business and Management*. https://doi.org/10.1093/acrefore/9780190224851.013.63

Michael Quinn Patton. (2002). *Qualitative research and evaluation Mehtods* (3rd ed.). Sage Publications, Inc.

Ngai, G., & Chan, S. C. F. (2019). Engaging Teachers in Teaching Service-Learning Subjects: Critical Issues and Strategies. In D. T. L. Shek, G. Ngai, & S. C. F. Chan (Eds.), *Service-Learning for Youth Leadership: The Case of Hong Kong* (pp. 309–322). Springer. https://doi.org/10.1007/978-981-13-0448-4_18

Oakes, W. (2014). Engineering Service-learning: A Model for Preparing Students for Engineering Practice while meeting Needs of the Underserved. *Journal of Engineering Education Transformations*, *27*(4), 46. https://doi.org/10.16920/ijerit/2014/v27i4/53300

Oakes, W., Duffy, J., Jacobius, T., Linos, P., Lord, S., Schultz, W. W., & Smith, A. (2002). *Service-learning in engineering*. *2*, F3A-F3A.

O'Meara, K. (2008). *Motivation for Faculty Community Engagement: Learning from Exemplars*.

O'Meara, K., & Niehaus, E. (2009). *Service-Learning Is... How Faculty Explain Their Practice*. 16.

Radhakrishnan, D., Kandakatla, R., & Delaine, D. A. (2022). Participatory Approach to Engineering Service Learning Programs—Quality Framework for the Implementation of Unnat Bharat Abhiyan Program in Indian Engineering Institution. *2022 IEEE Global Engineering Education Conference (EDUCON)*, 863–872. https://doi.org/10.1109/EDUCON52537.2022.9766735

Richard, F. D., Berkey, B., & Burk, H. M. (2022). *Motivation and Orientation: Faculty Perspectives on Development and Persistence in Service Learning and Community Engagement*. *14*(1).

Sylvan, L., & Becker, R. (2022). Implementing Service-Learning Projects for the First Time During the Pandemic: Two Instructors' Inspiration, Motivation and Goals. *Collaborations: A Journal of Community-Based Research and Practice*, *5*(1), 10. https://doi.org/10.33596/coll.93

Copyright statement

Research in Engineering Education Symposium-2024
Jan 4-6, 2024 - Hubballi, India

The effects of a humanities-based engineering course on engineering students' sociotechnical thinking

Siddhant Sanjay Joshi[a], Kirsten A. Davis[a], and Lori A Czerwionka[b].
School of Engineering Education, Purdue University[a], School of Languages and Cultures, Purdue University[b],
Corresponding Author Email: joshi110@purdue.edu

Abstract

Context

As the world becomes connected and globalized, engineering problems become more complex and multi-dimensional. To solve these problems, engineers require *sociotechnical thinking*, which involves addressing both the technical and contextual aspects of a problem and understanding the interconnections between these considerations. However, engineering programs traditionally emphasize technical thinking, resulting in a lack of sociotechnical thinking during problem-solving.

Purpose or Goal

Humanities-informed engineering education is one pedagogical approach that has shown promise in supporting engineering students' development of sociotechnical thinking skills. Our study explored how enrollment in a one-semester humanities-informed engineering course is related to the development of sociotechnical thinking.

Methods

We administered the Energy Conversion Playground (ECP), a scenario-based assessment of sociotechnical thinking, to three groups at the start and end of the semester: students in the humanities-informed engineering course, other students from engineering majors, and students from humanities majors. Students' performance on each dimension of the ECP (Technology, People, and Broader Context) was compared using a 3x2 mixed ANOVA design.

Outcomes

Our results reveal that students enrolled in the course were more likely to discuss social considerations compared to the two groups not enrolled in the course. Students in the course also improved over the semester in their discussion of social and contextual considerations while the other two groups did not.

Conclusions

The results of the study indicate that interdisciplinary training in humanities and engineering can help engineering students engage in sociotechnical thinking during problem-solving. Moreover, the results for the engineering control group reiterate previous findings that there may be a lack of emphasis on social and contextual aspects in traditional engineering education. Therefore, future research should focus on development of pedagogical frameworks and assessments on sociotechnical thinking.

Keywords—*Sociotechnical thinking, scenario-based assessment, interdisciplinary engineering education.*

I. INTRODUCTION

Problems that engineering students are trained to solve in classrooms are well-defined and closed-ended (Jonassen et al., 2006), and often decontextualized from contextual influences (McGowan & Bell, 2020). In practice, however, problems that engineers encounter are complex, ill-structured, and situated in social contexts, and thus sociotechnical by nature (Leydens & Lucena, 2018). Leydens & Lucena (2018) suggest that the disconnect between classroom and workplace problems is a consequence of technical-social dualism (Faulkner, 2000, 2007) where engineering students are taught to prioritize technical aspects and minimize contextual aspects during problem-solving (Swartz et al., 2019). In the real world however, engineering problems involve a complex interplay between technical and contextual aspects (Kaur & Craven, 2022; Leydens et al., 2018; McGowan & Bell, 2020; Trevelyan, 2014b, 2014a). For instance, engineers are tasked to handle sociotechnical problems such as provide access to clean water, ensure privacy and cybersecurity of people, improve urban infrastructure, make energy sustainable, accessible, and economical, etc. To tackle these sociotechnical problems, engineering students and practitioners require sociotechnical thinking, which can be conceptualized as the ability to address both technical and contextual dimensions of a problem (Mazzurco & Daniel, 2020) and the ability to understand the interconnections between these dimensions (Davis et al., 2021). However, the development of sociotechnical thinking has not historically been emphasized in engineering education.

Humanities-informed engineering education is one approach that has shown promise for developing engineering students' sociotechnical thinking ability (Davis et al., 2021). Humanities-informed engineering course involves teaching students to approach engineering problems through the lens of both engineering and humanities disciplines. It is an interdisciplinary course that uses sociotechnical case studies to develop students ability to consider contextual aspects and broader considerations associated with engineering problems (Davis et al., 2021).

The purpose of the current study was to investigate whether a humanities-based engineering course can develop

 https://doi.org/10.52202/073963-0016

sociotechnical thinking abilities in engineering students as assessed using a scenario-based assessment (Mazzurco et al., 2014; Mazzurco & Daniel, 2020) and to compare the ability of the students in the class to that of other groups of students. We addressed the following research questions (RQ):

1. How do the scores on a scenario-based assessment of sociotechnical thinking compare between students enrolled in a humanities-informed engineering course and control groups of engineering and humanities students?
2. How do the scores on a scenario-based assessment of sociotechnical thinking change from pretest to posttest for the students in the course group and control groups?

Our study responds to ongoing calls for equipping students with sociotechnical thinking skills (Cech, 2013; Leydens et al., 2018; Mazzurco & Daniel, 2020; Swartz et al., 2019; Trevelyan, 2014b). Our findings show that interdisciplinary approaches like humanities-informed engineering may provide opportunities for developing required sociotechnical thinking skills. Further, we reiterate previous findings by showing that engineering students do not often prioritize contextual considerations while solving problems. Finally, we urge educators to develop frameworks and assessment methods that help foster and measure the sociotechnical abilities of engineering students. These targeted efforts will enable universities to prepare engineers capable of addressing complex sociotechnical problems.

II. Literature Review

The problems that engineers face today are complex and consist of various technical and contextual aspects that are interconnected with each other (Kaur & Craven, 2022; Leydens et al., 2018; McGowan & Bell, 2020; Trevelyan, 2014b, 2014a). As a result, engineering problems and solutions exist within a complex sociotechnical space (Adams et al., 2011) indicating that these problems cannot be addressed solely by consideration of technical factors (Leydens & Lucena, 2018). In addition to the technical aspects, prior research on engineering practice shows that engineers are required to consider contextual factors such as stakeholder's needs, economic, political, legal, and environmental aspects of engineering problems (Bucciarelli, 1994; Cross, 2021, 2023; Jonassen et al., 2006; Petroski, 2011) during problem-solving. Therefore, engineers will be better prepared to address problems when they understand both the technical and contextual aspects and their interdependencies (Currie & Galliers, 1999; Davis et al., 2021; Grohs et al., 2018).

Despite engineering problems being sociotechnical in nature, research has shown that a culture of disengagement exists in engineering education that prepares students to prioritize technical aspects over contextual aspects during problem-solving and considers societal concerns tangential to engineering practice (Bardzell & Bardzell, 2013; Cech, 2014; Pawley, 2009; Riley, 2008; Stevens et al., 2014). For instance, Cech (2014) found that students' understanding of the societal consequences of technology solutions declined over the course of their engineering education. This indicates that as students' progress from the first year to the final year of engineering, their consideration of contextual factors like public welfare decreases. Furthermore, as students engage in solving linear, well-defined (Jonassen et al., 2006), and often decontextualized (Erickson et al., 2020; McGowan & Bell, 2020) problems over the course of their engineering education, they tend to discount contextual factors during problem-solving (Stevens et al., 2014). To prepare engineers for addressing sociotechnical problems, engineering education must be centered around teaching skills that encourage learners to value contextual aspects during problem-solving.

Sociotechnical thinking is a skill that enables engineers to understand the complex interconnections between the technical and contextual factors of a problem (Hoople & Choi-Fitzpatrick, 2020). Furthermore, with sociotechnical thinking skills, engineering students can discern how and why technical factors are co-dependent on contextual factors during problem-solving (Swartz et al., 2019), embrace the sociotechnical complexities of a problem (Cech, 2013), and think about how their decision making may impact the society as a whole (McGowan & Bell, 2020). Therefore, it is important to support the development of sociotechnical thinking skills in engineering education.

Research has explored a variety of approaches to develop students' sociotechnical thinking skills. For example, Reynante (2022) investigated how sociotechnical thinking skills of engineering students can be developed by exposing them to a two-course community-engaged engineering program. They found that students enrolled in the program shifted their emphasis from being solely focused on technical aspects to accounting for relevant contextual factors along with technical factors while solving engineering problems. Prior research by Frank (2010) suggests that multidisciplinary educational experiences may also encourage engineers to consider contextual aspects while solving problems. Additionally, Bucciarelli & Drew (2015) proposed a Liberal Studies in Engineering degree to expose students to social complexities and implications of their work through a humanities perspective. Similarly, other studies have shown that multidisciplinary education can help students learn to solve complex engineering problems (Bornasal et al., 2018; Jesiek et al., 2017; Stevens et al., 2014).

For our study, we chose to focus on the humanities which is known to develop professionals' abilities needed to solve socially contextualized problems (Benneworth, 2015). Further, a humanities based perspective improves ability to consider unintended consequences of engineering on the society (Fila et al., 2014) and prepare engineers to keep society central to their problem solving (Hynes & Swenson, 2013). Therefore, we developed a one-credit humanities-informed engineering course. Our objective through this course was to integrate

https://doi.org/10.52202/073963-0016

engineering and the humanities and, thus, examine the integration of contextual and technical aspects during engineering problem-solving. We exposed students enrolled in the course to engineering case studies that had integrated technical and contextual aspects. Students engaged with the case studies by participating in activities like role-playing, discussions, reflections, in-class readings, and brainstorming in groups. Through these activities, students discussed various approaches to analyzing and solving engineering problems from an integrated contextual and technical perspective. For more information on the course, please refer to Davis et al., (2021).

III. Methods

To address our research questions, we used quantitative research methods. We implemented a scenario-based assessment in a pre/posttest study design at a large midwestern university in the USA.

TABLE I
SOCIOTECHNICAL THINKING ABILITIES (DEPENDENT VARIABLES) ASSESSED BY THE ECP SCENARIO
(REVISED BY JOSHI ET AL., 2023; DEVELOPED BY MAZZURCO & DANIEL, 2020).

Dimensions	Definitions and Key Characteristics
Technology	Considerations focused on four technical categories: • *Inputs or constraints to the technology:* Power requirements, time of operation, cost, materials, safety, climate, people as a source of energy, etc. • *Functionality:* Efficiency, feasibility, ease of operation, maximum power generated, friction, storage of energy, functioning of components, alternative techs to meet the same goals, ability to generate the needed energy output, and so forth. • *Long-term technological considerations:* Maintenance, repairs, spare parts, upgrades, etc. • *Additional considerations:* Durability, Focus on system safety/equipment safety; people as part of the larger system; funding, budget, cost of maintenance and operation, etc.
People	Considerations focused on stakeholders' needs, desires, expertise, and degree of participation in the design process (e.g., listening to the community, hearing their voices, collaborating with them in the design process). *Additional considerations*: Focus on the safety of people; the willingness of people participation, and the influence of people on the playground system
Broader Context	Considerations focused on four contextual categories: • *Local norms*: Social norms, culture, gender/ethnic/power dynamics, religious views, etc. • *Ethics and law*: Regulations, standards, laws, moral and ethical issues. • *Other socio-material contexts*: Built environment, impact on the natural environment, local economy, education system. • *Additional considerations*: Political aspects (under ethics and law), Profitability, and Ability to own or produce the technology in a financial sense.

A. Participants

In Spring 2021 and Spring 2022, 38 undergraduate students from various engineering majors self-enrolled in a one-credit course called Humanities-Informed Engineering Projects (class group) (see course description in section II). In addition, we collected data from 62 undergraduate students who did not enroll in the course: 32 engineering students from different engineering majors (engineering group) and 30 students from different humanities majors (humanities group). All students in the course participated in the assessment as part of the class. Students in the engineering and humanities groups volunteered to participate after receiving a recruitment email distributed to known engineering and humanities contacts and using a snowball approach. They received gift certificates for their participation. The Purdue University IRB approved this investigation.

B. Data Collection

All participants responded to a scenario-based assessment at two points in time: the start (pretest) and end (posttest) of the semester. We collected data using an online questionnaire, and only the students who completed both the pretest and posttest assessments were included in this study. Upon collecting the data, we deidentified student responses prior to scoring them.

1) ECP scenario-based assessment.

To assess students' sociotechnical thinking, we used the Energy Conversion Playground (ECP) scenario-based assessment developed by Mazzurco & Daniel (2020). The ECP assessment measures sociotechnical thinking along three dimensions (see Table 1): Technology, People, and Broader Context. These dimensions explore both the technical and contextual aspects of defining and solving a problem. Student responses to the scenario-based assessment were scored on a scale of 0 to 3 for each dimension as per the rubric given in Mazzurco & Daniel (2020). For more details on the scenario, see Fig 1.

In developing countries, energy production is one of the most critical problems. Resources and technologies to produce energy are not often available. Thus, human power conservation systems might be used to power small appliances. Imagine you and your team are assigned to a design project in partnership with a Non-Governmental Organization of a developing country. The NGO needs a low-cost power system that can generate enough energy for the lights of a primary school. One of the members of your team suggests using a merry go round, seesaw, and swing to produce energy that can be converted to electricity for the lights.

Question: What considerations do you need to take into account to solve the problem described in the scenario? List and describe all the constraints and justify their inclusion

Fig. 1. Energy Conversion Playground scenario developed by Mazzurco & Daniel, (2020) .

We chose to use a scenario-based assessment because these assessments allow some insight into students' thinking and may more directly measure students' abilities than traditional self-report assessments (Davis et al., 2023). Furthermore, scenario-based assessments can be used as an instructional tool as well as an assessment tool (Davis et al., 2023). The course we studied for this project used case studies to teach students to

https://doi.org/10.52202/073963-0016

consider and integrate both technical and contextual factors (i.e., practice sociotechnical thinking) while problem-solving. Because the ECP is a scenario-based assessment that measured this skill, it aligned well with the objectives of the course and research study. In addition, earlier studies that compared two scenario-based assessments and found that the ECP may be better suited for assessing sociotechnical thinking over a semester-long course when compared to another similar instrument (Joshi et al., 2022).

C. Data Analysis

To compare students' performance on sociotechnical thinking across the three groups, we examined participants' pre and posttest scores on the ECP scenario-based assessment. We deidentified the data by removing participant information such as names, majors, and emails before analyzing the students' responses. Three researchers from our team scored all student responses to the scenario using the ECP rubric (see Mazzurco & Daniel, 2020) and then met to discuss the scores until they reached a mutual consensus on the final score for each response.

Next, we conducted three 3 x 2 mixed ANOVA analyses to answer the two research questions. The advantage of using mixed ANOVA is that it allows analyses of both the within-and between subject variables (i.e., changes over time and changes between groups) (Frey, 2018). For this study, as we wanted to analyze changes over time and changes between groups, a mixed ANOVA was suitable for our study.

Group	Time	
	Pretest	Post test
Treatment: Class	Technology	Technology
Control I: Engineering	Technology	Technology
Control II: Humanities	Technology	Technology

Group	Time	
	Pretest	Post test
Treatment: Class	Broader Context	Broader Context
Control I: Engineering	Broader Context	Broader Context
Control II: Humanities	Broader Context	Broader Context

Group	Time	
	Pretest	Post test
Treatment: Class	People	People
Control I: Engineering	People	People
Control II: Humanities	People	People

Fig. 2. 3X2 Mixed ANOVA Matrix for each dependent variable: Technology, People, and Broader Context

As shown in Fig. 2, the dependent variable for each ANOVA is one of the three dimensions of the ECP instrument (Technology, People, and Broader Context). Across all three analyses, the between-subjects factor was 'group-membership' in one of the three student groups: class group, engineering group, and humanities group. The within-subjects factor was 'time,' consisting of two levels: pretest and posttest. We checked that our data met the assumptions of this analysis, as described in the following sections.

D. Assumptions for 3x2 Mixed ANOVA

1) Normality

Our results of the normality test indicated that no sociotechnical thinking dimensions across the three groups were normally distributed. Hence, the non-normality of the data for the three groups could impact proceeding hypothesis testing as the mean may not appropriately represent the distribution. However, the central limit theorem suggests that for a random sample size greater than 30, the standard sample mean converges to a normal distribution with a mean equal to the sample mean (Islam, 2018). Hence, we satisfied the normality assumption as the sample size of all groups is greater than 30.

2) Homogeneity of Variances:

Next, we calculated the variances for pretest and posttest dimensions of sociotechnical thinking using Levene's test of homogeneity of variances (See Table II). We found significant results for the Broader Context Pretest score. For the other dimensions of sociotechnical thinking at pre and posttest, the homogeneity of variances is non-significant. This indicates that we can assume equal variances for the data except for Broader Context pretest score. The results of the study for Broader Context pretest score dimension will thus need to be interpreted based on the assumption of unequal variances.

TABLE II
TESTS OF HOMOGENEITY OF VARIANCES

		LEVENE STATISTIC	DF1	DF2	P
TECHNOLOGY PRETEST	BASED ON MEAN	.309	2	97	.735
TECHNOLOGY POSTTEST	BASED ON MEAN	.943	2	97	.393
PEOPLE PRETEST	BASED ON MEAN	2.363	2	97	.100
PEOPLE POSTTEST	BASED ON MEAN	1.723	2	97	.184
BROADER CONTEXT PRETEST	BASED ON MEAN	4.740	2	97	.011*
BROADER CONTEXT POSTTEST	BASED ON MEAN	2.051	2	97	.134

* = p <.05

3) Sphericity of our data

For this study, as we have two levels of within-subjects variable, there is only one set of differences (pretest vs. posttest) and hence, sphericity is not an issue (Field, 2013). Therefore, we can assume that the assumption of sphericity has been met.

E. Limitations

The ECP scenario-based assessment used in this paper is not developed based on the humanities-informed engineering framework from Davis et al., (2021). Therefore, the pedagogical framework used for fostering sociotechnical thinking is not the same as that used for developing the assessment instrument. However, there are overlapping contextual constructs between the framework and assessment instrument. Further, given the sample size for each group, it may not accurately represent the interactions with a high statistical power.

https://doi.org/10.52202/073963-0016

IV. Results

In this section, we describe the descriptive statistics of our pre/post data followed by interpretation of results in response to each research question.

A. Descriptive Statistics

The data for this study are pre/post scores of three sociotechnical thinking dimensions for three student groups (Table III).

TABLE III
Descriptive Statistics

Dimension	Group		Pretest		Post Test	
		N	M	SD	M	SD
Technology	Class	38	2.42	0.56	2.29	0.73
	Humanities	30	2.23	0.77	2.30	0.65
	Engineering	32	2.50	0.67	2.47	0.51
	Total	100	2.39	0.68	2.35	0.64
People	Class	38	0.84	0.89	1.32	0.84
	Humanities	30	1.27	1.06	0.70	0.75
	Engineering	32	0.44	0.72	0.62	0.66
	Total	100	0.84	0.94	0.91	0.82
Broader Context	Class	38	0.42	0.55	0.87	0.84
	Humanities	30	0.33	0.55	0.30	0.60
	Engineering	32	0.22	0.42	0.34	0.54
	Total	100	0.33	0.51	0.53	0.73

From Table III, we observe that the mean scores show an increase from the pretest to posttest for People and Broader Context dimensions of sociotechnical thinking for the class group. The students from the other two groups did not change much in their scores of Technology, People, and Broader context from the pretest to the posttest. This trend was expected as these students did not receive any training on sociotechnical thinking between the two assessment times. In addition, the mean scores of the class group students decreased on the dimension of technology from the pretest to the posttest as students focused more on contextual considerations (like People and Broader Context). Observing the spread of the data, we can infer that the posttest scores seem to have a greater spread for the majority of the sociotechnical thinking dimensions compared to the pretest data.

B. Research question 1: How do the scores on a scenario-based assessment of sociotechnical thinking compare between students enrolled in a humanities-informed engineering course and control groups of engineering and humanities students (not enrolled in the course)?

We used between-subjects ANOVAs for each of the three dimensions of sociotechnical thinking to address research question 1. The between-subjects ANOVAs test the difference between the three groups while ignoring the time variable. That is, this analysis considers the pre- and posttest scores of each group together when comparing across groups. We found no statistically significant difference in scores for the Technology dimension across the three groups ($F(2,97) = 5.714, p < 0.247, \eta_p^2 = .28$). However, there were statistically significant differences for the People ($F(2,97) = 5.714, p < 0.05, \eta_p^2 = .128$) and Broader Context ($F(2,97) = 6.521, p < 0.05, \eta_p^2 = .119$) dimensions, which indicated that students from the three groups had some differences on these two dimensions of sociotechnical thinking.

Next, we used Tukey's post-hoc tests to determine which pairs of student groups had statistically significant differences in their scores for the People and Broader context dimensions (see Table IV). For the Broader Context dimension, we found a statistically significant difference between the ECP scores for the class group and both the engineering control group ($p < 0.01$) and the humanities control group ($p < 0.05$), where the class group had higher scores than both groups. For the People dimension, we found a statistically significant difference between the ECP scores for the class group and the engineering control group ($p < 0.01$), where the class students scored higher on average. We also found a difference between the two control groups for the People dimension, where the humanities students scored higher than the engineering students ($p < 0.05$).

TABLE IV
Post Hoc analysis -Tukey's for Between-Subjects effects People and Broader Context

Measure	(I) Group	(J) Group	Mean Difference (I-J)	Std. Error	P.
People	Class	Humanities	.10	.155	.811
	Class	Engineering	.55	.152	.001***
	Humanities	Engineering	.45	.161	.017*
Broader Context	Class	Humanities	.33	.114	.014*
	Class	Engineering	.36	.112	.005**
	Humanities	Engineering	.04	.119	.952

* = p<.05, ** is p<.01, and *** is p<.001

C. Research Question 2: How do the scores on the scenario-based assessment of sociotechnical thinking change from pretest to posttest for the students in the course group and the two control groups?

We used within-subjects ANOVAs for each of the three dimensions of sociotechnical thinking to address research question 2. Within-subjects ANOVAs can explain (i) whether there is a difference between the pre-and posttest scores for all the students together (across all three groups) (ii) the interaction effect between time and group-membership i.e., how much do differences in scores on the sociotechnical thinking dimensions between the three groups change over time (pretest to posttest).

When considering all the students together, our within-subjects ANOVAs identified no statistically significant differences in the ECP scores over time for the Technology ($F(1,97) = 0.147, p = .702, \eta_p^2 = 0.002$) and People ($F(1,97) = 0.087, p = .769, \eta_p^2 = 0.001$) dimensions of sociotechnical thinking. For the Broader Context dimension, there was a statistically significant increase in the students' scores over time ($F(1,97) = 5.359, p < 0.05, \eta_p^2 = .052$).

https://doi.org/10.52202/073963-0016

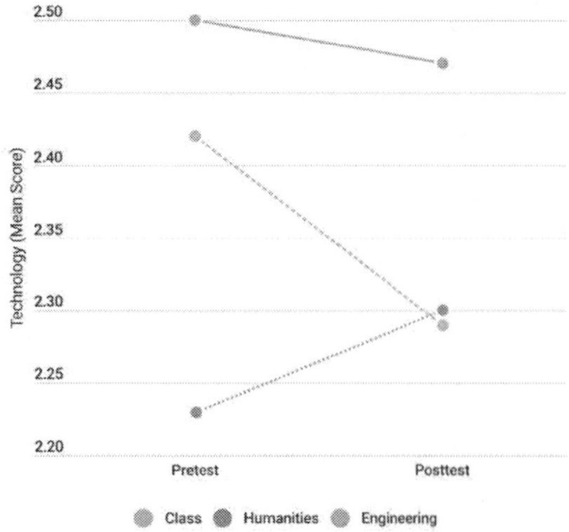

Fig. 3. Change in Means from Pretest to Post Test for Technology dimension.

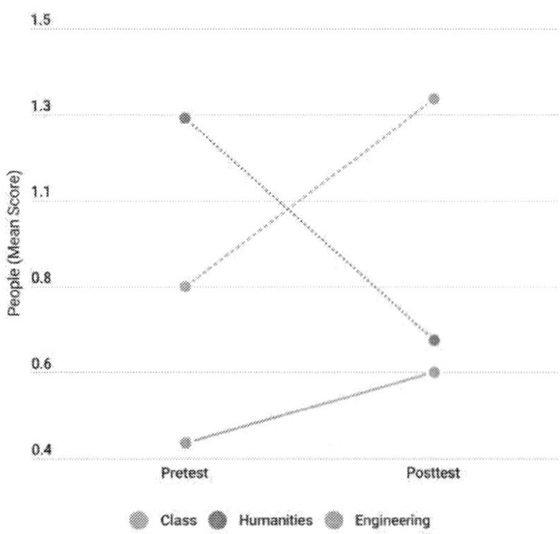

Fig. 4. Change in Means from Pretest to Post Test for People dimension.

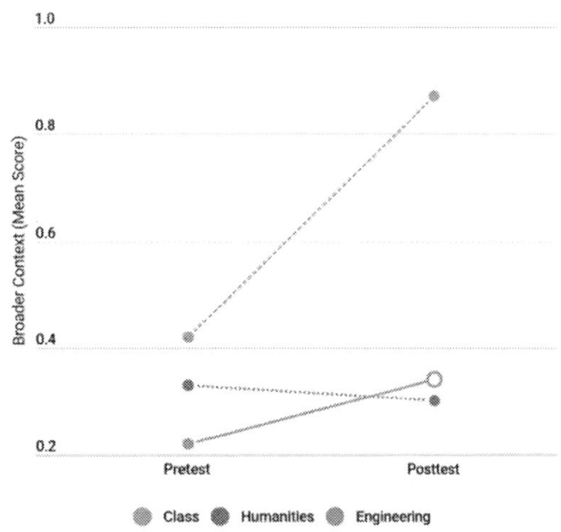

Fig.5. Change in Means from Pretest to Post Test for Broader Context dimension.

When we considered the interaction effect between time and group membership, we found no statistically significant effect for the Technology dimension. However, for the People ($F(2,97) = 8.308, p < 0.001$, $\eta_p^2 = .146$) and Broader Context ($F(2,97) = 3.469, p < 0.05, \eta_p^2 = .067$) dimensions, there was a statistically significant interaction effect between time and group membership. This effect indicated that the scores for some of the groups changed at different rates from the pretest to the posttest for these dimensions.

To further explore the impact of the interaction effect, we observed how the means vary at the pretest (Time 1) and posttest (Time 2) for all three student groups (see Fig. 3, Fig. 4, Fig. 5.). In Figure 3, we can see that there is little change in the Technology dimension scores over time for the three groups. This observation aligns with our within-subjects results suggesting that there is an interaction effect between time and group membership for only People and Broader Context dimensions. For the People dimension (Figure 4), we notice that students from the class group showed a large increase, the engineering group showed a slight increase, and the humanities showed a considerable decrease in their scores over time. Though the decrease may not seem much, it is large as the scale of scoring varies from 0–3. Similarly, in Figure 5 we see that the class group students showed a large increase in their scores over time on the Broader Context dimensions while the engineering group students showed a slight increase, and the humanities group students showed a slight decrease. Overall, the students in the class showed the most increase amongst the three groups over time on the dimensions of People and Broader Context.

V. Discussion

The purpose of this study is to compare the performance of students enrolled in humanities-informed engineering class with students from engineering and humanities majors who did not enroll in the class to investigate (i) how the scores between three groups compare on dimensions of sociotechnical thinking (ii) how the scores of the three groups change over the course of a semester. To compare the student performance, a scenario-based assessment was administered to three groups of students at the start and the end of Spring 2021 and Spring 2022 semesters. To answer the research questions, we conducted three 3 x 2 mixed factorial ANOVAs, one for each sociotechnical thinking dimension — Technology, People, and Broader Context.

In response to RQ1, we found that the class group scored higher on the Broader Context dimension than both control groups and higher on the People dimension than the engineering control group. The humanities control group also scored higher than the engineering control group for the People dimension. In response to RQ2, we found that student scores increased over time for the dimension of Broader Context. Additionally, there were differences between the groups on the People and Broader Context dimensions from pretest to posttest.

Our results suggest that completing the humanities-informed engineering course may improve engineering students' sociotechnical thinking by helping them to consider contextual aspects while problem-solving. This result aligns with Frank's (2010) study which found growth in systems thinking abilities with interdisciplinary education. Our comparison between the two control groups also support previous findings that humanities students are more likely to consider human aspects of problems because their education integrates various social and contextual factors (Benneworth, 2015). At the same time, the lack of change in Broader Context scores for the engineering group students supports previous findings that traditional engineering education largely prioritizes technical aspects (Trevelyan, 2007) and discounts contextual aspects (Cech, 2014; Faulkner, 2007; Paul et al., 2022; Riley, 2008).

Future research could focus on ways in which the traditional engineering curriculum can integrate contextual aspects The results of our study suggest that interdisciplinary training in humanities can improve engineering students' sociotechnical thinking, but only a limited number of students take this interdisciplinary course. The rest of the engineering education curriculum still focuses on teaching technical dimensions rather than social dimensions (Cech, 2013; Faulkner, 2000; Pawley, 2009), despite technical engineering decisions responding to and influencing the society. This lack of focus on social and contextual aspects limits engineering students' sociotechnical thinking abilities and thus, they marginalize contextual aspects while solving problems (Riley, 2003, 2008; Stevens et al., 2014). To overcome this challenge, a broader curricular shift is necessary to enable all engineering students to develop sociotechnical thinking abilities. Rather than having an elective course on this topic, all engineering courses could integrate contextual topics related to the course content. This approach could support students' development of sociotechnical thinking to a higher degree because they would see relevant content across the curriculum.

Building on this suggestion, future research could explore the effect of the duration of interdisciplinary or sociotechnical interventions on students' development of sociotechnical thinking. The current work emphasizes that brief exposure, through a semester-long course, shows positive results for including contextual aspects in engineering problem solving. However, training over an extended time may be more effective in producing gains or those gains may be more stable. These unexplored questions are important in understanding how to prepare engineering graduates to maintain a sociotechnical focus rather than a purely technical one. Furthermore, as integrating contextual and technical dimensions is an important component of engineering, future research should continue to explore how experiential learning opportunities with socially embedded experiences (like service learning) administered over different periods impact sociotechnical thinking in engineering students.

Along with these interventions, researchers can also explore the development of instruments to assess sociotechnical thinking in students and practicing engineers. Currently, limited tools like the ECP scenario (Mazzurco & Daniel, 2020), Abeesee scenario (Grohs et al., 2018), and Lake Urmia Vignette (Davis et al., 2020) are available to assess students' sociotechnical thinking abilities. Additional research is needed to identify ways to effectively assess sociotechnical thinking abilities in real-world settings (e.g., service learning or internship programs). It may also be useful to explore pedagogical frameworks that can simultaneously be used to teach as well as assess sociotechnical thinking.

VI. Conclusion

Our study explored the sociotechnical thinking skills of three groups of students over the course of a semester. We administered the ECP scenario-based assessment to students enrolled in humanities-informed engineering course and two control groups: engineering students and humanities students. Our results indicate that students from the course saw greater increases on their assessment scores over time when compared to the two control groups. These findings suggest that interdisciplinary education can help foster sociotechnical thinking abilities in engineering students. Future research could explore pedagogical frameworks and assessments on sociotechnical thinking and observing the effectiveness of these techniques over an extended period.

Acknowledgements

We would like to thank Gabriel Rios-Rojas for his help in teaching the humanities informed engineering course. We would also like to thank Elisa Camps Troncoso and Nguyen Huu Gia Bao for their help in scoring the ECP student responses.

References

Adams, R., Evangelou, D., English, L., De Figueiredo, A. D., Mousoulides, N., Pawley, A. L., Schiefellite, C., Stevens, R., Svinicki, M., Trenor, J. M., & Wilson, D. M. (2011). Multiple Perspectives on Engaging Future Engineers. *Journal of Engineering Education, 100*(1), 48–88. https://doi.org/10.1002/j.2168-9830.2011.tb00004.x

Bardzell, J., & Bardzell, S. (2013). What is "critical" about critical design? *Proceedings of the SIGCHI Conference on Human Factors in Computing Systems*, 3297–3306. https://doi.org/10.1145/2470654.2466451

Benneworth, P. (2015). Tracing how arts and humanities research translates, circulates and consolidates in society.. How have scholars been reacting to diverse impact and public value agendas? *Arts and*

Humanities in Higher Education, *14*(1), 45–60. https://doi.org/10.1177/1474022214533888
Bornasal, F., Brown, S., Perova-Mello, N., & Beddoes, K. (2018). Conceptual Growth in Engineering Practice. *Journal of Engineering Education*, *107*(2), 318–348. https://doi.org/10.1002/jee.20196
Bucciarelli, L. L. (1994). *Designing engineers*. MIT press.
Bucciarelli, L. L., & Drew, D. E. (2015). Liberal studies in engineering – a design plan. *Engineering Studies*, *7*(2–3), 103–122. https://doi.org/10.1080/19378629.2015.1077253
Cech, E. A. (2013). The (mis) framing of social justice: Why ideologies of depoliticization and meritocracy hinder engineers' ability to think about social injustices. *Engineering Education for Social Justice: Critical Explorations and Opportunities*, 67–84.
Cech, E. A. (2014). Culture of disengagement in engineering education? *Science, Technology, & Human Values*, *39*(1), 42–72.
Cross, N. (2021). *Engineering design methods: Strategies for product design*. John Wiley & Sons.
Cross, N. (2023). *Design thinking: Understanding how designers think and work*. Bloomsbury Publishing.
Currie, W., & Galliers, B. (1999). *Rethinking Management Information Systems: An Interdisciplinary Perspective*. OUP Oxford.
Davis, K. A., Ghaffarzadegan, N., Grohs, J., Grote, D., Hosseinichimeh, N., Knight, D., Mahmoudi, H., & Triantis, K. (2020). The Lake Urmia vignette: A tool to assess understanding of complexity in socio-environmental systems. *System Dynamics Review*, *36*(2), 191–222. https://doi.org/10.1002/sdr.1659
Davis, K. A., Grote, D., Mahmoudi, H., Perry, L., Ghaffarzadegan, N., Grohs, J., Hosseinichimeh, N., Knight, D. B., & Triantis, K. (2023). Comparing Self-Report Assessments and Scenario-Based Assessments of Systems Thinking Competence. *Journal of Science Education and Technology*. https://doi.org/10.1007/s10956-023-10027-2
Davis, K. A., Joshi, S. S., Czerwionka, L., Montalvo, F. J., Rios-Rojas, G. O., Tort, J., Marston William, J., & Nauman, E. (2021). Integrating the humanities with engineering through a global case study course. *Journal of International Engineering Education*, *3*(1), Article 4.
Erickson, J., Claussen, S., Leydens, J., Johnson, K., & Tsai, J. (2020). Real-world Examples and Sociotechnical Integration: What's the Connection? *ASEE Annual Conference Proceedings*. ASEE Annual Conference proceedings, Virtual Conference.
Faulkner, W. (2000). Dualisms, Hierarchies and Gender in Engineering. *Social Studies of Science*, *30*(5), 759–792. https://doi.org/10.1177/030631200030005005
Faulkner, W. (2007). `Nuts and Bolts and People': Gender-Troubled Engineering Identities. *Social Studies of Science*, *37*(3), 331–356. https://doi.org/10.1177/0306312706072175
Field, A. (2013). *Discovering statistics using IBM SPSS statistics*. sage.
Fila, N. D., Hess, J., Hira, A., Joslyn, C. H., Tolbert, D., & Hynes, M. M. (2014). The people part of engineering: Engineering for, with, and as people. *2014 IEEE Frontiers in Education Conference (FIE) Proceedings*, 1–9. https://doi.org/10.1109/FIE.2014.7044106
Frank, M. (2010). Assessing the interest for systems engineering positions and other engineering positions' required capacity for engineering systems thinking (CEST). *Systems Engineering*, *13*(2), 161–174. https://doi.org/10.1002/sys.20140
Frey, B. B. (2018). *The SAGE Encyclopedia of Educational Research, Measurement, and Evaluation*. SAGE Publications, Inc. https://doi.org/10.4135/9781506326139
Grohs, J. R., Kirk, G. R., Soledad, M. M., & Knight, D. B. (2018). Assessing systems thinking: A tool to measure complex reasoning through ill-structured problems. *Thinking Skills and Creativity*, *28*, 110–130. https://doi.org/10.1016/j.tsc.2018.03.003
Hoople, G. D., & Choi-Fitzpatrick, A. (2020). Drones for good: How to bring sociotechnical thinking into the classroom. *Synthesis Lectures on Engineers, Technology, and Society*, *9*(1), i–148.
Hynes, M., & Swenson, J. (2013). The Humanistic Side of Engineering: Considering Social Science and Humanities Dimensions of Engineering in Education and Research. *Journal of Pre-College Engineering Education Research (J-PEER)*, *3*(2). https://doi.org/10.7771/2157-9288.1070
Islam, M. R. (2018). Sample size and its role in Central Limit Theorem (CLT). *Computational and Applied Mathematics Journal*, *4*(1), 1–7.
Jesiek, B. K., Trellinger, N., & Nittala, S. (2017). Closing the practice gap: Studying boundary spanning in engineering practice to inform educational practice. *2017 IEEE Frontiers in Education Conference (FIE)*, 1–9. https://doi.org/10.1109/FIE.2017.8190503
Jonassen, D. H., Strobel, J., & Lee, C. B. (2006). Everyday Problem Solving in Engineering: Lessons for Engineering Educators. *Journal of Engineering Education*, *95*(2), 139–151. https://doi.org/10.1002/j.2168-9830.2006.tb00885.x
Joshi, S. S., Davis, K. A., Czerwionka, L., Camps Troncoso, E., & Montalvo, F. J. (2022, June). A Comparison of Two Scenario-Based Assessments of Systems Thinking. *Paper Presented at 2022 ASEE Annual Conference Content Access*. In 2022 ASEE Annual Conference Content Access, Minneapolis, MN.
Kaur, M., & Craven, L. (2022). Systems Thinking: Practical Insights on Systems-Led Design in Socio-Technical

Engineering Systems. In *Handbook of Engineering Systems Design* (pp. 189–217). Springer.

Leydens, J. A., Johnson, K., Claussen, S., Blacklock, J., Moskal, B. M., & Cordova, O. (2018, June 23). Measuring Change over Time in Sociotechnical Thinking: A Survey/validation Model for Sociotechnical Habits of Mind. *Paper Presented at 2018 ASEE Annual Conference & Exposition*. 2018 ASEE Annual Conference & Exposition, Salt Lake City, Utah. https://doi.org/10.18260/1-2--30794

Leydens, J. A., & Lucena, J. C. (2018). Social Justice is Often Invisible in Engineering Education and Practice. In *Engineering Justice: Transforming Engineering Education and Practice* (pp. 45–66). IEEE. https://doi.org/10.1002/9781118757369.ch1

Mazzurco, A., & Daniel, S. (2020). Socio-technical thinking of students and practitioners in the context of humanitarian engineering. *Journal of Engineering Education, 109*(2), 243–261. https://doi.org/10.1002/jee.20307

Mazzurco, A., Huff, J., & Jesiek, B. (2014). The Energy Conversion Playground (ECP) Design Task: Assessing How Students Think About Technical and Non-technical Considerations in Sustainable Community Development. *International Journal for Service Learning in Engineering*, *9*(2), 64–84. https://doi.org/10.24908/ijsle.v9i2.5585

McGowan, V. C., & Bell, P. (2020). Engineering Education as the Development of Critical Sociotechnical Literacy. *Science & Education*, *29*(4), 981–1005. https://doi.org/10.1007/s11191-020-00151-5

Paul, R., Lönngren, J., & Berge, M. (2022). Breaking down dualisms in engineering classrooms: How emotions can support engineering problem solving. *2022 IEEE Frontiers in Education Conference (FIE)*, 1–3. https://doi.org/10.1109/FIE56618.2022.9962596

Pawley, A. L. (2009). Universalized Narratives: Patterns in How Faculty Members Define "Engineering." *Journal of Engineering Education*, *98*(4), 309–319. https://doi.org/10.1002/j.2168-9830.2009.tb01029.x

Petroski, H. (2011). *The essential engineer: Why science alone will not solve our global problems*. Vintage.

Reynante, B. (2022). Learning to design for social justice in community-engaged engineering. *Journal of Engineering Education*, *111*(2), 338–356. https://doi.org/10.1002/jee.20444

Riley, D. (2003). Employing liberative pedagogies in engineering education. *Journal of Women and Minorities in Science and Engineering*, *9*(2).

Riley, D. (2008). *Engineering and Social Justice*. Springer International Publishing. https://doi.org/10.1007/978-3-031-79940-2

Stevens, R., Johri, A., & O'Connor, K. (2014). Professional Engineering Work. In A. Johri & B. M. Olds (Eds.), *Cambridge Handbook of Engineering Education Research* (pp. 119–138). Cambridge University Press. https://doi.org/10.1017/CBO9781139013451.010

Swartz, M., Leydens, J., Walter, J., & Johnson, K. (2019). Is sociotechnical thinking important in engineering education?: Survey perceptions of male and female undergraduates. *Paper Presented at 2019 ASEE Annual Conference & Exposition*. 2019 ASEE Annual Conference & Exposition, Tampa, FL. https://doi.org/10.18260/1-2--33030

Trevelyan, J. (2007). Technical Coordination in Engineering Practice. *Journal of Engineering Education*, *96*(3), 191–204. https://doi.org/10.1002/j.2168-9830.2007.tb00929.x

Trevelyan, J. (2014a). *The making of an expert engineer*. CRC Press.

Trevelyan, J. (2014b). Towards a theoretical framework for engineering practice. *Engineering Practice in a Global Context: Understanding the Technical and the Social*, 33–60.

Piloting Case-Based Instruction in Immunology Course for Enhancing of Cognitive Learning

Shivalingsarj Desai[a], Preeti Patil[b], Sharanappa Achappa[a]and Basavaraj Hungund[a]
[a]Department of Biotechnology, KLE Technological University, Hubballi-580 031, India
[b]Center for Engineering Education Research, KLE Technological University, Hubballi-580031, India
Corresponding Author Email: desaisv@kletech.ac.in

Abstract

Context

Immunology as a course occupies corner stone in the domain of Biotechnology due to its relevance and applications. The significance of the course lies in its critical applications like vaccines, diagnosis and organ transplantation. Learning immunology at undergraduate level is often a challenging task for students since it involves integration of diverse concepts and coherent thinking. In this context, the present study of employing case-studies was undertaken for undergraduate students of Biotechnology engineering studying the course of immunology.

Purpose or Goal

The key objectives of the pedagogical intervention were to enhance the understanding of the fundamental immunological concepts and to induce cognitive thinking amongst the students leading to attain higher levels of cognitive dimensions of Bloom's taxonomy. At a broader level, the exercise aims to aid the increase the retention and comprehensive understanding of the concepts.

Methods

Selected clinical-case studies from " Case Studies in Immunology- A Clinical Companion" by Geha and Notrangelo was used as instrument for the intervention. 'Cooperative learning model was practiced for the activity and rubrics-based assessment was practised .

Outcomes

The exercise of case-based instruction was found to be instrumental in enhancing the cognitive learning of the course amongst the students. Also the depth of the case-studies gave them an immersive experience and enabled them to think critically and analyze the cases to arrive at appropriate solutions for the questions posed at the end of the case-study.

Conclusion

The activity was instrumental in achieving key elements of cognitive and knowledge dimensions of the cognitive process of learning. Further iterations of the exercise would be helpful in more effective implementation from the faculty perspective.

Keywords— Case-based learning; Cooperative learning; Immunology. Key-terms-based assessment

I. INTRODUCTION

A About Immunology and its associated challenges

Immunology is the study of body's defense system and its functions. It deals with physiological functioning of the immune system in states of both health and disease as well as malfunctions of the immune system in immunological disorders like allergies, hypersensitivities, immune deficiency, transplant rejection and autoimmune disorders. Though immunology is a fascinating subject, yet arguably it is considered to be a challenging course both from teaching and learning perspectives. It involves complex interactions between innate and adaptive systems which are distinct yet interrelated.

The learning challenges of immunology for students involves comprehending the complex terminologies and think coherently about a physiological system that is so anatomically disseminated. Immunology requires students to learn the meaning of new words, and rapidly apply them to build a knowledge base and answer complex questions. Despite the course being practically relevant, it needs serious and conscious efforts from students end to internalize the concepts. Teaching immunology is challenging since it requires students to integrate knowledge derived from pre-requisite courses like Microbiology, Cell biology, biochemistry, anatomy and genetics.

B Case-Based Learning in Immunology

Conventional didactic lectures wherein students are passive learners are known to have a very minimal impact on students learning and acquisition of knowledge. They often fail to stimulate students' cognitive skills of higher order like analytical, reasoning skills and their problem-solving ability.
In this context, the present study of Case-Based-Learning (CBL) was practised for Immunology course of IV semester undergraduate students of Biotechnology with an objective to enhance the learning effectiveness of immunological concepts. Case-based learning is a student-centred active learning method that facilitates the students' learning. The pedagogical intervention evokes students' interest, promote their learning, and engage them in active discussion in solving a clinical

problem related to immunology. The CBL enables the students to apply the concepts learnt from theory to real-life scenarios reflected in the clinical case-studies using inquiry-based learning methods. This enriches their cognitive skills of higher order like critical and analytical thinking, (Tayem, 2013; Zhang et al, 2013). In the first place, the case studies in CBL though are comprehensive in nature, yet are loosely structured and hence require the students to recollect and comprehend the concepts from pre-requisite courses and articulate the concepts learnt. CBL brings in a sense of life-long learning amongst the students (McEnerney, 1999).

CBL is known to engage the students in solving complicated questions and enable to attain higher order cognitive skills by thinking cognitively about means of approaching the feasible solutions. (Stranford et al, 2020). In context of Bloom's taxonomy, generally "remember" and "understand" are regarded as lower order cognitive skills that need only a minimum level of understanding, the third level "Apply" is considered as transitional whereas the application of knowledge and critical thinking are higher-order cognitive skills (HOCS) that require deep conceptual understanding (Zoller, 1993). Students often have difficulty performing at these higher levels (Zoller, 1993; Bransford et al., 2000; Bailin, 2002).

Case-Based Learning (CBL) has gained recognition in medical education for fostering active engagement and critical thinking (Thistlethwaite et al., 2012; Parmelee et al., 2012). In the realm of immunology, CBL proves beneficial, demonstrating enhanced understanding and retention of intricate concepts among students (Radcliffe, Lester, & Perera, 2019).

Research suggests that CBL contributes to cognitive learning enhancement, particularly in medical and health science education contexts (Thistlethwaite et al., 2012; McLean et al., 2016). Comparative studies highlight the superiority of CBL over traditional lecture-based instruction in immunology, showing better long-term retention and application of knowledge (Thistlethwaite et al., 2012; Parmelee et al., 2012). Moreover, CBL positively influences student engagement and motivation, creating a more favorable learning experience (Graffam, 2007). Case-Based Instruction effectively integrates basic science principles with clinical applications in immunology courses, providing students with a comprehensive understanding of the subject (Radcliffe et al., 2019; Graffam, 2007).

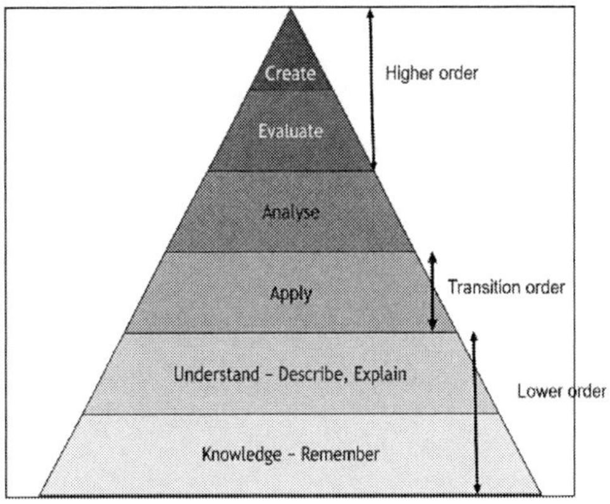

Fig.1. Levels of Bloom's taxonomy and order of learning

C Objectives of the study

The key objectives of the present study involving CBL are:

1. to incite the the students to understand the context, problem and analyse the facts of the case-scenario involved.
2. to enable the critical thinking skills in ambiguous case-scenarios, decision making, propose the possible solutions and chose the best feasible one and
3. to encourage co-operative learning amongst the team members.

II. METHODOLOGY

A Group Dynamics and target students

Engineering undergraduate students of IV semester Biotechnology were chosen for the CBL of Immunology course. Student groups (n=12) each comprising four members were formed for the CBL exercise. The groups were heterogeneous in terms of gender, merit and location. The exercise was a group activity and each group was assigned with a case-study for the purpose.

B Pedagogical instrument and reference material

Case-based instruction also referred to as case-based learning (CBL) was the pedagogical tool employed for the activity. "Case Studies in Immunology- A Clinical Companion" (VI Edition) authored by Raif Geha and Luigi Notarangelo regarded as the gold standard for clinical cases in Immunology was used as the reference study material for the activity. The rationale behind using the reference material was the clinical cases were complex and diagnostic in nature, mimic the real-world scenario, underscore the practical applications of Immunology and encompassed the test questions which were critical in nature.

C Cooperative learning- Write and Share activity

Cooperative learning wherein students work in groups towards a common course assignment was practiced with Write and Share activity. As part of Write and Share activity, the assigned topic was classified into different parts and each student of the

https://doi.org/10.52202/073963-0017

group was to study the assigned part in-depth and write the concept in their own words. All the members of the group were to meet, narrate their assigned part of the topic and discuss the topic in detail. The underlying objective was to bring-in individual responsibility, in-depth understanding, enhance attainment levels of students, develop positive relationships, create a learning community that values diversity and team coherence.

A template sheet for Write and Share activity consisting of topic and sub-topic and the concept chosen by the students amongst the group members was provided for brining in the clarity and uniformity in presentation of the activity.

D Focused Group Discussion (FGD) with faculty

The learning of the student groups was facilitated by scheduling periodical focused group discussion with the members wherein the case-study topic assigned was discussed at length. The students could clear the doubts and find means of approaching the questions for possible solutions. This complimented their efforts of self-study of the topic and ingrained the conceptual understanding.

E Choice of case studies

Twelve case studies were hand-picked from the reference book and assigned to the student groups keeping in view the level of difficulty, understanding of the undergraduate students, course content and diversity of the case-scenarios. Accordingly the following is the list of case studies

TABLE I

LIST OF IMMUNOLOGY TOPICS OF CASE STUDIES

Sl. No.	Topic of Case Study	Concepts covered
1	Allergic Asthama	Regulation of Immune-response and Immune tolerance
2	Drug-Induced Serum Sickness	
3	Hyper IgE Syndrome	
4	Deficiency of the C8 Complement Component.	Immunological disorders
5	Acquired Immunodeficiency Syndrome	
6	Multiple Sclerosis	
7	Rheumatoid Arthritis	
8	Adenosine Deaminase Deficiency	
9	Hemolytic Disease of the Newborn	
10	MHC Class I Deficiency	Transplantation immunology
11	MHC Class II Deficiency	
12	Graft versus Host Disease	

E Assessment

A two-level assessment was performed to evaluate the understanding of the case-studies by the students. The first review was oral on one-to-one basis with student members of the group. The key objective of the review was to assess the broad understanding and basic tenets of the topic assigned. This assumed significance in the light of the case studies being complex and the questions were analytical in nature which demanded a thorough knowledge of the basic concepts. All the students were expected to have a sound know-how of the concept with a threshold to attempt the test questions.

The second review was through a written report submitted as a group activity by the students. The report comprised responses to the questions which were critical and evaluative in nature. The students were expected to propose possible solutions with one best alternative. Given the descriptive nature of the report it needed a detailed study into the case study with reading between the lines and comprehending the information with an objective of attempting the questions which were critical in nature.

An aid in the form of key-word based assessment was practiced which proved to be an efficient solution for evaluating electronically (Mahmud et al, 2020). This was coupled with the descriptive evaluation of the written reports to assess the comprehensive understanding. The assessment was rubrics-based and inputs from different external experts working in the area of immunology was taken for the assessment.

F Insights into a Model Case Study

A model case study entitled “ Allergic Asthama was assigned to a students' group. The case study dealt with a case Frank Morgan-a 14-year boy with chronic asthama and rhinitis. Allergic Asthama is a Type-I hypersensitive chronic allergic disease caused by adaptive immune response to an inhaled antigen.

The following are sample questions posed as part of the CBL activity.

1. Explain the failure of Frank's asthma to improve despite the frequent use of bronchodilators, and his response to steroid therapy.

Keyword answers: Sensitized mast-cell, Degranulation, histamine, cytokines,

2. Eosinophilia is often detected in the blood and in the nasal and bronchial secretions of patients with allergic rhinitis and asthma. What is the basis for this finding?

Keyword answers: allergic rhinitis, Monteux test, IgE. Leishmans stain.

3. Frank developed wheezing on several occasions after taking the nonsteroidal anti-inflammatory drugs (NSAIDs) aspirin and ibuprofen (Motrin). Explain the basis for these symptoms.

Keyword answers: asthma exacerbation, cyclooxygenase inhibition, Polypoid hypertrophy.

https://doi.org/10.52202/073963-0017

Essentially the questions seek answers in the light of diagnosis, immune-response of the patient and rationale for the therapy administered by the physician.

III RESULTS AND DISCUSSION

A Graduate attributes and Course Outcomes

The CBL activity was instrumental in addressing some of the course outcomes outlined in the course. Also few graduate attributes expected of an undergraduate Biotechnology graduate could be addressed which would have been difficult to meet by routine methods of delivery and assessment . The following Table 2 summarises the competencies, performance indicators and course outcomes attained a s a result of CBL implementation.

TABLE II

GRADUATE ATTRIBUTES AND COURSE OUTCOMES

Competency	Performance Indicators	Course Outcomes addressed
1.2 - Demonstrate the competence in basic sciences	1.2.1 - Apply laws of basic science to an engineering problem	1.Analyze the mechanism of Cell-mediated Immunity, Major Histocompatibility Complex and Phagocytosis (L4)
1.4 - Demonstrate competence in domain knowledge of Biotechnology	1.4.1 - Apply knowledge of molecular biotechnology to solve conceptual engineering problems	2. Apply the principles of Complement system, cytokines, immune tolerance and hypersensitivity reactions in immune responses (L4) 3. Apply the concepts of autoimmunity & immunodeficiency and their associated disorders (L4)

A total of twenty marks were apportioned for the CBL activity with ten marks each for review 1 and 2. Among the total of fifty students, 15% 65% and 20% students scored in the range of 10-13, 14-16 and 17-20 marks. Majority of the students fairing average (14-16 marks out of 20) can be attributed to the facilitation of learning by cooperative learning. Given the criticality of the questions based on the complex case scenario this would have been difficult otherwise. Nevertheless the activity gave a rigor of learning experience for the students.

B Key experiences from the CBL activity

The students expressed that their depth of understanding enhanced due to reading and re-reading of the concepts. The write and share activity of cooperative learning not only honed the technical writing skills but also made them responsible learners amongst the group members. They appreciated the practical relevance of the immunology course in day-to-day life. The inclusion of external experts as stake-holders for evaluation brought in new thought processes and objectiveness to the activity. The course instructor in the process of facilitation had an overall knowledge of the all the concepts dealt.

From the faculty's perspective, studies indicate a positive attitude towards CBL in immunology courses, with many instructors recognizing the value of real-world cases in enhancing student learning (McLean et al., 2016). However, challenges exist, including the need for effective case selection, time constraints, and potential resistance from traditionalists (Parmelee et al., 2012; McLean et al., 2016).

In assessing CBL in immunology, diverse strategies such as case-based exams, group discussions, and reflective essays are proposed, showcasing a range of evaluation methods (Thistlethwaite et al., 2012). Moving forward, further research is crucial to explore optimal strategies for implementing CBL in immunology courses, considering variations in student backgrounds, institutional contexts, and the evolving landscape of medical education (Radcliffe et al., 2019).

III. CONCLUSION

The study on Case-Based Learning (CBL) in Immunology for fourth-semester Biotechnology engineering undergraduates demonstrated several key achievements. The primary objectives of the study were successfully met, including the stimulation of students' understanding of case scenarios, fostering critical thinking skills in decision-making, and promoting cooperative learning within diverse teams. The selection of "Case Studies in Immunology- A Clinical Companion" as the reference material proved effective in presenting complex, diagnostic, and real-world scenarios, aligning well with the pedagogical tool of CBL.

The implementation of cooperative learning through the Write and Share activity facilitated individual responsibility, in-depth comprehension, and positive relationships among group members. Focused Group Discussions (FGD) with faculty members further enriched students' conceptual understanding, providing a platform for clearing doubts and refining problem-solving approaches. The choice of twelve diverse case studies ensured a comprehensive coverage of difficulty levels, aligning with the undergraduate students' understanding and course content.

The assessment methodology, involving both oral reviews and written reports, provided a robust evaluation of students' comprehension. The results indicated that the CBL activity effectively addressed course outcomes and graduate attributes, contributing to a meaningful learning experience.

In summary, the study demonstrated that CBL in Immunology, when integrated with cooperative learning strategies, diverse case studies, and robust assessment methods, can effectively enhance students' critical thinking, problem-solving skills, and overall understanding of complex subject matter. The positive feedback from students and the alignment with course outcomes underscore the value of CBL as a pedagogical approach in higher education.

ACKNOWLEDGEMENT

The authors thank Dr. Ashok S. Shettar, Honorable Vice Chancellor., Dr. B.S. Anami, Registrar and Dr. Vijayalakshmi M., Director Center for Engineering Education, KLE Technological University, Hubballi, Karnataka for their support and encouragement in undertaking the study.

REFERENCES

Al Mahmud, T., Hussain, M. G., Kabir, S., Ahmad, H., & Sobhan, M. (2020). A keyword based technique to evaluate broad question answer script. In Proceedings of the 2020 9th International Conference on Software and Computer Applications. 167-171.

Bailin,S.(2002).Critical thinking and science education. Science & education,11, 361.

Bransford, J., Brown, A. L., and Cocking, R. (2000). How People Learn: Brain, Mind, Experience, and School, Washington, DC: National Academies Press.

Ciraj, A. M., Vinod, P., & Ramnarayan, K. (2010). Enhancing active learning in microbiology through case based learning: Experiences from an Indian medical school. Indian Journal of Pathology and Microbiology, 53(4), 729-733.

Graffam, B. (2007). Active learning in medical education: strategies for beginning implementation. Medical teacher, 29(1), 38-42.

McEnerney, K. (1989). Cooperative learning as a strategy in clinical laboratory science education. Clinical Laboratory Science, 2(2), 88-9.

McLean, S. F. (2016). Case-based learning and its application in medical and health-care fields: a review of worldwide literature. Journal of medical education and curricular development, 3, JMECD-S20377.

Parmelee, D., Michaelsen, L. K., Cook, S., & Hudes, P. D. (2012). Team-based learning: a practical guide: AMEE guide no. 65. Medical teacher, 34(5), e275-e287.

Shigli, K., Aswini, Y. B., Fulari, D., Sankeshwari, B., Huddar, D., & Vikneshan, M. (2017). Case-based learning: A study to ascertain the effectiveness in enhancing the knowledge among interns of an Indian dental institute. The Journal of the Indian Prosthodontic Society,17(1), 29-34.

Stranford, S. A., Owen, J. A., Mercer, F., & Pollock, R. R. (2020). Active learning and technology approaches for teaching immunology to undergraduate students. Frontiers in public health, 8, 11-17.

Tayem, Y. I. (2013). The impact of small group case-based learning on traditional pharmacology teaching. Sultan Qaboos University Medical Journal, 13(1), 115-120.

Thistlethwaite, J. E., Davies, D., Ekeocha, S., Kidd, J. M., MacDougall, C., Matthews, P., ... & Clay, D. (2012). The effectiveness of case-based learning in health professional education. A BEME systematic review: BEME Guide No. 23. Medical teacher, 34(6), e421-e444.

Zhang, Z., Liu, W., Han, J., Guo, S., & Wu, Y. (2013). A trial of patient-oriented problem-solving system for immunology teaching in China: a comparison with dialectic lectures. BMC medical education, 13(1), 1-8.

Zoller, U. (1993). Are lecture and learning compatible? Maybe for LOCS: Unlikely for HOCS. Journal of Chemical Education, 70(3), 195-198.

ACKNOWLEDGEMENT

The authors thank Dr. Ashok S. Shettar, Honorable Vice Chancellor., Dr. B.S. Anami, Registrar and Dr. Vijayalakshmi M., Director Center for Engineering Education, KLE Technological University, Hubballi, Karnataka for their support and encouragement in undertaking the study.

Copyright Statement

Transforming the Journal of Engineering Education Transformations

Sohum Sohoni[a] and Sandeep R Desai[b]

Milwaukee School of Engineering Department of Electrical Engineering and Computer Science, Milwaukee School of Engineering, Milwaukee USA[a], Department of Automobile Engineering K.E.S's Rajarambapu Institute of Technology Shivaji University, Kolhapur, India[b]

Corresponding Author Email: sohoni@msoe.edu

Abstract

Context

A reputable and reliable outlet for publication is an important component of building a research ecosystem. Currently, the Engineering Education Research community in India is at a nascent stage, and lacks a world-renowned outlet for publication. The Journal of Engineering Education Transformations (JEET), was established in 1985 to serve as an outlet for sharing narratives of educational transformations at engineering institutions in western India. For the past six years, the journal has itself undergone a transformation, from publishing a small set of case studies, to publishing peer-reviewed articles that range from engineering education research, to practice and even policy. As a result of this transformation, the number of submissions to the journal have skyrocketed over the last few years indicating the EER community's faith in the journal's quality and practices. This article aims to describe JEET's transformation and provides details of its inner workings including training programs such as a mentored reviewer program. The journal and the EER ecosystem in India, have a long way to go, and a discussion on JEET is necessary to engage the EER community.

Purpose or Goal

The goal behind submitting this article is to have a frank and open conversation on how operating procedures could be improved at the journal, how the EER community in India can benefit from adopting the journal and enabling its success, and to solicit innovative ideas on how the journal can best serve the needs of a growing ecosystem of engineering education researchers and scholarly teachers in India. A secondary goal is to involve the global EER community at large to help JEET in having an impact and presence beyond India.

Methods

The paper takes a simple narrative approach, with the journal editors presenting the history and growth of the journal, supported by statistics on number of submissions, time to review, time to publish etc.

Outcomes

The paper will showcase the journey of the journal from being a repository for the occasional case study to a Scopus-indexed journal that accepts papers on the scholarship of teaching and learning.

Conclusion

JEET has established strong practices for peer review and quality control. It seeks more engagement from budding engineering education researchers in India for participating in peer-reviews. It will also benefit from international engagement. Having become a Scopus-indexed journal, JEET serves to elevate the EER community in India and needs participation from all stakeholders to take it to the next level.

Keywords—journal; capability-building; peer-review; publication outlet.

I. INTRODUCTION

ONE of the goals of the research in engineering education is the transformation of engineering education through a variety of experiments in teaching-learning and related areas. The Journal of Engineering Education Transformations (JEET) is a scholarly journal committed to the advancement of theory, research, and practice in the field of engineering education. Published from India, the world's largest hub of engineering education, JEET was identified for a critical role in the capability building for EER in India (Sohoni et al., 2017). Although based out of India, it is international in its scope, inviting scholars and experts from across the globe to share their theoretical insights and innovative practices for the enhancement and transformation of engineering education. JEET is a peer-reviewed journal made available in both print and online versions. A double-blind peer-review process performed by experts in the field ensures that the highest standards in scholarship are maintained.

The Objectives of JEET are:

- To provide a world-class platform for publishing original research in engineering education.
- To provide a forum for sharing innovative practices for imparting engineering education.

https://doi.org/10.52202/073963-0018

- To provide a forum for sharing innovative strategies for combating issues unique to engineering education in India and abroad.
- To foster international collaboration and discourse for the betterment of different aspects of engineering education.

In 2017, JEET has embarked on a transformation from a journal that published a handful of case studies to a Scopus-indexed journal that publishes research in engineering education, including the scholarship of teaching and learning (Sohoni, 2018). Through a targeted campaign with the support of the Indo-Universal Collaboration for Engineering Education (IUCEE), aided by the fact that it became Scopus indexed, the journal saw a tremendous surge in the number of articles submitted per year as shown in Table 1.

TABLE I
TREND IN NUMBER OF SUBMISSIONS RECEIVED PER YEAR

Year	Submissions
2017	23
2018	200
2019	300
2020	Unknown due to change in platform
2021	Unknown due to change in platform
2022	401
2023	519 (as of November)

As expected, the editorial team was overwhelmed by the deluge of submissions and has been expanding the number of associate editors as shown in Table 2 (JEET, 2023) as well as the number of reviewers over the past several years (Sohoni, 2021).

TABLE II
SCALING OF TOTAL NUMBER OF AEs

Year	Number of Associate Editors
2017	4
2018	5
2019	9
2020	15
2021	26
2022	30
2023	27

In this article, the operating procedure of the Journal of Engineering Education Transformations (JEET) and initiatives taken by the journal to improve its operations, relevance and ranking are discussed in detail. The performance indicators of the journal, the factors affecting the performance indicators and improvement in the performance of the journal over the past six years are also discussed.

II. STANDARD OPERATING PROCEDURE FOR PEER REVIEW

JEET follows the double-blind peer review process. To ensure the quality of peer review and quality of the manuscripts

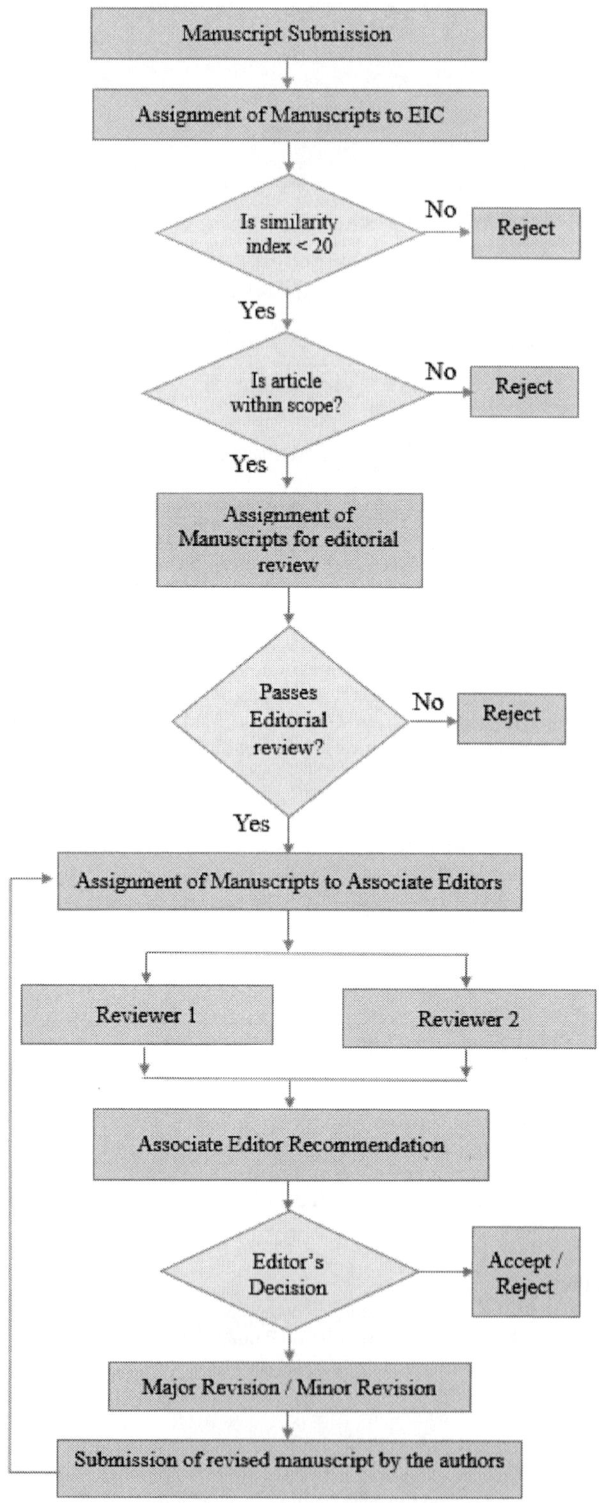

Fig. 1. Peer-review process.

https://doi.org/10.52202/073963-0018

published in the JEET, the rigorous peer review process is designed and implemented. Figure 1 shows the flow chart to indicate the journey of the manuscripts from its submission to final decision. An online peer review tool (Manuscript Communicator) is used as the editorial management system. The manuscripts undergo a plagiarism check and only those manuscripts having similarity less than or equal to 20% are considered for the editorial review. Technical (non-EER) manuscripts or the manuscripts beyond the scope of the journal and the manuscripts having unacceptable similarity are rejected before review. Over the past few years, the large volume of out-of-scope submissions have been a significant challenge and have required additional efforts on the editorial team's part. The manuscripts recommended by the editorial review team are further assigned to the Associate Editors to assign them to at least two reviewers for peer review. The reviewers after accepting the invitation to review are expected to submit the review reports and review recommendation within two months. After at least two reviews are received, Associate Editors send their recommendation to Editor-In-Chief. The Editor-In-Chief reviews the review reports and Associate Editor's recommendation and communicates the decision to the authors. The decision would be accept / reject / major revision / minor revision. The accepted manuscripts are sent to the production team and they are considered ready for publication after proof reading. In case of the manuscripts with a revision decision, authors are asked to send the revised manuscripts and rebuttal within one month. The revised manuscripts submitted by the authors again undergo the plagiarism check and are then assigned to the respective Associate Editors and reviewers for the revision review. The same peer review process is repeated till manuscript reaches the final decision.

III. Performance of the Journal

A. Performance Evaluation Parameters

The performance of a journal is evaluated on the basis of the performance indicators such as h index, Scopus indexing, impact score, Best Quartile, Overall Ranking etc. The performance is evaluated using quantitative data as well as using qualitative measures. The achievement of the performance indicators for any Journal depends on evaluating the stakeholders such as reviewers, Associate Editors and Editorial Board members. The performance of a journal also depends on impact of the articles published on society, time to first decision, time to final decision, acceptance rate etc. (O'Rourke, 2015). Gangan Prathap (2012) discussed various indicators that are used as the basis to rank Journals. He discussed various quality and quantity indicators such h-index, g-index, p-index, eigen factor score and article influence and their relationships. This section provides data on these metrics for JEET.

B. Initiatives to Elevate Journal Performance

The Editors and Editorial Board members are responsible to ensure achievement of the target time for the first decision, target time for the final decision, quality of the publication, controlling acceptance rate, declining the manuscripts before review to ensure that the journal is not overloaded with unsuitable submissions, increasing database of the reviewers, increasing number of Associate Editors, and effective communication with the authors, reviewers, Associate Editors, administrator and staff, online peer review management team, and production team. Since its inception, the Journal of Engineering Education Transformations (JEET) has taken variety of initiatives to improve the Journal's performance. As a result, JEET has achieved steady and gradual improvement in the performance over the period of time in terms of quality of reviews, quality of publication, indexing, and impact score etc. Following are the initiative taken to control the factors affecting the performance of the Journal.

1) Mentored Reviewer Program

Inspired by the success (Benson et al 2021, Jensen et al, 2022) of the mentored reviewer program conducted by the Journal of Engineering Education (JEE) (Journal of Engineering Education Mentored Reviewer Program, 2019), JEET embarked on its own version of it in May of 2020 (Hattingh et al. 2021). Given the volume of submissions received by JEET and the corresponding scaling up of Associate Editors, the program was modified to provide additional mentoring for AE's prior to the start of the program. This mentoring was done in the form of two one-hour workshops on consecutive weekends. This helped greatly in building a community of practice (Wenger, 1998) for the AEs themselves. Each AE was then asked to pick a reviewer for whom the AE would serve as a mentor. Each mentor-mentee pair was then tasked with collaboratively reviewing two papers with sessions held in between reviews to facilitate discussion and reflection on the review process. Overall, the program has had a positive effect on the quality of reviews, but challenges exist in terms of scaling it to the needs of JEET. An unanticipated positive outcome of this program was the realization of how lonely the peer-review process is, and how important it is to create some social interaction around it. However, the initial enthusiasm around wanting to continue the mentored reviewer program and providing opportunities for interaction has unfortunately faded at this point. A follow-up study to gather more detailed data from participants is currently underway.

2) Increase in number of Associate Editors

The number of Associate Editors that were 4 in 2017 are now 30 (Table 2), to take care of increased submissions. The involvement of Associate Editors from within and outside India has helped the journal to reach to authors and reviewers at international level.

3) Involvement of Global EER Community

In an effort to bring in global perspectives and international standards to JEET, 7 Associate Editors from outside of India were added between 2017 and 2022. The goal is to continue to grow this number with the help of Research in Engineering Education Network (REEN). In 2021, the journal was also added to the list of EER journal maintained by REEN on its website (REEN 2023). The journal has also become available on the SCI list, an internationally recognized listing.

4) Increase in Database of Reviewers

There has been significant increase in number of reviewers over the last few years and contribution from the reviewers outside India has also been increased in the recent past. The involvement of increased number of reviewers has helped to strengthen the peer review process and reach the final decision within expected time. However, the journal still faces a tremendous challenge in terms of responsiveness of reviewers and review quality.

5) Evaluation and Change in the Online Peer Review System

An effective and efficient peer review tool plays an important role in ensuring that the standard operating procedure is in place for manuscript submission, peer review and publication. The performance evaluation of the online peer review tool was done and recommendations were given to improve the submission and peer review process. The revised version of online peer review tool was installed to help the authors, editors, associate editors, reviewers and production managers. Based on the user experience, several enhancements were suggested to the system, many of which were included in a new version of the system.

C. Achieved Performance Indicators

The Journal was indexed in Scopus in the year 2018 and since then the Journal has been indexed in Scopus continuously. The best quartile for this journal is Q3 and this journal has an h-index of 9. The best quartile Q3 indicates the middle-low position, next 25% Journal title after Q2 fall under this category (between 50% to 75%). The ISSN of Journal of Engineering Education Transformations journal is 23941707, 23492473. The overall rank of Journal of Engineering Education Transformations is 18694. According to SCImago Journal Rank (SJR), this journal is ranked 0.210. SCImago Journal Rank is an indicator, which measures the scientific influence of journals. It considers the number of citations received by a journal and the importance of the journals from where these citations come. SJR acts as an alternative to the Journal Impact Factor (or an average number of citations received in last 2 years). Figure 2 shows improvement in SCImago Journal Ranking (SJR) of the Journal for the past four years.

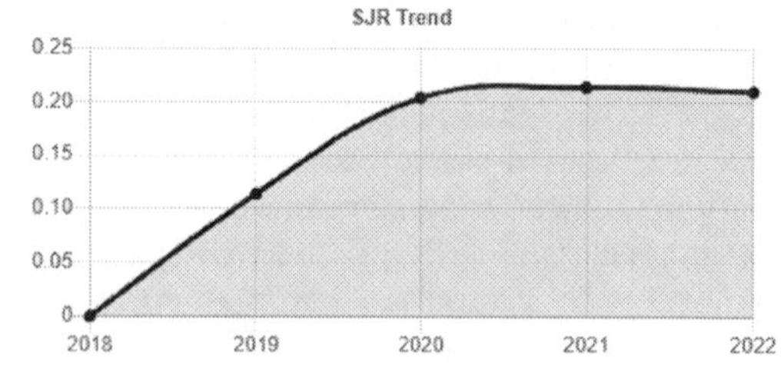

Fig. 2. Improvement in SCImago Journal Ranking since 2018.

The impact score (IS), also denoted as Journal impact score (JIS), of an academic journal is a measure of the yearly average number of citations to recent articles published in that journal. It is based on Scopus data. The impact score (IS) 2022 (which is computed in 2023 as per its definition) of Journal of Engineering Education Transformations is 0.79 and has increased by 0.3, an approximate percentage improvement of 61.22% when compared to the preceding year 2021. The highest and the lowest impact index or impact score of this journal are 0.79 (2022) and 0.00 (2018), respectively, in the last 5 years. Moreover, its average IS is 0.39 in the previous 5 years. Figure 3 shows improvement in Impact Score (IS) of the Journal for the past four years. Although the trend is in the right direction, the absolute numbers are low, and efforts are underway to increase these numbers, including training faculty in India in EER and promoting the journal to a worldwide audience.

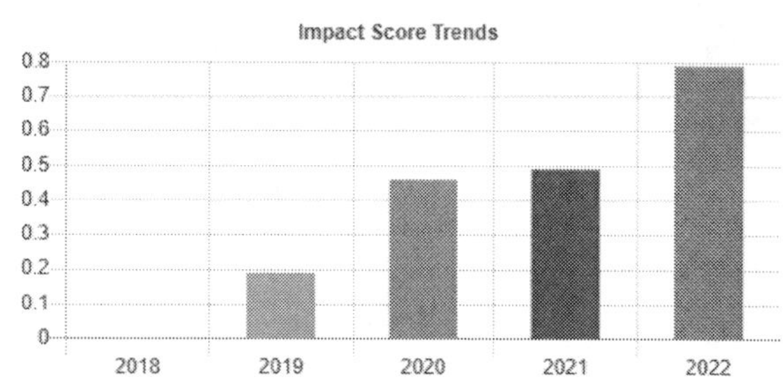

Fig. 3. Improvement in SCImago Jouranal Impact Score.

Figure 4 shows trend of overall ranking for the Journal of Engineering Education Transformations (JEET).

D. Reviewer and Associate Editor Grading System

The role of the reviewers is crucial to ensure quality of the manuscripts published by the Journal. The evaluation of reviewers is necessary to achieve the target time of final decision and publication as well as to ensure quality of publication through quality of reviews. The number of reviews completed in a year, number of times the reviewers declined the review, quality of the review report submitted by the reviewers,

https://doi.org/10.52202/073963-0018

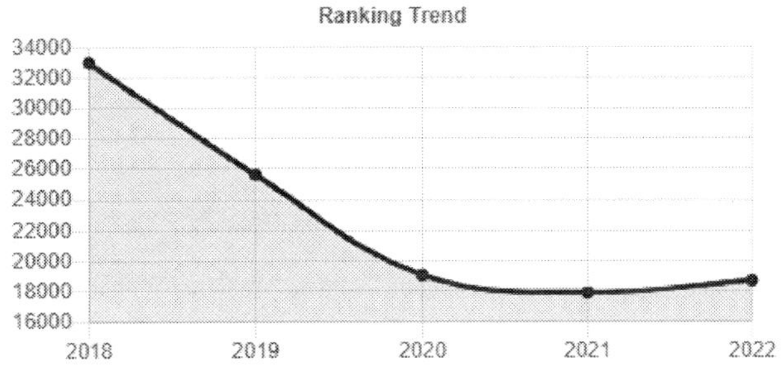

Fig. 4. Improvement in journal ranking.

timely submission of the review, contribution of the reviewers for number of years and acceptance rate are the parameters considered for evaluating the reviewers.

The Journal of Engineering Education Transformations (JEET), as a future step, has planned to design and implement reviewer grading system to evaluate the performance of the reviewers and to award them based on their contribution and rating. The objective of such an initiative is to further increase the reviewer database by motivating best reviewers to join JEET reviewers' team. The initiative to further increase reviewers' database and quality of reviews will help to reduce the time to decision as well as quality of the manuscripts published in JEET.

Similarly, Associate Editors will be evaluated on the basis of the number of manuscripts completed, follow up with the reviewers, timely submission of the AE's recommendation and attempts to ensure quality of the reviews assigned to them.

IV. CONCLUSIONS

The goal of this paper was to discuss the operating procedures of JEET, describe how the EER community in India can benefit from adopting the journal and enabling its success, and to solicit innovative ideas on how the journal can best serve the needs of a growing ecosystem of engineering education researchers and scholarly teachers in India. JEET has established strong practices for peer review and quality control. However, the journal can greatly benefit from the involvement of the international EER community on all fronts- article submission, peer-review process as well as readership. Thus, this paper may well be viewed as a call to the global EER community for participation in JEET.

REFERENCES

Benson, L., Bates, R. A., Jensen, K., Lichtenstein, G., Watts, K., Ko, M., & Albayati, B. (2021, January). Building research skills through being a peer reviewer. In American Society for Engineering Education.

Hattingh, T., Sohoni, S., Agrawal, A., Desai, S., & Joshi, S. (2021). An exploration of capacity development of journal reviewers through a mentored reviewer program. REES AAEE 2021 conference: Engineering Education Research Capability Development. Engineers Australia. https://search.informit.org/doi/10.3316/informit.354412078738540

JEET, 2023. Editorial Board. https://journaleet.in/editorial-board Accessed September 19, 2023.

Jensen, K., Bates, R., Lichtenstein, G., Watts, K., Ko, E., & Benson, L. Building a Community of Mentors in Engineering Education Research Through Peer Review Training. 2022 Annual American Society of Engineering Education (ASEE) Conference and Exposition

Journal of Engineering Education Mentored Reviewer Program, 2019. https://sites.asee.org/wied/journal-of-engineering-education-mentored-reviewer-program/ Accessed September 19, 2023

O'Rourke, K. (2015). https://www.csescienceeditor.org/article/by-the-numbers-evaluation-strategies-to-improve-journal-performance/ Accessed September 19, 2023.

REEN (2023) Research in Engineering Education Network. https://reen.co/eer-journals/ Accessed November 26, 2023.

Sohoni, S. Craig, S. D., Vedula, K. (2017, June). A Blueprint for an Ecosystem for Supporting High Quality Education for Engineering, Journal of Engineering Education Transformations, 30 (4) pp 58-66. DOI 10.16920/jeet/2017/v30i4/114107.

Sohoni, S. (2018, October). Editorial, Journal of Engineering Education Transformations, 32 (2) pp 4-6. DOI 10.16920/jeet/2018/v32i2/139489.

Sohoni, S. (2021, January). Editorial, Journal of Engineering Education Transformations, 32 (2) pp 5-6. DOI 10.16920/jeet/2021/v34i3/157714.

Wenger, E. (1998). Communities of practice. Learning, meaning, and identity. Cambridge, UK: Cambridge University Press.

Enablers and Barriers for Engaging in Engineering Education Research in India

Pallavi Meshram[a], Rucha Joshi[a*], Rohit Kandakatla[b], Shreel Prasad[a], Mayanglambam Pooja Devi[a], and Ashish Agrawal[c]
Plaksha University, Mohali, Punjab, India[a], KG Reddy College of Engineering and Technology, Hyderabad, Telangana, India[b], College of Engineering Technology, Rochester Institute of Technology, Rochester, New York, USA[c]
Corresponding Author Email: rucha.joshi@plaksha.edu.in

Abstract

Context
India needs to improve its Engineering Education Research (EER) to address its unique educational requirements and align with the visionary goals set forth in the National Education Policy (NEP) 2020. To truly follow the NEP 2020 vision, we need to understand and tackle the challenges stopping us from diving deeper into EER.

Purpose or Goal
Engineering faculty in India are one of the key stakeholders as they can engage in EER in alignment with the vision of NEP 2020. However, an extremely low participation of engineering faculty in EER remains a major concern. Therefore, this work aims to systematically explore the barriers to engaging in EER in India and suggest potential directions to overcome these barriers via dialogue with stakeholders about enablers.

Methods
A qualitative approach was used in this study to gather insights and perspectives from various stakeholders regarding the barriers to engaging in EER in India and potential enablers. Semi-structured interviews were conducted with engineering education researchers, and leaders at engineering education institutes in India. Collected data was analyzed using thematic analysis and specific themes are reported.

Outcomes
This study identifies barriers and enablers to facilitate high-quality EER in India, provides recommendations for institutional and governmental policies, and motivates individuals to implement practical solutions. A few of the barriers we found include a lack of institutional support for EER, bridging research and teaching practice in academia, disciplinary silos, lack of direction, lack of recognition of EER as a field of inquiry, a small community of EER researchers, and lack of funding. Some enablers found include motivation to advance EER in India, institutional recognition, acknowledgment of EER by the engineering community, capacity building and collaborations, technology-based research, and NEP 2020.

Conclusion
This analysis provides valuable guidance for future research and policy initiatives aimed at enhancing engineering education in India. The findings can be utilized by stakeholders at both individual and systemic levels to drive positive change within the EER ecosystem in India.

Keywords— Barriers; Enablers; Challenges; Engineering education research, EER in India

I. INTRODUCTION

ENGINEERING Education Research (EER) is a multifaceted field that goes beyond traditional education and research boundaries. It encompasses reforming, implementing, and exploring interdisciplinary aspects of engineering education. EER aims to tackle worldwide issues in higher engineering education, such as attracting students, fostering skills across disciplines, and managing intricate knowledge (Borri & Maffioli, 2008). This interdisciplinary domain involves engineering, science, social science, and educational psychology researchers. By incorporating advancements in engineering education and learning sciences, EER seeks to improve teaching methods and prepare well-rounded graduates capable of making societal impacts (National Academies of Sciences, 2018).

India with its rich history of academic excellence and its burgeoning role in the global technology arena, presents a unique backdrop for EER. Performing quality EER could help the nation's academic institutions be at the forefront of producing world-class engineers. However, there are only limited systematic comparisons available for EER across various national and cultural settings. This holds true for India as well, where EER has a limited presence in both historical and global contexts. This limitation hinders its ability to be widely disseminated and recognized within the broader academic community. We realize that the journey of integrating EER into the Indian academic fabric has its own set of challenges and opportunities that need to be identified and resolved. India has taken steps towards incorporating the advancements towards

https://doi.org/10.52202/073963-0019

engineering education in the form of NEP 2020, for which the current status is now in its implementation phase. Between 2020 and 2023, the government conducted webinars and informational sessions across various institutions.

Specifically, in the domain of EER, there is a noticeable gap in discussions related to the historical foundations of the engineering profession. A limited number of individuals in this field genuinely delve into historical aspects in their research or pedagogical approaches, as noted by (Wisnioski, 2015). Previous unsuccessful initiatives in the field, which have led to a sense of caution, coupled with misunderstandings or misconceptions surrounding the concept of the Scholarship of Teaching and Learning (SoTL), can also serve as significant deterrents or obstacles that discourage individuals from engaging in EER. For a new researcher, the challenges of understanding theoretical frameworks and establishing the credibility of qualitative research remain large obstacles (Smith & McGannon, 2018). In a study conducted by (Streveler et al., 2015), which aimed to familiarize Ph.D. students with the realm of EER, the participants recognized that the most conceptually demanding elements included understanding the theoretical framework, grappling with qualitative and mixed-method approaches, and establishing the credibility of qualitative research. Despite the complexities, the students managed to navigate these challenges by leveraging their existing engineering knowledge and past experiences. They connected these insights to their educational experiences or various facets of their professional engineering roles.

Considering the above-mentioned facts, this paper aims to provide a comprehensive understanding of the barriers and enablers influencing engagement in EER in present-day India. Drawing from insights by researchers, leaders, and other stakeholders, we have explored the intricate dynamics at play, offering insights that can guide future endeavors in this critical field in India.

II. LITERATURE REVIEW

While a longstanding interest has persisted in enhancing and documenting engineering education, the formal recognition of EER as a discipline-specific area within education research remains a relatively recent development compared to research in conventional engineering disciplines that have been around for many decades (Council, 2012; Johri & Olds, 2014; Lohmann & Froyd, 2010). New entrants into EER often originate from faculty roles within engineering or related domains, driven by a shared ambition to enhance educational practices (Duschl & Bybee, 2014). As a result, the field of EER has traditionally emphasized two key areas: one is centered on expanding participation, and the other concentrates on enhancing competencies (Feiman-Nemser, 1989). However, while EER has introduced innovative pedagogical methods, the intricate nature of engineering work across its various subfields demands a more nuanced perspective (Buckley et al., 2023).

Engineering education can be seen as a multilevel system comprising four main levels: the student level, the institutional personnel level, the institutional culture and industry level, and the societal level (Klein et al., 2019). As the field of engineering education evolves, researchers are increasingly emphasizing an understanding of the individual student's journey, from their initial motivations and emotions to their progression through various educational phases. However, beyond the student level, institutional dynamics play a pivotal role. The personnel involved in education, their roles, expertise, and perspectives significantly shape the education process. Moreover, the institutional culture, closely tied to industry trends and demands, greatly influences curriculum design and pedagogical approaches, shaping the directions for EER (Burbules & Torres, 2000). Navigating the newly forming EER landscape is complex because many researchers in the EER domain traditionally began as engineering faculty in conventional engineering fields like civil engineering, mechanical engineering, etc., and later shifted to research within EER (Kamp, 2020).

However, due to limited structural support on how to conduct EER and a lack of established novice-expert relationships, there's a growing need to train the new researchers to conduct research that directly impacts students and stakeholders, as current EER work is being critiqued for risks of becoming isolated and less relevant (Dart et al., 2023). To make EER more relevant to the needs of the local community, modern educational context, and changing ecosystems, barriers must be recognized and solved using an enabling ecosystem.

A. *Barriers*

Researchers typically shift to EER post-technical engineering studies, facing challenges merging their training with the distinct demands of educational research. Formal training pathways, especially in postgraduate studies, are not well-established, leading to a lack of clear standards and practices (Gardner & Willey, 2018). Further, this area of research underscores several "institutional" challenges, encompassing time limitations, inadequate institutional backing and growth prospects, insufficient funding, and a perception of undervaluation of engineering education research. These hindrances are well-documented in prior literature (Brodie et al., 2011; Haigh et al., 2011; McKinney, 2002; Wankat et al., 2002), prompting consideration of strategies to overcome these issues and stimulate increased engagement in education research among engineering academics. Another major obstacle mentioned in the literature for EER is the considerable job-related difficulties experienced by faculty members. Limited scholarship support and incentives prevent them from fully engaging in EER despite their motivation. This makes it challenging for them to effectively balance their roles as both

 https://doi.org/10.52202/073963-0019

educators and researchers (Ko et al., 2021). The departmental organization of universities, structured around subjects, sometimes hinders the creation of interdisciplinary teaching environments needed for holistic projects and skill enhancement. Tackling these issues is vital for advancing EER in India (Valero, 2022). It is observed that faculty at institutions with doctoral programs significantly value reduced teaching loads as an enabler for motivating research output (Chen et al., 2010). Similarly, in another study it was noted that the factors contributing to success of EER initiatives include proactive leadership support, adequate resource allocation, enthusiastic staff involvement, and access to valuable conceptual frameworks (Wenger et al., 2011). Importantly, the study emphasizes that the supportive elements outweigh the challenges, indicating a positive outlook for the advancement of EER when these factors are effectively leveraged.

B. Enablers

Researchers (Borrego & Bernhard, 2011) (Beddoes et al., 2010) identify the significance of global research collaborations and how it is equally important to adapt EER to local contexts. Researchers and educators should make their work understandable across various institutions and countries, considering cultural and educational differences (Jesiek, Borrego, & Beddoes, 2010). However, it's crucial to understand that not all findings can apply everywhere due to cultural and educational differences (Beddoes et al., 2010). It is important for EER researchers to show impact for gaining traction in the field (Mehta & Berdanier, 2019), review the need for incorporation of EER in the educational landscape, and emphasize the need for curriculum adjustments to address the implementation challenges. It brings out the necessity of integrating EER to tackle emerging issues in curriculum design, teaching methods, expertise development, and diversity for Additive Manufacturing context.

Various research articles have delved into the assessment and progression of EER as an independent field (Borrego & Bernhard, 2011; Jesiek et al., 2009) (Borrego and Bernhard, 2011; Jesiek, Newswander, and Borrego, 2009). Numerous studies have explored the evolution of EER within diverse settings, encompassing the U.S.A. (Lohmann & Froyd, 2010) (Froyd and Lohmann, 2014), Portugal (van Hattum-Janssen et al., 2015) (Sorby et al., 2014; van Hattum-Janssen, Williams, and Nunes de Oliveira, 2015), Ireland (Sorby et al., 2014), Australia and New Zealand (Godfrey & Hadgraft, 2009) (Godfrey and Hadgraft, 2009), Europe (Bernhard, 2018) (Bernhard, 2018), and three Nordic countries (Edström et al., 2018) (Edström et al., 2016). Additionally, research has analyzed EER in a global context (Streveler & Smith, 2010) (Jesiek, Borrego, and Beddoes, 2010a, 2010b; Streveler and Smith, 2010). However, there remains a gap in terms of investigating EER within the Indian context, especially with respect to enabling and hindering factors. The absence of such studies in India leaves an information gap regarding the barriers, recognition, institutional support, funding, and career pathways for EER researchers in the country. This dearth underscores the need for targeted research initiatives in India to better understand and address the challenges and potential solutions related to the advancement of EER in the Indian engineering education.

III. METHODS

The literature review serves as the foundation for developing the interview protocol by providing a comprehensive understanding of the historical context, key concepts, emerging trends, and challenges in EER. Here, we systematically investigated data obtained from interviews in India by following Braun et al. thematic analysis (Braun & Clarke, 2006). Initially, we carefully selected our stakeholders, comprising key figures in engineering education, including the Deans, Vice-chancellors, Teaching and learning center heads, EER researchers, and Practitioners. To guide our interviews with these stakeholders, we crafted a questionnaire that probed their views on "Evolution of Engineering Education in India in terms of their impressions of current state, factors shaping it, and most pressing challenges"; "Engineering Education in India next two decades, including factors driving it, and role of NEP", and "Potential Directions for EER in India". The detailed questionnaire is given in (Appendix I). Subsequently, we invited participants (India-over) who possessed relevant experience within the Indian engineering education context or had engaged in research related to it. Ultimately, we conducted semi-structured interviews with a total of 18 participants from institutes across India, including 6 institutional leaders, 8 EER researchers, 2 practitioners, and 2 heads of teaching-learning centres. Interviewees consisted in total of 4 female participants and 14 male participants. Interviews were transcribed using the transcription tool available in Microsoft Teams application. Our analysis utilized an exploratory coding method, as advocated by Creswell (Cresswell et al., 2012), to systematically uncover significant themes and patterns within the transcripts. These themes included barriers and enablers for EER in India, providing valuable insights into the challenges and opportunities within the field. Throughout the analysis, we maintained the credibility and trustworthiness of our findings through continuous dialogue and resolution of disagreements among the authors. This comprehensive methodology allowed us to gather structured insights from the stakeholders and perform a rigorous analysis of the interview data, contributing to a deeper understanding of EER in the Indian context. Overall, 7 themes emerged in barriers and 6 themes emerged in enablers. Illustrative quotes are given for themes, with the pseudonyms in place to protect privacy of stakeholders.

IV. RESULTS

The analysis of transcripts revealed the key themes that act as barriers and enablers for EER in India. Participants talked of

https://doi.org/10.52202/073963-0019

many problems including lack of institutional support, people not recognizing EER as a field of inquiry, teaching overload, shortage of time, lack of training, lack of working opportunities, lack of funding, lack of proper research questions framing, challenges in ecosystem, lack of opportunities, lack of practice, absence of dedicated centre, etc. However, some participants talked positively of motivation for educational research, and how factors such as connecting with different researchers, having people acknowledge significance of EER, conferences and appropriate platforms to disseminate findings, technology enabled research, administrative policies to enable creativity, faculty development programs, mentorship, and ecosystem enabling new initiatives in EER- can be enabling for EER in India. In what follows, we systematically describe our findings categorized in themes of "barriers" and "enablers" emerged at personal, institutional and policy level as discussed by various participants mentioned in pseudonyms to protect their privacy.

A. Barriers

1) Theme: Lack of Institutional support for research activities

Diversity and administrative variations among Indian universities create inconsistencies and challenges in coordinating and standardizing research efforts across institutions. Many universities are privately owned or have restricted access, making it difficult to establish collaborations or engage in research activities. As Dr. Rajesh Patel a researcher says, *"majority of the institutions if you see here are all mass education and academic oriented. Now slowly we must build a research culture"*. Most institutions being referred to are primarily focused on providing mass education and are geared toward academic-oriented programs. This implies that their primary emphasis has been on teaching and delivering educational content to a broad audience. However, Dr. Rajesh Patel emphasizes the need for a shift in perspective and priorities. He suggests that it's time to gradually cultivate a "research culture" within these institutions. The wide range of administrative styles and ownership structures creates complexities and challenges in attempting to navigate and penetrate these institutions for the purpose of EER. A engineering education researcher Mr. Ishaan said, *"... research (in engineering education) itself is not a priority across institutions in India,... like I mean outside (of a couple institutes) that promote intuitively those things within their system, I don't know of many other institutions that prioritizes (engineering educational) research and structurally that's been how you know education has been modelled."*

2) Theme: Convergence challenge: Bridging research and teaching practice in Indian academia

Participants revealed the feeling that experimental research with an emphasis on conducting experiments and analyzing the resulting data is valued more than qualitative research in engineering education in India. For the fewer ones doing research in engineering education, the nature of research varies from one individual to another, depending on their interests and the specific problems they aim to address in the country, however once concern raised was that the translation of the EER into practice is not happening to the scale desired. There might be even a disparity between the research conducted and its practical application within the same institute or across different institutes. For example. Mr. Darshan a EER researcher said, *"you know what they are researching, maybe the research is there, but may not be always applied by the same institute or by even other institutes, right?"*. Also, Professor Bhattacharya said, *"There is another big conflict that I see, research versus teaching. There is so much that one is expected to do and India being predominantly having institutions with teaching focus, there is a dichotomy you see among teachers."* In brief, he highlighted EER and effective teaching as two different areas that caused pressure in striking balance and exceling in both areas amongst faculty.

3) Theme: Disciplinary silos: Hurdles to multidisciplinary collaborations

The faculty's resistance to embracing collaborative efforts across disciplines is highlighted as a barrier to advancing research and innovation in their context. While intrinsic motivation to be better faculty, to learn and conduct EER, etc. might be observed at some places, there is largely an absence of extrinsic motivators. The traditional academic reward structure in many institutions prioritizes research in technical fields over EER. Many of the faculty members hold the belief that there is no value or appreciation for collaborations between different disciplines, such as engineering, arts, management, and education, thus deterring their motivation to do EER. As Dr. Raman an EER practitioner states, *"unless and until there is motivation from an individual who has been entering education, these conversations typically require encouragement."* Other faculty may perceive these additional conversations as burdensome, especially when they are already heavily committed to teaching and administrative duties. In addition to the above, a lack of incentives, recognition, and support for EER endeavors further hinder interdisciplinary collaboration.

4) Theme: Small community of engineering education researchers in India

Lack of a supportive and collaborative research community is challenging due to the fewer representation of faculty within it. Many faculty faculty associated with EER community choose to remain with their parent discipline, and not fully dive in as EER is not recognized as separate at many institutions. For example, Head of Teaching Learning Center, Professor Ram said, *"A number of associated faculty who kind of were associated with the center... and but they continue to be in their parent disciplines"*. Professor Sham said, *"I wish there was like more we could do and like at least I would be interested in like collaborations and things like that. But I don't think we're there yet."* This leads to difficulties in finding individuals with whom researchers can openly discuss their challenges, successes, and another valuable knowledge. As a result, this limited community interaction becomes a barrier to sharing information and mutually benefiting from each other's experiences and insights. Also, researcher Arun adds, *"There are very less people with whom the researchers can share their problems or success or any other knowledge that they have which might be useful mutually in so that is another barrier."* Many researchers also mentioned that due to the limited small community of the people, EER is struggling to attract sufficient attention, funding, and recognition compared to larger academic disciplines.

5) Theme: Lack of recognition for EER as a field of inquiry

EER has not been given significant priority or recognition in various aspects within the academic and research community. The acceptance of EER as a distinct discipline encounters significant obstacles, primarily due to the abundance of teaching professionals within engineering faculties who may view themselves as pedagogical experts. Unlike conventional engineering fields, where peers might more readily acknowledge expertise, EER researchers frequently confront doubt concerning the necessity for evidence and the applicability of their discoveries. Establishing credibility becomes notably challenging when EER delves into the practical aspects of education, as colleagues might believe their own expertise suffices. For example. Head of the Teaching Learning Center, Professor Ram here expressed, *"See one of the thing is acceptance. You know, acceptance as a discipline is challenging for any engineering education research because there are everybody you know is a instructor okay. So there are 700 colleagues who do who teach like 3 courses a year and you are one of them. So now suddenly if you say, okay, I know how to teach and I know the pedagogy and you don't know or something or come, I will tell you that doesn't work well. So and a lot of people have a, you know, practitioners understanding of what it is to do education research, you know...so we have to keep talking to them about these levels of engineering education research and to say that okay, this is Level 3 actually we are trying to do at level 5 and you know what exactly is the difference. And a lot of times you know the need for evidence or the generalizability of findings. All those things are often questioned."* This credibility dilemma is particularly conspicuous in the domains of disciplinary and teacher education research, where numerous professionals perceive themselves as authorities, rendering the acknowledgment of EER's distinctive contributions an enduring hurdle. Additionally, the participants emphasize that the recognition of the importance of this research area by individuals and institutions could potentially overcome the barriers created by the lack of funding. This includes performance appraisals, where EER is not emphasized as much as technical research areas. For example, a researcher Mr. Mohan said, *"The weight that is given to the engineering education research. Whether it is in terms of, uh, yeah, the performance appraisals in the formal criteria, wherever the faculty is appraised. Uh, those have not been given priority."* Additionally, the publication opportunities and platforms for EER are comparatively limited when compared to those available for technical research. Due to the limited recognition and priority given to EER, many researchers in India may not be familiar with important practices which could potentially hinder the quality and effectiveness of their research efforts.

6) Theme: EER in India lacks strategic focus and direction

One recurrent theme is the ambiguity surrounding research direction. For example, Dr. Harpreet an EER reseacher mentioned *"Are we kind of like recreating the way things happened at like pretty where US origin, or like, are we trying to figure out what is the need in this case and are we kind of using so, you know, like what are the student aspirations, the needs, the needs of the community, right?"*. Additionally, EER often seems relegated to a secondary or retirement pursuit rather than a primary focus, impacting its perceived importance and the expectations associated with it. For example, Dr. Raghav also, an researcher said, "I care, but I have unfortunately seen this trend, I saw to get engaged in engineering education research and it's almost like a… you know, a retirement kind of a thing that, you know, I know that I don't need to teach or, you know, be doing technical research. I can safely retire and do ER." Altering these attitudes and perspectives poses a considerable challenge.

On a different note, the gap between industry requirements and the development of academic programs was also recognized as a key area needing research and enhancement in the field. For example, Dr. Meenakshi a dean mentioned, *"I always feel like there is not much connection between what the industry wants and what's being developed...I feel like that's something that would be interesting to see and I see a lot of like opportunity and potential in that to be able to kind of bridge that gap or think about research in that area".*

7) Theme: Lack of funding for EER

The lack of sufficient research funding is a significant barrier to advancing EER in India. Adequate funding emerges as a critical factor for advancing EER, while success stories from other regions serve as valuable models. The speakers express a sense of frustration and limitation due to the insufficient funding available for this field compared to other countries like the US. For example, Dr. Rao, an institutional TLC leader and researcher mentioned, *"Money speaks. You look at US right there is so much funding for engineering education research. It is it is unbelievable. I mean, I am in awe of how much money NSF is providing [for EER]."* Another participant leader indicated how it could be frustrating to find no funding even if faculty might be trained for excellence in research in the area. For example, Dr. Gupta an EER practitioner said, *"It is unfortunate that you know, in terms of research money, we don't have enough research money. NITs today have the kind of faculty members, the younger lot in NIT. They are all very well trained. These faculty members come from good schools, and they're trained in, you know, top 80s or abroad. You know, these are well trained faculty members, but they don't have [funding]. They go back to NIT, and they don't [find support] they find that there is no research infrastructure. So, they struggle, they struggle, right…and, you know, in research, research is such a thing that if you were not on that train for, three years after that, you can't catch the train. You know it's gone, right?"* These examples highlight the need for funding and support for the growth and recognition of EER.

Thus, the barriers encountered by stakeholders in India encompass a lack of a supportive and collaborative research community, insufficient recognition of EER, unfamiliarity with vital research practices, administrative variations among universities, resistance to interdisciplinary collaboration, and limited research funding. These challenges collectively hinder the progress of EER in the country. However, there is potential for growth and improvement of EER through heightened awareness, comprehensive training initiatives, and increased recognition, as elaborated below.

B. Enablers

1) Theme: Motivation to advance engineering education in India

Motivation emerges as a crucial enabler within Engineering Education Research (EER), driving engagement, innovation, and progress. The motivation to engage in EER also originates from a sense of responsibility towards sculpting the students and a desire to contribute to India's educational advancement. For example, Professor Pandit says, *"I am an educator and I think the want and the desire to do engineering research stemmed out of the need to want to be a better [educator]."* Collaborative initiatives, like conferences and workshops, motivated by passionate practitioners, foster engagement, and dialogue within the EER ecosystem. Global collaborations, nurtured by a shared motivation to enhance research quality and impact, enrich the field by exchanging expertise and methodologies. This motivation is exemplified in the initiatives undertaken by individuals like Professor Rohan, who established platforms like Engineering Education Trust and engineering education journals, inspiring active participation. Furthermore, a motivated drive to enhance teaching methodologies and incorporate experiential learning underlines the commitment of faculty members to evolve pedagogical approaches. In essence, motivation emerges as the driving force propelling active contributions, global connections, innovative initiatives, and continuous improvement in teaching practices, collectively shaping the future of EER.

2) Theme: Institutional recognition of EER activities

Institutional support plays a significant role in enhancing the growth and development of EER by encouraging faculty engagement with EER. By recognizing EER achievements in faculty appraisals and career progression, institutions could validate the significance of this field and motivate educators to actively contribute to it. Such recognition not only benefits individual researchers but also elevates the status of EER within the academic community. It also facilitates collaboration and networking within the EER community. For example, Dr. Kalawati an EER researcher said, *"Institutional support first of all comes from the leadership and the leadership has to prioritize in terms of what is it that they need and be aware of the needs as well."* By organizing seminars, workshops, and conferences focused on EER, institutions can foster a vibrant ecosystem where researchers exchange ideas, share insights, and collaborate on joint projects. This collaborative environment not only enriches research quality but also helps in disseminating findings effectively.

3) Theme: Acknowledgment of EER by engineering community in India

While the initial lack of recognition from both individuals and institutions presented a notable barrier, it's crucial to underscore that acknowledging the importance of EER has the potential to evolve into a potent enabler. Recognizing and valuing EER can help in its development and impact in shaping more effective practices in engineering education. Hence, acknowledgment is an enabler because it emphasizes the importance of changing prevailing attitudes towards EER. By altering the perception that EER is somehow less significant or less serious than technical research, there is an opportunity to create a more supportive environment that recognizes the value and impact of EER. Professor Prakash mentions, *"Research can be done on what should be the appraisal for faculty, HR related. What will be the best method to do the appraisal of faculty so that they continuously learn then?"*. This shift in mindset can motivate researchers and institutions to invest more in EER, ultimately fostering its growth and impact in India's educational landscape.

4) Theme: EER Community for Capacity Building and Collaborations

The network plays a pivotal role in enhancing EER endeavors through its multifaceted contributions. Consciously promoting networking facilitates knowledge exchange, enabling global collaborations, offering publication avenues, and fostering active participation in conferences and workshops. Additionally, the alumni network creates an enriched environment for research growth. This platform not only addresses the potential isolation that can arise from the institution's small size but also serves as a medium for maintaining connections, sharing experiences, and engaging in insightful discussions. This collaborative atmosphere is nurtured by the regular interactions maintained with alumni, who bring diverse expertise from various industries and research fields. For example, Teaching and Learning center head- Dr. Ram said, *"Since we are small and people are likely to get isolated, so what we have very consciously done is to set up a very strong alumni network. So, we keep interacting with our alumni. They come and give guest lectures in our courses and then we do have meetings with them".* Moreover, the network's impact extends to education as well, with former students providing guest lectures that infuse real-world insights, industry trends, and relevant case studies into the learning experience. This dynamic transfer of knowledge becomes especially valuable for research aimed at bridging academia and industry.

5) Theme: Technology based research

Technology-based research serves as a significant enabler within the field of EER. It catalyzes transformative shifts in teaching and learning methodologies. The realization that technology is not merely a tool for course transmission, but also a medium for innovative research, propels institutions to explore its diverse applications. For example, researcher Dr. Rishabh says, *"I'm always interested in using technology in interesting ways. That's one of the things that I always look for in the work that I do. Like what can we do in this kind of interesting moment where technology has become more accessible."* - showing his excitement towards technology and his acceptability towards the same in EER. He expresses a strong interest in utilizing technology in creative ways and conveys excitement about its potential applications. Similarly, if encouraged, the integration of technology with pedagogical innovation can drive institutions to contemplate the initiation of programs like PhDs to further investigate these intersections. The inclination towards technology is also reflective of a forward-looking approach, seeking to harness its accessibility and potential to redefine educational spaces. By leveraging technology creatively, educators can cultivate engaging learning environments and foster a sense of belonging. This visionary stance fosters a departure from traditional gap-filling approaches and embraces future-oriented perspectives. The fusion of technology and pedagogy not only transforms educational ecosystems but also extends its impact to institutional rankings. The ability to virtually conduct interviews or classroom sessions is a testament to technology's role in facilitating remote learning experiences. Overall, technology-based research introduces dynamic possibilities, reshaping teaching paradigms, enhancing engagement, and amplifying the role of institutions in driving educational advancements.

6) Theme: National Education Policy (NEP) 2020

As universities shift their focus towards research, there is a concern that attention to teaching-learning might diminish. Striking a balance between teaching and research is crucial, and this is where policies come into play at both institutional and national levels. Policies can guide how much emphasis is given to research while still valuing effective teaching. Policies can be enabling to support those who want to teach core engineering subjects, at the same time incentivize to pursue research in engineering education. This two-fold approach aligns with the interest of external factors like governments, who see value in research to inform policies and educational practices. This alignment validates the efforts of engineering education researchers. The national education policy's focus on employability skills and multidisciplinary education serves as an enabler in EER by promoting curriculum enhancements and interdisciplinary approaches to better prepare engineering students and promote holistic development. For example, Professor Chattopadhyay an institutional leader says, *"the next big thing that is happening is national education policy where, you know, it talks about employability skills of the undergraduate engineers and multidisciplinary education leading to holistic development of individuals."* With the backing of policies and growing enthusiasm, institutions are encouraged to implement such initiatives. Collaborations like Indo Universal Collaboration for Engineering Education (IUCEE) further contribute to creating platforms for these efforts. These enablers collectively can pave the way for a thriving an impactful EER ecosystem in India.

The enablers` collectively show the path towards enhancing the quality and effectiveness of engineering education with a plus hand of overcoming the barriers mentioned there as well. The proactive promotion of a research culture, adequate funding support, collaborative networks, innovative teaching practices, the incorporation of technology, and effective dissemination mechanisms collectively contribute to the advancement of EER.

V. DISCUSSION

The themes presented in the results section can be categorized into three levels – personal, institutional, and policy level, as shown in figure 1. At personal level, it's crucial to acknowledge that the barriers faced by engineering education researchers in India are multifaceted and demand strategic attention for the field to thrive. These challenges encompass a limited collaborative community due to minority representation, unfamiliarity with vital research practices, resistance to interdisciplinary collaboration, and the interdisciplinary nature of EER. Despite these challenges, many Indian faculty members recognize the intrinsic value of EER and are self-motivated to become better teachers, which drives them to conduct EER, as mentioned for other global counterparts in literature (Kittur et al., 2020). Faculty members, regardless of their backgrounds, are actively engaging in this field, showcasing the potential of EER to revolutionize engineering education and foster inclusivity.

Moving to the institutional level, we find that support from institutions is pivotal. This includes active encouragement for research in the engineering education domain (Kandakatla et al., 2018), funding opportunities, recognition in appraisal policies, and the acceptance of publications in the EER domain as equivalent to other engineering disciplines for promotion. Additionally, a balanced focus on both teaching and research

https://doi.org/10.52202/073963-0019

initiatives, rather than being solely teaching-centric, is crucial to help the EER community grow in India and ensure that all faculty members can contribute effectively.

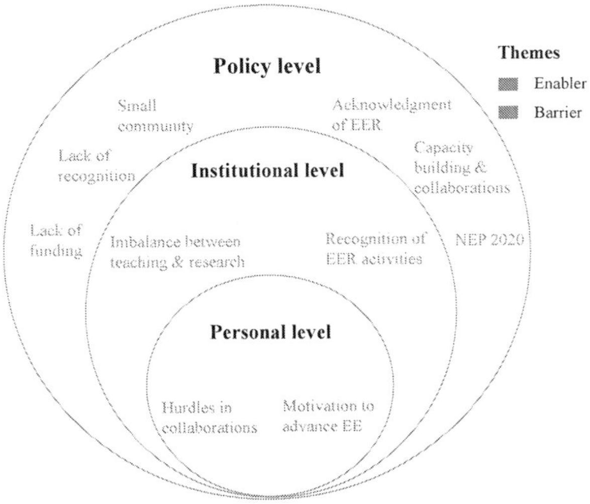

Figure 1: Barriers (orange) and enablers (green) at Personal, Institutional and Policy level identified from data analysis.

Collaborations through professional societies and international partnerships, provide avenues for networking, knowledge transfer, and joint research efforts, enhancing the overall impact of engineering education research and promoting a diverse and inclusive research community (Delaine et al., 2018). Early efforts in this direction, such as blended certification programs focused on building EER capacity among engineering faculty in India (Nagabhushan & Sohoni, 2020), highlight the commitment to inclusivity and capacity building.

To foster Engineering Education Research (EER) in India at the policy level, a comprehensive strategy must address various vital themes. First and foremost is the imperative recognition of EER's significance within academia, supported by incentives and acknowledgment at both policy and institutional levels. Second, the need to cultivate a larger EER community necessitates policies that encourage networking and collaboration among practitioners across institutions. Communities of practice particularly have been reported to enable growth of ecosystems that could contribute to large-scale change (Kandakatla & Palla, 2020). Third, addressing the scarcity of funding opportunities for EER projects requires dedicated resources and grants. Furthermore, identifying, and prioritizing EER focus areas, guided by the National Education Policy (NEP) of 2020, can ensure relevance and growth. Lastly, facilitating faculty transitions from traditional engineering roles to EER-focused positions through capacity-building efforts is crucial. These approaches, as emphasized by (Vijaylakshmi et al., 2022), collectively promote the development and impact of EER in India, ultimately contributing to enhanced engineering education quality, equity, and national development objectives.

These findings underscore that addressing the multifaceted barriers to EER in India requires action at personal, institutional, and policy levels. A concerted effort from all stakeholders, including institutions, policymakers, and researchers, is necessary to overcome these challenges and leverage enablers for the advancement of EER in India. The key takeaway message for all stakeholders is the need for community, collaboration, focused efforts, capacity building, recognition, and support for EER to improve the quality of engineering education in the country.

VI. CONCLUSION

In conclusion, our study's findings, while aligning with existing knowledge in several aspects, offer specific insights into the barriers and enablers of EER within the Indian context. Our findings emphasize the multifaceted nature of the challenges confronting EER in India, encompassing conceptual clarity, perceptions, financial support, and alignment with industry needs. A call for proposals with specific themes that have priority could remove this ambiguity in defining and prioritizing EER efforts in India. This comprehensive analysis of the challenges and opportunities within the Indian EER landscape contributes to a deeper understanding of the field, providing valuable guidance for future research and policy initiatives aimed at enhancing engineering education in the country. Our discussion effectively bridges these findings with the existing literature, emphasizing the critical need to address these barriers and harness the identified enablers at personal, institutional, and policy levels to promote EER and elevate the quality of engineering education in India. In a resource-constrained, and infrastructure-constrained environment of a developing country of India, EER can provide benefit to Indian researchers to make their mark using the unique Indian engineering educational landscape and many unexplored research areas. However, many barriers that are currently hindering this effort, include lack of funding, lack of institutional support in recognizing EER at par with other research areas in engineering, minimal awareness about suitable approaches and correct research methodologies for EER, lack of focus in research questions specific to India's educational system, missing a supportive ecosystem, incentives, platforms for disseminating high-quality EER and a poor mindset for appreciating EER engagement of stakeholders.

Overall, funding, fostering collaborations within and outside India with people doing EER, cross-disciplinary collaborations and learning from diverse colleagues facilitated via conferences, industry engagement to drive technology-based research in engineering education, external validation, and a shifting mindset could be additional pivotal forces propelling

https://doi.org/10.52202/073963-0019

EER's growth and enhancing its contribution to India's educational landscape. However, it is important to acknowledge the limitations of this study. The research sample consisted of a relatively small group of 18 participants, all of whom were from India. Furthermore, the gender diversity within the sample was limited, with only 4 female participants. This restricted sample size and gender imbalance may impact the generalizability of the findings to a broader population. In future EER studies, efforts should be made to include a more diverse and representative sample, both in terms of numbers and demographic characteristics, to enhance the robustness of research outcomes.

ACKNOWLEDGMENT

The authors declare that they have no conflict of interest. Authors would like to thank all the participants including the Teaching and Learning Center Heads, Engineering education researchers, Practitioners, and Leaders of various engineering institutes for giving their insights and time in the interview. We would also like to thank Plaksha University for providing us technical support with Microsoft Teams software with transcription feature for conducting the interviews.

REFERENCES

Beddoes, K. D., Jesiek, B. K., & Borrego, M. (2010). Identifying Opportunities for Collaborations in International Engineering Education Research on Problem- and Project-Based Learning. *Interdisciplinary Journal of Problem-Based Learning*, *4*(2). https://doi.org/10.7771/1541-5015.1142

Bernhard, J. (2018). Engineering Education Research in Europe–coming of age. In *European Journal of Engineering Education* (Vol. 43, Issue 2, pp. 167–170). Taylor & Francis.

Borrego, M., & Bernhard, J. (2011). The Emergence of Engineering Education Research as an Internationally Connected Field of Inquiry. *Journal of Engineering Education*, *100*(1), 14–47. https://doi.org/10.1002/j.2168-9830.2011.tb00003.x

Borri, C., & Maffioli, F. (Eds.). (2008). *Re-engineering Engineering Education in Europe* (1st ed., Vol. 39). Firenze University Press. https://doi.org/10.36253/978-88-8453-676-1

Braun, V., & Clarke, V. (2006). Using thematic analysis in psychology. *Qualitative Research in Psychology*, *3*(2), 77–101. https://doi.org/10.1191/1478088706qp063oa

Brodie, L., Bullen, F., & Gibbings, P. (2011). Developing an engineering education research culture. *2011 IEEE Global Engineering Education Conference (EDUCON)*, 214–219.

Buckley, J., Wallin, P., Matemba, E., Power, J., Mohanty, A., & Bombaerts, G. (2023). The Future of Engineering Education Research. In *International Handbook of Engineering Education Research* (pp. 711–729). Routledge.

Burbules, N. C., & Torres, C. A. (2000). *Globalization and education: Critical perspectives*. Routledge.

Chen, Y., Nixon, M. R., Gupta, A., & Hoshower, L. (2010). Research productivity of accounting faculty: An exploratory study. *American Journal of Business Education (AJBE)*, *3*(2), 101–115.

Council, N. R. (2012). *A framework for K-12 science education: Practices, crosscutting concepts, and core ideas*. National Academies Press.

Cresswell, K. M., Worth, A., & Sheikh, A. (2012). Integration of a nationally procured electronic health record system into user work practices. *BMC Medical Informatics and Decision Making*, *12*, 1–12.

Dart, S., Seniuk Cicek, J., & Sohoni, S. (2023). REES AAEE special issue on engineering education research capability development: Introduction by guest editors. *Australasian Journal of Engineering Education*, *28*(1), 2–7. https://doi.org/10.1080/22054952.2023.2231768

Delaine, D. A., Cukierman, U., Kandakatla, R., Morell, L., & DeBoer, J. (2018). Leveraging Professional Networks for an Equitable, Smart Society—A Case Study on the International Federation of Engineering Education Societies. In M. E. Auer & K.-S. Kim (Eds.), *Engineering Education for a Smart Society* (pp. 237–252). Springer International Publishing.

Duschl, R. A., & Bybee, R. W. (2014). Planning and carrying out investigations: An entry to learning and to teacher professional development around NGSS science and engineering practices. *International Journal of STEM Education*, *1*(1), 1–9.

Edström, K., Kolmos, A., Malmi, L., Bernhard, J., & Andersson, P. (2018). A bottom-up strategy for establishment of EER in three Nordic countries–the role of networks. *European Journal of Engineering Education*, *43*(2), 219–234.

Feiman-Nemser, S. (1989). *Teacher preparation: Structural and conceptual alternatives*. National Center for Research on Teacher Education.

Gardner, A., & Willey, K. (2018). Academic identity reconstruction: The transition of engineering academics to engineering education researchers. *Studies in Higher Education*, *43*(2), 234–250. https://doi.org/10.1080/03075079.2016.1162779

Godfrey, E., & Hadgraft, R. (2009). Engineering education research: Coming of age in Australia and New Zealand. *Journal of Engineering Education*, *98*(4), 307–308.

Haigh, N., Gossman, P., & Jiao, X. (2011). Undertaking an institutional 'stock-take'of SoTL: New Zealand university case studies. *Higher Education Research & Development*, *30*(1), 9–23.

Jesiek, B. K., Newswander, L. K., & Borrego, M. (2009). Engineering education research: Discipline, community, or field? *Journal of Engineering Education*, *98*(1), 39–52.

Johri, A., & Olds, B. M. (2014). *Cambridge handbook of engineering education research*. Cambridge University Press.

Kamp, A. (2020). Navigating the landscape of higher engineering education. *Education*, *2*, 115ce70ecb98.

Kandakatla, R., Goldenstein, A., Evenhouse, D. A., Berger, E. J., Rhoads, J. F., & Deboer, J. (2018, June 23). *MEERCat: A Case Study of How Faculty-led Research Initiatives Gave Rise to a Cross-departmental Research Center with Potential to Inform Local Policy*. 2018 ASEE Annual Conference & Exposition. https://peer.asee.org/meercat-a-case-study-of-how-faculty-led-research-initiatives-gave-rise-to-a-cross-departmental-research-center-with-potential-to-inform-local-policy

Kandakatla, R., & Palla, A. (2020). Role of Community of Practice (CoP) to Facilitate Change in STEM Instructional Practices through Faculty Development Programs. *2020 IFEES World*

Engineering Education Forum - Global Engineering Deans Council (WEEF-GEDC), 1–5. https://doi.org/10.1109/WEEF-GEDC49885.2020.9293660

Kittur, J., Coley, B. C., & Kellam, N. N. (2020). Understanding how novice Indian faculty engage in engineering education research. *2020 ASEE Virtual Annual Conference Content Access.*

Klein, C., Lester, J., Rangwala, H., & Johri, A. (2019). Learning analytics tools in higher education: Adoption at the intersection of institutional commitment and individual action. *The Review of Higher Education*, *42*(2), 565–593.

Ko, M., Mirabelli, J. F., Barlow, A. J., Jensen, K. J., & Cross, K. J. (2021). Faculty motivations and barriers for engineering education research. *American Society for Engineering Education Annual Conference*.

Lohmann, J., & Froyd, F. (2010). Chronological and ontological development of engineering education as a field of scientific inquiry. *Second Meeting of the Committee on the Status, Contributions, and Future Directions of Discipline-Based Education Research, Washington, DC. Available: Http://Www7. Nationalacademies. Org/Bose/DBER_Lohmann_Froyd_October_Paper. Pdf.*

McKinney, K. (2002). The scholarship of teaching and learning: Current challenges and future visions. *Remarks Presented at the Ceremony to Install the Cross Chair in the Scholarship of Teaching and Learning at Illinois State University. Accessed May, 10*, 2010.

Mehta, P., & Berdanier, C. (2019). A Systematic Review of Additive Manufacturing Education: Toward Engineering Education Research in AM. *2019 ASEE Annual Conference & Exposition Proceedings*, 32006. https://doi.org/10.18260/1-2--32006

Nagabhushan, P., & Sohoni, S. (2020). An Introductory Course on Research in Engineering Education- An experiment in training faculty in India. *2020 IEEE Frontiers in Education Conference (FIE)*, 1–7. https://doi.org/10.1109/FIE44824.2020.9273918

National Academies of Sciences, E. (2018). *The integration of the humanities and arts with sciences, engineering, and medicine in higher education: Branches from the same tree*.

Smith, B., & McGannon, K. R. (2018). Developing rigor in qualitative research: Problems and opportunities within sport and exercise psychology. *International Review of Sport and Exercise Psychology*, *11*(1), 101–121.

Streveler, R. A., Brown, S., Herman, G. L., & Montfort, D. (2015). Conceptual change and misconceptions in engineering education: Curriculum, measurement, and theory-focused approaches. In *Cambridge handbook of engineering education research* (pp. 83–102). Cambridge University Press.

Streveler, R. A., & Smith, K. A. (2010). From the margins to the mainstream: The emerging landscape of engineering education research. *Journal of Engineering Education*, *99*(4), 285–287.

Valero, M. (2022). Challenges, difficulties and barriers for engineering higher education. *Journal of Technology and Science Education*, *12*(3), 551. https://doi.org/10.3926/jotse.1696

van Hattum-Janssen, N., Williams, B., & de Oliveira, J. N. (2015). Engineering education research in Portugal, an emerging field. *International Journal of Engineering Education*, *31*(2), 674–684.

Vijaylakshmi, M., Kandakatla, R., Baligar, P., Joshi, G., & Shettar, A. (2022). Design of Research Canvas to align Research Efforts at Engineering Education Research Centre in India. *2022 IEEE Global Engineering Education Conference (EDUCON)*, 844–849. https://doi.org/10.1109/EDUCON52537.2022.9766458

Wankat, P. C., Felder, R. M., Smith, K. A., & Oreovicz, F. S. (2002). The scholarship of teaching and learning in engineering. *Disciplinary Styles in the Scholarship of Teaching and Learning: Exploring Common Ground*, 217–237.

Wenger, E., Trayner, B., & De Laat, M. (2011). *Promoting and assessing value creation in communities and networks: A conceptual framework*. Rapport.

Wisnioski, M. (2015). What's the use? History and engineering education research. In *Journal of Engineering Education* (Vol. 104, Issue 3, pp. 244–251). Wiley Online Library.

APPENDIX

Appendix I - Interview questionaire.docx

Copyright statement

Enhancing Conceptual Understanding of Electricity and Magnetism: Experiences from an Indian University

Babulakshmanan Ramachandran[a] ; Yogesh Velankar[b], Michael Smith [a,c], Sheh Lit Chang [d] ; Krithi Krishna[a];

[a] Department of Physics, Amrita School of Physical Sciences, Amrita Vishwa Vidyapeetham, Amritapuri Campus, Kollam, Kerala, 690525, India

[b]Amrita School for Sustainable Futures, Amrita Vishwa Vidyapeetham, Amritapuri Campus, Kollam, Kerala, 690525, India

[c]School of Physics and Astronomy, University of Kent, Canterbury, United Kingdom, CT2 7NH

[d]School of Mathematics and Science, Singapore Polytechnic, 500 Dover Road, Singapore 139651

Corresponding Author Email id: ramachandranb@am.amrita.edu

Abstract

Context

Students studying for disciplines such as Electrical and Electronics engineering need a good foundation in Electricity and Magnetism (EM). Therefore, it is imperative to know the prior knowledge of students in this topic to address misunderstandings and misconceptions.

Purpose or Goal

Our motivation is to help students to improve their learning in EM in the post-covid era. We want to introduce suitable pedagogy and learning technology that helps address conceptual misconceptions and engages students.

Methods

We administered the conceptual inventory (CI) known as Brief Electricity and Magnetism Assessment (BEMA) to students at an Indian university studying for a physics degree. Students took BEMA in the first week of the semester (pre-test) and once at the end of the semester (post-test). We introduced the world class Pearson products, viz, Mastering Physics (MP) and Learning Catalytics (LC) and undertook a survey on MP and LC to understand the students' perspective.

Outcomes

We found that the scores from BEMA went from 24 % (pre-test) to 39 % (post-test) which is statistically significant with a large effect size (0.99). The results from the MP and LC surveys are positive which indicate that these tools were well received by students.

Conclusion

There is an improvement in the conceptual understanding of EM. MP and LC are important factors in the student learning of the subject. Students indicate that they find MP and LC quite useful. Also, the E-learning fatigue needs to be addressed to help students improve further. We intend to introduce BEMA and other CIs to engineering students through our engineering faculty colleagues and use tools such as MP and LC to help students build a solid foundation in EM and physics in general.

Keywords— Concept Inventory; Conceptual Understanding; Learning Technology

I. INTRODUCTION

THE physics courses in an engineering curriculum directly impact core engineering subjects. A deep conceptual understanding of physics is essential to build a strong Engineering background. Typical physics courses included in an engineering curriculum include Mechanics, Thermodynamics and Electricity and Magnetism. A course on Electricity and Magnetism (EM) is usually introduced in the first-year engineering and physics degree programs as it forms a basis for science and technology. Usually, students study EM in the second semester as a calculus-based physics course after going through an introduction to classical mechanics and some courses in mathematics in the first semester.

Unlike mechanics where students are familiar to situations that they can relate to daily life such as moving objects and forces, EM is perceived to be abstract, difficult, and often confusing (Chabay & Sherwood, 2006). In EM, students are expected to grasp a level of abstraction in concepts such as field, potential and flux. These abstract concepts involve in-depth mathematical description requiring differential and integral calculus in ways that students may have not had an experience previously. As a student progresses to higher years, subsequent courses depend on the concepts learnt in EM. Therefore, it is important for students to understand the fundamental concepts in EM thoroughly. This is possible if we can supplement the traditional lectures with suitable educational technology to enhance the learning of EM as well as engage them in a meaningful way thereby improving the quality of education.

When students enter university after high school education, research has shown that they have misconceptions and misunderstandings in physics (Halloun & Hestenes, 1985; Trowbridge & McDermott, 1980; McDermott et al., 1987). This led to the development of concept inventories to test student understanding (Lindell et al., 2007). A Concept Inventory (CI) is a research-based assessment instrument that probes students' understanding of a particular concept or a set of concepts that can help measure the effectiveness of teaching approaches (Porter et al., 2014). Concept inventories are intended to assess understanding and they assess different things from conventional exams (Sands et al., 2018).

Typically concept inventories contain multiple-choice questions. These questions are a result of a systematic study of the misconceptions and misunderstandings in a subject. The incorrect choices in multiple choice items are called distractors, based on the common student misconceptions (Sadler et al., 2009). Concept inventories are tests usually taken twice within the instruction sequence and do not usually part of summative assessments. They fulfil several requirements for assessing learning gain. These requirements include Validation through research, Standardization by using the same test on different students in different institutions to allow for meaningful comparisons of different students' understanding of the concept in question and Longitudinal in the sense of the same test used at two different points in time to allow for a meaningful assessment of gain by the students (McGrath et al., 2015).

The Force Concept Inventory is one of the earliest CI to be developed in the topic of force and motion and has been administered to thousands of students over the years (Hestenes et al., 1992). Apart from physics, there are several concept inventories in different disciplines such as chemistry, engineering and biosciences that are administered in universities across the globe. Examples include the biology concept inventory (Garvin-Doxas et al., 2007), and the astronomy diagnostics test (Hufnagel, 2002). Currently, there are around 60 CIs in different topics in physics and astronomy for various introductory and upper-level topics in physics and astronomy.

For Electricity and Magnetism (EM), one of the most widely used CI is the Brief Electricity and Magnetism Assessment (BEMA) (Ding et al., 2006). Many studies have used BEMA and it has been found very useful to instructors to help them evaluate their pedagogy and curriculum and understand students' level of conceptual understanding of EM (Kohlmeyer et al. 2009; Pollock, 2009; Ding et al., 2006).

II. Motivation and Aim

We would like to know how we can help enhance the conceptual understanding of EM. In the post-covid era, we are not aware of many studies on conceptual understanding of EM. Therefore, we would like to know how students in the post-covid era fare in conceptual understanding of EM using BEMA. Studies has shown that research-based teaching methods, such as interactive engagement, lead to improvements in students' gains when compared to traditional lectures (Hake, 1998). Therefore, we have incorporated Mastering Physics and Learning Catalytics in our class that enhances and engages students in their learning of EM.

In this paper we aim is to answer the following:

1. What is the impact of introducing Mastering Physics and Learning Catalytics on the conceptual understanding in EM?
2. What are the students' perspectives of Mastering Physics and Learning Catalytics?

III. Methodology

In this study, we chose students in a physics degree program at an Indian university taught by one of us. There was only one class. The students mainly went through the online mode of learning during their final years of high school due to Covid pandemic. There were 29 students who joined in September 2022. In the first semester they studied courses such as Mechanics, Chemistry and Mathematics. They continued in the second semester face-to-face starting in January 2023 where they studied EM among other subjects. This 4-credit course on EM was taught 4 hours per week for a semester spanning around 16 weeks.

Students were introduced to the world class Pearson products, viz, Mastering Physics (MP) and Learning Catalytics (LC) when they came to campus in the first semester for their Mechanics course. These products came with the textbook "University Physics with Modern Physics" by Hugh D Young and Roger A Freedman, 15th edition. We used the same book for EM as well in the second semester. The university purchased student licenses.

Mastering Physics is an online homework, tutoring and assessment system that has been used worldwide for several years with much success. It was developed at Massachusetts Institute of Technology (Lee et al., 2008). It was conceived because it was not feasible for instructors to sit down with every student on an individual basis. The system offers students hints and/or answer-specific feedback to address their misconceptions. It helps students when they encounter learning obstacles and gives them the individual coaching, they need to overcome those obstacles. Diagnostics features in MP allows instructors to know the learning misconceptions and difficulties faced by students which need to be addressed during the face-to-face session. MP can be integrated into Learning Management Systems such as Blackboard, Canvas and Moodle. Also, MP has an Adaptive Follow-Up feature. Based on each student's performance, Adaptive Follow-Up assignments provide additional coaching and targeted practice as needed to help the

https://doi.org/10.52202/073963-0020

students. MP contains the e-Textbook with video tutor solutions and demonstrations and offers several different question types such as ranking, drawing, and keying in equations. It also contains the widely used PhET simulations which aid immensely in the learning of Physics. The grade book in MP contains scores, time taken and difficulty level for the questions set in the assignments which can be used to track student progress. To get started with MP and LC, students were first given an orientation by Pearson which consisted of a detailed explanation of the products.

We set 8 assignments in MP for the semester. These assignments were on topics such as Static Electricity, Gauss's law, Circuits, Magnetism and Electromagnetism. Each assignment consisted of different question types such as multiple-choice questions, numerical questions and ranking based questions. These were selected by us based on the world-wide statistics provided by MP regarding the median time taken as well as the difficulty level of the questions as perceived by students from all over the world. Each assignment was assigned for 2-3 weeks to be completed outside class hours. The time spent on an assignment was around 1-2 hours. We allowed students to take up to 6 attempts per question with a 3 % penalty imposed for each attempt taken. This ensured that students were motivated to do the problems without worrying too much on the scores. We provided the opportunity for students to think about solving problems with interest and enthusiasm instead of just submitting the assignments only for getting good grades. The assignments constituted around 12 % of continual assessment component (the other components were mid-term exam, end semester exam and other continual assessments).

Learning Catalytics is an interactive classroom tool developed at Harvard and used in several disciplines all over the world. In this system, instructors can pose a variety of questions that help students recall ideas, apply concepts, and develop critical thinking skills. Students can submit their responses through their smartphones, tablets, or laptops. The real-time display of student responses allows instructors to immediately address any student misconceptions. Instructors can adjust their teaching approach as well as initiate peer-to-peer discussion, collaboration, and communication. There are 18 question types in LC such as composite sketch, hotspot, expression, and direction which allows the deployment of a wide range of questions to engage and motivate students. This has a direct impact on their learning. We deployed LC in the classroom especially as a revision before exams. This helped students refresh their concepts and they were excited to see features such as sketching answers for questions. The real time feedback gave the instructor an insight into the misconceptions which were addressed immediately.

Students went through BEMA during week 1 (pre-test) at the start of the semester (before MP and LC were deployed) and nearly at the end of the semester in week 15 (post-test). BEMA consists of 31 multiple choice questions to be taken in 45 min. The items in BEMA cover the core concepts of EM such as electrostatics, direct current circuits, magnetostatics, and Faraday's Law. We administered a survey on MP and LC to understand the students' perspectives and feedback.

IV. Findings

There is an improvement of 15 % in the average post-test BEMA (39 %) score of students when compared to the average pre-test BEMA score (24 %) and is statistically significant at 5 % level with an effect size of 0.99. We also found that 86 % of students improved their scores in the post-test. The results are shown in Table I and Table II.

TABLE I
STATISTICS OF BEMA ANALYSIS

Number of students	Average pre-test score (%)	Average post-test score (%)	SD (pre-test)	SD (post-test)
29	24	39	11	17

TABLE II
RESULTS OF BEMA SCORES

Number of students	p-value (at 5 % level)	Effect size	Normalized gain
29	3×10^{-6}	0.99	0.19

We also found the Normalized gain to be 0.19. Normalized gain is a quantitative way of understanding the effectiveness of a course in promoting conceptual understanding (Hake, 1998) and is given by $\langle g \rangle = (\langle Post \rangle - \langle Pre \rangle)/(100 - \langle Pre \rangle)$.

The students' survey consisted of questions consisting of 5-point Likert scale as well as open ended to understand their perspective on MP and LC. The Likert scale being such that a rating of 1 being Strongly disagree to 5 being Strongly agree.

The result of the survey is as follows:

TABLE III
RESULTS OF MP AND LC SURVEY

Question	Average score out of a max 5
The number of assignment questions in MP is appropriate.	3.67
The assignment questions in MP make me aware of how much I have learnt.	4.17
MP is user friendly.	3.33
MP is an effective tutoring system.	3.83
LC is easy to use.	3.73
LC is very engaging.	3.70
I would like MP and LC for other physics topics	3.63

https://doi.org/10.52202/073963-0020

The following are a sample of the responses to the open-ended questions:

Can you elaborate on positive aspects of Mastering Physics and Learning Catalytics that helped you?

"Helped to understand concepts on a practical basis and not only theoretical. Was effective in fixing my mistakes if there were any, some problems were intuitive, whereas some needed some brain power. The steps and hints helped me to progress through questions."

"Mastering physics and Learning Catalytics help to understand the topics very well."

"It helps to properly assess myself and helps to realize and understand my weak topics."

"Mastering physics has many questions that makes you understand the concepts better."

"Helps to know how much I have learnt."

Can you suggest some improvements that can be done in Mastering Physics and Learning Catalytics?

"Upload videos related to certain the topic which should be short and easy to understand."

"It is a bit tricky to input answers in mobile."

"No improvements needed."

"Nothing in particular."

Any other comments/feedback on Mastering Physics and Learning Catalytics.

"Overall, it was a decent experience using Mastering Physics and Learning Catalytics. Just a few tweaks here and there would make the experience smoother and more fluid."

"Excellent material for learning."

"I loved it".

"It is really good."

V. DISCUSSION

We find that the increase in BEMA post-test score is statistically significant. The effect size is large, and this implies that the difference in the post and pre-test score is important. We think that the MP and LC have contributed in a positive way that is reflected in the improvement of the scores and the student survey. Research has shown that activities that promote Interactive Engagement contribute to concept building (Hake, 1998) compared to just traditional lectures. Content taught in the lectures in the classroom is supplemented and enhanced by MP and LC. During online learning, it is important to have activity-based pedagogy that can engage students in a meaningful way which can involve tools such as virtual labs and simulations (Nedungadi et al., 2015; Nedungadi et al., 2017; Nair et al., 2015; Achuthan & Murali, 2017; Raman et al., 2015; Chandrasekhar et al., 2020). Therefore, we think that the features in MP and LC help in addressing abstract topics such as fields by visualization, video lab demonstrations as well as giving detailed explanation to students which is useful in increasing the conceptual understanding of topics. We also realize that courses in semester one such as mechanics and some calculus-based math courses need to be mastered well. Since we introduced MP and LC in the first semester, students benefited from the experience in mechanics before they embarked on EM and could get used to MP and LC to take on new abstract ideas and concepts.

Our study also opens doors for more research in this topic. We observe that the normalized gain is only 0.19. Moreover, even though there is a 15 % increase in the post-test scores, in other related studies, in the pre-covid era, the improvements are much higher. For example, a 35 % increase in the post-test score is noticed by Pollock (Pollock, 2008) while a 20 % increase is reported by Eaton (Eaton et al. 2019). One reason for not achieving a higher score in the post-test may be due to E-learning fatigue. This E-learning fatigue happens when students feel a sense of overload due to constant use of technology, creating mental and physical dynamics that result in less efficient and uncomfortable learning (Reed, 2022). We know that students in our study spent a considerable amount of time learning online due to Covid 19 before they started university. Hence, E-learning fatigue may have set in for many students in exploring features in MP such as the Dynamic Study Modules (DSM). The DSM tool in MP consists of many conceptual questions in every topic that require respondents to think as well as indicate their confidence level before submitting their responses. We think that while the assignments were very helpful, spending time on other features in MP needs to be highlighted to students.

Based on the MP and LC survey (Table III), students' perspectives are quite positive. From the open-ended responses, students seem to benefit from the fact that MP and LC give feedback on how much they have learnt and where they lag. Such a system where students know their learning gaps is useful so that students can take timely appropriate actions. This will have a bearing on their subsequent learning as a stronger

https://doi.org/10.52202/073963-0020

foundation in topics will pave way to a deeper learning of future topics and courses with confidence.

The data for instructors from MP and LC helps directly to address the learning of students on a personalized basis. MP gives a variety of statistics such as average assignment scores, time taken by students and difficulty level perceived by students (Figs. 1-3). Similarly, LC gives valuable feedback in real-time in the class where students misconception can be immediately addressed (Fig 4). Learning Analytics is an emerging area in education (Chan et al., 2017). We see immense potential to do Learning Analytics using MP and LC that can benefit students early in the semester. For example, if a student is taking too much time to do questions in a particular topic, instructor can discuss with the student (Fig 2). On the other hand, if a student takes a very short time to do a multiple-choice question and takes a lot of attempts, the student has not put the required effort and is trying his/her luck to obtain the right answer. Again, the instructor can intervene and help the student realize the importance of hard work and encourage the student to spend time thinking about physics problems.

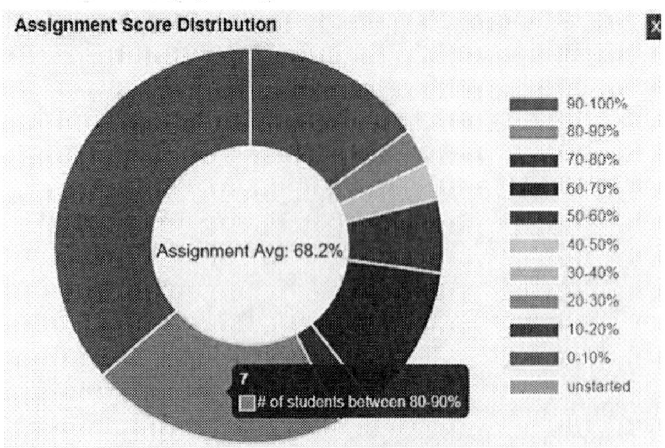

Fig 1. An example of an assignment score.

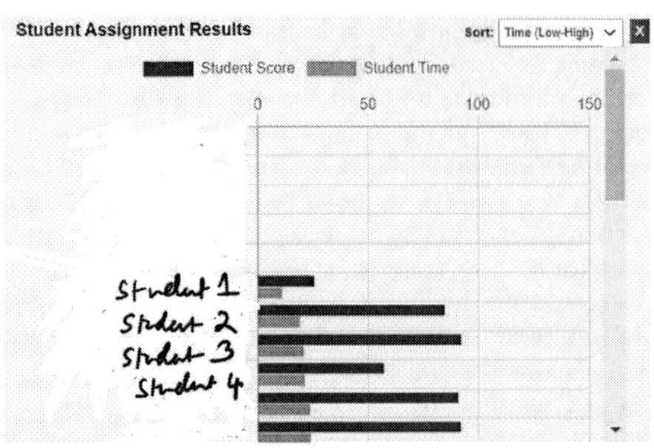

Fig 2. Student scores and time spent.

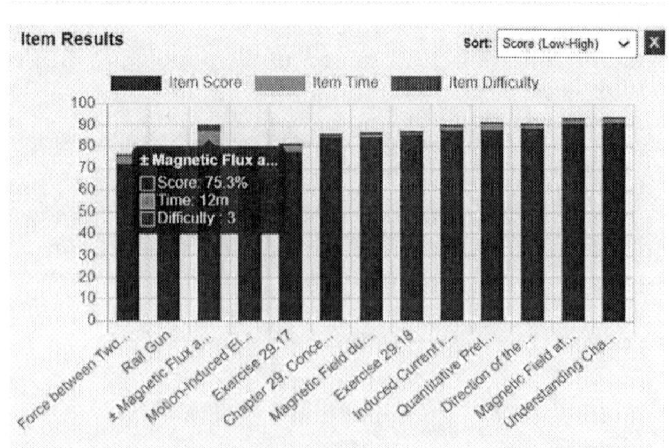

Fig 3. Item score, time and difficulty level perceived by students.

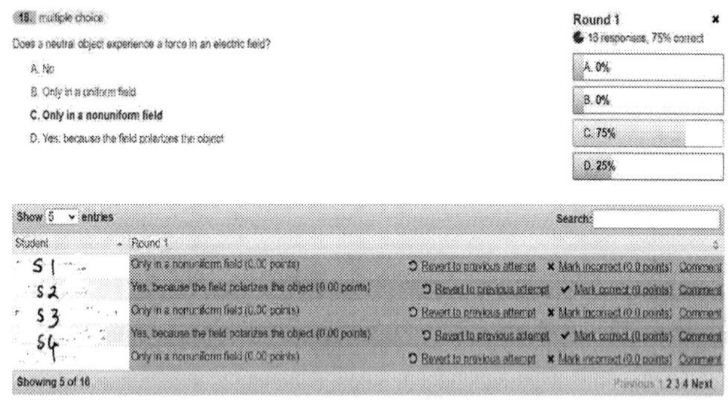

Fig 4. LC question and responses.

VI. Conclusion and Future Work

We find that there is an increase in conceptual understanding of Electricity and Magnetism at an Indian University which we found through the concept inventory, BEMA. The post-test scores of students showed an average of 15% increase compared to pre-test scores that are statistically significant with a large effect size. We have used Mastering Physics and Learning Catalytics in our course which we think played an important role in improving the conceptual understanding of EM. We have also shared that the data from MP and LC can be used to do Learning Analytics to benefit students. The students were quite positive in their responses to MP and LC. They find MP and LC useful and engaging. They also find that it helps them to understand how much they have learnt. We also discussed the potential of MP and LC to give us meaningful data to perform Learning Analytics.

We think our study needs to be conducted with more students so that we can be sure of our conclusions. Hence, we are going to continue with MP and LC with the new cohorts of students. We recognize that students were diligent in submitting the assignments. However, we need to encourage students to utilize other tools in MP such as the Dynamics Study Module which

https://doi.org/10.52202/073963-0020

can enhance their conceptual understanding. We hope to include these tools in our future curriculum formally so that students take these seriously and overcome E-learning fatigue. We hope to share our experience with the engineering faculty in our university on the concept inventories, MP and LC. This will help engineering students build a strong base in physics for a successful learning in their discipline. We would like to use Learning Analytics and conduct a more in-depth study to further improve the learning of physics and engineering. Another project is to continue our investigation by doing an in-depth item analysis in BEMA. We aim to conduct interviews with students and perform a qualitative study. We would extend our study to other topics such as mechanics and administer concept inventories such as Force Concept Inventory to understand the challenges faced by students in mechanics.

ACKNOWLEDGEMENT

We thank Mata Amritanandamayi for guidance encouragement. We thank the students who participated in this study.

REFERENCES

Achuthan, K., & Murali, S.S. (2017). Virtual lab: an adequate multi-modality learning channel for enhancing students' perception in chemistry. Advances in Intelligent Systems and Computing.574, 419-433.

Chabay, R. & Sherwood, B. (2006). Restructuring the introductory electricity and magnetism course, Am. J. Phys. 74(4), 329-336.

Creswell, J. W. (2012). Educational research: Planning, conducting, and evaluating quantitative and qualitative research (4th ed). Pearson.

Chan, E., Lai, S.B., Lim, M., Soh, Y.Y., & Tan-Yeoh, A.C. (2017). Piloting Learning Analytics to Support Differentiated Learning via LearningANTS. Proceedings from Learning Analytics and Knowledge, Vancouver, British Columbia, Canada.

Chandrashekhar, P., Prabhakaran, M., Gutjahr, G., Raman, R.,& Nedungadi, P. (2020). Teacher Perception of OLabs Pedagogy. in Fourth International Congress on Information and Communication Technology, Singapore.

Ding L., Chabay R., Sherwood B., & Beichner R. (2006). Evaluating an electricity and magnetism assessment tool. Phys. Rev. ST Phys. Educ. Res., 2, 010105.

Eaton,P., Johnson,K., Frank, B., & Willoughby, S. (2019). Classical test theory and item response theory comparison of the brief electricity and magnetism assessment and the conceptual survey of electricity and magnetism. Phys. Rev. ST Phys. Educ. Res., 15, 010102.

Garvin-Doxas K., & Klymkowsky M. W., Elrod S. (2007). Building, using, and maximizing the impact of concept inventories in the biology education: a meeting report. CBE Life Scbiology,7–282.

Hufnagel, B. (2002). Development of the Astronomy Diagnostic Test, The Astronomy Education Review, Issue 1(1), 47-51.

Hestenes, D., Wells, M., & Swackhamer, G. (1992). Force Concept Inventory. The Physics Teacher, 30, 141–158.

Hake, R.R. (1998). Interactive-engagement versus traditional methods: A six-thousand student survey of mechanics test data for introductory physics courses. Am. J. Phys. 66(1), 64–74.

Halloun, I.A. & Hestenes, D. (1985). Common Sense Concepts About Motion. American Journal of Physics, 53 (11), 1056-1065.

Kohlmyer, M.A., Caballero, M.D., Catrambone, R., Chabay, R.W., Ding, L., Haugan, M.P., Marr, M.J., Sherwood, B.D., & Schatz. (2009). Tale of two curricula: The performance of 2000 students in introductory electromagnetism, Phys. Rev. Phys. Educ. Res. 5, 020105.

Lee, Y.J., Palazzo, J.P., Rasil, W., & Pritchard, D.E. Measuring Student Learning With Item Response Theory. (2008). Physical Review, 4(1).

Lindell, R.S., Peak, E., & Foster, T.M. (2007). Are they all created equal? A comparison of different concept inventory development methodologies. IP conference proceedings, Syracuse, New York 883, 14–17.

McDermott, L.C., Rosenquist, M.L. & van Zee E.H. (1987). Student difficulties in connecting graphs and physics: Examples from kinematics, in Am. J. Phys., 55, 503-513.

McGrath, C., Guerin, B., Harte, E., Frearson, M., & Manville, C. (2015). Learning gain in higher education. Santa Monica, CA: RAND Corporation.

Nair, B., Sasidharakurup, H., Radhamani, R., Kumar, D., Nizar, N., & Achuthan, K.(2015). Assessing students and teachers' experience on simulation and remote biotechnology virtual labs: a case study with a light microscopy experiment. Proceedings of 2nd International Conference on e-Learning, e-Education and Online Training.

Nedungadi, P., Malini, P., & Raman, R. (2015). Inquiry based learning pedagogy for chemistry practical experiments using OLabs", Advances in Intelligent Systems and Computing, 320, 633-642, 2015.

Nedungadi, P., Prabhakaran, M., & Raman, R. (2017). Benefits of Activity Based Learning Pedagogy with Online Labs (OLabs) in 5th IEEE International Conference on MOOCs, Innovation and Technology in Education (MITE).

Pollock, S.J. (2009). Longitudinal study of student conceptual understanding in electricity and magnetism, Phys. Rev. Phys. Educ. Res. 5, 020110.

Pollock, S.J. (2008). Comparing student learning with multiple research-based conceptual surveys: CSEM and BEMA. AIP Conf. Proc. 1064, 171–174 (2008).

Porter, L., Taylor, C., & Webb, K. (2014). Leveraging open-source principles for flexible concept inventory development. Proceedings of the conference on innovation & technology in education, Uppsala, Sweden,(243-248).

Raman, R., Achuthan, K., Nedungadi, P., Diwakar, S., & Bose, R. (2015). The VLAB OER experience: modeling potential-adopter students' acceptance. IEEE Transactions on Education, vol. 57, 235-241.

Reed, H.C., (2022). E-Learning Fatigue and the Cognitive, Educational, and Emotional Impacts on Communication Sciences and Disorders Students During COVID-19. https://doi.org/10.1044/2022_PERSP-22-00049.

Sadler, P.M., Coyle, H., Miller, J.L., Cook-Smith, N., Dussault, M., & Gould, R.R. (2009). The astronomy and space science concept inventory: Development and validation of assessment instruments aligned with the K-12 national science standards. Astronomy Education Review, 8(1).

Sands, D., Parker, M., Hedgeland, H., Jordan, S., & Galloway, R.(2018). Using concept inventories to measure understanding, Higher Education Pedagogies, 3(1), 60-69.

Trowbridge, D & McDermott, L. (1980). Investigation of Student Understanding of the Concept of Velocity in One Dimension. American Journal of Physics. Vol. 48. No.12.

Copyright Statement

https://doi.org/10.52202/073963-0020

Digital Games to Develop Empathy in Students: A Scoping Review

Radhika Amashi, Preethi Baligar, and Vijayalakshmi M
KLE Technological University
radhika.amashi@kletech.ac.in

Abstract

Context

Empathy is developed in learning contexts where the learner is familiar, similar, and proximal to the protagonist, which often poses a challenge in designing learning environments. Although technology can bridge the gap between learner and the context, research in this area is still in its infancy. Games and virtual reality, is frequently used to develop empathy and empathetic behaviors in medical, nursing, and allied health students. In line with this, Researchers suggest that digital games can aid educational institutions in fostering empathy among students.

Purpose or Goal

This study aims to identify the games, the crucial elements that influence game design for empathy development, and how the elements map to the three dimensions of empathy: cognitive, affective and behavioral.
Also, we present simple scenarios where the game can be redesigned to develop competency for engineering problem-solving by situating participants in different case studies, past, present, and futuristic.

Methods

This proposed study is a scoping review of the Scopus database for articles between 2013 and 2023. Based on the five-step methodological framework, 17 articles were included in the review.

Outcomes

It is observed that narration, immersion, and interactivity play important roles in empathy development while supporting other game elements like role play, decision making, character identification, rules/tasks and challenges to develop empathy at different dimensions (cognitive, affective and behavior).

Conclusion

Combined with others, game elements like narration, immersion, and interaction contribute to different dimensions of empathy development. Further research is needed to establish a precise correlation between these elements and empathy dimensions, but the framework offers valuable guidance for game designers and researchers in this area.

Keywords— Empathy, digital, virtual, simulation, video games, technology-enhanced learning, gamification.

I. INTRODUCTION

Empathy is a social phenomenon defined as "the ability to sense other people's emotions, coupled with the ability to imagine what someone else might be thinking or feeling and compassionately take appropriate action" (Knezek et al., 2022), (Wulansari et al., 2020). Irrespective of the domain, empathy manifests itself in three dimensions: cognitive, affective, and behavioral. Cognitive empathy, also known as perspective-taking, involves consciously trying to understand another person's emotional state or point of view. Emotional empathy is the unconscious emotional response to someone else's emotions. The third dimension called behavioral or motivational or compassionate empathy, refers to the action in response to someone else's feeling (Boltz et al., 2015) (López-Faican & Jaen, 2023). Empathy has recently been promoted as the desirable outcome (Knezek et al., 2022), leading researchers and instructors to develop curricular interventions for fostering empathy in engineering education (Preethi B et al., in press).

The relationship between empathy and the use of technology has been researched since 1980. This association is complex as the development or decline of empathy due to the use of Technology depends on how much and in what way the Technology is used (Knezek et al., 2022). Among other technological tools, online chat, video chat, simulations, and video games (Wulansari et al., 2020) have demonstrated the potential to promote empathy as they let players become immersed in the situation and participate.

Digital games, once primarily seen as sources of entertainment, are now recognized as powerful tools that can engage players on emotional, cognitive, and social dimensions (Yusoff et al., 2018) (Wulansari et al., 2020). By immersing players in diverse and complex virtual worlds, digital games offer unique opportunities to experience different perspectives, confront challenging scenarios, and practice empathetic decision-making in a safe environment (López-Faican & Jaen, 2023). Empathy is developed in learning contexts where the learner is familiar, similar, and proximal to the protagonist, which often poses a challenge in designing learning environments.

In today's increasingly interconnected and diverse world, engineers are tasked with solving technical challenges and understanding and addressing the needs, perspectives, and emotions of a wide range of stakeholders. While researchers are designing frameworks, courses, and interventions to develop empathy (Walther et al., 2017), integrating digital games to foster empathy in engineering education holds great promise and is at its infancy.
By immersing engineering students in virtual scenarios, they must navigate complex ethical dilemmas, consider the impacts of their designs on various communities, and make empathetic decisions. This helps engineers contribute positively to society

https://doi.org/10.52202/073963-0021

by creating solutions that align with the diverse needs and emotions of the people they serve.

As technology advances and game designers harness the potential of immersive storytelling and dynamic character interactions, researchers highlight that digital games can support educational institutions to develop empathy in students(Chan et al., 2023) (Yusoff et al., 2018). Therefore, this study aims to explore different games and how digital games are used to develop empathy in undergraduate students through a scoping review. Section II discusses the scoping review process followed for this study. Section III discusses the synthesis and discussion of the study followed by the conclusion and future scope.

II. METHODOLOGY

(Arksey & O'Malley, 2005) five step methodological framework is used for the scoping review.

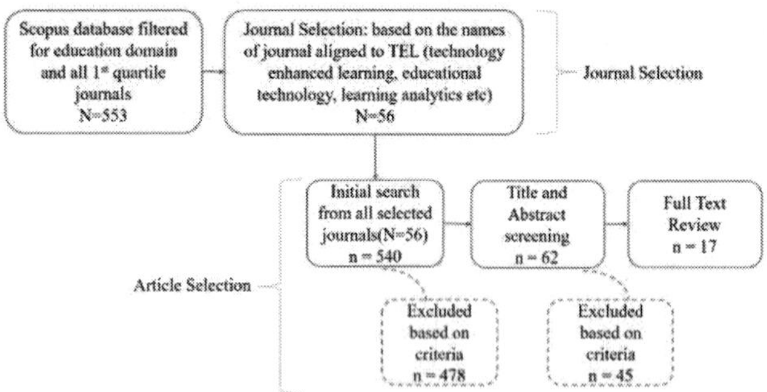

Fig. 1. Selection Process

1) Identifying the research question: This study aims to answer the following research questions:
   - *RQ1: What games are designed to develop empathy?*
   - *RQ2: What elements of game design are important to promote the development of empathy in students?*
   - *RQ3: How do the elements of game design map to the three dimensions of empathy: cognitive, affective, and behavioral?*
2) Identifying relevant studies: The authors followed the steps below to identify the suitable studies. Fig.1 shows the Flow chart for searching and filtering process.
   a. Top 25% of journals on technology-enhanced learning and its related journals indexed in the Scopus database were selected, resulting in 56 journals.
   b. Search String: All these selected journals were searched with the string empath* AND (digit* OR simulat* OR video* OR virtual*) AND game* for 2013 to 2023, resulting in a total of 540 articles.
3) Study selection: The selection of articles was done in two stages. The first stage selection was done by reading the paper titles and the abstract. The article was selected or rejected based on inclusion-exclusion criteria. The second stage selection was done by reading the full paper and charting the data.

Inclusion: The criteria for inclusion in this study encompass a range of factors crucial as listed below:

1. The studies must describe the design of games for empathy development by focusing on either of the mechanisms for empathy development: perspective-taking, responsible decision-making, empathetic concern, socio-cognitive skills, and socio-emotional skills
2. The games described in the studies must be grounded in existing game design theories.
3. The games must include elements like immersion and narration.
4. Furthermore, studies related to the development of empathy within pre-service teachers and teachers, and within the contexts of fields/courses related to literacy, history, drawing, and addressing contemporary issues like cyberbullying are considered.
5. The game must be technology-based.

Exclusion:

1. Papers situated within the domain of medicine or relevant fields are excluded, as the primary focus is on empathy using games within an engineering educational context.
2. Constructs such as curiosity, interests, intercultural competence, and gender equality, fall outside the defined parameters and are thus not considered.
3. Research involving participants classified as specially challenged students is excluded.
4. Studies centered on gamification without technological integration are excluded, given the emphasis on game elements like immersion and narrative.
5. Finally, the utilization of robots or Artificial Intelligence (AI) for Social Emotional Learning (SEL) competencies is excluded, as this study specifically explores the efficacy of Augmented Reality (AR) and Virtual Reality (VR) games.

Collectively, these exclusion criteria ensure that the selected research aligns closely with this study's intended focus and objectives.

4) Charting the data: The data from the articles is charted into an Excel sheet to identify different contexts, scope of studies, research questions, game design elements, a technology used for game design, definitions of empathy and its components, and so on.
5) Collating, summarizing, and reporting the results: The collated data is analyzed and reported in the paper's results section.

III. RESULTS AND DISCUSSION

This section presents and discusses the results of the study. Figure 2 shows the increasing trend of articles on technology-

https://doi.org/10.52202/073963-0021

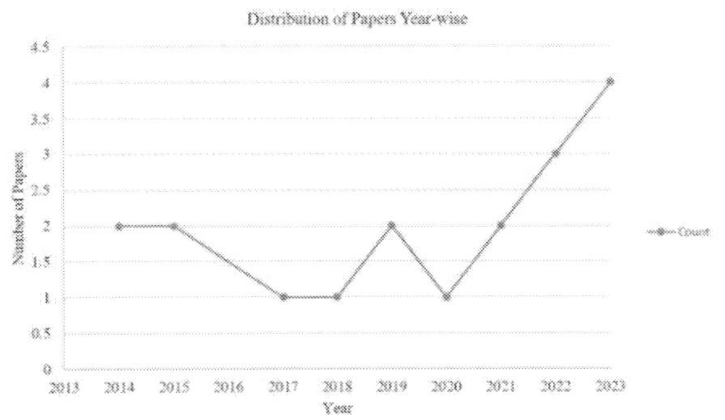

Fig. 2. Year-wise count of paper

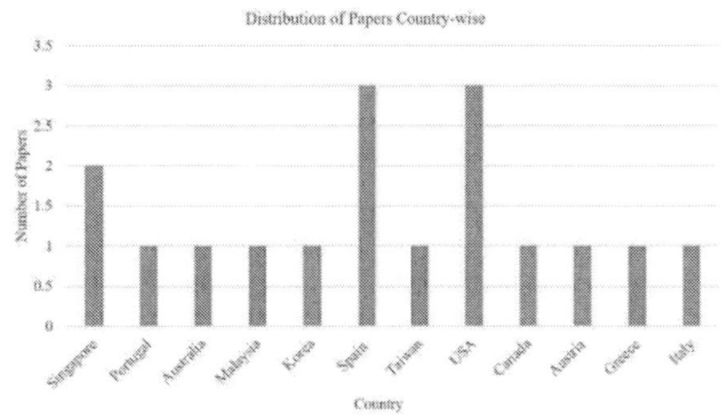

Fig. 3. Country-wise count of paper

based empathy development from the year 2020. This coincides with the Covid-19 pandemic, during which all educational institutions were forced to leverage technology for teaching and learning. Figure 3 shows the distribution of papers across different countries. The USA, Spain, and Singapore have used gaming for empathy development.

A. What games are designed to develop empathy and their suitability for engineering problem-solving?

This research question is answered through Table I, which shows names and descriptions of different games that have the potential to develop empathy in students. A total of 13 games were identified, where one of them (Rector-Aranda & Raider-Roth, 2015) did not mention the level of empathy being catered through the games.

The diverse selection of games discussed in this study offers a nuanced perspective on using digital platforms to cultivate empathy across various age groups and contexts. Synthesis of this study shown in Table I also highlight that empathy development through games has been focused on participants aged 6 to 22.

The game "Com@Viver" is an example of how games can intentionally evoke empathy (Ferreira et al., 2021) by immersing players in a school context and presenting them with cyberbullying scenarios and bystander responses. The game aims to stimulate emotional engagement and empathetic concern among adolescents. This game presents pointers for developing scenarios of real-world problem for engineering problem solving. These scenarios are interdisciplinary, complex and ill-structured problems that can be situated in different countries and cultures to highlight that engineering is not the context-free application of technical knowledge (Hoeborn & Bredtmann, 2012).

Similarly, "Conectado" (Calvo-Morata et al., 2020) takes an educational approach, simulating a high school student's experience with bullying and harassment (Calvo-Morata et al., 2020). It fosters empathy in adolescents by emphasizing seeking help, making choices to change outcomes, and teaching strategies to prevent victimization and bystander involvement. Additionally, this game is valuable for engineering ethics and sustainability development, as it encourages students to address dilemmas in practice (Voss, 2013). It presents scenarios from different stakeholder perspectives and environments, prompting students to assess how solutions impact people, the planet, and profit.

The results also highlight that empathy-building extends beyond contemporary contexts. Games like "The Battle of Yiwei" (Chan et al., 2023) and "Kokoda VR" (Calvert & Abadia, 2020) take players into historical narratives, offering opportunities to empathize with characters' challenges and struggles, effectively fostering historical empathy, which is to understand the ideas, feelings, experiences, decisions, and acts of people in the past within certain historical circumstances (Karn, 2023). As these games focus on historical empathy, the players can be placed in several historical case studies related to the engineering profession, like the first bridge/dam construction, computers, to driverless cars, along with the constraints of that time, to enable players to appreciate the growth and progress of engineering and the technological, social and economical affordances that they need today to make rapid advances (Young et al., 2021).

Games such as "Why Did Baba Yaga Take My Brother?" (Muravevskaia & Gardner-McCune, 2023) demonstrate how virtual reality can provide an interactive and imaginative platform for instilling empathy and problem-solving skills in children.

Furthermore, the game "EmpathyAR" creatively utilizes augmented reality to provide tasks that involve helping virtual characters in distress (López-Faican & Jaen, 2023). By encouraging players to seek solutions and aid these characters, the game prompts individuals to step into others' shoes and experience empathy through action, offering a unique approach to empathy development. This game appears apt for courses that focus on disaster management in engineering as they focus on

The "Mysterious Museum" game adeptly tackles both cognitive and affective empathy, with a primary emphasis on nurturing cognitive empathy by fostering an understanding and acceptance of differing perspectives. This concept, referred to as self-other differentiation, plays a pivotal role in enhancing empathy development and positively impacting one's ability to

https://doi.org/10.52202/073963-0021

TABLE I

SUMMARY OF GAMES FOR EMPATHY AND TECHNOLOGY USED

S.No	Name of the Game	Description	Technology/Platform	Age Group
1	Unknown (Tan et al., 2022)	Game depicts an incidence of bullying in school in which the protagonist experienced being physically harassed and victimized online.	Immersive Virtual Environments	15
2	Com@Viver (Ferreira et al., 2021)	The game immerses players in a school context, interacting with profiled social agents to organize a field trip, including cyberbullying scenarios with bystander responses, aiming to evoke empathy and empathic concern based on theoretical models.	Web based interface	12 to 14
3	The Walking Dead (PC game) and The Last of Us (PS3 game) (Toh & Lim, 2022)	The Walking Dead (PC game) and The Last of Us (PS3 game) both feature protagonists navigating post-apocalyptic worlds alongside young companions (Clementine in TWD and Ellie in TLOU), one set in a zombie-infested universe and the other in a world transformed by a mutated fungal infection.	PC game	15 to 24 (youths)
4	Kokoda VR and the Kokoda 360° video. (Calvert & Abadia, 2020)	Kokoda VR offers students a immersive narrative experience, placing them at the heart of the historic Kokoda Track campaign during World War Two, condensing key events while maintaining historical accuracy and alignment with the Australian curriculum.	Virtual Reality	12 to 18 (high school)
5	Mysterious Museum game(Jeon et al., 2023)	Player solves various puzzles based on ambiguous images and three-dimensional models.	Virtual Reality/Unity and XR Interaction Toolkit	22 to 32
6	Conectado(Calvo-Morata et al., 2020)	The video game simulates a high school student's experience facing bullying and harassment over five days, emphasizing the importance of seeking help, altering dialogue choices to impact outcomes, while also teaching strategies to prevent victimization, bullying, and bystander complicity.	Unity 3D	12 to 17
7	The Battle of Yiwei (Chan et al., 2023)	This game situated in the context of the 1895 Yiwei War in Taiwan, where players assume the role of a militia leader fighting against the Japanese army alongside numerous Taiwanese people.	Online Role-Playing Game/Gather Town and Google Jam board	22 to 55
8	EmpathyAR (López-Faican & Jaen, 2023)	Game takes players through series of tasks where the player must locate and provide assistance to different individuals, each with unique needs and potential solutions, such as seeking support from friends, offering medical aid, or securing resources.	Augmented Reality/Unity Engine	12 to 15
9	"Why Did Baba Yaga Take My Brother?" (Muravevskaia & Gardner-McCune, 2023)	Players must watch their younger brother, but a magical swan geese abduction sends them on a quest through an enchanted forest to rescue him.	Virtual Reality	6 to 9
10	JCAT and Place Out ofTime simulation (POOT) (Rector-Aranda & Raider-Roth, 2015)	This game is a web-mediated simulation designed for middle school classrooms where students take on roles of various characters throughout the world, history, and literature to address an imaginary court case.	Simulation	11 to 14
11	Nintendogs (Tsai & Kaufman, 2014)	Player pet a dog and use various items that can be found or purchased, such as balls, frisbees, toys, and grooming supplies, all designed to keep the dog happy.	Simulation Video Game	9 to 11
12	Path-Out(Wulansari et al., 2023)	This is an autobiographical adventure game where players follow the journey of Abdullah Karam, a Syrian artist who escaped the 2014 civil war. Disguised as a Japanese RPG, the game offers surprises, challenges, and humor,	Video Game/Unity 3D	18 to 22
13	Unknown (Di Tore, 2014)	In the first task, players must identify the man whose perspective is displayed while changing viewpoints. In the second task, they select the correct viewpoint window with one man present, and the third task involves choosing a park area based on a hidden man's perspective shown at the top.	Video Game/Unity 3D	Unknown

empathize with others (Jeon et al., 2023).

"Path-Out" is an autobiographical game that employs narrative elements to immerse players in the protagonist's life story. It utilizes gaming elements to provide players with a firsthand experience of the journey of a young Syrian artist who escaped the civil war in Syria (Wulansari et al., 2023). This game and "Mysterious Museum" game appears suitable and can be used to communicate the need for human-centred design. Give a design say, a coffee mug or a cell phone, and the participant needs to redesign it for a person with upper limb disabilities. They can also be used to track the daily lives of people suffering from disabilities and develop products to ease their everyday challenges.

JCAT and its counterpart, Place Out of Time (POOT), are educational simulations for schools, fostering skills like critical thinking, empathy, and communication. One scenario presented

TABLE II
GAME ELEMENTS USED TO DEVELOP EMPATHY

S.No	Name of the Game	Level of Empathy	Game Elements
1	Unknown (Tan et al., 2022)	Cognitive, Affective and Behavior	Scenario, Role Play, Decision Making
2	Com@Viver (Ferreira et al., 2021)	Cognitive	Interaction, Feedback
3	Mysterious Museum game (Jeon et al., 2023)	Perspective taking	Scenario, Control, Interaction
4	Conectado (Calvo-Morata et al., 2020)	Behavioral and Affective	Immersion, Character Identification,, Player`s Choice, Decision Making, Scenario
5	The Battle of Yiwei (Chan et al., 2023)	Affective Connection, and Perspective Taking	Player`s choice, Character Identification, Planning, Feedback and Suggestion, Challenges, Immersive and Interactive
6	Empathy AR (López-Faican & Jaen, 2023)	Cognitive, Affective Behavior	Play, Rules, Scenario Interface, Tasks
7	"Why Did Baba Yaga Take My Brother?" (Muravevskaia & Gardner-McCune, 2023)	Cognitive, Affective and Behavior	Story, Interaction, Challenge, Hints, Character Identification
8	JCAT and Place Out of Time simulation (POOT) (Rector-Aranda & Raider-Roth, 2015)	NA	Research their characters (Character Identification), Post biographies, Speeches and Comments as their character (Tasks/Rules), Role Play.
9	Nintendogs (Tsai & Kaufman, 2014)	Cognitive and affective	Player`s choice, Interaction, Rewards
10	Path-Out (Wulansari et al., 2023)	Cognitive and Affective	Story, Player`s Choice, Role Play, Character Identification), and Tasks.
11	Unknown (Di Tore, 2014)	Perspective taking	Interface, Tasks, Feedback

to player were simulated trial on reparations for SS. St. Louis passengers' descendants denied entry to the U.S. Another was on the debate over religious ornamentation in French schools, with Jewish and Muslim students advocating for their right to wear religious headwear (Rector-Aranda & Raider-Roth, 2015). This game can be used to integrate real engineering projects and case studies into the simulation which require the context of multiple perspectives (Murray et al., 2019), which has been presented here for the context of religious headwear.

"Nintendogs" is a real-time pet simulation video game by Nintendo, played on the DS (Dual Screen) console with a touchscreen and microphone. Players care for a virtual dog by petting it, using items, and teaching it commands. The game emphasizes grooming, feeding, and playing with the dog. Players can go on walks, to the park, and engage in activities like disc-catching and agility trials. Contests are the main way to earn in-game currency. This virtual pet game serves as a unique platform for children to develop empathy by experiencing and responding to the virtual pet's needs, fostering emotional connections, and understanding of caring behaviors (Tsai & Kaufman, 2014). For engineering, the context of virtual pets can be extended for virtual laboratories (Potkonjak et al., 2016) to develop competency in preventive maintenance of engineering artefacts like cars, equipment, solar panels, etc.

The diverse range of games examined in this discussion collectively demonstrates the potential of digital platforms to foster empathy among different age groups and contexts.

B. What elements of game design are important to promote the development of empathy in students

This research question is addressed by describing various elements used in game design. Table II shows the list of different elements used in the game design for empathy development.

1) Scenario/Story and Narration: A scenario in a game refers to a specific situation or context in which players must make decisions and take actions (Calvo-Morata et al., 2020; Ferreira et al., 2021). It sets the stage for the gameplay and often presents challenges that players need to overcome (Jeon et al., 2023). Story/Plot: The story or plot of a game encompasses the narrative, characters, and events that unfold as players progress (López-Faican & Jaen, 2023). A compelling story can enhance players' emotional engagement and empathy. Players immersion is based on how well the story/plot or scenario is narrated (Muravevskaia & Gardner-McCune, 2023; Tan et al., 2022; Wulansari et al., 2023).
2) Character Identification: is the emotional connection players establish with the characters in the game, particularly the protagonist or central figures. It's about players empathizing or feeling a strong bond with these characters. Character identification is when players mentally align themselves with the in-game characters' perspectives, emotions, and motivations. This can lead to a deeper understanding of different viewpoints. Role-play in games involves players taking on specific roles or characters within the game's narrative or setting (Rector-Aranda & Raider-Roth, 2015; Tan et al., 2022; Wulansari et al., 2023). They assume these roles' identity, characteristics, and behaviors during gameplay. The distinction between role-play and character identification lies in the active engagement of players. While role-play centers around assuming a character's role and making decisions, character identification focuses on establishing an emotional bond with the game's characters. (Calvo-Morata et al., 2020; Chan et al., 2023; Muravevskaia & Gardner-McCune, 2023; Wulansari et al., 2023).

https://doi.org/10.52202/073963-0021

3) Decision Making: Decision-making in games refers to players evaluating their options, weighing potential consequences, and ultimately choosing from those available to them. Decisions often significantly impact the game's progression or outcome (Calvo-Morata et al., 2020; Tan et al., 2022). Player's choice refers to the options or actions offered to players within a game. It encompasses a range of possibilities that players can select from, and these choices may or may not have immediate or long-term consequences (Calvo-Morata et al., 2020; Chan et al., 2023; Tsai & Kaufman, 2014; Wulansari et al., 2023). The player's choice encompasses all available options and actions in the game, including those with minimal consequences. Decision-making focuses specifically on evaluating options and selecting one with significant consequences.
4) Interaction: Interaction in games refers to the ways players engage with the game environment, characters, objects, and other players (Chan et al., 2023; Yusoff et al., 2018). It can include actions like movement, communication, and manipulation of in-game elements (Muravevskaia & Gardner-McCune, 2023). An intuitive and user-friendly interface can enhance the gaming experience and interaction (Di Tore, 2014; Tsai & Kaufman, 2014).
5) Feedback: Feedback in games informs players about their actions and progress. Positive feedback reinforces desired behaviors, while negative feedback helps players adjust their strategies(Chan et al., 2023; Di Tore, 2014; Ferreira et al., 2021).
6) Challenges: Challenges in games refer to obstacles or tasks that players must overcome to progress. They can include puzzles, enemies, and tasks that require problem-solving (Chan et al., 2023; Ferreira et al., 2021; Muravevskaia & Gardner-McCune, 2023).
7) Tasks/Rules: Tasks in games are specific actions or objectives that players need to complete. Tasks can vary in complexity and contribute to the overall gameplay (Di Tore, 2014; López-Faican & Jaen, 2023).
8) Immersion refers to how players feel fully engaged and absorbed in the game world. Elements such as graphics, sound, and narrative can contribute to creating an immersive experience (McMahan, 2004).
   In each of these games, the incorporation of specific game elements aligns with different dimensions of empathy, creating a holistic and immersive experience that engages players on cognitive, affective, and behavioral front

While there are various game design frameworks like Mechanics Dynamics and Aesthetics (MDA) (Hunicke et al., 2004), and Learning Mechanics and Game Mechanics (LM-GM) (Arnab et al., 2015) which highlight on categorizing game elements for mechanics, dynamics, and aesthetics, however, in this study, the eight game elements are presented without any categorization due to the scope of the study being limited to understanding different game elements and their relationship with empathy dimensions. There is a scope for categorizing these eight elements into game world and the mechanics.

C. How do the elements of game design map to the three dimensions of empathy: cognitive, affective and behavioral?

Table II presents a comprehensive overview of different games, their associated dimensions of empathy, and the specific game elements they incorporate to foster empathy in players.

1. Cognitive Empathy:

Game Elements: Narration, Immersion, Decision making, Role Play, Interaction, Feedback

Cognitive empathy involves understanding another's perspective and feelings (Jeon et al., 2023). Game elements that enable players to make decisions, interact with character stories through hints and feedback (Muravevskaia & Gardner-McCune, 2023), and take on the role of characters can stimulate cognitive empathy(Jeon et al., 2023).

A well-constructed **narration** can help players understand and appreciate the perspectives, motivations, and thoughts of in-game characters. By immersing players in the characters' stories and dilemmas, narration can enhance their ability to cognitively empathize with these virtual individuals.

Narration combined with **decision-making** becomes a powerful tool for enhancing cognitive empathy. When players are presented with complex choices that involve ethical considerations and moral dilemmas, they are encouraged to think deeply about the consequences of their decisions.

Along with narration and decision-making, **interactive dialogues and conversations** allow players to engage directly with characters, enabling them to explore different perspectives and understand the cognitive aspects of empathy. Through interactions, players can ask questions, seek information, and engage in meaningful discussions, fostering a deeper understanding of the characters' thoughts and viewpoints. While narration and decision-making contribute to fostering cognitive empathy, interaction, feedback, and role play enhance the experience by adding a layer of direct engagement.

Role-play with narration, interaction, immersion, decision making, and interactive experience can promote cognitive empathy by encouraging players to think and act as their in-game counterparts, gaining insights into the characters' cognitive processes.

How cognitive empathy manifests in game design?

The game "Mysterious Museum" (Jeon et al., 2023) narrative focuses on the robot's journey to develop its empathetic ability. It highlights the theme of ambiguity and the significance of grasping diverse viewpoints to enhance cognitive empathy. Players assume the role of the robot, making choices that directly impact their understanding of others' perspectives. They decide which images and objects to explore, how to interpret them, and how to approach perspective-related challenges. Cognitive empathy is fostered through interactions with the virtual world. Players use a joystick and controller to navigate, interact with 2D images and 3D objects, and make decisions by selecting options and manipulating objects. These interactions encourage players to consider alternative viewpoints and broaden their cognitive empathy skills.

The game "The Battle of Yiwei,"(Chan et al., 2023) begins with a stage of familiarization where players are introduced to historical characters and context through narration. Throughout the game, players are required to make decisions related to strategic planning and historical events. Players can select historical characters such as President Su, Vice-President Su, or Branch President Chen. This choice allows players to embody these characters and make decisions from their perspectives, enhancing their ability to empathize with the roles and responsibilities of these historical figures. Interactions with real-person NPCs, such as Battalion Officer Su, provide players with guidance, explanations, and feedback. These interactions facilitate the transfer of historical knowledge and support players in understanding the consequences of their decisions.

In the game "Nintendogs" (Tsai & Kaufman, 2014) the progression of the virtual dog's growth, behavior, and achievements becomes a narrative of the player's relationship with their pet. Players constantly make decisions that impact their virtual pet's well-being. They must decide when and what to feed the dog when to groom it, which activities to engage in, and how to participate in contests. These decisions require players to consider the dog's needs and preferences, encouraging them to think empathetically about what would be best for their virtual pet's happiness and health. Nintendogs encourage role-playing as players take on the role of a pet owner. They are responsible for the dog's care, training, and overall well-being. Through interactions, they can pet, feed, groom, train, and play with the dog, fostering cognitive empathy as players learn to understand their pet's cues and emotional states.

The game "Com@Viver" (Ferreira et al., 2021) strongly emphasizes cognitive empathy, encouraging players to interact with characters and provide feedback. By engaging players in interactions that necessitate understanding characters' thoughts and emotions, the game facilitates the development of cognitive empathy. Through these interactions, players are challenged to decipher emotional cues and respond appropriately, enhancing their ability to perceive and understand the feelings of others.

2. Affective Empathy:

Game Elements: Narration, Character Identification, Challenges and Immersion

Affective empathy is about sharing and feeling the emotions of others (Jeon et al., 2023). Game elements that encourage players to emotionally connect with characters through identification, immersion, and choices can evoke affective empathy. Emotional storytelling can be a potent tool for enhancing affective empathy.

In addition to narration, immersion and interaction, **character identification** facilitated by realistic and relatable characters is another driver of affective empathy. When players form strong emotional bonds with in-game characters, they are more likely to feel and share in the characters' emotions. This is the key difference between character identification and role play.

Immersive game worlds and environments contribute significantly to affective empathy when players are fully immersed in the game's setting. Players become more emotionally invested in the characters and their struggles.

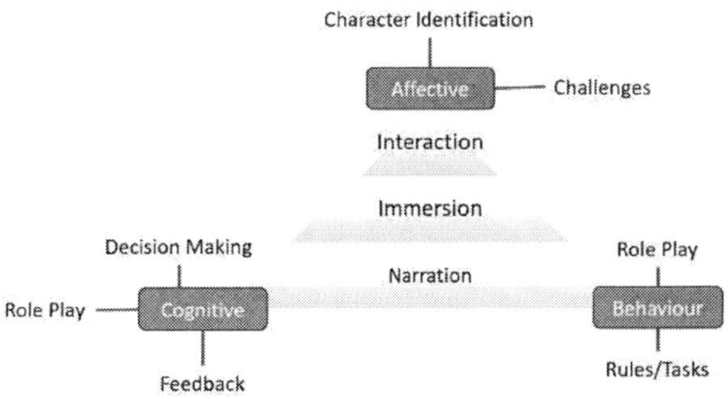

Fig. 4. Relationship between game elements and levels of empathy: Empathy Dimension and Game Elements (EDGE) framework

Challenges in games often evoke emotional responses from players. The frustration, excitement, and satisfaction that arise from overcoming challenges are primarily tied to emotions rather than cognitive processes. Therefore, the challenge element typically categorized within the affective dimension in mapping game elements to the empathy dimension framework.

How affective empathy manifests in game design?

The player in "Conectado," (Calvo-Morata et al., 2020) assumes the role of a high school student who experiences bullying and harassment, allowing for deep character identification. Immersion in the game is achieved through the player's ability to make choices that impact the story's progression and outcome. As players witness the protagonist's experiences of bullying and frustration, they are likely to feel empathy for the character's emotional suffering. The game effectively utilizes character identification, challenges, and immersion to create an emotionally engaging experience.

In "The Battle of Yiwei," (Chan et al., 2023) the real-person NPC's(Non Player Character) appeared in period-appropriate attire and their interactions with learners in a historically accurate tone add an important dimension of affective empathy. This allows players to better relate to the characters and events of that era while engaging in cognitive and strategic tasks within the game.

The game mechanics in "Why Did Baba Yaga Take My Brother?"(Muravevskaia & Gardner-McCune, 2023) require players to perform empathy actions, such as hugging characters, to progress and complete the game. This direct involvement in demonstrating empathy reinforces the emotional connection between the player and the characters. The game imposes a challenge to the player to rescue her little brother who is being kidnapped.

Nintendogs (Tsai & Kaufman, 2014) game allows players to name their dog and use the built-in microphone to interact with it by calling it and teaching it commands. The process of naming and personalizing the virtual pet fosters character identification. The challenge game elements, is demonstrated in the form of contests to earn money and conduct training to earn trainer points.

3. Behavioral Empathy:

Game Elements: Narration, Immersions, Interaction, with Tasks/Rules, Role Playing

https://doi.org/10.52202/073963-0021

Behavioral empathy involves demonstrating understanding through actions (Boltz et al., 2015). **Interactions** that require players to respond empathetically, **follow rules** that mirror empathetic behaviors, and **perform tasks** to engage in prosocial tasks for learning can promote behavioral empathy (Muravevskaia & Gardner-McCune, 2023). Interactions encompass players' choices and actions within the game world. When players are given the agency to make decisions that reflect empathy and compassion, they are more likely to exhibit behavioral empathy by taking actions that benefit others within the game. **Role-playing** situations where players adopt characters' roles encourage empathy-driven actions (Tan et al., 2022).

How behavioral empathy manifests in game design?

In the game (Tan et al., 2022), the goal was to engage participants in a thought-provoking and immersive experience. This experience would encourage them to consider the complexities of social issues (Social and Income Inequality in Singapore and Bullying Faced by Young Singaporeans). The overarching goal was to generate discussions, and potentially develop solutions to the dilemmas presented in the scenarios.

Conectado (Calvert & Abadia, 2020) encourages players to ask for help and offer help to address the bullying issue. It fosters behavioral empathy by emphasizing the need for players to take action, seek help, and offer support to combat bullying, mirroring the behaviors of empathetic individuals.

In the game (Muravevskaia & Gardner-McCune, 2023) players' actions and interactions with the virtual characters reflect behavioral empathy. Specifically, in the strategies used by children to find their VR brother, such as actively listening to characters, asking questions, and performing empathy-driven actions like hugging characters.

Relationship between game elements and dimensions of empathy

The relationship between dimensions of empathy and game elements is complex and multi-faceted as shown in Figure.4 which we refer to Empathy Dimension and Game Elements (EDGE) empathy dimensions framework. Game elements of narration, immersion, and interaction play a central role in any game environment to foster empathy. At the core, narration, immersion, and interaction combined with decision-making, role-play, and feedback enable players to develop cognitive empathy. Similarly, narration, immersion, and interaction combined with character identification, and challenges can collectively contribute to the development of affective empathy. Narration, immersion, and interaction combined with tasks/rules and role play develop behavioral empathy in players. The interplay between these elements creates a rich and immersive gaming experience that has the potential to foster empathy on multiple dimensions, from understanding different perspectives to sharing emotional connections and engaging in prosocial behaviors within the game world.

The relationship between game elements and dimensions of empathy varies based on how the researcher interprets it. This aspect remains open for further exploration and investigation to uncover concrete evidence supporting a direct and clear correspondence between game elements and dimensions of empathy. The framework proposed in this paper lays the foundation and provides directions for researchers, practitioners, and game designers interested in designing games for the development of empathy.

IV. CONCLUSION

This scoping review has examined the multifaceted relationship between digital games and the development of empathy in students. The exploration of this intricate connection has revealed that digital games, ranging from simulations to virtual reality experiences, hold significant potential as tools for nurturing empathy across diverse age groups and educational contexts. The study has presented a diverse array of games designed to cultivate students' empathy. These games span various educational contexts, from addressing contemporary issues like cyberbullying to immersing players in historical narratives. Games like "Com@Viver", "Conectado", "The Battle of Yiwei," and "Nintendogs" serve as illustrative examples of how different games target various dimensions of empathy development, catering to different age groups and learning objectives.

The empathy developed through these games can be valuable when working on engineering projects that affect communities and society as a whole. By incorporating these games into the engineering curriculum, students can not only develop technical skills but also become more empathetic and socially conscious engineers, better equipped to address the complex challenges of our interconnected world

The research has highlighted how game design elements develop empathy. Game elements such as narration, decision-making, role-play, character identification, interaction, feedback, challenges, tasks/rules, and immersion play crucial roles in fostering empathy. Through these elements, players can engage with virtual scenarios, empathize with in-game characters, and make decisions that reflect empathetic behaviors. Elements such as narratives, immersive environments, and meaningful interactions are fundamental to creating an educational game for developing emotional connection between players and the game world. While specific game elements within the game and these basic game elements help develop different dimensions of empathy.

Lastly, EDGE framework for mapping game design elements and dimensions of empathy is also presented which opens up opportunities for testing the game elements-empathy dimensions framework.

By analyzing the intricate interplay of game elements and their impact on different dimensions of empathy, this study provides valuable insights for educators, game designers, and researchers seeking innovative approaches to cultivate empathy skills in today's digitally-driven educational landscape.

The essence of engineering lies in solving problems (Passow & Passow, 2017). Engineers solve design problems, decision-making problems, troubleshooting, and systems analysis; each calls for different cognitive processes (D. Jonassen et al., 2006;

D. H. Jonassen, 2000). This opens up opportunities to create problem contexts using technology-enabled game-based learning environments to develop different levels of knowledge, as pointed out by (Anderson & Krathwohl, 2001) factual, conceptual, procedural, and metacognitive knowledge within the context of engineering problem-solving.

As technology advances and game designers harness the power of immersive storytelling, integrating digital games into educational institutions holds promise for nurturing empathy in students of all ages and backgrounds. This study presents further avenues for exploration:

1. What is the relationship between role play, decision making, feedback game elements and cognitive empathy?
2. What is the relationship between character identification, challenges game elements in game and affective empathy?
3. What is the relationship between role play, rules /tasks game elements and behavioral empathy?

Limitations: Due to lack of time and effort, the scope of this study was limited to selection of papers from 2013 to 2023 year and only the journals in the Scopus database. However, it could have been extended to other databases and extending the time span for 10 more years would have helped gain a broader perspective.

REFERENCES

Anderson, L. W., & Krathwohl, D. R. (2001). *A taxonomy for learning, teaching, and assessing: A revision of Bloom's taxonomy of educational objectives: Complete edition*. Addison Wesley Longman, Inc.

Arksey, H., & O'Malley, L. (2005). Scoping studies: Towards a methodological framework. *International Journal of Social Research Methodology*, *8*(1), 19–32.

Arnab, S., Lim, T., Carvalho, M. B., Bellotti, F., De Freitas, S., Louchart, S., Suttie, N., Berta, R., & De Gloria, A. (2015). Mapping learning and game mechanics for serious games analysis. *British Journal of Educational Technology*, *46*(2), 391–411.

Boltz, L. O., Henriksen, D., Mishra, P., & Group, D.-P. R. (2015). Rethinking technology & creativity in the 21st century: Empathy through gaming-perspective taking in a complex world. *TechTrends*, *59*(6), 3–8.

Calvert, J., & Abadia, R. (2020). Impact of immersing university and high school students in educational linear narratives using virtual reality technology. *Computers & Education*, *159*, 104005. https://doi.org/10.1016/j.compedu.2020.104005

Calvo-Morata, A., Rotaru, D. C., Alonso-Fernandez, C., Freire-Moran, M., Martinez-Ortiz, I., & Fernandez-Manjon, B. (2020). Validation of a Cyberbullying Serious Game Using Game Analytics. *IEEE Transactions on Learning Technologies*, *13*(1), 186–197. https://doi.org/10.1109/TLT.2018.2879354

Chan, H.-Y., Liu, S.-W., & Hou, H.-T. (2023). Interacting with real-person non-player characters to learn history: Development and playing behavior pattern analysis of a remote scaffolding-based situated educational game. *Interactive Learning Environments*, 1–21.

Di Tore, P. A. (2014). Perception Of Space, Empathy And Cognitive Processes: Design Of A Video Game For The Measurement Of Perspective Taking Skills. *International Journal of Emerging Technologies in Learning (iJET)*, *9*(7), 23. https://doi.org/10.3991/ijet.v9i7.3718

Ferreira, P. C., Veiga Simão, A. M., Paiva, A., Martinho, C., Prada, R., Ferreira, A., & Santos, F. (2021). Exploring empathy in cyberbullying with serious games. *Computers & Education*, *166*, 104155. https://doi.org/10.1016/j.compedu.2021.104155

Hoeborn, G., & Bredtmann, J. (2012). Serious games as an interdisciplinary approach in Engineering degree courses. In *GIEE 2011: Gender and Interdisciplinary Education for Engineers* (pp. 169–182). Brill.

Hunicke, R., LeBlanc, M., Zubek, R., & others. (2004). MDA: A formal approach to game design and game research. *Proceedings of the AAAI Workshop on Challenges in Game AI*, *4*(1), 1722.

Jeon, H., Jun, Y., Laine, T. H., & Kim, E. (2023). Immersive virtual reality game for cognitive-empathy education: Implementation and formative evaluation. *Education and Information Technologies*. https://doi.org/10.1007/s10639-023-11840-3

Jonassen, D. H. (2000). Toward a design theory of problem solving. *Educational Technology Research and Development*, *48*(4), 63–85.

Jonassen, D., Strobel, J., & Lee, C. B. (2006). Everyday problem solving in engineering: Lessons for engineering educators. *Journal of Engineering Education*, *95*(2), 139–151.

Karn, S. (2023). Historical Empathy: A Cognitive-Affective Theory for History Education in Canada. *Canadian Journal of Education*, *46*(1), 80–110.

Knezek, G., Christensen, R., & Gibson, D. (2022). Empathy: How Can Technology Help Foster Its Increase Rather Than Decline in the 21st Century? In *Social and Emotional Learning and Complex Skills Assessment: An Inclusive Learning Analytics Perspective* (pp. 51–73). Springer.

López-Faican, L., & Jaen, J. (2023). Design and evaluation of an augmented reality cyberphysical game for the development of empathic abilities. *International Journal of Human-Computer Studies*, *176*, 103041. https://doi.org/10.1016/j.ijhcs.2023.103041

McMahan, A. (2004). Immersion, Engagement, and Presence: A Method for Analyzing 3-D Video Games. In *The Video Game Theory Reader*. Routledge.

Muravevskaia, E., & Gardner-McCune, C. (2023). Designing a Virtual Reality Empathy Game framework to create empathic experiences for children. *International*

Journal of Child-Computer Interaction, *35*, 100561. https://doi.org/10.1016/j.ijcci.2022.100561

Murray, J. K., Studer, J. A., Daly, S. R., McKilligan, S., & Seifert, C. M. (2019). Design by taking perspectives: How engineers explore problems. *Journal of Engineering Education*, *108*(2), 248–275.

Passow, H. J., & Passow, C. H. (2017). What competencies should undergraduate engineering programs emphasize? A systematic review. *Journal of Engineering Education*, *106*(3), 475–526.

Potkonjak, V., Gardner, M., Callaghan, V., Mattila, P., Guetl, C., Petrović, V. M., & Jovanović, K. (2016). Virtual laboratories for education in science, technology, and engineering: A review. *Computers & Education*, *95*, 309–327.

Rector-Aranda, A., & Raider-Roth, M. (2015). `I finally felt like I had power': Student agency and voice in an online and classroom-based role-play simulation. *Research in Learning Technology*, *23*. https://doi.org/10.3402/rlt.v23.25569

Tan, M. C. C., Chye, S. Y. L., & Teng, K. S. M. (2022). \textquotedblIn the shoes of another\textquotedbl: Immersive technology for social and emotional learning. *Education and Information Technologies*, *27*(6), 8165–8188. https://doi.org/10.1007/s10639-022-10938-4

Tsai, Y.-F., & Kaufman, D. (2014). Interacting with a Computer-Simulated Pet: Factors Influencing Children's Humane Attitudes and Empathy. *Journal of Educational Computing Research*, *51*(2), 145–161. https://doi.org/10.2190/EC.51.2.a

Voss, G. (2013). Gaming, texting, learning? Teaching engineering ethics through students' lived experiences with technology. *Science and Engineering Ethics*, *19*(3), 1375–1393.

Walther, J., Miller, S. E., & Sochacka, N. W. (2017). A model of empathy in engineering as a core skill, practice orientation, and professional way of being. *Journal of Engineering Education*, *106*(1), 123–148.

Wulansari, O. D. E., Pirker, J., & Guetl, C. (2023). Analysis of Empathic Experiences in Video Games with Microblogging. *International Journal of Emerging Technologies in Learning (iJET)*, *18*(03), 83–95. https://doi.org/10.3991/ijet.v18i03.36291

Wulansari, O. D. E., Pirker, J., Kopf, J., & Guetl, C. (2020). Video Games and Their Correlation to Empathy. In M. E. Auer, H. Hortsch, & P. Sethakul (Eds.), *The Impact of the 4th Industrial Revolution on Engineering Education* (pp. 151–163). Springer International Publishing. https://doi.org/10.1007/978-3-030-40274-7_16

Young, M. F., Bassett, L., Burkey, D. D., Streiner, S., & Reed, J. B. (2021). Let's play! Gamifying engineering ethics education through the development of competitive and collaborative activities. *ASEE Annual Conference*.

Yusoff, Z., Kamsin, A., Shamshirband, S., & Chronopoulos, A. T. (2018). A survey of educational games as interaction design tools for affective learning: Thematic analysis taxonomy. *Education and Information Technologies*, *23*(1), 393–418. https://doi.org/10.1007/s10639-017-9610-5

Copyright Statement

Enhancing Learning and Engagement in Computer Organization and Architecture through Hands-On Activity

Aruna S. Nayak [a], Namrata D. Hiremath[a], Umadevi F.M [a] , Preeti B. Patil[b]
School of Computer Science & Engineering, KLE Technological University, Hubballi[a]
Centre for Engineering Education Research, KLE Technological University, Hubballi[b]
Corresponding Author Email: arunan@kletech.ac.in.

Context

The educational phenomenon being studied in this paper is the integration of hands-on activities in the teaching and learning of Computer Organization and Architecture (COA), a foundational subject in computer science and engineering. The paper aims to explore how incorporating hands-on experiences can enhance the learning process, reinforce complex concepts and develop problem-solving and critical thinking skills.

Purpose or Goal

Given the intricate nature of COA, which encompasses a multitude of theoretical concepts and models, it becomes imperative to provide students with immersive hands-on activities that serve to enhance their comprehension of these complexities. Moreover, active engagement in practical exercises not only enriches the learning trajectory but also imparts students with invaluable tangible knowledge. This effectively narrows the gap between theoretical understanding and practical application.

Methods

The study compared two student groups: one focused on individual processor modules called as control group (CG) and the other, on designing a complete processor which was the experimental group (EG). Both groups aimed to develop critical thinking, problem-solving, teamwork, communication and technical skills. The research assessed skill development using rubrics and employed statistical analysis to compare the two groups. This paper details the EG's activity and skill attainment. In this hands-on activity, approximately 250 students in EG were divided into teams and tasked with designing and simulating a processor datapath based on specific design requirements (Hamacher, V. C., Vranesic, Z. G., Zaky, S. G., Vransic, Z., & Zakay, S. (1996), Computer organization, McGraw-Hill). The activity emphasized effective instructional strategies, teamwork and scaffolded learning experiences, offering step-by-step guidance and opportunities for exploration and experimentation. Pre- and post-activity conceptual assessments were conducted to measure students' understanding of concepts. Data analysis explored the impact of teamwork and collaboration on knowledge acquisition, problem solving and critical thinking (Koppikar, Vijayalakshmi, Mohanachandran & Shettar, 2022).

Outcomes

Performance statistics revealed that a significant number of students improved their understanding of COA concepts through the hands-on activity, as evidenced by increased average scores in COA assessments.

The hands-on approach boosted student engagement, motivation and interest, as they actively applied theoretical concepts. Additionally, team activities facilitated lifelong skills, knowledge exchange, promoting peer learning and concept clarification.

Conclusion

Students demonstrated proficiency in applying COA principles and techniques through the hands-on activity. They excelled in designing and simulating computer architectures, analyzing performance metrics and optimizing system components, showcasing their practical knowledge. The interactive and experiential nature of the activity provided a holistic learning experience, equipping students with valuable skills for success in the field of COA.

Keywords— Addressing modes; critical thinking; datapath design; hands-on activity; instruction set architecture; problem solving; processor design.

I. INTRODUCTION

COMPUTER Architecture and Organization are integral subjects in the field of Computer Science & Engineering. They lay the groundwork for understanding the hardware aspects of computing systems, including processors. This course also lays a sound foundation for the learning of further courses like microcontroller & embedded systems, operating systems, system software, principles of compiler design and so on. Given COA's complexity with various theoretical concepts, practical activities are essential for enhancing understanding. Furthermore, active participation in hands-on activities not only enhances students' learning journey but also equips them with invaluable practical knowledge (Erdil, Bowlyn & Randall, 2021). This reduces the gap between theory and practice. However, engaging students and encouraging their interest in these hardware courses remains a critical concern. This scenario prompted the course instructors to consider the following research questions for their study:

1. What are the most effective instructional strategies or techniques within a hands-on activity framework that facilitate learning and understanding of COA concepts?

https://doi.org/10.52202/073963-0022

2. How does collaboration and teamwork in a hands-on activity contribute to students' learning outcomes in COA?

The objectives of the activity, includes

- Reinforcing COA concepts
- Applying theoretical knowledge to practical implementation (Chen, Huang, Lin, Chang, Lin, Lin, Hsiao, 2020),
- Honing problem-solving, and critical-thinking skills (Clausen & Andersson, 2019)
- Improving team building,
- Refining tool usage
- Allowing students to gain valuable insights into the inner workings of computer systems.

The activity begins with an introduction to the importance of COA in computer science and engineering (Nayak, Hiremath, Umadevi & Garagad, 2021). Students are familiarized with the key components of a processor, its role in executing instructions, and its significance in overall system performance. The hands-on activity (Erdil et al, 2021) on building a simple processor in the context of COA was conducted for approximately 250 students of second-year undergraduate Computer Science and Engineering (CSE) at a Technological University in Karnataka, India.

II. LITERATURE SURVEY

An exhaustive exercise was done to survey the existing literature where similar ideas were proposed. This gave the course teachers an in-depth understanding of the gaps existing where similar activities were conducted. An attempt has been made to address these gaps which led to the formulation of research questions mentioned in the previous section. In the work (Alqadi & Malhis, 2007) authors propose a structured methodology for imparting practical knowledge of processor design to students. The authors recognize the challenges faced by universities in developing countries when it comes to teaching advanced computer engineering topics due to limited resources, outdated equipment and lack of access to cutting-edge technologies. Therefore, the paper aims to address these challenges by presenting an approach that can be implemented with relatively modest resources. The authors in this paper (Nayak & Vijayalakshmi, 2013) share their experiences in teaching the COA course, highlighting the teaching methodologies and strategies used to effectively convey complex concepts to students. This may include lectures, hands-on labs, assignments, and assessments. The authors of the paper (Hiremath, Umadevi, Meena, 2018) provide insights into the realm of COA tutorials. The paper delves into the benefits of COA tutorials for students while also addressing the potential hurdles that educators might encounter in this context. The paper (Blackburn, Villa-Marcos, Williams, 2018) underscores the significance of using simulation software as a preparatory tool to enhance student readiness and competence in laboratory-based practical sessions, ultimately contributing to a more effective and enriching learning experience. The paper (Clausen & Andersson, 2019) adopts PBL method and discusses how to develop crucial employability skills such as critical thinking, problem-solving, teamwork and communication. These skills were seen as valuable assets for their future careers. Students felt better prepared to tackle challenges they might encounter in their future professions. The approach discussed in (Erdil, 2019) allows students to engage in practical implementation and experimentation, which enhances their understanding, engagement and real-world application of COA principles. The study in paper (Chen et al, 2020) emphasizes the significance of experiential learning in augmenting traditional classroom instruction. It states that by integrating virtual reality based hands-on activities, educators can bridge the gap between theoretical understanding and practical application. The study in (Rini, Adisyahputra, Sigit, 2020) involves a research design that includes pre-tests and post-tests to measure changes in students' critical thinking abilities after undergoing the proposed instructional intervention. The paper (Kamerikar, Patil & Watharkar, 2020) discusses on higher-order skills that include critical thinking, problem-solving, creativity and other abilities that are valuable for students' academic and professional development. The paper (Nayak et al, 2021) discusses the implementation of project-based learning (PBL) to enhance the teaching of COA. This approach emphasizes practical, hands-on experiences to deepen students' comprehension of COA concepts. The paper underscores the effectiveness of PBL in COA education, highlighting its potential to engage students and improve learning outcomes. The study conducted by authors in (Erdil et al, 2021) on the other hand, emphasizes the value of hands-on learning in Computer Organization & Architecture (COA) education. The authors designed interactive workshops where students were introduced to fundamental COA topics through hands-on activities. The authors (Siddamal & Despande, 2021) advocate that through collaborative initiatives, students engage in practical projects that mirror real-world scenarios, enabling them to develop problem-solving, teamwork, and critical thinking skills. The paper (Patil & Karikatti, 2022) discusses various assessment techniques and strategies tailored for PBL contexts. It involves strategies for evaluating project work, teamwork, problem-solving abilities and other skills that are cultivated through PBL approaches. The paper (Koppikar et al, 2022) mainly focuses on conducting the post-test effectively and carrying out an extensive analysis of the student's performance. In alignment with the various studies cited, our paper explores the implementation of a practical oriented activity in the COA course, emphasizing hands-on experiences to enhance students' understanding in addition to imparting lifelong skills.

III. METHODOLOGY

This study aims to foster a profound understanding of computer organization and architecture, encompassing both theory and practical applications, by engaging students in the hands-on task of designing and constructing a basic processor It was conducted for approximately 250 students of second-year undergraduate Computer Science and Engineering (CSE) at a Technological University in Karnataka, India. Prior to conducting the hands-on activity for knowledge acquisition and other significant skills (Erdil et al, 2021), several preparations were made to ensure a successful and effective learning experience for students. The course instructors indulged in rigorous brain storming sessions to meticulously plan, implement and assess the outcome of the activity.

A. Control Group(CG) vs. Experimental Group(EG):

The activity involved the assessment of skill attainment among two sets of students: CG and EG with 67 and 68 teams respectively. The CG participated in course projects that centered on the creation of individual processor modules such as Booth's multiplier, array multiplier, barrel shifter, carry lookahead adder among others, while the EG tackled the more

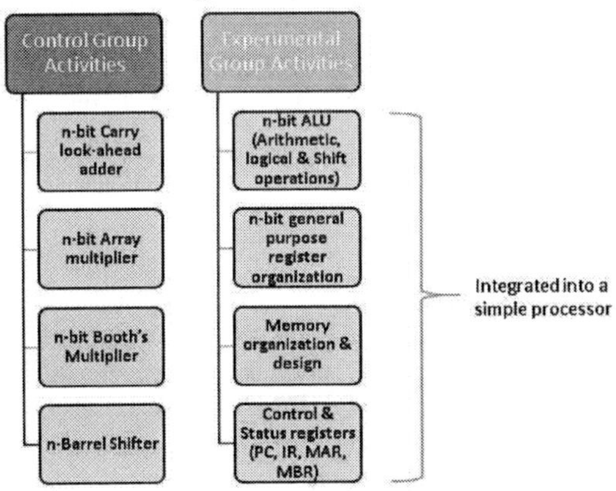

Fig 1. Activity List

holistic task of designing an entire processor. As Fig 1. depicts, the CG involved design of individual modules within the processor, which were not holistically integrated. This shortcoming was overcome in the EG. Both groups used Logisim as simulation tool. The CG and EG of students were from consecutive cohorts, with a one-year interval between them. The CG approach offered in-depth knowledge of a specific module enabling skill specialization, whereas the EG approach nurtured skills in module integration and system-level thinking. The EG were assessed for both individual modules and the overall processor design (Alqadi & Malhis, 2007). The CG had limited collaboration due to individual module focus, while the EG necessitated extensive collaboration for seamless integration. The core focus of the study was to assess the attainment of skills among students in both groups and assessment questions to both groups were set at the same difficulty levels. Skills included critical thinking, (Cáceres, Nussbaum & Ortiz, 2020) problem-solving (Clausen & Andersson, 2019), teamwork, communication, and technical competence in processor design. Rubrics based assessment was employed to consistently evaluate and measure students' skill development in both groups. This structured approach ensured fairness and accuracy in skill assessment (Patil & Karikatti, 2022). The study utilized statistical analysis methods to objectively compare the level of skill development between the control and EG. This paper focuses on the detailed description of the activity carried out for the EG.

B. Activity Planning:

Relevant concepts pertaining to the theoretical aspects of COA was imparted to students both in theory and laboratory sessions. Accordingly the entire activity was rolled out at the beginning of the semester as detailed below.

1) Learning Objectives: Clear and specific learning objectives for the activity were defined.

- Students should be able to demonstrate comprehensive knowledge acquisition by applying theoretical concepts such as Instruction Set Architecture (ISA), data path elements, control unit signals to design and develop the processor architecture.
- Students should be able to enhance their problem-solving, critical thinking (Clausen & Andersson, 2019), team building and communication skills by identifying design challenges, troubleshooting, optimizing performance and collaborative problem-solving (Siddamal & Despande, 2021).

These objectives were then aligned with the overall course goals to ensure that they are measurable. The activity alignment with the broader curriculum and learning outcomes of the course was ensured so that it complemented and reinforced the theoretical concepts covered in COA lectures (Nayak et al, 2021). The structure and sequence of the activity was meticulously planned. The scope of the processor design project, the level of complexity, and the required resources (such as software tools and materials) was identified and defined. The activity was finally integrated into the course timeline guided by following step-by-step process:

- *Pre-Activity Review:* A pre-activity review session was conducted to refresh students' understanding of relevant COA concepts and foundational knowledge (Clausen & Andersson, 2019). This review ensured that all participants are adequately prepared for the activity (Rini et al, 2020).
- *Resource Preparation:* Students were familiarized with the necessary resources, materials and digital logic simulator software tool (LogiSim). Necessary guidelines for tool usage and access to relevant resource materials were provided.
- *Group Formation:* Students were organized into small groups to foster collaboration and teamwork. The diversity of skills and backgrounds within each group to promote knowledge sharing and equitable contributions was considered referring to humanmetrics.com personality test with each team comprising of 4 students. Teams were

presented with the project scope and objectives, which involved building a simple processor capable of executing basic instructions. The processor's architecture, instruction set and supported operations were defined.

- *Communication with students:* The learning objectives, expectations and guidelines for the activity were communicated to the students well in advance. An overview of the project scope and the resources available for their design process was provided.
- *Documentation:* Throughout the activity, teams maintained detailed documentation of their design choices, implementation steps, challenges encountered and solutions (Hiremath et al, 2018). The documentation served as a record of their learning journey and a crucial part of the assessment.
- *Assessment Plan:* A clear assessment plan was charted out to evaluate students' knowledge acquisition, problem-solving and critical thinking skills (Clausen & Andersson, 2019) & (Cáceres et al, 2020) team collaboration and communication skills during the activity.

C. Activity Implementation:

During the lab sessions, students were actively engaged in the design and implementation of simple computer building blocks. Some of the specific tasks included:

- *Address Decoders:* Students were tasked with designing and implementing address decoders that enabled the selection of specific memory locations or peripheral devices based on address inputs.
- *Memory Design:* Memory design is a crucial component in computer architecture, and this lab activity provided students with practical experience (Erdil, 2019) in selecting specific memory location for read/write operation.
- *Multiplexers & ALU:* Multiplexers are essential in data selection and routing within digital circuits for data selection and manipulation of arithmetic & logical operations in ALU within processors.
- *Instruction life cycle:* An instruction goes through following phases during its life cycle in the processor.

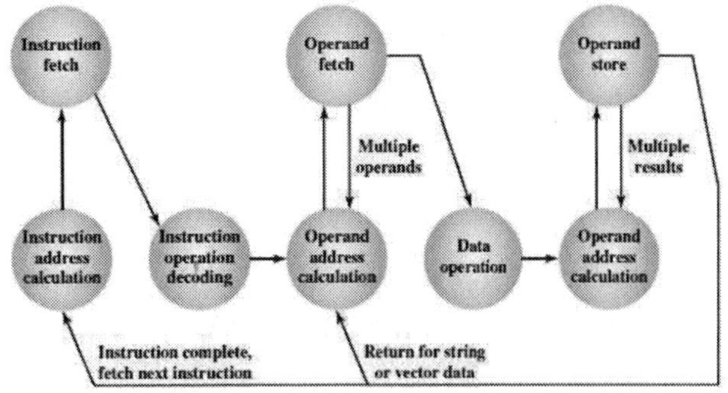

Fig.2. Instruction life cycle state diagram

The instruction life cycle state diagram as depicted in Fig.2, refers to the series of steps performed by a computer's central processing unit (CPU) to process and execute a single instruction (Stallings, W. (Ed.) (2010). Computer organization and architecture: designing for performance. Pearson Education India). It is a continuous process that occurs repeatedly as the CPU executes a sequence of instructions from a program. Each instruction goes through these stages in a sequential manner. By following the instruction life cycle, the CPU can efficiently process instructions and carry out the tasks specified by the computer program, enabling the computer to perform complex computations and operations.

Guided by instructors, teams embarked on the design and implementation of their processors. This involved taking critical design decisions, carefully considering factors like instruction set design, data path, control unit, and memory organization. The following were the guidelines floated to students to carry out the activity:

- ➢ Design and simulate a processor, which can perform load/store, arithmetic & logical operations on a set of data.
- ➢ Design for a Harward architecture, separate code memory & separate data memory to store program instructions and operands respectively.
- ➢ Include all the control & status registers like program counter (PC) (to hold address of instruction), memory address register, memory buffer register, and processor status word and instruction register (IR). Include a register file of 16 registers (R0—R15).
- ➢ Include a data memory and code memory of suitable size.
- ➢ Fetch the instruction using the contents of PC and update PC.
- ➢ Decode the instruction from IR.
- ➢ Fetch the operands (wherever applicable).
- ➢ Execute the operation.
- ➢ Write the result back in the destination (wherever applicable).
- ➢ Each team needs to implement the problem statement using 8-bit/16-bit/32-bit for 1-address/2-address/3-address format for given addressing mode. The addressing modes to be implemented are:
  - ✓ Direct addressing
  - ✓ Indirect addressing
  - ✓ Register addressing
  - ✓ Immediate addressing

- *Troubleshooting and Optimization*: As teams progressed, they encountered challenges typical in real-world processor design (Alqadi & Malhis, 2007). They involved in troubleshooting the issues, optimizing their design for performance, and ensuring

https://doi.org/10.52202/073963-0022

proper functionality. Once the implementation was completed, teams conducted rigorous testing and validation to ensure that their simple processors functioned correctly and executed instructions accurately.

D. Assessment Strategy:

The COA activity was assessed based on several criteria as shown in Fig 3.

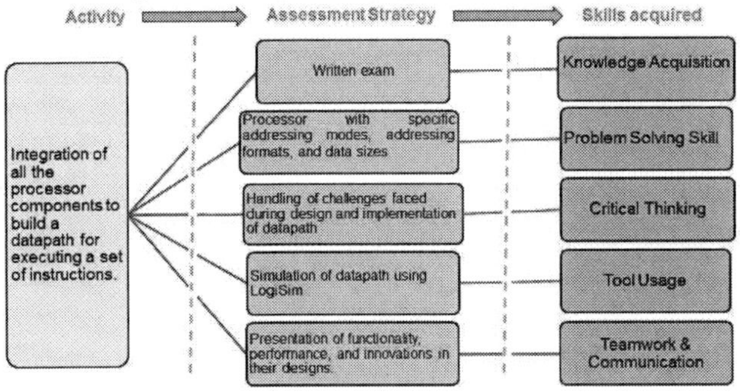

Fig.3. Activity at a glance

Knowledge acquisition: This criterion focused on measuring how well students grasped and absorbed the theoretical concepts and practical knowledge (Alqadi & Malhis, 2007) related to COA (Erdil, 2019). Students were evaluated before and after the activity to gauge their comprehension of theoretical concepts. For the pre-activity assessment a survey questionnaire was set covering all the theoretical concepts (Erdil et al, 2021) required for the conduct of the activity. Likewise the post-activity assessment (Erdil et al, 2021) was conducted using similar survey instrument (Rini et al, 2020). The functionality of the simple processor and its ability to execute instructions correctly were primary evaluation criteria. Additionally, teams were assessed on the performance metrics achieved through optimization efforts (Patil & Karikatti, 2022).

Team work and communication skills: A qualitative assessment was conducted by the course instructors where each team presented their final processor design to the class. They showcased the functionality, performance and innovations in their designs. Additionally, teams submitted a comprehensive report detailing their entire design process, including challenges faced and lessons learned (Hiremath et al, 2018). Teams were evaluated on their ability to work collaboratively, communicate effectively and contribute to the collective learning experience (Siddamal & Despande, 2021).

Problem solving, practical application and tool usage: Each individual in a team was assessed for problem solving (Kamerikar, Patil & Watharkar, 2020) practical application and tool usage skill through a single question as follows:

You have designed a processor with specific addressing modes, addressing formats and data sizes. The processor specifications dictate that you need to perform the following sequence of operations:

i. *Read two operands based on the processor's addressing mode, addressing format and number of bits.*
ii. *Perform addition on these operands and complement the result.*
iii. *Logically AND the result obtained in step ii. with another operand (read in the same way as mentioned in i. above).*
iv. *Store the final result in the specified memory location.*

Each team member was expected to write an assembly language code snippet for the above given problem statement. Convert it to machine level language and store it in the code memory of their designed processor. Walk through the steps taken to implement this sequence of operations in the processor's microarchitecture and ensure efficient execution as per the process of instruction life cycle mentioned in Fig. 1.

Critical Thinking skills: A set of questions as mentioned below was administered to each individual student in the team to assess their critical thinking.

i. *Why is it important to carefully select the addressing mode and format for each operand?*
ii. *Explain how the processor handles data size mismatches during complex arithmetic & logical operations.*
iii. *How do you ensure that the operands and the result are read from and stored in the correct memory locations?*
iv. *Are there any trade-offs or compromises you had to make in designing the processor to execute this sequence of operations efficiently?*
v. *How do you maximize the number of instructions executed per clock cycle?*

These questions were designed to challenge students to delve into the intricacies of the processor's design, consider the implications of various decisions and apply their knowledge to practical scenarios. Their responses provided insights into their ability to analyze complex situations evaluate options and synthesize solutions related to COA concepts and hence enabling measurement of critical thinking skills (Cáceres et al, 2020).

E. Assessment Rubrics:

Rubrics for assessment were used to provide a structured and transparent way to evaluate students' performance based on specific criteria. It ensured consistency in evaluation and helped both students and instructors to understand the expectations for each aspect of the activity. Rubrics as shown in Table I were

https://doi.org/10.52202/073963-0022

effectively used to measure various skills and competencies, including knowledge acquisition, critical thinking & problem-solving to mention a few among others. Similarly other acquired skills such as practical application, tool usage, team work & communication were also assessed through suitable rubrics.

TABLE I
ASSESSMENT RUBRICS

	Excellent	Good	Average
Knowledge Acquisition	• Demonstrates a deep understanding of COA concepts related to the simple processor design. • Accurately explains the principles and components involved in building a basic processor. (8-10M)	• Shows a solid understanding of most of the COA concepts relevant to the simple processor design. • Explains the principles and components with few inaccuracies. (4-7M)	• Displays some understandin g of COA concepts, but with significant gaps in knowledge. (0-3 M)
Problem-Solving Skills	• Demonstrates exceptional problem-solving skills, effectively analyzing and resolving complex issues related to processor design. • (8-10M)	• Displays strong problem-solving skills, effectively resolving most issues encountered during the design process. • (4-7M)	• Shows some problem-solving ability, but struggles to address certain challenges. (0-3 M)
Critical Thinking	• Demonstrates exceptional ability to identify underlying issues and challenges in the COA scenario. • Skillfully evaluates strengths and weaknesses of processor design options • Applies COA concepts to a real-world scenario and justifies their application. (8-10M)	• Identifies key aspects of the problem but lacks depth and thoroughness. • Demonstrates basic ability to evaluate evidence but lacks depth in evaluation. • Demonstrates basic application of COA concepts to a scenario but may lack clear relevance. (4-7M)	• Attempts to analyze problems, but lacks clear understandin g of problem analysis. • Attempts to evaluate evidence but struggles to present clear insights. • Attempts to apply COA concepts to a real-world scenario but lacks clear justification. (0-3 M)

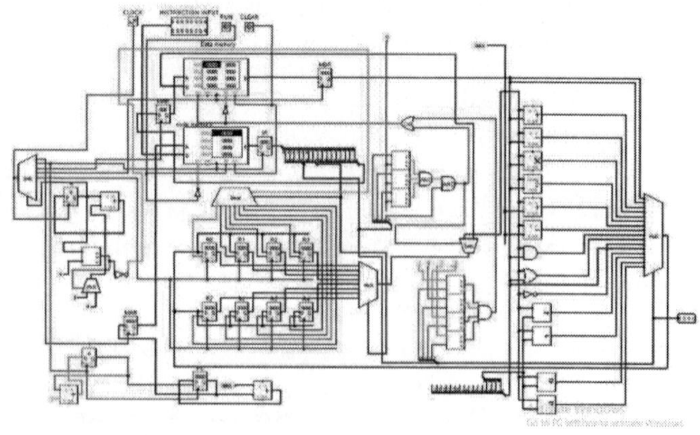

Fig.4. Sample Processor design done by a student team

Fig.4 depicts the processor designed by a one of the student teams as part of their course activity using LogiSim simulation (Blackburn, et al, 2018) tool.

IV. RESULTS & ANALYSIS

Statistical analysis was performed to assess skill acquisition and the influence of the study on both the control and experimental groups. A comprehensive breakdown of the analysis pertaining to the initial research question is presented below.

To address the first research question on effective methods to find best instructional strategies to facilitate learning, a quantitative research analysis was conducted. Initially, a descriptive statistical analysis was carried out to gain insights into the central tendency and variability of the data. Measures such as mean and standard deviation were calculated. Additionally, Shapiro wilk, skewness, and kurtosis were examined to assess the normality of the data distribution. As the data did not follow a normal distribution, non-parametric tests were deemed appropriate. Descriptive statistical analysis was performed to understand the distribution of data and non-parametric tests, specifically the Mann-Whitney U test and Wilcoxon signed-rank test, were conducted to assess the differences and identify the most effective instructional strategies or techniques within a hands-on activity framework that facilitates learning and understanding of COA concepts (Nayak & Vijayalakshmi, 2013). The dataset used in this analysis comprises student performance metrics from a CG and and EG. Performance data includes scores from assignments and activities. The Mann-Whitney U test as shown in Fig. 4 was employed to assess whether there were any statistically significant differences in performance between the CG and EG. This test is suitable for comparing two independent groups when the assumption of normal distribution is not met. Wilcoxon signed-rank test was performed within each group to assess any significant differences in performance before and after the intervention within each group.

https://doi.org/10.52202/073963-0022

As per the descriptive statistical analysis data presented in TABLE II, The CG had a sample size of N = 269 and the EG had a sample size of N = 273. There were three variables that were studied and they are Knowledge acquisition (KA), Problem-solving skills (PSS), and Critical Thinking Skills (CTS).

TABLE II
DESCRIPTIVE STATISTICS

	N	Mean	Std. Deviation	Skewness		Kurtosis	
	Statistic	Statistic	Statistic	Statistic	Std. Error	Statistic	Std. Error
EG_2KA	269	7.00	1.946	-1.104	0.149	0.759	0.296
EG_2PA	269	7.36	1.806	-0.599	0.149	-0.354	0.296
EG_2PSS	269	6.25	2.090	-0.792	0.149	0.114	0.296
EG_2CTS	269	5.90	2.365	-0.456	0.149	-0.622	0.296
EG_SCORE	269	62.56	17.080	-0.871	0.149	0.202	0.296
CG_2KA	273	6.14	1.284	-0.220	0.147	0.005	0.294
CG_2PA	273	6.26	1.637	-0.592	0.147	-0.134	0.294
CG_2PSS	273	5.46	1.723	-0.664	0.147	0.102	0.294
CG_2CTS	273	4.91	1.950	-0.301	0.147	-0.871	0.294
CG_SCORE	273	56.24	12.646	-0.799	0.147	0.129	0.294
Valid N (listwise)	269						

As per the data analysis report the mean (M) and standard deviation (SD) for KA before intervention for the CG was (M =6.14, SD = 1.28) lower when compared to (M =7.00, SD = 1.94) after intervention for the EG. Additionally, the (M =5.46, SD = 1.723) for PSS for CG was lower than (M =6.25, SD = 2.09) PSS in EG. Similarly, the M and SD for CTS for CG was (M =4.91, SD = 1.95) lower when compared to the (M =5.90, SD = 2.36) of EG. Overall it can be observed that the mean value for EG is higher than the CG indicating that the students in the EG have performed more effectively in KA, PSS, and CTS assessments when compared to the students in the CG. It is interesting to observe that the students in EG had (M =62.56, SD = 17.08) when compared to CG (M =56.24, SD = 12.646) indicating that EG students performed better than CG students in the end semester assessment for the course. The skewness and kurtosis indicated that the data for all the variables was not normally distributed. To second the data another Shapiro-Wilk normality test was done and the report is in TABLE. III. The data indicated that the $p<0.05$ was statistically significant rejecting the hypothesis that the data is normally distributed violating the assumption of normality. Homogeneity of variance was conducted and the $p<0.05$ violating the homogeneity of variance.

TABLE III
SHAPIRO-WILK

	Statistic	df	Sig.
EG_KA	0.926	269	0.000
EG_PSS	0.920	269	0.000
EG_CTS	0.939	269	0.000
EG_2KA	0.879	269	0.000
EG_2PA	0.926	269	0.000
EG_2PSS	0.925	269	0.000
EG_2CTS	0.947	269	0.000
CG_KA	0.944	269	0.000
CG_PSS	0.934	269	0.000
CG_CTS	0.960	269	0.000
CG_2KA	0.945	269	0.000
CG_2PA	0.933	269	0.000
CG_2PSS	0.933	269	0.000
CG_2CTS	0.939	269	0.000

Since the data was not normally distributed a nonparametric analysis was conducted to see which are the most effective instructional strategies or techniques within a hands-on activity framework that facilitates learning and understanding of COA concepts (Nayak & Vijayalakshmi, 2013). Initially, the Wilcoxon Signed-Ranks Test was performed to assess the impact of hands-on activity in improving students' understanding of the concept. When studying EG for the KA variable, the test indicated that student's scores on the post-test with (Mdn = 8.0) were statistically significantly higher than pre-test scores (Mdn = 6.0) Z = 13.87, p = 0.00. The PSS variable also had a statistically significantly higher value on post-test (Mdn = 6.0) compared to pre-test (Mdn = 7.0) Z = 12.731, p = 0.00. Similarly, CTS improved in students from pre-test (Mdn = 5.0) to post-test (Mdn = 6.0) Z = 11.398, p = 0.00. The CG indicated increase in students performance and the score was higher in pre-test compared to post-test. TABLE IV represents the data. Overall the results indicated that regardless of pre-test or post-test performance students performed significantly better in the EG compared to the CG.

TABLE.IV
WILCOXON SIGNED RANKS TEST

	Statistic (Z)	*Mdn*	Sig.
EG_KA	13.875	6.00	0.000
EG_PSS	12.731	6.00	0.000
EG_CTS	11.398	5.00	0.000
EG_2KA	13.875	8.00	0.000
EG_2PSS	12.731	7.00	0.000
EG_2CTS	11.398	6.00	0.000
CG_KA	11.450	5.00	0.000
CG_PSS	9.199	5.00	0.000
CG_CTS	6.479	4.00	0.000
CG_2KA	11.450	6.00	0.000
CG_2PSS	9.199	6.00	0.000
CG_2CTS	6.479	5.00	0.000

A second analysis was conducted to observe a comparison between EG and CG to observe whether the student's performance improved between the two. A Mann-Whitney U test as shown in Fig. 5, was conducted to examine the differences in students' performance on the final test between the EG (Mdn = 66.0) and the CG (Mdn = 58.0). TABLE V shows the report. The Mann-Whitney U statistic was U = 2.586, p <0.000, indicating a statistically significant difference between the groups. Thus we can suggest that the performance of the EG

https://doi.org/10.52202/073963-0022

and CG differs significantly. The students in the EG performed well on the test compared to the CG.

TABLE. V
TEST STATISTICS [a]

	Score
Mann-Whitney U	2.586E4
Wilcoxon W	6.326E4
Z	-5.959
Asymp. Sig. (2-tailed)	0.000

a. Grouping Variable: Group

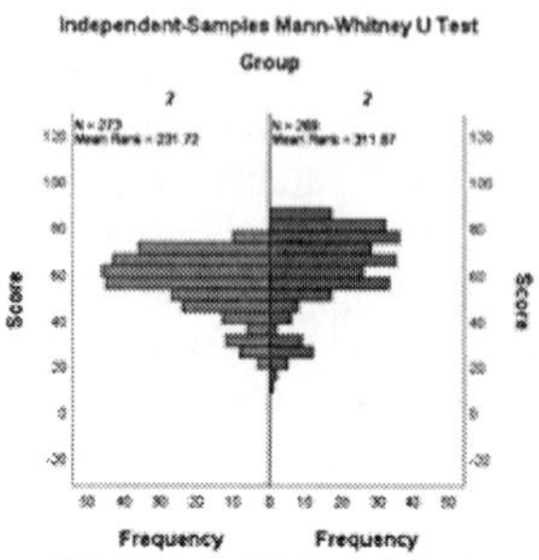

FIG.5.Mann-Whitney U test

The second research question, on how collaboration and teamwork in a hands-on activity contribute to students' learning outcomes in COA, was tackled through a comprehensive and qualitative approach that involved aggregating the average scores obtained from both peer-review assessments and evaluations conducted by course instructors. Each team member participated in the assessment process by evaluating their fellow team members using a meticulously designed form. This form encompassed a range of criteria including reliability and responsiveness, quality of work, contribution to ideas, team communication, time management, collaboration, and the overall contribution to the project. In tandem with this, the course instructors also undertook the evaluation process, employing identical parameters to assess the students' performances. The responses thus received were quantified mapping it to range of marks. This multi-faceted evaluation strategy ensured a comprehensive and well-rounded assessment of each team member's contributions and performance..

V. CONCLUSION

In conclusion, the COA activity demonstrated its effectiveness in fostering a comprehensive understanding of computer organization and architecture principles. By engaging students in hands-on processor design and implementation, the activity successfully bridged the gap between theoretical concepts and practical applications. The collaborative nature of the activity not only enhanced teamwork skills but also facilitated knowledge exchange among peers. Furthermore, the assessment outcomes showcased improved problem-solving abilities and critical thinking skills among participants. The activity's holistic approach encompassed various facets of COA, including processor design, memory hierarchy, and instruction execution. As evidenced by the statistical analysis, the activity positively impacted both the CG and EG, affirming its value in promoting skill attainment and overall learning outcomes. This COA activity serves as a model for integrating practical experiences into theory-based subjects, paving the way for a more enriching and effective educational journey.

REFERENCES

Alqadi, R. A., & Malhis, L. M. (2007). A Systematic Approach for Building Processors in a Computer Design Lab Course at Universities in Developing Countries. Information Technology Journal, 6(4), 497-508.

Blackburn, R. A., Villa-Marcos, B., & Williams, D. P. (2018). Preparing students for practical sessions using laboratory simulation software.

Clausen, H. B., & Andersson, V. (2019). Problem-based learning, education and employability: a case study with master's students from Aalborg University, Denmark. Journal of Teaching in Travel & Tourism, 19(2), 126-139.

Chen, J. C., Huang, Y., Lin, K. Y., Chang, Y. S., Lin, H. C., Lin, C. Y., & Hsiao, H. S. (2020). Developing a hands-on activity using virtual reality to help students learn by doing. Journal of Computer Assisted Learning, 36(1), 46-60.

Cáceres, M., Nussbaum, M., & Ortiz, J. (2020). Integrating critical thinking into the classroom: A teacher's perspective. Thinking Skills and Creativity, 37, 100674.

Erdil, D. C. (2019). Teaching hands-on computer organization and architecture using single-board computers. Journal of Computing Sciences in Colleges, 34(6), 163-164.

Erdil, D. C., Bowlyn, K. N., & Randall, J. (2021, July). A Hands-on Learning Approach to Introducing Computer Organization and Architecture to Early-college Students. In 2021 ASEE Virtual Annual Conference Content Access.

Hiremath, N. D., Umadevi, F. M., & Meena, S. M. (2018). Tutorial on Computer Organization and Architecture-Advantages and Challenges. Journal of Engineering Education Transformations, 31(Special Issue).

Kamerikar, U., Patil, M. S., & Watharkar, S. R. (2020). Project Based Learning-An Innovative approach to enhance higher order skills. Journal of Engineering Education Transformations, 33(3), 92-97.

Koppikar, U., Vijayalakshmi, M., Mohanachandran, P., & Shettar, A. (2022, March). Effectiveness of introducing concept-wise questions through post-tests in ensuring student learning. In 2022 IEEE Global Engineering Education Conference (EDUCON) (pp. 894-900). IEEE.

Nayak, A. S., & Vijayalakshmi, M. (2013, December). Teaching Computer System Design and Architecture course—An experience. In 2013 IEEE International Conference in MOOC, Innovation and Technology in Education (MITE) (pp. 21-25). IEEE.

Nayak, A. S., Hiremath, N. D., Umadevi, F. M., & Garagad, V. G. (2021). A hands-on approach in teaching computer organization &

Proceedings of REES 2024 KLE Technological University, Hubli, India, Copyright ©, Aruna S. Nayak , Namrata D. Hiremath, Umadevi F.M, Preeti B. Patil, Enhancing Learning and Engagement in Computer Organization and Architecture through Hands-On Activity 2023.

architecture through project based learning. Journal of Engineering Education Transformations, 34, 742-746.

Patil, P., & Karikatti, G. (2022). Effective Assessment Strategies for Project-Based Learning. Journal of Engineering Education Transformations, 36.

Rini, D. S., Adisyahputra, D. V. S., & Sigit, D. V. (2020). Boosting student critical thinking ability through project based learning, motivation and visual, auditory, kinesthetic learning style: A study on Ecosystem Topic. Universal Journal of Educational Research, 8(4), 37-44.

Siddamal, S. V., & Despande, A. M. (2021, November). Project-Based Learning to Enhance Employability and Entrepreneurship: A Collaborative Approach. In 2021 World Engineering Education Forum/Global Engineering Deans Council (WEEF/GEDC) (pp. 381-386). IEEE.

Stallings, W. (Ed.) (2010). Computer organization and architecture: designing for performance. Pearson Education India.

Hamacher, V. C., Vranesic, Z. G., Zaky, S. G., Vransic, Z., & Zakay, S. (1996). Computer organization, McGraw-Hill.

GLOSSARY

Abbreviation	Meaning
CG	Control Group
EG	Experimental Group
KA	Knowledge Acquisition
PSS	Problem-solving skills
CTS	Critical Thinking Skills
M	Mean
SD	Standard Deviation

https://doi.org/10.52202/073963-0022

Impact of Early PBL Course on Final-Year Engineering Project Work

Chetan C Jadhav[a], Preeti Basavaraj Patil[a], Ramesh Kurbet[a], Nandish Humbi[a], Praveen V Goggal[a], Vani Yelamali[a], and Shivprasad M Channangi[a]
KLE Technological University, Hubballi[a]
Corresponding Author Email: chetan.jadhav@kletech.ac.in

Abstract

Context

Conventional teaching methods mostly rely on teacher centric approach and doesn't provide interdisciplinary knowledge, with that practice it is difficult for students to survive in the corporate world. In this paper a study on comparison of engineering projects completed with and without project-based learning knowledge has been done by collecting the feedback from passed out engineering students.

Purpose or Goal

To prove the importance of mechatronic skills and project management knowledge in developing quality engineering projects during the final year.

Methods

In this paper a semi structured qualitative analysis is attempted to assess the quality of final-year project work with and without early project-based learning knowledge. First, we have framed two research questions and 11 interview questions and invited the students to KLE Technological university campus through email. Data collection was done through in-person interviews.

Outcomes

By the end of study, we hypothesized that the skillsets developed by the students who have attended first year engineering exploration course, help them in creating a quality project compared to that of students who did not learn the course. hence, we can say that developing mechatronic skills at the early stage of engineering exploration program will help them on understanding and producing high-quality project work.

Conclusion

This study has proven that PBL has more impact on producing quality projects compared to conventional learning. The interactions amongst the students amply illustrated the demand for project-based learning courses to improve the quality of final-year projects as well as the importance of more such learning during their first year of engineering.

***Keywords*—** Project based learning; Mechanism; programming; Project management.

I. INTRODUCTION

CURRENTLY graduate engineers are facing various industry expectations, including technical competence, problem-solving skills, communication skills, teamwork, adaptability, initiative, professionalism, safety awareness, willingness to learn, industry tools knowledge, time management, and business awareness. Employers expect engineering graduates to have a solid foundation in their field and be able to solve real-world problems. Additionally, they expect graduates to communicate complex technical information clearly, collaborate effectively, and adapt to new technologies and trends. Familiarity with industry-specific software and tools is also crucial, as is time management and understanding the broader business context. Hence it is necessary to train undergraduate students in multidisciplinary skills during the college days.

Traditional learning approaches in engineering face disadvantages such as limited practical application, inadequate problem-solving skills, lack of interdisciplinary integration, limited exposure to modern tools and technologies, insufficient project management skills, fixed curricula, limited soft skills development, slow feedback loop, and lack of real-world constraints. These drawbacks hinder student's ability to effectively manage engineering projects, adapt to industry demands, and address real-world constraints.

To address these issues, educational institutions are embracing experiential and project-based learning approaches, incorporating real-world projects, internships, industry collaborations, and modern teaching methods. By embracing these approaches, students can better prepare for real-world engineering projects and develop the necessary skills for success in the field. Hence it is necessary to understand how training students with multidisciplinary skills at their first semester will help them in future projects. therefore, in this study we have analysed how greatly the project-based learning will impact and enhance the student's performance during their final-year projects work, by comparing two set of students with and without PBL course. The details of literature survey are explained in the next section.

https://doi.org/10.52202/073963-0023

II. Literature Survey

Research has shown that project-based learning can significantly improve student learning outcomes and engagement in various subjects. In research on the introduction of a project-based learning strategy in an undergraduate course on new product development, Zancul et al., (2017) focused on this issue. Their study sought to assess how well this pedagogical strategy improved the learning results and engagement of students. The results of this study showed that adding project-based learning strategies to the curriculum significantly improved the course. In the context of introductory engineering design courses, Carpenter et al., (2016) investigated the application of project-based learning. Their research emphasized this strategy's beneficial effects on student engagement. The study revealed that by incorporating real-world projects into the curriculum, students were more motivated and actively involved in their learning. This essay highlights the potential advantages of project-based instruction in introductory engineering courses. In specifically, in the area of requirement engineering, Daun et al., (2016) discussed their experiences with project-based learning in academic settings. Their research provided understandings into how real-world business examples might be incorporated into the classroom setting to improve student comprehension and implementation of course content. This experience report provides insightful advice on how to effectively apply project-based learning in engineering education. In an academic engineering course, Frank et al., (2003) investigation focused on the application of the project-based learning strategy. Their in-depth analysis covered not only the theoretical underpinnings but also the practical elements of incorporating project-based learning into the curriculum. In order to adapt project-based learning for engineering instruction, a core grasp of the concept is provided in this work. Uziak (2016) looked into the use of a project-based learning strategy inside an engineering programme. The study examined the beneficial effects of this educational strategy, highlighting improved learning opportunities and higher levels of student involvement. The paper discusses the potential advantages of project-based learning for both educators and students studying engineering. In a technical institution in India, Patil et al., (2022) looked into the learning preferences of first-year undergraduate engineering majors. Although not specifically about project-based learning, the information in this paper about the various learning styles of engineering students might help teachers when creating and implementing project-based learning activities. Multimodal machine learning was used by Joshi et al.,(2022) to predict student's performance. Although this research is not specifically relevant to project-based learning, it may provide useful insights into evaluating the outcomes and efficacy of project-based learning strategies. The essential elements of project-based learning in K–12 scientific education was examined by Markula & Aksela (2022) The principles and traits presented in this work may still be applicable to understanding how project-based learning can be implemented successfully in engineering education, despite the fact that the educational level is different. It offers a more comprehensive viewpoint on teaching methods. Project-based learning facilitates an integrated experience, as mentioned by Shet et al., (2015) This essay focuses on the all-encompassing advantages of this teaching strategy, such as how information and abilities from other fields are integrated. It helps us comprehend how project-based learning in engineering education can change the way students learn. The idea of developing an integrated learning experience within curriculum threads through mini-projects was first proposed by Mudenagudi et al., (2015) This strategy entails including smaller-scale projects in the curriculum, which is beneficial for engineering programmes in particular. The article presents suggestions for how project-based learning can be modified to fit various course formats. Mini-projects were described as a transformative teaching and learning process by Jadhav & Patil (2015) This study investigates an alternative viewpoint on incorporating real-world projects into engineering education, which, while comparable to project-based learning but may provide instructors with a variety of student engagement tactics. A study on troubleshooting within an online circuit modeling platform was carried out by Humbi et al., (2022) as a component of a blended project-based learning course. The troubleshooting procedure used by first-year undergraduate students is the specific topic of the paper. This study sheds light on the difficulties and teaching opportunities related to practical engineering activities. A freshman engineering course's design and evolution was discussed by Baligar et al., (2019) as part of their shared collaborative experience. Although not directly relevant to project-based learning, this study provides insightful information about curriculum development and cutting-edge instructional strategies. Teachers can incorporate project-based learning components by learning about the design and development of engineering courses.

According to Prince, M. J., & Felder, R. M. (2006) Science and engineering courses are traditionally taught by starting with lectures on fundamental concepts and having students apply what they have learned. This research, however, examines other methods that start with observations, questions, issues, or case studies, such as inquiry learning, problem-based learning, and discovery learning. With varied degrees of instructor assistance, these approaches enable students to gain knowledge and skills in context, encouraging more in-depth study and critical thinking. There is compelling evidence that supports these inductive approaches above conventional deductive teaching, despite the fact that support for them differs. Empirical studies and brain research support the use of inductive methods in

 https://doi.org/10.52202/073963-0023

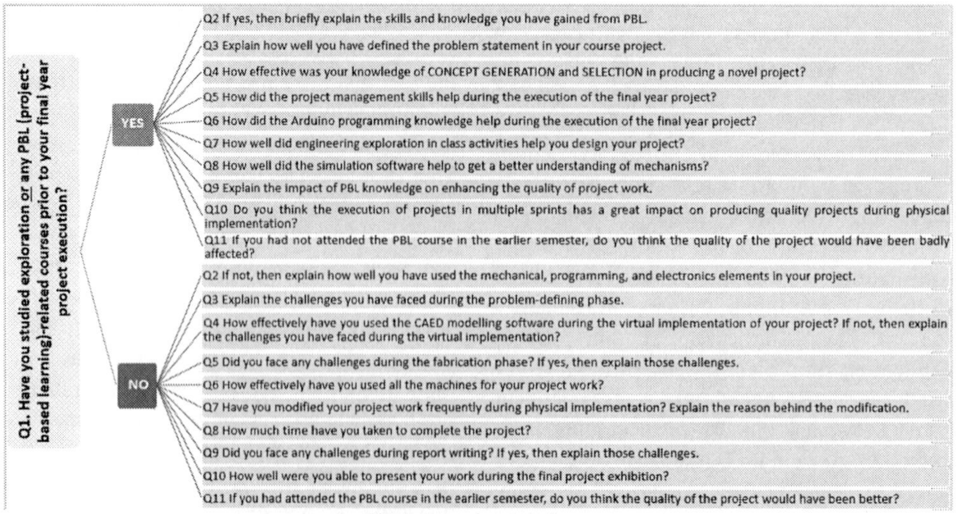

Fig. 1. Interview questions

education, which are consistent with ideas such as constructivism. They support students in acquiring critical thinking and self-directed learning abilities and promote meaningful learning as opposed to mindless memorizing. But using an inductive technique alone does not ensure improved learning outcomes. Student resistance and discontent might result from poor implementation. Students may find it difficult to take on more responsibility for their education, which could result in subpar performance and unfavorable reviews. In order to use inductive teaching effectively, teachers need to get familiar with best practices, such as giving students enough support at first and reducing it gradually as they gain confidence. Teachers need to prepare for and use tactics to deal with opposition from students.

When taken as a whole, these publications present a substantial and varied body of research on project-based learning in engineering education. They offer important insights into its application, effects, and possible advantages for both students and educators. Best practices in engineering pedagogy can be informed by these findings and additional study can be guided by them.

To address all the above discussed issues, we conducted a semi structured qualitative analysis by conducting in person student's interview at our college campus. To collect the student's feedback, we have created two research questions, which are as follows.

R1: What challenges did students encounter when working on projects without having taken a PBL course?

R2: How did PBL training in the early stages of engineering assist students to deal with challenges while working on their upcoming projects?

The first research question will aid in obtaining the perspectives of students who have not previously taken any PBL courses before the execution of final-year engineering projects. Second question will help to get the feedback of students those who have completed engineering exploration or any other PBL course prior to final-year project execution.

About the engineering exploration course that was made available to first-year students at KLE Technical University Hubli in 2015. Students are learning project management and mechatronics in this course.

III. Methodology

The significance of how mechatronic and project management information obtained during the first-year exploration course will help in generating a quality engineering project in the final-year is investigated using a semi structured qualitative analysis method. The study has been conducted at KLE Technological university for the graduated students who have finished their degrees in engineering during the last four years. The sampling and data collection process is as follows.

A. Sampling Process

We have selected two types of participants for our study. Population A consists of graduated engineering students who have learned engineering exploration course, which is introduced during their first-year academics and they have gained the mechatronic and project management skills.

Population B consists of students who were taught using traditional methods without having any prior experience to engineering exploration or any PBL course.

Population A has the sample size of 5 students and population B has the sample size of 6 students.

B. Data Collection Process

We have collected the data from the graduated engineering students by informing through email. The data collection mode was through in-person one to one interview at our KLE Technological university campus. The following techniques were used to gather data for this study:

1) Informed consent

Prior to participation, informed consent paperwork outlining the study's goals, data collection techniques, and participant's rights will be given to all chosen students. They will have the choice to accept or reject the offer to participate.

2) Voice recording process

During the interview process we have asked the questions and recorded the students feedback using live transcriber application software. This software will help to convert voice to word.

3) Privacy and considerations

Data was coded anonymously and participant identities were kept private. Access to personally identifiable data was only be available to the research team.

4) Interview questions

For two Research questions as mentioned in the methodology section, we framed 11interview questions each to both the Population A and B, so that students will be able to give detailed insight of their experience while conducting the final-year projects w.r.t Problem definition, concept generation, concept selection, Virtual implementation, physical implementation, report writing. These questions are framed in such way that students will give their genuine feedback without any manipulations. block diagram of all 11 interview questions is depicted in Fig. 1.

C. Data Analysis

We have used MAXQDA software to analyze the data collected from the students. The detailed analysis of students without early PBL education are done and presented in TABLE I by mentioning questions, themes and quotations. We have coded the student response using the pseudonyms a1, b1...... v1.The detailed analysis of students with early PBL education are done and presented in TABLE II by mentioning questions, themes and quotations. We have coded the student response using the pseudonyms a2, b2............ v2.

IV. Results and Discussion

The findings of our qualitative study on the influence of early Problem-Based Learning (PBL) knowledge on the quality of final-year projects are presented in this section. The analysis included looking at the student's feedback. Our primary goal was to identify patterns, themes, and trends regarding how early PBL exposure affected project outcomes. The detailed results of this qualitative study are as follows.

A. Improved Problem Definition and Identification

According to our data, students who were exposed to PBL early on showed a more advanced comprehension of problem description and identification in their final-year projects. They were better at identifying the fundamental problems and situating their work within a wider context. They were able to develop clear project objectives and research questions thanks to their early understanding, which ultimately resulted in more focused and pertinent projects.

B. Enhanced Analytical and Critical Thinking

Students that participated in PBL in their early years of instruction demonstrated improved analytical and critical thinking abilities. They were better able to evaluate existing material critically and synthesize it, which allowed them to create initiatives with more thorough theoretical underpinnings. This resulted in more rigorous investigation and analysis, which had a favourable effect on the general caliber of their capstone projects.

C. Increased Communication and Collaboration

PBL encourages a collaborative learning environment, and students who engaged in PBL activities from the start of their academic careers improved their ability to communicate and work in teams. During the final-year project phase, this was demonstrated by their capacity to work productively with their classmates, faculty advisors, and external stakeholders. The successful completion of interdisciplinary initiatives and the integration of many views that enriched the project outputs were made possible by improved teamwork.

D. Increased Creativity and Innovation

PBL early on encouraged student's inventiveness and creativity. In their final-year projects, they were more likely to investigate unusual ways and solutions, leading to original and cutting-edge contributions to their respective fields. This creative problem-solving was especially clear in projects that dealt with complicated, real-world issues, demonstrating the beneficial effects of early PBL knowledge on the cultivation of innovative problem-solving skills.

E. Increased Project Management Capabilities

Early PBL experience students showed exceptional project management abilities. Their project planning and execution, deadline adherence, and resource management were all more well-organized. The timely completion of excellent final-year projects was directly attributed to the enhanced project management.

In summary, our qualitative analysis offers strong evidence in favour of the beneficial influence of early PBL knowledge on the caliber of senior project outcomes. The results indicate that early exposure to PBL improves problem definition and identification of problems, analytical and critical thinking skills, teamwork and communication skills, creativity, and project management abilities, all of which contribute to the overall excellence of final-year projects. These findings highlight the value of using PBL at an early age in order to support student's academic and professional development.

https://doi.org/10.52202/073963-0023

TABLE I
SUMMARY OF STUDENTS ANALYSIS WITHOUT EARLY PBL KNOWLEDGE

Questions	Themes	Quotations
Q1: Have you studied exploration or any PBL (project-based learning)-related courses prior to your final-year project execution?	Lack of PBL Training Authors	"I had to rely solely on my mechanical, programming, and electronics knowledge because I had no prior PBL training."a1 "Without PBL, I had to figure out how to integrate mechanical, programming, and electronics elements on my own."b1
Q2: If not, then explain how well you have used the mechanical, programming, and electronics elements in your project. Keywords	Utilization of Knowledge	"I had to effectively utilize my existing knowledge to make up for the lack of PBL training."c1 "I had to apply what I learned in my coursework to make sure that my project included mechanical, programming, and electronics elements."d1
Q3: Explain the challenges you have faced during the problem-defining phase.	Problem Formulation Challenges 3	"Defining the problem was a major challenge, as I lacked structured guidance that PBL could have provided."e1 "Without PBL, it was difficult to gather client requirements and convert them into a clear problem statement."f1
Q4: How effectively have you used the CAED modelling software during the virtual implementation of your project? If not, then explain the challenges you have faced during the virtual implementation?	Challenges in Virtual Implementation	"Without prior PBL training, I had trouble using CAED modelling software for virtual implementation. "g1 "Virtual implementation was difficult without PBL because I had to use difficult software tools alone."h1
Q5: Did you face any challenges during the fabrication phase? If yes, then explain those challenges.	Challenges in Fabrication Phase	"Fabrication presented a number of challenges, particularly with regard to sourcing materials and coordinating the assembly process."i1 "I had trouble fabricating the project components because there was no PBL-based guidance."j1
Q6: How effectively have you used all the machines for your project work?	Challenges in Machine Utilization	"Without PBL, I found it difficult to use machines for my project work because I had little experience with them."k1 "Machine utilization was difficult because I had to pick up new skills on the job and get used to using different equipment."l1
Q7: Have you modified your project work frequently during physical implementation? Explain the reason behind the modification.	Frequent Project Modifications	"I found myself frequently changing the project because, at first, I didn't have clarity on certain aspects due to the absence of PBL guidance."m1 "Without PBL, I encountered unforeseen issues during physical implementation and had to make several modifications. n1
Q8 How much time have you taken to complete the project?	Project Completion Time	"The project took longer than expected to complete because I had to spend extra time troubleshooting and addressing challenges."o1 "It took me longer to complete the project without PBL because of the learning curve and overcoming challenges independently."p1
Q9 Did you face any challenges during report writing? If yes, then explain those challenges.	Challenges in Report Writing	"Without PBL, I had difficulty structuring and effectively presenting my project findings in the report."q1 "Without PBL, I had difficulty structuring and effectively presenting my project findings in the report."r1
Q10 How well were you able to present your work during the final project exhibition?	Project Presentation	"Without PBL experience, I found it a little difficult to articulate and present my work effectively during the final project exhibition."s1 "I think I could have improved my presentation skills if I had PBL-based training earlier."t1
Q11 If you had attended the PBL course in the earlier semester, do you think the quality of the project would have been better?	Hypothetical Impact of PBL Training	"I believe that if I had taken PBL courses earlier, the quality of my project would have been significantly better."u1 "PBL training would have given me useful skills and direction that could have improved the overall quality of my project"v1

https://doi.org/10.52202/073963-0023

TABLE II
SUMMARY OF STUDENTS ANALYSIS WITH EARLY PBL KNOWLEDGE

Questions	Themes	Quotations
Q1: Have you studied exploration or any PBL (project-based learning)-related courses before executing your final-year project?	PBL Education	"Yes, I had the chance to take courses related to PBL, which greatly enhanced my project experience."a2 "I benefited from prior PBL training, which was essential in the development of my final-year project."b2
Q2: If yes, then briefly explain the skills and knowledge you have gained from PBL.	Skills and Knowledge Acquired	"I developed my problem-solving, teamwork, and project management skills through PBL."c2 "PBL taught me how to programme Arduino, design mechanisms, and use simulation tools, all of which were extremely helpful for my project."d2
Q3: Explain how well you have defined the problem statement in your course project.	Effective Problem Statement Definition	"PBL training gave me the abilities to communicate with clients, collect requirements, and formulate a precise problem statement."e2 "By utilizing PBL principles of client interaction and need analysis, I was able to define the problem statement effectively."f2
Q4: How effective was your knowledge of CONCEPT GENERATION and SELECTION in producing a novel project?	Facilitating Concept Generation and Selection	"PBL emphasized methods such as concept scoring and brainstorming, which significantly aided in generating and choosing innovative project concepts,"g2 "The concept generation and selection methods used by PBL helped to generate creative and workable project ideas."h2
Q5: How did the project management skills help during the execution of the final-year project?	Project Management Skills	"PBL gave me the project management skills I needed to successfully manage my final-year project's resources, time, and team dynamics."i2 "I credit the project management abilities I developed through PBL training for the efficient execution of my project."j2
Q6: How did the Arduino programming knowledge help during the execution of the final-year project?	Benefits of Arduino Programming Knowledge	"The incorporation of electronic components in my final-year project was greatly facilitated by my proficiency in Arduino programming, gained through PBL."k2 "My project's functionality was improved by the ease with which I was able to incorporate sensors and actuators thanks to PBL's focus on Arduino programming."l2
Q7: How well did engineering exploration in class activities help you design your project?	In class activities	"Engineering exploration activities in PBL courses gave me real-world experience and improved my ability to design complex projects," m2 "Engineering Exploration's hands-on activities helped me better understand mechanisms and apply that knowledge to my final-year project,"n2
Q8: How well did the simulation software help to get a better understanding of mechanisms?	Benefits of Simulation Software	"The introduction of simulation software in PBL training gave me a clear understanding of mechanisms and allowed me to fine-tune the design of my project."o2 "Using simulation software was a key step in my project's development process, ensuring better clarity and precision in mechanisms."p2
Q9: Explain the impact of PBL knowledge on enhancing the quality of project work.	Enhancing Project Quality	"PBL knowledge elevated the quality of my project by allowing me to deliver a more refined solution, make informed decisions, and choose the right components."q2 "The thoroughness of my project execution and the final output show the impact of PBL knowledge on project quality."r2
Q10: Do you think the execution of projects in multiple sprints has a great impact on producing quality projects during physical implementation?	Benefits of Multiple Sprints	"Using PBL lessons learned, the project was implemented in multiple sprints, allowing for better control, problem-solving, and overall higher project quality during physical implementation."s2 "PBL's approach of carrying out projects in multiple sprints played a crucial role in ensuring that the project components integrated seamlessly."t2
Q11: If you had not attended the PBL course in the earlier semester, do you think the quality of the project would have been badly affected?	Hypothetical Effects of PBL Training Absence	"I feel that my project's quality would have suffered without PBL training, especially in terms of problem definition, concept selection, and project management."u2 "The lack of PBL training would have had a negative impact on the overall quality and execution efficiency of my project."v2

https://doi.org/10.52202/073963-0023

V. Conclusion

By the end of study, we came to conclude that the skillsets developed by the students who have attended first year engineering exploration course, help them in creating a quality project compared to that of students who did not learn the course. hence, we can say that developing mechatronic skills at the early stage of engineering exploration program will help them on understanding and producing high-quality project work.

Hence this study has proven that PBL has more impact to enhance the student's abilities in areas like reading technical publications, creating, developing, and simulating the mechanisms, organizing their work, effective communication and presentation, self-assurance, and teamwork.

Future Implications

To enhance the qualitative analysis, we propose expanding the participant pool to encompass a more diverse range of students from various institutions and backgrounds. An in-depth look at skill development would be possible with a longitudinal study that follows students from their initial exposure to PBL to their final-year projects. The effects can be directly compared and isolated by including a control group that hasn't had any early PBL exposure. Furthermore, adding quantitative data collected through surveys and standardized tests will support the qualitative findings obtained. Focusing on experiences and viewpoints, in-depth interviews and reviews by external experts should be more systematic. The analysis will gain depth and richness by taking socioeconomic and cultural issues into account and by creating a feedback loop with students.

References

Carpenter, M. S., Yakymyshyn, C., Micher, L. E., & Locke, A. (2016, June). Improved student engagement through project-based learning in freshman engineering design. In 2016 ASEE Annual Conference & Exposition.

Daun, M., Salmon, A., Weyer, T., Pohl, K., & Tenbergen, B. (2016, April). Project-based learning with examples from industry in university courses: an experience report from an undergraduate requirement engineering course. In 2016 IEEE 29th International Conference on Software Engineering Education and Training (CSEET) (pp. 184-193). IEEE.

Frank, M., Lavy, I., & Elata, D. (2003). Implementing the project-based learning approach in an academic engineering course. International Journal of Technology and Design Education, 13, 273-288.

Uziak, J. (2016). A project-based learning approach in an engineering curriculum. Global Journal of Engineering Education, 18(2), 119-123.

Patil, P. B., Vijayalakshmi, M., Basawaraj, & Humbi, N. (2022, November). Learning Styles of First-Year Undergraduate Engineering Majors at an Indian Technological University. In 2022 IEEE IFEES World Engineering Education Forum-Global Engineering Deans Council (WEEF-GEDC) (pp. 1-5). IEEE.

Y. Joshi, K. Mallibhat and V. M. (2022). Students Performance Prediction Using Multimodal Machine Learning, 1-5.

Markula, A., & Aksela, M. (2022). The key characteristics of project-based learning: how teachers implement projects in K-12 science education. Disciplinary and Interdisciplinary Science Education Research, 4(1), 1-17.

Shet, R. M., Iyer, N. C., Nissimgoudar, P. C., & Ajit, S. (2015). Integrated experience: Through project-based learning. In Proceedings of the International Conference on Transformations in Engineering Education: ICTIEE 2014 (pp. 479-486). Springer India.

Mudenagudi, U., Shettar, A., Meena, M., Nalini, C., Sujata, K., Desai, B. L., & Banakar, R. M. (2015). Creating an Integrated Learning Experience Within Curriculum Threads Through Mini-projects. In Proceedings of the International Conference on Transformations in Engineering Education: ICTIEE 2014 (pp. 359-367). Springer India.

Jadhav, H. S., & Patil, S. N. (2015). Mini projects: a new concept of transformation of teaching-learning process. In Proceedings of the International Conference on Transformations in Engineering Education: ICTIEE 2014 (pp. 89-94). Springer India.

Humbi, N., Patil, P. B., & Vijayalakshmi, M. (2022). Troubleshooting in an online circuit simulation platform: a study on the troubleshooting process by first-year undergraduate students in a blended project-based learning course. In EDULEARN22 Proceedings (pp. 2295-2300). IATED.

Baligar, P., Kavale, S.M., M., K., Joshi, G., & Shettar, A.S. (2018). Engineering Exploration : A Collaborative Experience of Designing and Evolving a Freshman Course. 2018 World Engineering Education Forum - Global Engineering Deans Council (WEEF-GEDC), 1-5.

Prince, M. J., & Felder, R. M. (2006). Inductive teaching and learning methods: Definitions, comparisons, and research bases. Journal of Engineering Education, 95(2), 123–138.

Zancul, E. D. S., Sousa-Zomer, T. T., & Cauchick-Miguel, P. A. (2017). Project-based learning approach: improvements of an undergraduate course in new product development. Production, 27.

https://doi.org/10.52202/073963-0023

Evaluating the effectiveness of MBSR on Engineering Students using PSS & STAI

Sneha Varur; Vishwanath P Baligar
K L E Technological University, Vidyanagar, Hubballi – 580031
sneha.varur@kletech.ac.in, vpbaligar@kletech.ac.in

Abstract

Context

Engineering students embark on a demanding journey in higher education, facing challenges that can potentially impact their mental health and overall well-being. Often, these students are less inclined to seek help when they experience stress, which may be triggered by factors such as academic pressure, a heavy workload, and time constraints. This can pose significant risks to their mental health.

Purpose

The primary objective of this study was to assess the stress levels experienced by engineering students, implement a stress-reduction intervention, and subsequently evaluate its effectiveness in reducing stress.

Methods

A quasi-experimental design with pretest and post-test assessments was used in the study. Stress levels were measured using established scales like Perceived Stress Scale (PSS) and State-Trait Anxiety Inventory(STAI) on a sample of 50 engineering students (n=50). A Mindfulness- based stress reduction (MBSR) program was then introduced, and data was collected before and after the implementation of a stress reduction intervention.

Outcomes

The results indicated that there was a significant decrease in the overall mean of PSS and STAI score after participating in the MBSR program.

Conclusion

We conclude that the use of MBSR is effective in reducing stress among engineering students

Keywords— Stress, Engineering students, MBSR, PSS, STAI, Paired t-test

I. INTRODUCTION

In today's fast-paced world, the pursuit of innovation and financial success has become a relentless race. The allure of technological advancements and wealth generation often overshadows the importance of leading a stress-free and contented life. This phenomenon is particularly evident among students, who, under the pressure of societal expectations and academic demands, grapple with stressor that compromise their well being (Ban et al., 2022).

The contemporary era is marked by an unprecedented emphasis on innovation and money-making. Technological marvels are consistently reshaping our lives, promising convenience and advancement. While this is undoubtedly remarkable, the relentless pursuit of innovation can lead to a state of perpetual discontent. In our relentless quest for the 'next big thing, we often overlook the simple joys that life has to offer. Happiness is often elusive in this whirlwind of innovation and financial gain.

The pressure that students face today is emblematic of these societal phenomena. The modern education system places immense demands on students. Student stress can be due to heavy academic workload, complex subjects, time management challenges, competition, isolation, performance anxiety, lack of resources, career concerns, health issues, financial strains and difficulties in group projects.

Throughout its history, engineering has maintained its reputation as a challenging and rigorous academic discipline. The competitive nature of the curriculum often leads to high levels of stress among students. Recognizing that a certain level of stress is an inherent aspect of college life, it can actually contribute positively to a student's academic and personal accomplishments when managed appropriately (Oyewobi et al., 2020). Nevertheless, elevated and persistent stress levels can have adverse effects on students' mental well-being, potentially resulting in conditions such as depression, anxiety, and various forms of psychological discomfort (Negi & Khanna, 2019).Furthermore, stress serves as an early indicator of potential mental health issues among undergraduate students (Acharya et al., 2018).

Approximately 10.7% of the global population experiences daily mental health challenges, including 3.4% who contend with anxiety disorders (Auerbach et al., 2018). This phenomenon is even more pronounced among college students, where as of 2018, an estimated 35.3% of students worldwide grappled with mental health disorders, with anxiety disorders affecting 23.6% of this population (Dalvi et al., 2023).

https://doi.org/10.52202/073963-0024

Under stress, students may adopt unhealthy coping mechanism, including procrastination, isolation, cheating, self harm and negative self talk. Ignoring physical health, over committing and avoiding support systems or professional help are also common maladaptive responses. There are several practices that create a holistic approach to stress reduction and MBSR is one among them.

Our paper focuses on capturing the stress levels for a set of students and apprehending certain techniques to overcome stress.

II. RELATED WORK

There are numerous scientifically tested programs that have demonstrated their effectiveness in alleviating stress among students. Some of them are Mindfulness-Based Cognitive Therapy (MBCT), Mindfulness-Based Cognitive Therapy for Children (MBCT-C), Mindfulness-Based Relapse Prevention (MBRP), Dialectical Behavior Therapy (DBT), Acceptance and Commitment Therapy (ACT), Mindful Self-Compassion (MSC), Cultivating Emotional Balance (CEB), Search Inside Yourself (SIY), iRest (Integrative Restoration), Breath works Mindfulness for Stress and Chronic Pain etc. In the midst of various options, we have selected MBSR due to its simplicity, making it easy for students to comprehend and practice.

Lee (2012) provided an in-depth overview of the Perceived Stress Scale (PSS) and its application. The PSS comprises 10 inquiries probing feelings and thoughts over the past month, gauging the level of unpredictability and overwhelm in an individual's life. Respondents rate their responses to stressful situations on a scale of 0 to 4, where 0 signifies 'Never' and 4 signifies 'Very often.' The cumulative scores yield an overall assessment of perceived stress, with individual scores ranging from 0 to 40. Higher scores correspond to elevated levels of perceived stress. Table I shows for the score interpretation.

Spielberger and colleagues (1983) were the inventors of STAI. STAI was designed to offer reliable and concise self-report assessments for calculating two types: state anxiety and trait anxiety also called as A-State and A-Trait sub-scales respectively. Table II shows the scale of scores and Table III shows the differences between the two types ofSTAI.

Yusufov and colleagues (2019) analyzed 43 global studies on stress reduction interventions for college and graduate students. Found that various approaches effectively reduce anxiety and perceived stress. The cognitive-behavioral program, social support, the coping skills work well for stress, while relaxation training and mindfulness help with anxiety. Short and long-term interventionsshow positive effects.

TABLE I
RANGE OF SCORES FOR PSS

Stress score	Levels
Range(0,13)	Meaning stress is low
Range(14,26)	Meaning stress is moderate
Range(27,40)	Meaning stress is high

TABLE II
RANGE OF SCORES FOR STAI

Stress score	Levels
Range(20,37)	Low anxiety
Range(38,44)	Moderate anxiety
Range(45,80)	High anxiety

TABLE III
SUMMARY DIFFERENCE BETWEEN A-STATE & A-TRAIT

A-State	A-Trait
Expresses current emotion state.	Expresses specific anxiety symptoms over a period of time
Aims to assess the intensity of feelings such as tension, nervousness, worry, and apprehension at a specific moment in time	This scale gauges individual variations in anxiety proneness, which means a person's general inclination to worry, particularly in situations that pose a potential threat to their self-esteem.
It tends to rise in situations like physical or psychological stress. Situations like electric shock, watching distressing films, delivering a speech, or receiving negative feedback on performance.	It will be stable over the time and less affected by physical or psychological stress.
Scores are elevated prior to surgical procedures and gradually decline during the post-surgical recovery period.	Tends to rise when faced with situations involving threats to their ego or self-esteem

Peterson and colleagues (1992) examined the Efficacy of a Meditation- Based Stress Reduction Program for Treating Anxiety Disorders. This research aimed to assess the impact of a mindfulness meditation group program on individuals with anxiety disorders. Out of 22 participants, all met the criteria for generalized anxiety disorder after undergoing a structured clinical interview.

Furthermore Hsieh and colleagues (2012) research indicate that this group significantly alleviate indicators of panic, stress and sustain these improvements for individuals with anxiety and other disorders. Hsu and colleagues (2014) proved similar theory in their work.

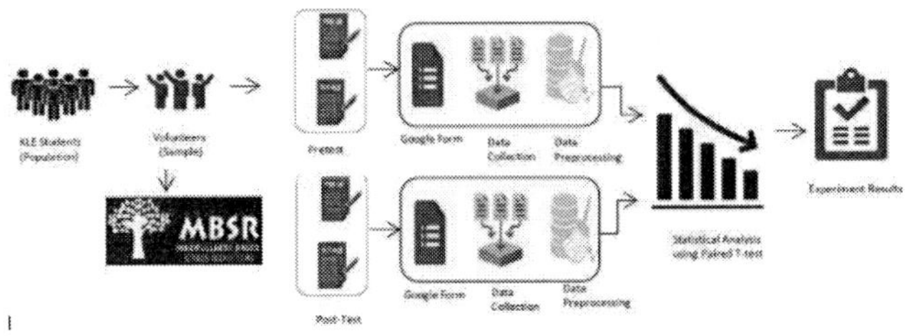

Fig. 1. System Architecture Diagram.

III. PROPOSED METHODOLOGY

A. *Ethical Considerations*

As per KLE University mandated informed consent form was shared with each student, which comprehensively outlined the details of this experimental study. It was made explicitly clear that their data would be gathered in an anonymous and confidential manner, solely for the purpose of the study. We conducted assessments exclusively on relevant components, and students had the freedom to discontinue their participation at any time of their choosing.

B. *Random Sampling*

The population under consideration was all students of KLE University, while the randomly volunteered students were selected participants who constituted the sample. A simple random sampling with a fixed sample size of 50 (Acharya et al., 2013).

C. *Paired T-test*

The choice of hypothesis testing method hinges on both the nature of the data and the specific research inquiry at hand.

The paired t-test and unpaired t-test serve as distinct statistical tools for comparing means across two groups. This is employed when two measurements are taken on the same individuals, typically before and after an intervention while unpaired t-test is used for comparing the means of two separate groups (Hsu & Lachenbruch, 2014).

Given the characteristics of our dataset, the paired t-test aligns more closely with our research objectives, and thus, we have chosen to it. The proposed hypotheses for the paired t-test are framed as one-tailed tests, specifically testing for a decrease in scores from pretest to post-test as shown below:

Null Hypothesis (H0): The population mean of the post-test scores (μpost) is greater than or equal to the population mean of the pre-test scores (μpre) $H0$: μpost >= μpre Alternative Hypothesis (H1):The population mean of the post-test scores (μpost) is less than the population mean of the pre-test scores (μpre) $H1$:μpost < μpre

This setup is specifically interested in whether there is a significant decrease in scores from the pretest to the post-test. Keep in mind that this means we are not considering the possibility of an increase or no change in scores. It's a one-tailed that is test focused on a decrease.

D. *Research Question and Hypothesis*

Research Question: Does participating in a MBSR program lead to a significant reduction in stress levels among students?

We have devised two hypotheses to address our research inquiry

I Hypothesis: Participation in the MBSR program will lead to a statistically significant decrease in the mean PSS score among the group of 50 students, keeping the level of significance 0.05

Null Hypothesis (H0): "Participating in the MBSR program will result in either a greater than or equal to the mean PSS score among the group of 50 students (μ_after >= μ_before).
Alternative Hypothesis (H1): Participating in the MBSR program will result in a statistically significant decrease in the mean PSS score among the group of 50 students (μ_after < μ_before).

II Hypothesis: Participation in the MBSR program will lead to a statistically significant decrease in the mean STAI score among the group of 50 students, keeping the level of significance 0.05

Null Hypothesis (H0): "Participating in the MBSR program will result in either a greater than or equal to the mean STAI score among the group of 50 students (μ_after >= μ_before).
Alternative Hypothesis (H1): Participating in the MBSR program will result in a statistically significant decrease in the mean STAI score among the group of 50 students (μ_after < μ_before).

E. *Stress assessments*

Prior to initiating our research, we engaged in consultations with an esteemed psychologist who offered valuable insights regarding stress detection and different stress reduction programs. Among the various standard stress assessments available, the PSS & STAI were chosen based on their established effectiveness in accurately gauging stress levels and anxiety traits among the students.

F. Data Collection

The process began by distributing links to the PSS and STAI tests to the students. They were instructed to complete both assessments and provide a screenshot of their results. Subsequently, a Pretest survey form was generated and sent out, prompting students to update their scores along with the attached screen-shots as evidence. This information was documented in the initial survey. Following this, the students were engaged in an eight-week MBSR program. Subsequently, students were asked to reattempt the PSS and STAI tests and submit their results through a Post-test survey form for recording.
Pretest link - https://forms.gle/AnZSsXWTYHpVhETc6
Post-test link - https://forms.gle/ipJVXntcvCURshvDA
Fig. 1. Shows the flow chart of our work.

G. Data Preprocessing

The data from the Pretest and Post-test forms was downloaded as a csv (comma separated values) file format that serves as a dataset to perform analysis. Following the data collection, we carried out several data preprocessing steps (Garcia et al., 2015).

1. ***Data Cleaning:*** Examined the dataset for any missing values. In the instances where students have not provided values for certain attributes, we have employed appropriate imputation methods. For numerical data, the mean or median was used, while for categorical data, the mode method has been used to ensure data completeness.
2. ***Data Integration:*** Different naming conventions emerged as issues. To ensure uniformity throughout the dataset, we standardized naming conventions. Additionally, to maintain data integrity, duplicate records were identified and removed from the dataset.

H. Exploratory Data Analysis (EDA)

In addition to the standard tests, we also asked students to answer some simple questions in the survey forms to gain some meaning full insights by performing EDA (Tukey & J. W, 1977) (Hartwig et al., 1979)

Question 1: In which of these specific events do you experience stress?

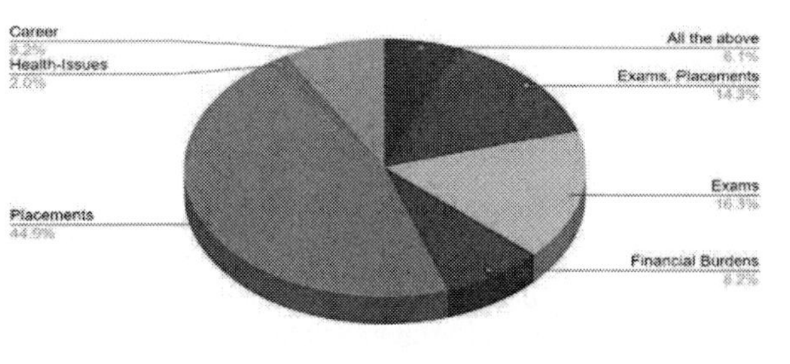

Fig. 2. Percentage of stressful events

Question 2: Did the test provide a valuable perspective?

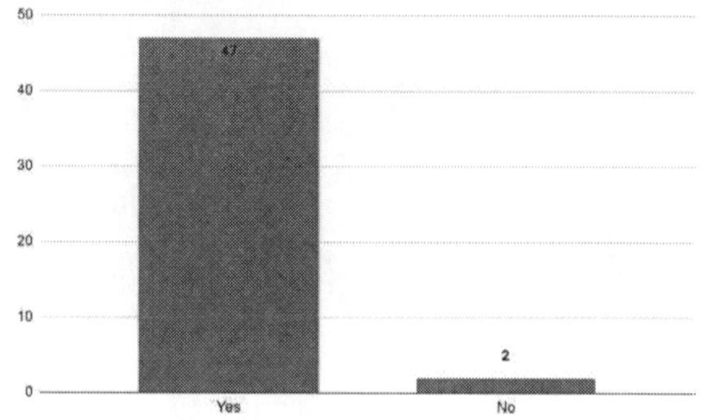

Fig. 3. Perspective about the tests

IV. RESULTS AND DISCUSSION

We first work on hypothesis I that is based on PSS score and then hypothesis II that is based on STAI score. The STAI scores are of two types A-State and A-Trait, so we perform paired t-test twice on STAI score. Table shows the resulting descriptive statistics score. Table IV shows the resulting values.

TABLE IV
DESCRIPTIVE STATISTICS USING PASS AND STAI

	PSS		A-State STAI		A-Trait STAI	
	Pretest	Post-test	Pretest	Post-test	Pretest	Post-test
Mean	19.56	16	45.52	43.52	44.68	42.22
Standard deviation	10.154	9.36	11.168	6.347	8.87	7.985
Variance	103.10	87.7	124.72	40.295	78.83	63.76
Sample Size	50	50	50	50	50	50

For PSS: The paired t-test was conducted on a PSS scores consisting of pretest and post-test scores from 50 individuals. With the values shown in table mean, standard deviation and variance for both before and after MBSR revealed a significant decrease in scores, with a mean difference of -3.56 and standard deviation of the differences of approximately 4.669. The calculated t-statistic was -5.394 with 49 degrees of freedom. At a significance level of 0.05, the critical t-value for a one-tailed test was approximately -1.676. As the calculated t-statistic was much farthest than

https://doi.org/10.52202/073963-0024

the critical t-value, we reject the null hypothesis. This provides strong evidence to suggest a significant decrease in scores from the pretest to the posttest. These findings indicate that the intervention or treatment implemented likely had a meaningful impact on the measured outcome.

For A-State anxiety :
The paired t-test was conducted on STAI A-state scores with all details shown in TABLE 4. The analysis unveiled a significant decrease in scores, with a mean difference of -2 and a standard deviation of the differences of approximately 6.093. The calculated t-statistic was -2.32 with 49 degrees of freedom. At a significance level of 0.05, the critical t-value for a one-tailed test with 49 degrees of freedom was approximately -1.676. Since the calculated t-statistic was more farthest than the critical t-value, we reject the null hypothesis. This implies strong evidence supporting a significant decrease in scores from the pre to post. The outcome indicate that the intervention or treatment likely had a meaningful impact on the measured outcome.

For A-Trait anxiety :
The paired t-test was conducted on STAI A-Trait containing pretest and posttest scores from 50 individuals. The pretest scores had a mean, standard deviation values shown in the table. The analysis revealed a significant decrease in scores, with a mean difference of -2.46 and a standard deviation of the differences of approximately 4.202. The calculated t-statistic was -4.143 with 49 degrees of freedom. At a significance level of 0.05, the critical t-value for a one-tailed test with 49 degrees of freedom was approximately -1.676. As the calculated t-statistic was substantially more farthest than the critical t-value, we reject the null hypothesis. This strongly suggests a significant decrease in scores from the pretest to post. The outcome indicate that the intervention or treatment likely had a meaningful impact on the measured outcome.

Therefore we conclude that in both hypothesis I and II the Null hypothesis is rejected and Alternate hypothesis is accepted, that means the post-test scores have reduced compared to pretest. This answers our research question that MBSR does help to reduce the stress for engineering students.

V. CONCLUSION AND FUTURE SCOPE

In summary, our findings indicate that engineering students commonly experience a notable degree of stress. Introducing initiatives like MBSR has shown promise in alleviating this stress. Nevertheless, it's important to acknowledge that programs like MBSR can come with a significant financial investment. As we look ahead, our goal is to pioneer a cost-effective stress reduction app tailored specifically for our students. This innovative solution aims to offer all the advantages of MBSR without imposing any financial burden.

Various other stress reduction programs, under medical supervision can be used to assess the impact of stress level.

REFERENCES

Ban, N., Shannon, H., Wright, C. J., Miller, M. E., Hargis, L. E., Usher, E. L., ... & Wilson, S. A. (2022, January). Identifying common perceived stressors and stress-relief strategies among undergraduate engineering students. In *ASEE annual conference exposition*.

Iraola-Real, I., Nolberto-Quispe, L., Soto, A. C. Z., & Del Rosario, H. (2020, December). Predictors of the Test Anxiety in Systems Engineering and Chemical Engineering Students. In *2020 IEEE Congreso Bienal de Argentina (ARGENCON)* (pp. 1-4). IEEE.

Oyewobi, L. O., Bolarin, G., Oladosu, N. T., & Jimoh, R. A. (2020). Influence of stress and coping strategies on undergraduate students' performance. *Journal of Applied Research in Higher Education*, *13*(4), 1043-1061.

Negi, A. S., Khanna, A., & Aggarwal, R. (2019). Psychological health, stressors and coping mechanism of engineering students. *International Journal of Adolescence and Youth*, *24*(4), 511-520.

Acharya, L., Jin, L., & Collins, W. (2018). College life is stressful today–Emerging stressors and depressive symptoms in college students. *Journalof American college health*, *66*(7), 655-664.

Auerbach, R. P., Mortier, P., Bruffaerts, R., Alonso, J., Benjet, C., Cuijpers, P., ... & Kessler, R. C. (2018). WHO world mental health surveys international college student project: prevalence and distribution of mental disorders. *Journal of abnormal psychology*, *127*(7), 623.

Dalvi, A., Kapur, P., Ashtekar, R., Gorugantu, A., & Siddavatam, I. (2023, March). Transforming College Counseling: An In-House Solution for Student Mental Wellness. In *2023 2nd International Conference for Innovation in Technology (INOCON)* (pp. 1-5). IEEE.

Lee, E. H. (2012). Review of the psychometric evidence of the perceived stress scale. *Asian nursing research*, *6*(4), 121-127.

Spielberger, C. D., Gorsuch, R. L., Lushene, R., Vagg, P. R., & Jacobs, G. A. (1983). Manual for the State-Trait Anxiety Inventory; Palo Alto, CA, Ed. *Palo Alto: Spielberger*.

Nigussie, S., Belachew, T., & Wolancho, W. (2014). Predictors of preoperative anxiety among surgical patients in Jimma University specialized teaching hospital, South Western Ethiopia. *BMC surgery*, *14*,1-10.

Yusufov, M., Nicoloro-SantaBarbara, J., Grey, N. E., Moyer, A., & Lobel, M. (2019). Meta-analytic evaluation of stress reduction interventions for undergraduate and graduate students. *International Journal of Stress Management*, *26*(2), 132.

Peterson, L. G., & Pbert, L. (1992). Effectiveness of a meditation-based stress reduction program in the treatment of anxiety disorders. *Am J Psychiatry*, *149*(7), 936-943.

Hsieh, P. H., Sullivan, J. R., Sass, D. A., & Guerra, N. S. (2012). Undergraduate engineering students' beliefs, coping strategies, and academic performance: An evaluation of theoretical models. *The Journalof Experimental Education*, *80*(2), 196-218.

https://doi.org/10.52202/073963-0024

Hsu, H., & Lachenbruch, P. A. (2014). Paired t test. *Wiley StatsRef: statistics reference online*.

Acharya, A. S., Prakash, A., Saxena, P., & Nigam, A. (2013). Sampling: Why and how of it. *Indian Journal of Medical Specialties*, *4*(2), 330-333.

García, S., Luengo, J., & Herrera, F. (2015). *Data preprocessing in data mining* (Vol. 72, pp. 59-139). Cham, Switzerland: Springer InternationalPublishing.

Tukey, J. W. (1977). *Exploratory data analysis* (Vol. 2, pp. 131-160).

Hartwig, F., & Dearing, B. E. (1979). *Exploratory data analysis* (No. 16). Sage.

https://doi.org/10.52202/073963-0024

Exploring shifts in engineering students' understanding of engineering work during their co-op experiences

Siddhant Sanjay Joshi[a]; Kirsten A. Davis[a]
School of Engineering Education, Purdue University[a]
Corresponding Author Email: joshi110@purdue.edu

Abstract

Context

Engineering students often struggle when they enter the workforce after graduation due a lack of workplace readiness. Cooperative (co-op) programs are experiential educational programs that provide workforce experience prior to graduation, improve employability, provide exposure to workplace culture, and help bridge the disconnect between engineering work and academia.

Purpose or Goal

Well-designed co-op programs enhance student learning by providing a balance of challenge and support. To effectively support students, it is helpful to understand how their perceptions and experiences of engineering work shift during the co-op. Therefore, we aim to understand: How does students' understanding of engineering work shift during their co-op rotations?

Methods

During Fall 2022, 639 engineering students from a large public U.S. university enrolled in the co-op program. 295 students from the co-op program participated in an online survey where we asked them to reflect on their progress toward their goals during their co-op. We analysed this reflection data using a two-step approach: open coding followed by thematic analysis.

Outcomes

We found that students' perceptions of engineering work changed as they gained industry-specific knowledge and skills, dealt with open-ended, ill-structured problems, and worked in teams. Our analyses also revealed that students improved their technical skills, adapted to workplace cultures, took ownership and initiative on projects, and acquired interpersonal skills to accomplish their tasks.

Conclusion

Our study identifies how understanding students' perception of engineering work can inform the design of experiential learning programs. We also provide insights on how co-op programs can ease students' transition to the workforce by considering skill development but also changes in understanding of what engineering work entails. Our results can help educators support students in experiential learning programs that bridge the gap between industry and academia.

Keywords*—co-op education, thematic analysis, student experience.*

I. INTRODUCTION

There is a growing demand for engineers capable of developing innovative solutions to address the grand challenges that lie ahead of us (Mote et al., 2016). To address these grand challenges, it is important for universities to develop competent engineers that are workplace ready and prepared to solve these problems (Ehlert & Orr, 2020). However, prior research shows that engineering students lack workplace readiness and often struggle after they enter workforce (Brunhaver et al., 2018; Huff et al., 2016). Additionally, positive academic achievement in engineering may not necessarily translate into required workplace readiness (Parsons et al., 2005). Therefore, it is important for engineering students to acquire workplace experience prior to graduation.

Co-op programs are experiential learning programs that help prepare engineers for workplace readiness and develop their required competencies (Ramirez et al., 2016). Additionally, these experiences also help students understand what engineering work is and provide opportunities to learn about engineering as a profession (Brunhaver et al., 2018). Therefore, we investigated how co-op experiences helped students shift in their understanding of what engineering work entails. Furthermore, this study will also help with intentional design of experiential learning programs so that students are better prepared for a career in industry upon graduation.

II. LITERATURE REVIEW

When engineering students graduate, they typically have a narrow understanding of the activities involved in engineering practice (Brunhaver et al., 2018). Most students perceive engineering to be technically oriented and based on concepts and equations they learned in classrooms (Anderson et al., 2010; Trevelyan, 2010). In reality, engineers also require professional competencies like communication skills, teamwork skills, creativity, lifelong learning, etc. (ABET Engineering Accreditation Commission, 2020). As a result, there is a disconnect between the expectations of engineering education and practice. This misalignment can cause engineering graduates to feel unprepared and face difficulties when transitioning from academia to industry (Huff et al., 2016; Korte et al., 2015; Trevelyan, 2014, 2019). To minimize this misalignment, it is important to study the workplace preparedness of engineering students prior to their graduation.

Prior work suggests that experiential learning strategies like project-based learning (Bielefeldt et al., 2010), and service-learning (Bielefeldt et al., 2013) help bridge this gap by preparing students for a career in engineering. For instance, Huff et al. (2016) found that enrollment in service-learning

https://doi.org/10.52202/073963-0025

programs helped graduates bridge the gap between industry and academia, acquire workplace experience, and develop professional skills required to work in industry after graduation.

Other experiential learning techniques like Cooperative education (co-op) and internships also help bridge the gap by providing hands-on industry experience (Gardner & Motschenbacher, 1997; Jin & Fabretto, 2019; Main et al., 2020). Co-op programs are defined as experiential learning opportunities that help integrate classroom education with planned and supervised work experience in an industry setting (Garavan & Murphy, 2001). Co-op experiences can improve students' chances of employability by providing them with technical and professional skills, industry experience, and exposure to the workplace culture (Ramirez et al., 2016). In addition, students' participation in co-op programs can positively impact their academic performance (Blair et al., 2004) and ABET learning outcomes (Parsons et al., 2005).

In summary, previous research suggests that co-op programs offer a variety of employability benefits and help engineering students develop skills that will prepare them for engineering practice. Moving beyond skill development, however, co-op experiences have the potential to shift students' perceptions of engineering work (Brunhaver et al., 2018). Less research has focused on this outcome, despite the fact that such perceptions may inform students' career choices and persistence in engineering after graduation (Gilmartin et al., 2018). Based on experiences like co-op, students may make an assessment of their preparedness for engineering workforce (Baytiyeh & Naja, 2012; Martin et al., 2005). Students may also develop a more realistic understanding of the expectations that employers have from working engineers. Understanding students' perceptions of engineering work can help educators better support students before, during and after their co-op experiences. Therefore, we addressed the following research question in our study: How do students' perceptions of engineering work shift during their co-op rotations?

III. Methods

To address our research question, we qualitatively analyzed students' open-ended survey responses on a survey completed at the start and end of their co-op programs. A qualitative research methodology was chosen for this study because we wanted to understand student perceptions and beliefs about engineering work without limiting their potential responses through the use of closed-ended survey questions.

A. Participants

In Fall 2022, 639 students from various undergraduate and graduate majors enrolled in an experiential learning course. Engineering students are required to take this course while completing a semester-long co-op rotation in an engineering company (See Table I for breakdown as per major for 295 students who completed at least one of the course surveys). 33.56% students in this study were completing their first co-op, whereas others had prior co-op experiences. In the co-op program at Purdue University, students gain practical experience before graduation by working with the same employer for 3-5 work sessions lasting 12-22 months. With every co-op term, students take on increasing responsible roles that determine their professional and personal goals for the rotation. Our project has been approved by the Purdue University IRB as an analysis of existing data because the survey responses were collected as part of the continual assessment for the experiential learning course.

B. Data Collection

An online survey was distributed as a class assignment at two points in time: the start of their co-op rotation and the end of their co-op rotation. At the start of the co-op rotation, we asked students to state their goals for the co-op experience, and at the end we asked them to reflect on their progress towards these goals. We asked students to reflect on their co-op rotation goals because reflection and goal-setting are best practices for supporting and assessing student learning during experiential learning (Chan, 2022). 295 Students responded to the following prompts in the online survey assignments:

TABLE I
Breakdown of Co-op students Who Completed At Least One Survey by major

Engineering Major	Undergraduate	Graduate	Percentage for major
Mechanical	83	22	35.59
Industrial	14	25	13.22
Biomedical	36	0	12.20
Aeronautical & Astronautical	28	1	9.83
Chemical	25	4	9.83
Electrical	17	0	5.76
Computer	13	0	4.41
Civil	8	0	2.71
Materials	8	0	2.71
Multidisciplinary	3	0	1.02
Biological	2	0	0.68
Environmental & Ecological	0	2	0.68
Engineering (undecided)	1	0	0.34
Manufacturing Technology	1	0	0.34
Mechanical Technology	1	0	0.34
Agricultural	0	1	0.34
Percentage	81.35	18.64	

1. Start of co-op survey: Set two professional growth or work-related goals for your next work experience/co-op rotation using the SMART (Specific-Measurable-Achievable-Relevant-Timebound) format. Include a metric which can be used to later measure the success of

achieving your goals.

2. End of the co-op survey: Reflect on your progress towards the goals you set at the beginning of your most recent experience/co-op rotation. Elaborate on any successes, challenges, new insights, or changes in goals. How have you grown professionally or personally to meet your goals? Reference your beginning of session survey responses sent to your email to recall your goals.

Prior to the start of co-op survey, students were given training on different ways they could set their goals for their co-op rotation. This training helped to ensure that students identified specific measurable goals that they could evaluate and reflect upon at the end of their co-op term. Students were encouraged to complete the training and online survey assignments to help them track their progress and get the most out of their co-op experiences. However, as it was a zero-credit course, students' responses were completely voluntary, and they did not receive any points for completing the training modules and survey assignments. Nevertheless, 295 students completed at least one of the surveys (pre or post) and 287 completed both out of 639 total co-op students. This represents a response rate of 44.9% for completing both surveys.

C. *Data Analysis*

To address our research question, we conducted a thematic analysis. The key advantage of using a thematic analysis is that it is driven by the research questions and helps identify major themes associated with the research questions from the data (Braun & Clarke, 2006). Because we wanted to focus on students' perceptions for this study, thematic analysis was a logical choice. It allowed the student voice to be centered in our analysis as the key themes were developed through detailed reading of the students' survey responses.

We followed the thematic analysis process described by Braun & Clarke, (2006). We began our analysis by familiarizing ourselves with the data and documenting memos and ideas commonly discussed in student reflections. In the next step, we generated an initial coding scheme by open-coding thirty randomly selected student reflections. We discussed this coding scheme to reach a consensus on the final codebook and the definition for each code. We then coded the remaining responses using the predefined coding scheme and identified the potential themes that emerged from our coding for each research question. We subsequently reviewed these potential themes to generate a final set of themes. We completed this analysis using Nvivo Release 1.

D. *Limitations*

The reflection assignment was not mandatory for students in the experiential learning course, which may have led to some of them not completing the survey. Additionally, as the survey consisted of multiple open ended and close ended responses (we analyzed only open-ended reflections), the time students spent on each question may be limited. As a result, some responses lacked detail and depth of reflection which may have limited the research team's understanding of the students' perceptions.

IV. RESULTS

The purpose of this study was to investigate how engineering students shift their perception of engineering work over the course of their co-op rotation. In this section, we present the results of our thematic analysis of 295 student reflections. We have summarized the themes and subthemes for our research question in Table II. The two main themes were *Nature of Engineering Work* and *Skills Needed to Succeed in the Workplace*. We present each theme in its own sub-section with descriptions of the sub-themes and representative quotes.

A. *Nature of Engineering Work*

The first major theme that emerged from our thematic analysis was the *Nature of Engineering Work*. This theme depicted students' perception of engineering work such as understanding the day-to-day work of engineers, the challenges they face, experiencing how engineering work is different from academic work, and developing insights into the requirements of engineering work. Our analysis revealed that while discussing this theme, students focused on seven major subthemes (organized from most common to least common): (1) *Acquire niche knowledge required by job role* (2) *Apply knowledge and skills gained from school/work to work projects.* (3) *Real-world problem solving* (4) *Hands-on work* (5) *Develop/work with engineering tools* (6) *Work in Teams* (7) *Developing documentation about processes and projects.*

TABLE II
THEMES AND SUBTHEMES PREVAILING FROM STUDENT REFLECTIONS

Theme	Subtheme*
Nature of Engineering Work	▪ Acquire niche knowledge required by job role. ▪ Apply knowledge and skills gained from school/work to work projects. ▪ Real-world problem solving ▪ Hands-on work ▪ Develop/work with engineering tools. ▪ Work in teams ▪ Developing documentation about processes and projects
Skills Needed to Succeed in the Workplace	▪ Upskilling/improving existing skills. ▪ Communication Skills. ▪ Take ownership or initiatives at work. ▪ Be a team player: Collaboration and Teamwork. ▪ Be organized: Scheduling and Time management. ▪ Be adaptable and flexible at workplace.

*Organized by most common to least common subtheme

1) *Acquire niche knowledge required by job role.*

The first subtheme, *acquire niche knowledge required by job role,* highlights that students felt a need to acquire industry-specific knowledge to complete their job responsibilities.

https://doi.org/10.52202/073963-0025

Students learned the fundamentals of various technical concepts during their education. However, upon entering the industry, they realized that they must acquire niche technical knowledge specific to their industry and position. Upon realizing this need, students took steps to gain knowledge that will help them complete their on-job tasks and responsibilities. One students explained the benefit of acquiring niche knowledge required by their job role said:

"During the 3rd rotation of my co-op, I spent time every week with the Demand Planning team to understand their processes, tools and systems. I also worked with them on the forecasting cycles from October to December. This helped me develop an understanding of Demand Planning and how its role is crucial in Monthly Production Planning at a plant level." [Par. 143]

Similarly, another student undergoing co-op discussed how challenges they faced during tasks helped them gain industry-specific knowledge. They said:

"My second goal was a "reach" goal, where I would need to put in a lot more work just for a chance of accomplishing it...The challenge of this goal was that it would require me to learn about our materials at a much more technical level and be able to vet ideas for their feasibility to be run. So, I did a lot of research, watched videos, talked to expert coworkers with lots of experience on the asset/ the materials and products it can make. I ended up deciding that it might be worth it to try running a wipe base sheet material." [Par. 22]

Thus, we observe that students had to identify areas in which they did not have specific technical knowledge and acquire that knowledge in order to fulfill the requirements of their jobs.

2) Apply knowledge and skills gained from school/work to work projects.

The second subtheme, *apply knowledge and skills gained from school/work to work projects*, refers to the application-oriented nature of engineering work. Engineering practice requires engineers to apply the concepts and skills they learned in school or the workplace to projects. As a result, students entering the co-op noticed that their job roles emphasized the application-oriented nature of engineering work. One student discussing how they applied their knowledge from engineering courses said:

"I applied all the techniques learned and templates developed during the preparation of my Lean Six Sigma exam and the project management course to create a training and communications plan to ease operations in my current company." [Par. 13]

This subtheme can also be considered as an extension of the previous subtheme because students often *apply* the knowledge they acquired during co-op to accomplish their job goals. One student described how they applied the knowledge gained from the industry to solve production issues by saying:

"Reflecting on my initial goals I think this was a successful experience since I was able to understand the process of battery manufacturing in depth and develop multiple root cause analyses that highlighted problems in the production line with faulty equipment that was generating many scrapped parts that only had false failure" [Par. 163]

These statements indicate how students emphasized the application-oriented nature of engineering work and how acquiring skills and knowledge about their tasks helped them apply what they learned.

3) Real-world problem solving

Our third subtheme focuses on how engineering workplace problems are different from problems that students encounter in the industry. During their co-op students came across different facets of engineering problems like ambiguity, detail-oriented, complexity, open-endedness, and dealing with failure. In addition, they had to learn to grapple with various constraints like deadlines, limited resources, changing priorities of goals, etc. This dynamic nature of engineering problems helped students become aware that real-world engineering problems are different than close-ended linear problems they encountered in their classrooms. Describing the differences between real-world problems and classroom problems, a student noted:

"The biggest challenges and change from college I had was not having all the answers with me. At school, we are given problems with answers. All the H.W, theory questions or engineering problem have an answer. However, while I was working on my project, I realized that a lot of things might not have an answer and it might take a lot longer to figure it out. Since I was so used to getting answers quickly, my biggest trouble was understanding that it is okay not to have all the answer and I need to continue to persevere and dig deeper to get to the bottom of the problem." [Par. 218]

Another student explained how engineering work problems involve challenges making these problems complex and dynamic. They said:

"The first was to run an entire experiment. I was able to complete this goal during my extension term, but I did run into some challenges. The experiment became a much more complex design that I had initially thought it out to be, so the maintenance and lab work became excessive and at some points overwhelming." [Par. 72]

Furthermore, students also discussed various constraints that must be accounted for while working on engineering projects. One of the students discussing time and hierarchical constraints and their influence during engineering problem-solving said:

"The challenges to achieving this goal mainly involved the timeliness of releases and iterations. I realized that in the industry, changes and releases are done quite slowly compared to projects I have worked on in the past due to the different levels of approval and management that deliverables must pass through. Accounting for this challenge, I was still able to mostly see the lifecycle of a deliverable that I was involved in the engineering of." [Par. 277]

Observing the reflections, we noticed that students encountered real-world problems, developed a sense of how these problems are different from classroom problems, and identified ways to work through challenges associated with the problems. Further, students' reflection also illustrates that real-

https://doi.org/10.52202/073963-0025

world problems are complex, have higher stakes and implications, and may involve external factors that are beyond their control. However, these reflections also indicate that an essential element of engineering work is to account for these factors while addressing problems.

4) Hands-on work

The fourth subtheme corresponds to the hands-on nature of engineering work. We define hands-on engineering work as participating actively and being involved *firsthand* in the operations of engineering work. Hands-on engineering work encompasses various participatory roles like working in the laboratory, conducting experiments, testing, process improvement, coding, etc.

Upon entering the co-op programs, several students found that much of engineering work is hands-on and requires active involvement by an individual to produce desired results. A student describing how their hands-on work will help improve the production line said:

"My main project involved the installation of modifications to the production line to accommodate both the current and new models running simultaneously and autonomously. The modifications that I have made will be long term changes to the production line and will likely make the accommodation of future models easier going forward." [Par 9.]

Another student who grew to realize the importance of the hands-on nature of engineering word mentioned:

"I didn't run into any challenges, but an insight that I made during these trials was that the manufacturing process, especially on machines as large as we have, is not as simple as pressing a few buttons to make the machine work. The machines require an entire team to properly operate and fix any hiccups that arise. This made me appreciate the more hands-on side of our business because without all the people at the mills with deep technical knowledge and experience in the business, our trials would not have been successful." [Par. 22]

From the above descriptions, we notice that students grew to understand that hands-on work is an integral part of engineering practice.

5) Develop/work with engineering tools.

The fifth subtheme, *develop/work with engineering tools,* depicts how co-op students recognized the role of computational tools in facilitating engineering work. A large part of engineering work today requires engineers to design and develop computational engineering tools and use these tools to solve a variety of multifaceted problems. Our analysis indicates that students frequently discussed using engineering tools to solve problems and accomplish the tasks assigned to them, and thus recognized the role of computational engineering tools in engineering work. One student explaining the role of computational tools in the industry said:

"My second goal is to become more familiar with as much of the engineering software as possible. This is a great opportunity to be able to experiment with common industry software that would typically be out of reach outside of an engineering workplace. By having access now, I will come into a new job with that experience" [Par. 144]

From the response, we notice that engineering students also recognize that computational software are used in the industry and many of these tools may not be available to learn in an academic environment. Thus, co-op programs also offer opportunities for students to learn and work with industry-specific computational tools.

6) Work in Teams

Our sixth subtheme suggests that working in teams is an inherent part of engineering work. Students during their co-op had to coordinate and work with different teams to complete their job tasks. Working in teams included reaching out to people from different teams, accounting for recommendations and insights from different teams while solving a problem, discussing, and working with people from their own team, etc. As students worked in teams, they understood that engineering work has many moving parts that can function well when teams collaborate and work together. Further, students also encountered challenges related to teamwork that impacted their project tasks and deliverables. One of the students discussing about their teamwork experience said:

"Halfway through the internship, an FDA audit initiated the development of a very high priority hierarchy that I was assigned to lead. This required coordination and collaboration with multiple teams, providing valuable experience in managing complex projects. However, the reliance on other teams also led to challenges, as some teams were unable to complete their documentation on time, ultimately hindering the completion of the project" [Par. 98]

Another student explained how working in teams contributes to the success of the company by saying:

"My second goal was to learn how post-market engineering impacts the company on the whole. I also was able to achieve this goal through speaking with my teammates as well as individuals on other teams. I can now see how tasks as simple as a print change are able to help clear backorder and promote continuity. Without the team I was a part of, thousands of dollars of inventory would be locked up or would continue to be made in a manner that is non-conforming." [Par. 54]

Therefore, we can say that though working in teams comes with its own set of challenges, it is an important component of engineering work that contributes significantly to the success of an organization.

7) Developing documentation about processes and projects

Our last subtheme reveals that engineering work requires students to understand, interpret, document, and communicate processes and projects they work on. As students worked on projects in cross-functional teams, they had to document the data (like problems, processes, anomalies, operating procedures, and findings) and communicate them to other departments in their organization. Students undergoing their co-op thus experienced that working with information and documenting it was an important component of engineering work. One student discussed how their main project of developing reports was useful to the organization. They said:

"My main project was to automate a process of creating Problem Follow-up Sheets to issue Design Investigation Reports more efficiently. I was able to save about 40-50% of the time to issue DIR's which is a success. However, it is still in the prototype phase so it can definitely be improved upon." [Par. 157]

Our analysis reveals that students identified various components that makeup engineering work. However, in order to succeed in their engineering tasks at work, they would need to acquire relevant skills. Our next major theme looks at different skills that students identified and learned which helped them prepare and succeed during their co-op.

B. Skills Needed to Succeed in the Workplace.

This theme explores the skills students believed were important for them to succeed at their workplace. Based on their perception of engineering work, students identified several skills that will help them successfully complete their job tasks and work in the industry. Therefore, this major theme, *Skills Needed to Succeed in the workplace,* is centered around the different professional and technical skills students described in their reflections. This theme on skills can be further divided into the following subthemes: (1) *Upskilling/improving existing skills* (2) *Communication Skills.* (3) *Take ownership or initiatives at work* (4) *Be a team player: Collaboration and Teamwork.* (5) *Be organized: Scheduling and Time management* (6) *Be adaptable and flexible at the workplace*

1) Upskilling/improving existing skills.

After observing and experiencing the nature of engineering work, students realized that they had to improve their technical skills and learn new industry/role-specific skills during their co-op. Our first major theme indicated that students had to acquire domain-specific knowledge or skills and apply them while working. As a part of this process, they needed to gain new skills and use these skills to accomplish the tasks assigned to them. Therefore, many students mentioned that their goals for co-op rotation were centered around gaining job-specific technical skills and using them to complete their tasks. A student describing industry-specific skills they learned said:

"One example of a new skill I learned was soldering. Up to this point, I have not had to ever solder anything, and I was given the opportunity to learn to this semester, so I did. I am still not great at it, but we all have to start somewhere. Another example of a practical engineering skill I learned this year is how to make engineering drawings. While we touch on this in the classwork I have taken thus far, there is a lot that is not covered that is needed if you are actually sending a part to get machined." [Par. 95]

Additionally, students also mentioned they worked to hone their existing technical or software skills as it helped them progress towards their work goals. One student explaining the benefit of working on their technical skills said:

"I was really focused on developing CAD experience, and I have been able to gain a lot of skills in using AutoCAD during this co-op. At first, it was difficult to navigate through the program and picture ideas in a 2D form, when being used to 3D. However, during this rotation, I was able to develop AutoCAD drawings of new workstations and flow in the factory, as well as communicating ideas on floor load ratings in the mezzanine using AutoCAD" [Par. 248]

From the above reflections, we notice that students worked on developing their existing skills and gained new skills that helped them comply with the requirements of their work.

2) Communication skills

Along with improving their technical skills, students also felt that being able to communicate effectively was an important skill in the engineering workplace. Our prior analysis suggests that communicating findings or information is a crucial part of engineering work. Furthermore, as engineers work in teams, they need to be able to communicate effectively with their colleagues and other different teams about their work and also form relationships with them. To work effectively in such settings, many students mentioned that they had to overcome their fear of public speaking by developing their communication skills. One student discussing about their public speaking abilities mentioned:

"I want to develop my presentation skills. I have never been strong at public speaking.... Moving on to presentation skills, not only was I tasked with giving a presentation to Delta operations leaders at the end of my rotation but also I would frequently speak in the bi-weekly update my manager gives to leaders detailing progress on the application. I would honestly say I did not have the best public speaking skills coming into college, but through my experiences at both Purdue and now Delta, I feel much more confident went giving presentations to a large audience" [Par. 152]

Another student mentioned the need to develop their communication skills when they interacted with people from different disciplines. People from other disciplines may not be aware of technical engineering jargon, which may require the engineer to communicate their work in simple language. Explaining a similar scenario, the student said:

"Another goal is to get better at communicating my work to non-technical people and people who are unaware of my work. I find it difficult to provide context and explain background information when talking about my work with colleagues. I would like to improve this as it a necessary skill for any engineer." [Par. 265]

In the context of developing their communication skills, students also indicated that asking questions was beneficial for them and helped them perform better at their workplace. One student realizing the value of asking questions said:

"In my first term, I struggled to be proactive and ask for help when facing a problem that I didn't understand. I was too concerned with appearing competent, and worried that my coworkers would think less of me if I struggled. In this term, I had more confidence in my abilities coming in, and did not feel

https://doi.org/10.52202/073963-0025

ashamed to ask for help and be proactive when planning my projects. This helped me to be more productive when working on my projects." [Par. 63]

Thus, we observe that students felt the need to improve their communication skills by pushing themselves to speak up, reflecting and recognizing the limitations of their existing communication skills, and being proactive at their workplace by asking questions.

3) Take ownership or initiative at work.

Though co-op students worked in teams, they found that they had to take initiative and ownership of the tasks they were assigned. Engineering work requires each individual to be responsible and accountable for their work, and hence, the individual needs to take ownership of their own task. This ability to take ownership proved to be a great learning experience for students as it helped them develop their leadership skills. Realizing the importance of taking ownership of a project, one student reflected:

"I have participated in quite a few multi-week projects that apply prior knowledge, so the length and nature of the project should not have been the emphasis of the goal. Instead, I should have challenged myself to INDEPENDENTLY complete a project that requires knowledge in the energy field. This would have not only pushed me to learn, but also to apply my learning." [Par. 6]

Another student who actively took ownership and leadership on tasks explained the importance of this skill. They said:

"My second goal was to develop my skills as a leader. I served as the co-op president this term and was able to hold co-op and co-op council meetings as well as be a point of contact for the co-ops on site. Over the course of the term, I learned to handle conflict, address uncomfortable issues, and make decisions on behalf of the group. It was a rewarding experience, and it allowed me to develop into a good leader." [Par. 205]

Thus, we noticed that actively taking ownership and initiative helped the students develop leadership skills and understand the intricacies of managing people and deliverables.

4) Be a team player: Collaboration and Teamwork

Our prior analysis revealed that a major component of engineering work required co-op students to work in teams. Upon realizing that teamwork is an integral part of engineering work, the students mentioned that they actively engaged in working on their teamwork skills and contributed to the company's goals by working with their teammates. We defined teamwork skills as the ability to work in teams, work with diverse sets of coworkers, and be open to feedback and suggestions from one's teammates. Reflecting on their experiences of teamwork and receiving feedback from colleagues, one student mentioned:

"Always share your projects with your colleagues; their constructive criticism will enable you to refine both your work and the project." [Par. 73]

Another student recognized the potential of teamwork skills and how discussing project issues with co-workers and supervisors helped provide insights about a problem. They said:

"After I failed on my first try, I reached out to co-workers and my manager who is more proficient than I am and set up zoom meetings where I explained the problem with as much depth as I could, and they offered up meaningful advice that took me a long way." [Par. 87]

Students also recognized how building relationships with their teammates was beneficial for them to gain confidence in their engineering work. One student said:

"I made a genuine attempt to connect more with my teammates and ask them for their guidance and help. I was able to get to know them a lot more personally and I think it made a huge impact on the quality of my work as I was able to work with a lot more knowledge and confidence knowing that people won't hold my mistakes over me." [Par. 29]

Reviewing these student responses, we observe that reaching out to co-workers and peers and collaborating with them in teams was valuable for students and helped them understand the importance of teamwork skills.

5) Be organized: Scheduling and time management.

To complete their designated tasks and comply with the constraints of the projects, students felt the need to develop a system that helps them organize and manage their workloads. In addition, students recognized that they needed to learn to prioritize their tasks without compromising the quality of their work. As a result, students employed various scheduling and time management strategies to stay on track for their projects and accomplish tasks on time. One student explaining their organization strategy said:

"I have made progress toward my goals that I set. One of my coworkers introduced me to how good OneNote is for organization of notes. I was able to use OneNote to keep all of my information organized, and keep track of different work I wanted to complete on a day-to-day manner which was one of my goals I set at the beginning." [Par. 124]

Therefore, we observe that students felt the need to develop new organization systems and gain skills that help them manage the requirements of their role.

6) Be adaptable and flexible at the workplace.

When students entered their co-op rotation, they had to adjust and adapt to the company culture and be open and flexible to changes and requirements of the tasks they were assigned. Furthermore, if students were working on their co-op abroad, they had to adapt to the local culture, learn a new language to communicate with their peers and adapt to a new working style based on the location of their company. One of the students who was completing their co-op abroad mentioned how they had to adjust to their new location and work culture. They said:

"My language skills are only improving as I continue to immerse myself living here, I'm growing accustomed and adapting to a new culture of working, and I'm ready to tackle travelling on business as well! Overall, I am quite proud of the progress I have made on the goals I set for myself, which were to improve my Spanish language skills, adapt to a new way of working, become comfortable with business travel, and become

https://doi.org/10.52202/073963-0025

more self-aware as to my professional interests and strengths/weaknesses" [Par. 31]

Another student discussed being flexible in the context of engineering work as deadlines and other uncertainties related to the project milestones can occur. This student said:

"My biggest take away from this goal would be that it is definitely wise to go in with this mentality to remain focused and diligent, yet the ability to be flexible is equally as important because the timing and schedule of project work can be shifted or put on hold to afford time for others." [Par. 179]

Therefore, we observe that students acquired the ability to be flexible and adaptable at their workplace during co-op rotation.

V. Discussion

Our study explored the experiences of engineering students while undergoing their co-op rotation to understand how they shift in their perceptions of engineering work. Our findings revealed two major themes (i) Engineering students had a better understanding of the *Nature of Engineering Work* after their co-op rotation (ii) Engineering students identified professional and technical *Skills Needed to Succeed in the Workplace*.

Our first major theme *Nature of Engineering Work* aligns closely with prior research findings on engineering practice. In particular, our subthemes on *acquiring niche knowledge required by job role* and *applying knowledge and skills gained from school/work to work projects* align with prior research on technical work (American Society for Engineering Education, 2020). Barley and Orr (1997) found that engineering practice requires acquisition and application of knowledge from formal education as well as through workplace peers. In addition to formal education and learning from peers, our theme also highlights that gaining and applying industry specific knowledge is possible by learning on the job using company training and other available resources in the organization.

Our subtheme of *real-world problem solving* echoes previous findings by suggesting that engineering problem solving is ambiguous, sociotechnical, ill-structured, and open ended with a lot of uncertainties (Downey et al., 2006; Jonassen, 2000). As a result, we suggest that engineering education must work on developing students with abilities to solve complex sociotechnical problems (Author et al., Year; Grohs et al., 2018; Smith et al., 2021). Further, our subtheme draws attention to bureaucratic aspects like layers of approval, resource constraints, changes in business decisions, etc. that influence engineering problem solving and decision making.

Finally, our subtheme *work in teams* reiterates that engineering is not just technical but has social components such as working in diverse teams (Trevelyan, 2007). Accordingly, efforts must be focused on exposing students to similar settings during engineering education through approaches like project-based learning and teamwork activities (Dym et al., 2005). Our subtheme also emphasizes that teamwork in an industry is different from university setting. Teamwork in the industry has unique challenges like collaborating across teams and adapting to other teams' working styles. These factors may also impact project success. Therefore, future work must try to expose students to similar real-world settings and explore how cross-team collaboration and competence can be developed.

Our next major theme, *Skills Needed to Succeed in the Workplace*, indicates that upon understanding the nature of engineering work, students identified specific professional and technical skills that they needed to develop. Companies and industries required students to gain job specific knowledge and thus, students discussed specific technical skills that they developed during the co-op experiences. Further, we noticed that students had to *upskill* their technical skillset as per the requirements of specific their role. This finding suggests that it is important to help students develop life-long self-learning abilities before going on their co-op experiences (and especially before graduation) so that they can adapt their skills for their specific work environment (Martin et al., 2005).

Along with technical skills, professional skills are an equally crucial component of engineering work (Baytiyeh & Naja, 2012; Brunhaver et al., 2018; Korte et al., 2015; Martin et al., 2005; Trevelyan, 2019). Prior research indicates that employers expect new graduates to co-ordinate and communicate with people from different disciplines, collaborate with peers and teams, and manage projects (Brunhaver et al., 2018). However, research also indicates that students often feel that they lack these professional skills upon graduation (American Society for Engineering Education, 2020; Baytiyeh & Naja, 2012; Martin et al., 2005). Our study revealed that co-op helped students acquired professional skills like *Communication skills*, *Be a team player: Collaboration and Teamwork*, and the ability to *Take ownership or initiative at work which* helped them meet the expectations of their employers. Additionally, our findings from the collaboration subtheme highlight that building relationships with colleagues and teammates was beneficial in improving student's workplace confidence and help them feel prepared for practice.

Our data also provided additional evidence of the multifaceted nature of engineering work. Students in our study began to recognize how technical and professional skills are integral to engineering work and how they will need to use all of these skills together as part of their engineering careers (Brunhaver et al., 2013; Martin et al., 2005). Therefore, we recommend that professional skills should not be taught in isolation from technical skills. Rather, these skills can be taught in tandem by embedding them in technical coursework and increasing student involvement in extracurricular activities and experiential learning programs (American Society for Engineering Education, 2020). Because co-op programs and similar experiential learning experiences require students to use professional and technical skills together, they can be important experiences in helping students feel prepared to enter the workforce. Building on our work, future research on co-op experiences could explore students' ability to integrate technical and professional aspects of engineering work and how life-learning abilities can be fostered in students before, during, and after their co-op programs.

ACKNOWLEDGMENTS

We would like to thank the Office of Professional Practice at Purdue University for co-operation in providing the data and program information. Special thank you to Jenny Strickland, Joseph Tort, Dr. Phillip Dunston, and Beata Johnson for their support developing the survey and sharing insights about its development.

REFERENCES

ABET Engineering Accreditation Commission. (2020). *ABET criteria for accrediting engineering programs*.

American Society for Engineering Education. (2020). *ASEE Corporate Member Council Survey for Skills Gaps in Recent Engineering Graduates* (p. 21). https://workforcesummit.asee.org/student-survey/

Anderson, K. J. B., Courter, S. S., McGlamery, T., Nathans-Kelly, T. M., & Nicometo, C. G. (2010). Understanding engineering work and identity: A cross-case analysis of engineers within six firms. *Engineering Studies*, *2*(3), 153–174. https://doi.org/10.1080/19378629.2010.519772

Barley, S. R., & Orr, J. E. (1997). *Between craft and science: Technical work in US settings*. Cornell University Press.

Baytiyeh, H., & Naja, M. (2012). Identifying the challenging factors in the transition from colleges of engineering to employment. *European Journal of Engineering Education*, *37*(1), 3–14. https://doi.org/10.1080/03043797.2011.644761

Bielefeldt, A. R., Paterson, K. G., & Swan, C. W. (2010). Measuring the value added from service learning in project-based engineering education. *International Journal of Engineering Education*, *26*(3), 535–546.

Bielefeldt, A. R., Paterson, K., Swan, C., Pierrakos, O., Kazmer, D. O., & Soisson, A. (2013). *Spectra of learning through service programs*. 23–1080.

Blair, B. F., Millea, M., & Hammer, J. (2004). The impact of cooperative education on academic performance and compensation of engineering majors. *Journal of Engineering Education*, *93*(4), 333–338.

Braun, V., & Clarke, V. (2006). Using thematic analysis in psychology. *Qualitative Research in Psychology*, *3*(2), 77–101. https://doi.org/10.1191/1478088706qp063oa

Brunhaver, S. R., Gilmartin, S., Grau, M., Sheppard, S., & Chen, H. (2013). Not All the Same: A Look at Early Career Engineers Employed in Different Sub-Occupations. *Paper Presented at 2013 ASEE Annual Conference & Exposition Proceedings*, 1–27. https://doi.org/10.18260/1-2--22315

Brunhaver, S. R., Korte, R. F., Barley, S. R., & Sheppard, S. D. (2018). Bridging the Gaps between Engineering Education and Practice. In R. B. Freeman & H. Salzman (Eds.), *U.S. Engineering in a Global Economy* (pp. 129–163). University of Chicago Press. https://doi.org/10.7208/chicago/9780226468471.003.0005

Chan, C. K. Y. (2022). *Assessment for Experiential Learning* (1st ed.). Routledge. https://doi.org/10.4324/9781003018391

Davis, K. A., Joshi, S. S., Czerwionka, L., Montalvo, F. J., Rios-Rojas, G. O., Tort, J., Marston William, J., & Nauman, E. (2021). Integrating the humanities with engineering through a global case study course. *Journal of International Engineering Education*, *3*(1), Article 4.

Downey, G. L., Lucena, J. C., Moskal, B. M., Parkhurst, R., Bigley, T., Hays, C., Jesiek, B. K., Kelly, L., Miller, J., Ruff, S., Lehr, J. L., & Nichols-Belo, A. (2006). The Globally Competent Engineer: Working Effectively with People Who Define Problems Differently. *Journal of Engineering Education*, *95*(2), 107–122. https://doi.org/10.1002/j.2168-9830.2006.tb00883.x

Dym, C. L., Agogino, A. M., Eris, O., Frey, D. D., & Leifer, L. J. (2005). Engineering Design Thinking, Teaching, and Learning. *Journal of Engineering Education*, *94*(1), 103–120. https://doi.org/10.1002/j.2168-9830.2005.tb00832.x

Ehlert, K., & Orr, M. (2020). Understanding How Co-op Students View their Learning. *2020 ASEE Virtual Annual Conference Content Access Proceedings*, 35421. https://doi.org/10.18260/1-2--35421

Garavan, T. N., & Murphy, C. (2001). The co-operative education process and organisational socialisation: A qualitative study of student perceptions of its effectiveness. *Education + Training*, *43*(6), 281–302. https://doi.org/10.1108/EUM0000000005750

Gardner, P. D., & Motschenbacher, G. (1997). Early Work Outcomes of Co-op and Non-co-op Engineers: A Comparison of Expectations, Job Level, and Salary. *Journal of Cooperative Education*, *33*(1), 6–24.

Gilmartin, S. K., Brunhaver, S. R., Chen, H. L., & Sheppard, S. D. (2018). Career Plans of Undergraduate Engineering Students: Characteristics and Contexts. In R. B. Freeman & H. Salzman (Eds.), *U.S. Engineering in a Global Economy* (pp. 49–86). University of Chicago Press. https://doi.org/10.7208/chicago/9780226468471.003.0003

Grohs, J. R., Kirk, G. R., Soledad, M. M., & Knight, D. B. (2018). Assessing systems thinking: A tool to measure complex reasoning through ill-structured problems. *Thinking Skills and Creativity*, *28*, 110–130. https://doi.org/10.1016/j.tsc.2018.03.003

Huff, J. L., Zoltowski, C. B., & Oakes, W. C. (2016). Preparing engineers for the workplace through service learning: Perceptions of EPICS alumni. *Journal of Engineering Education*, *105*(1), 43–69.

Jin, G., & Fabretto, C. (2019). "Journey to Success"-Helping International Co-op Engineering Students Succeed in the Workplace. *CSCE Annual Conference*, 1–10. https://csce.ca/elf/apps/CONFERENCEVIEWER/conferences/2019/pdfs/PaperPDFversion_181_0228031107.pdf

Jonassen, D. H. (2000). Toward a design theory of problem solving. *Educational Technology Research and Development*, *48*(4), 63–85. https://doi.org/10.1007/BF02300500

Korte, R., Brunhaver, S., & Sheppard, S. (2015). (Mis)Interpretations of Organizational Socialization: The Expectations and Experiences of Newcomers and Managers. *Human Resource Development Quarterly*, *26*(2), 185–208. https://doi.org/10.1002/hrdq.21206

Main, J. B., Johnson, B. N., Ramirez, N. M., Ebrahiminejad, H., Ohland, M. W., & Groll, E. A. (2020). A case for disaggregating engineering majors in engineering education research: The relationship between co-op participation and student academic outcomes. *International Journal of Engineering Education*, *36*(1), 170–185.

Martin, R., Maytham, B., Case, J., & Fraser, D. (2005). Engineering graduates' perceptions of how well they were prepared for work in industry. *European Journal of Engineering*

Education, 30(2), 167–180. https://doi.org/10.1080/03043790500087571

Mote, C. D., Dowling, D. A., & Zhou, J. (2016). The Power of an Idea: The International Impacts of the Grand Challenges for Engineering. *Engineering*, *2*(1), 4–7. https://doi.org/10.1016/J.ENG.2016.01.025

Parsons, C. K., Caylor, E., & Simmons, H. S. (2005). Cooperative Education Work Assignments: The Role of Organizational and Individual Factors in Enhancing ABET Competencies and Co-op Workplace Well-Being. *Journal of Engineering Education*, *94*(3), 309–318. https://doi.org/10.1002/j.2168-9830.2005.tb00855.x

Ramirez, N., Smith, S., Smith, C. F., Berg, T., Strubel, B., Ohland, M., & Main, J. (2016). From interest to decision: A comparative exploration of student attitudes and pathways to co-op programs in the United States and the United Kingdom. *International Journal of Engineering Education*, *32*(5A).

Smith, J. M., Lucena, J. C., Rivera, A., Phelan, T., Smits, K., & Bullock, R. (2021). Developing Global Sociotechnical Competency Through Humanitarian Engineering: A Comparison of In-Person and Virtual International Project Experiences. *Journal of International Engineering Education*, *3*(1), 5.

Trevelyan, J. (2007). Technical Coordination in Engineering Practice. *Journal of Engineering Education*, *96*(3), 191–204. https://doi.org/10.1002/j.2168-9830.2007.tb00929.x

Trevelyan, J. (2010). Reconstructing engineering from practice. *Engineering Studies*, *2*(3), 175–195. https://doi.org/10.1080/19378629.2010.520135

Trevelyan, J. (2014). *The making of an expert engineer*. CRC Press.

Trevelyan, J. (2019). Transitioning to engineering practice. *European Journal of Engineering Education*, *44*(6), 821–837. https://doi.org/10.1080/03043797.2019.1681631

Research in Engineering Education Symposium-2024
Jan 4-6, 2024 - Hubballi, India

Exploring the role of engineering skills-building activities amongst youth in rural Karnataka

Heather R. Beem; David Asante-Asare, and Pashington Obeng
Ashesi University, Ghana
Corresponding Author Email: hbeem@ashesi.edu.gh

Abstract

Context

Engineering skills-building activities, such as those found within the Creative Capacity Building (CCB) approach, have been used as a human capital-building approach to enabling developmental progress amongst rural communities.

Purpose or Goal

The Siddi (Afro-Indian) community can tend to view themselves as outsiders within broader Indian society. This study began delving into two research questions: 1) What constructs may be apt for describing Siddi youths' experience? 2) To what extent do self-perceptions change as Siddi youth engage in an engineering skills-building session?

Methods

This mixed-method exploratory study began with stakeholder interviews in a village and nearby town within Karnataka State. Coding was done to ascertain which constructs resonated well. A three-day CCB-inspired training was then deployed for three different groups of youth – a private-school in the town, a public-school in the town and hostel students in the village. Pre-post changes in Likert scale responses on a survey were analyzed using t-tests and Hedge's g tests to ascertain how interest in STEM, creativity and confidence levels were affected.

Outcomes

Interest emerged as a common first step in the Siddi youths' narrative. As a result of the training, interest in engineering increased with statistical significance and large effect size for all disaggregations. The public-school students performed objectively better than the private-school students during the training, however they reported lesser gains in self-perception.

Conclusion

Engineering skills-building can have a significant impact on rural youth, such as those from the Siddi community. Stereotype threat may however limit its manifestation in their self-perceptions.

Keywords—creative capacity building, design-build activities, Siddi community

I. INTRODUCTION

A. Background on the Siddi Community

Some rural villages in Karnataka State are the home of members of the Siddi (African-descent) community. As the authors support teaching and learning in an African high education institution, we sought to understand how engineering skills development, akin to what we currently offer in Africa, might impact members of this community in India, South Asia.

The Siddi people are descendants of Africans who have been in India for over 800 years as a result of enslavement and travels by merchants and artisans. They constitute a minority group, with a population of about 5,000 in Karnataka within a country of 1.4 billion. They live in clusters in parts of North Karnataka. They tend to only travel to the bigger towns and cities on market days and other times when they need to purchase items outside of their immediate communities. Otherwise, they live lives that are largely geographically and socially isolated from the broader society. Even amongst the surrounding villages where they live, they tend to be amongst the most marginalized communities. For example, even if Siddis live or settle next to Gowlis (shepherds), Lambanis (those originally responsible for providing firewood and construction), by virtue of pigmentation or skin color, the Siddis stand out even in those clustered communities where the others are also marginalized. The Siddis have been reclassified as a Scheduled Tribe by the Central Government of India. They are thus entitled to subsidized education, rationing of food items, and health care, although not all of them are still able to access these.

Based on the authors' previous interviews of Siddi community members, they tend to see themselves as outsiders whose forebears originated outside of South Asia. They tend to be self-conscious outsiders that point to how they have less resources than others. For example, Siddis who attend school outside of their immediate communities can exhibit a sense of shyness, withdrawal, and initial inability to advance as expected by the educational standards. In such a context where the school may be headed by Indians of larger society and different complexion, Siddi students may tend to cluster together. By so doing, they may either consciously or unconsciously reinforce their perceived differences – how they think of themselves in relation to how the larger society sees them. Their self-consciousness may sometimes inadvertently be reinforced by others they interact with in society. For example, staff and teachers in a school may unknowingly pass comments that Siddi children are not as capable as the others, which can in turn feed the misperception that they cannot perform at the same

https://doi.org/10.52202/073963-0026

level as others. Any intervention that manages to break down any perception of low self-esteem or self-consciousness as outsiders could hold value.

The aim of this study was to begin exploring the role of design-build activities on internal attitudes and external livelihood factors amongst the members of a community of African descent located in rural India. We hypothesize that training in design methodology and use of basic tools will empower the youth so that they can advance themselves in and out of school. By so doing, they in turn can inspire others since they can be role models for those who may not necessarily benefit directly from the intervention.

B. *Stereotype Threat*

Stereotype threat (Steele, 1997) is an effect studied in US education research and which originally focused mainly on the context of African American performance in school. It is the case where minority groups conform to the negative stereotypes they are viewed as, even on something like standardized exams. It is possible that this effect may have a strong influence here as well. For example, students from the majority group and attending a wealthier private-school likely receive messaging that they are better off than others. In that case, any positive input they receive may easily boost their self-perception. And the opposite can be true for those in the minority. One can either internalize stereotypes or reject them. But in either case, breaking the minority groups away from stereotypes likely requires a concerted effort.

C. *Engineering Skills-Building*

Designed and pioneered by MIT D-Lab, Creativity Capacity Building (CCB) is a workshop in which participants from a community co-create low-cost solutions using local resources. They learn the design process, basic fabrication skills and teamwork (MIT D-Lab, 2023). Since 2007, it has been deployed in dozens of countries spanning Africa (Taha, 2011), Southeast Asia (Drain, Shekar & Grigg, 2019) and beyond. The users have equally spanned a diverse set, such as refugees, mining community members, and farmers. Although quantifying the exact impact of CCB can seem elusive (McWilliam, Dawson, & Tan, 2008), many stories of its impact exist (Childs, 2017), and participants typically report an increase in confidence to solve challenges using local resources and feeling inspired to share the process with others (MIT D-Lab, 2023).

A shorter version of the full CCB training was offered in this intervention. Key elements of the 5 full-day program were offered within three 1.5-3 hour long sessions. This included one design challenge, fabrication skills practice, and one group design-build project.

In a similar vein, engineering skills-building has been deployed through a project-based learning approach in courses at Ashesi University in Ghana. First semester students were seen to significantly increase in self-efficacy after they engaged in all the elements described above, but over the course of one semester (Beem, 2021b). Evidence from these two approaches suggests the possibility of Siddi youth changing their self-perceptions as they engage in design-build activities.

D. *Author Positionality*

Author 1 is an engineering faculty at Ashesi University, where she has designed and led multiple design/project-based courses. She has also facilitated hands-on training sessions for pre-tertiary STEM teachers in multiple African countries. She is an American who has been living in Ghana for the last seven years. The trip described in this study constituted her first time spent in India.

Author 2 is a Ghanaian and a recent alumni of Ashesi University's engineering program. He served as the Teaching Assistant for Author 1's Introduction to Engineering course, in which he facilitated hands-on experiences for the students throughout the semester.

Author 3 is a Ghanaian who has been a humanities faculty in both US and Ghanaian institutions. For more than 20 years, he has supported various efforts in the rural community in this study. He has engaged closely with a group of nuns and priests whose organization has been based in the community for a similar amount of time.

Author 3 invited Author 1 to join him on one of his scheduled trips to this community and begin exploring if and how her work in Ghana could be extended to support this community. Authors 1 and 3 conducted a field visit to the community over a two-week period: July 17-31, 2019. This trip was aimed at building understanding of the context and gathering preliminary evidence on the hypothesis.

E. *Research Questions*

Given that this was Author 1's first trip to India, this work is intended to be exploratory in nature. The results were expected to enable the authors to identify any striking areas to focus their attention and thereafter design a narrower set of research questions in a follow-on study. For this exploratory study, two research questions (RQ) were used:

RQ 1: What constructs may be apt for describing Siddi youths' experience, vis-a-vis their position in society and future prospects?

RQ 2: To what extent do self-perceptions change as Siddi youth engage in an engineering skills-building session?

We hypothesize that aspirations and confidence will capture Siddi youths' experience, and we hypothesize that their interest in STEM, creativity and confidence will increase from the intervention.

II. METHODOLOGY

This work took the form of a mixed-methods exploratory study. A needs assessment was first conducted and then the training intervention was delivered. Surveys and interviews were conducted in English for a few parties (private-school

https://doi.org/10.52202/073963-0026

students and some adults) and the rest were conducted in Kannada. Nearby parties served as translators for the researchers.

A. Method 1 – Needs Assessment

The first few days were spent getting to know some of the many stakeholders in the diverse communities of study. This included Siddis (Indians of African descent), other Indians (Gowlis etc.), nuns, priests, convent staff, and more.

As part of the needs assessment, interviews and design challenges were carried out with a few of the stakeholder groups. This included the four (4) full-time staff at the local convent, five (5) Siddi male youth residing in the village, twenty one (21) of the girls (class 8-10) who reside in the village hostel at the convent, fifteen (15) students in the 7th grade class at the public school in the nearby town who also reside in the village hostel, and then twenty-one (21) students in the 9th grade class at the private school in the nearby town. Each of these groups were engaged separately. English was used for the interviews with the convent staff and the private school students, and Kannada was used with the other groups. A translator supported with real-time translation to English.

A "banana raise" design challenge served as an icebreaker and fun activity for each group to engage in, as they challenged themselves to stack bananas on two sheets of paper. Through a combination of group interviews and paper surveys, questions about their background, parents' work, and career aspirations were asked.

B. Method 2 – Training

For the three groups that went on to receive the training (the girls in the village hostel, the private-school students in the nearby town, and the public-school students in the nearby town), they also completed a written pre-survey. They were asked to rank, on a scale of 1 to 10 (where 10 is the highest), their interest in science, interest in engineering, how creative they think they are, and how confident they are to design and build their own product.

These groups were also asked to complete a 3-question creativity exercise. Based on the key components that researchers have tended to decompose creativity into, a creativity exercise (Beem & Jones, 2016) was administered here. Each question measures one of each of these components: fluency, which is the ability to generate quantities of ideas, flexibility, which is the ability to create different categories of ideas and to perceive an idea from different points of view, and elaboration, which is the ability to expand on an idea by embellishing it with details. Question 1 (fluency) asked participants to write down as many blue items they could possibly think of. The score was based purely on quantity – the higher the number of ideas, the higher their fluency. Question 2 (flexibility) asked participants to list as many unique use cases of a paper clip as possible. The score was based on the number of distinct responses given. Question 3 (elaboration) asked participants to write down their initials in the middle of a sheet of paper and then create as detailed of a drawing which included those initials as possible. The more details present in the drawing, the higher the mark. One minute was given for questions 1 and 2, and five minutes was given for question 3.

After finalizing logistical details, including scheduling and materials procurement, a 3-day training was administered to each of the three groups. One and a half hours were spent with the private-school students in the mornings, then three hours were spent with the public-school students in the evenings. After completion with these two groups, the same training was administered to the village hostel students for 3 hours each evening. An American university student served as a volunteer assistant throughout this intervention.

The training first led participants through skills building in electronics (learning to build a simple circuit), metalworking (cutting, filing, and snipping), and woodworking (cutting, hammering, filing, and drilling). The participants were placed in the same groups throughout. The participants were then guided to build a wooden box correctly sized to hold their circuit. Finally, participants were guided through a brainstorming activity, then tasked to draw at least one design per person of a lamp for their box, and then finally brought together in their group to finalize on a single design, plan for materials procurement (of which they were limited to whatever they could readily find around) and fabrication. Fig. 1 shows some examples of these training elements. The training can therefore be described as a prescribed set of skills-building tasks followed by an open-ended design prompt. Due to an immediate observation of students self-segregating by gender (with a physical distance between themselves), the author decided to mix the student groups across all classifications available in that particular group (gender and race).

Fig. 1. Training activities included (left) practicing fabrication techniques and (right) designing and assembling an LED-powered lamp

Each group had its distinct characteristics. The private-school in the nearby town was a new school, which had just recently opened, and it was the only English-medium school in a large radius. This school boasted of nice facilities, which came with higher school fees and an expectation of better educational

https://doi.org/10.52202/073963-0026

returns. No Siddi students were present in the class we engaged with here. In fact, less than 5 Siddi students were noticed during the authors' observation of the school during the period in question. Our training took place in their Physics Lab. Although this group was not the target group of interest, they served as a useful comparison. This training was provided to the Grade 9 students, who were 21 in number.

The public-school in the nearby town is located in close proximity to the private-school. The old school was founded in the 1990's, with the explicit goal of serving the marginalized communities in the area, including the Siddis. Some of the students live in the hostels on campus, and a group of 15 of the Grade 7 hostel students were selected by the staff for participation in the training. Their selection was based on their availability during the time period and aimed to include a mix of gender and race. The group make-up was 6 females, 9 males, 6 Siddis, and 9 Non-Siddis. Our training took place in an empty room in what appeared to be a temporarily abandoned Kindergarten building.

The final group of participants was the entire set of female students at the rural village hostel, spanning Grades 8-10. These were all females and they included 7 Siddis and 4 Non-Siddis. The economic background of these students is similar to that of the public-school students. Our training took place in an empty room in the convent building.

Upon completion of the training, the participants were asked to again rank their responses to the four questions asked at the beginning of the intervention. Pre-post differences were calculated across all questions and disaggregated across gender and race.

1) General Observations

General observations were noted by the lead author while the training took place and captured in journal entries. Select observations, such as those that repeated across multiple groups and/or were particularly striking, are mentioned in the Results.

2) Surveys

The pre-survey captured aspirations through an open-ended question "What kind of work do you hope to do as an adult?". Likert scale responses (1 to 10) were captured on their interest in science, interest in engineering, perceived confidence and perceived creativity levels. It also captured creativity levels, through the 3-measure creativity metric. On the post-survey, the same Likert scale questions were asked. An additional open-ended question "How do you feel after this program?" was added.

The open-ended responses were coded and then noteworthy and/or frequently-occurring codes were identified. Pre-post survey differences were calculated to determine any impact that the intervention had. Paired, two-tailed t-tests were used to determine statistical significance ($p < 0.05$). Hedge's g tests were used to measure the effect size of any statistically significant changes. The effect size was considered large if $|g| \geq 0.8$, medium if $|g| \geq 0.5$, and small if $|g| \geq 0.2$. All results were compiled in aggregate and also disaggregated across the training groups, gender, and race.

3) Focus Groups

The rural village hostel girls were not given the post-survey, but they were rather engaged through a focus group in an attempt to go deeper with the responses.

III. RESULTS

A. Method 1

1) Aspirations

The aspirations of the various stakeholders interviewed fit more or less into the categories one may expect globally. Many of those in the private-school aim to become doctors and engineers, whereas those in the public-school and village setting aim to become teachers and Catholic sisters. Hence they largely aim to be what they see. A few exceptions were seen. Some of the youth aspired towards professions beyond what their caste may seem to dictate. The Gowli females interviewed, for example, didn't cite wanting to become cowherders as their caste would prescribe. They rather aspired to become teachers, singers, and even a nurse.

The four convent staff members shared their original career aspirations, and all of them had, for one challenging reason or another, had to change course along the way. Where they had originally aspired to be a teacher or a software engineer, they had to shift and take up their current position. They displayed some sadness associated with that, but the fact that each of them has been living and working in the rural village for 7-20 years suggests that they have found new interest and fulfillment in this line of work. Also, they had clearly learned to work well together, as evidenced by their success in the design challenge. The ideas they came out with were unique compared to all other designs produced by other groups, and their teamwork was similarly strong.

All of this suggests that relying on career aspirations as a construct for describing how the Siddis view their place in society, may not be sufficient.

2) Confidence

In the course of data collection efforts, we asked four different translators to help translate "confidence". All struggled to do so – often using the English word itself and then filling in with examples in Kannada for how the translator interpreted it. This included, in one case, "competence to pass your exams".

Interaction with the male Siddi youth (in their 20's) of the

https://doi.org/10.52202/073963-0026

rural village brought out some other interesting aspects of Siddi life and perceptions here. The interviewees have currently found work as mechanics, welders, and cooks. They travel to the bigger cities to access more steady streams of income, despite the societal challenges they face as a result.

An attempt to determine how they view "confidence" led us to ask them how they would describe a farmer who persists in the face of adversity and continues working even when he faces a low-yield harvest. In responding, the participants kept bringing the conversation to interest – they would persist and go for something if they were interested in it. It then came out that half of them have created the first Afro-Indian rap group, and they are pursuing means of spreading their music. Interest and/or passion may be a necessary ingredient to help Siddis break free from negative perceptions.

B. Method 2

1) General Observations

The most immediate and obvious observation made when interacting with the students was that they self-segregated, especially along gender lines. They sat closely with those similar to themselves, chatting frequently and huddling together, but leaving a physical space between the girls and boys. The private-school students struggled to mix together even after multiple sessions. The public-school students, however, eventually mixed after prompting by the facilitators. And on the last day, during the design-build portion of the training, the public-school students began working together in their groups of mixed gender without any prompting.

The public-school students objectively performed better in the training than the private-school students– they worked more efficiently in their groups, they made it through the content more quickly, they broke less equipment. Secondly, the private-school students progressed too slowly to carry out the final session of the training – the one on design itself. Despite their prior experience working with electronics, the private-school students struggled to put their circuits together during the skills-building session. They got frustrated more easily and spent less effort troubleshooting than the public-school students.

The rural village girls' group was the easiest to work with, working well in their teams, keeping things in order, listening to instructions readily, and displaying strong motivation to complete the work. It is possible that as a single-gender group, it was easier for them to progress in their teams. Or perhaps simply being a smaller group made it easier. Or perhaps the fact that they are a small community living together and carrying out frequent programming together, made the teamwork easier as well.

Open-ended problems:

A few observations were made on how the various groups responded to being presented open-ended problems. Firstly, the private-school students, who had excelled in diligently writing down answers to any questions or prompts we wrote on the board, struggled to answer the creativity questions. In the elaboration prompt, where they were asked to draw a detailed picture around their initials, the private-school students disobeyed repeated commands to not talk to their neighbor and copy. Their teacher ended up going around the room, dictating potential drawing options they could produce. The other groups (public-school and rural village girls) also struggled to produce sufficient results on the creativity prompt. They, however, complained less and did obey the instructions to fill it out themselves. This does, however, question the validity of this exercise in measuring creativity levels, as opposed to capturing students' ability to write an exam efficiently.

TABLE I
SUMMARY OF SURVEY RESPONSES TO WHETHER STUDENTS HAD BUILT SOMETHING BEFORE AND IF YES, WHAT IT WAS

	Private-school	**Public-school**	**Girls Hostel**
Number (%) who had built something before	16/20 (80%)	14/15 (93%)	11/11 (100%)
Some examples cited	Hydraulic car using syringes, model volcano, science projects done in school	Paper-based objects: flower, boat, kite, etc.	Doll, flower, cloth, etc.

When it came to the open-ended design prompt of creating a lamp structure, the students flourished in producing ideas. They were all able to readily draw something. And they were excited to do so – there was an audible gasp of excitement when we announced to the public-school students that they were now to design their own lamp. This appears to have been one of the few times when students were allowed to engage in such an open-ended exercise, based on an observation that the artwork on display in the private-school was of the fill-in-the-outline style.

2) Surveys (Interest, Creativity, Confidence)

Pre-Survey Results:

Prior to the intervention, the private-school students reported higher confidence and perceived creativity than the public-school students. The public-school students reported an average creativity level of 6.9 and confidence level of 8.2, whereas the private-school students reported an average of 8.7 and 8.4, respectively. This is in spite of the fact that none of them had worked with such hardware tools before. Only a few of the students at the private-school cited having worked with some electronics before. And yet, similar to the dynamic captured at Ashesi, a private university in Ghana (Beem, 2021a), almost all survey respondents reported high values (between 8-10) for their confidence to design and build their own product prior to any training.

As seen in Table I, although the vast majority of students in each group cited experience with designing and building objects, the examples cited seemed to come more from school-based projects for the private-school students and more from home-based craft-type projects for the public-school and girls hostel groups. This could, however, also have been influenced by the translation given to each group. If the translator provided ideas for what types of things they could list, that would likely have affected their responses.

Fig. 2 captures the creativity results, in the form of box-and-whisker plots to describe the distribution of responses for each cohort. Each of the three measures are depicted separately. Single-factor ANOVA tests reveal a statistically significant difference between the cohorts for fluency and flexibility, but not for elaboration. For fluency, $F(2{,}43) = 37.97, p = 3.16E-10$, for flexibility $F(2{,}43) = 5.77, p = 6.01E-03$, and for elaboration $F(2{,}43) = 0.12, p = 8.87E-01$. The private-school cohort scored significantly higher in fluency than the other groups, and the hostel girls cohort scored significantly higher in flexibility than the other groups. T-tests conducted between women and men disaggregations for each of the three creativity measures revealed no statistically significant difference ($p > 0.05$). Hence there was no significant difference in how the students responded to this assessment, based on their gender.

Based on the observations made to how the private-school students responded to this exercise, one interpretation of this is that due to their experience practicing writing exams under pressure, they were better positioned to produce a large quantity of ideas. As mentioned above, the private-school students kept talking to each other as they completed the questions, and the teacher even supported this, which suggests that these students are used to operating in a competitive/high-achievement mode, even in this case where no grade would be administered.

It is striking that in spite of this, the hostel girls cohort scored significantly higher in flexibility. Perhaps because they were in an environment that they associate more as a relaxing one than at school, they were able to tap into divergent thinking more readily. The public-school students noticeably struggled to write things down on this exercise, which perhaps could be attributed to the timing – it was right after their play period, and it was later in the day. This raises the question of the extent to which this test is an ability of their actual creativity versus their ability to focus and execute on written prompts. The following is a quote from Author 1's journal entry on 25 July 2019: *"Looking around as the [public-school] students progressed through the electronics and the creativity exercises, I felt sad to see that all groups were fitting into their respective "positions" on all the tests I was administering. The Non-Siddi boys, then the Siddi boys, then the Non-Siddi girls, then the Siddi girls. In that order. Without even looking at the data yet, I could see based on how vigorously they were writing or not, that that's the order they were falling in. Are the metrics incorrect to capture their abilities? Or are the systemic barriers truly holding people in their expected positions? If the latter, then I want to understand what those truly are and observe how the design-build activities can level the playing ground."*

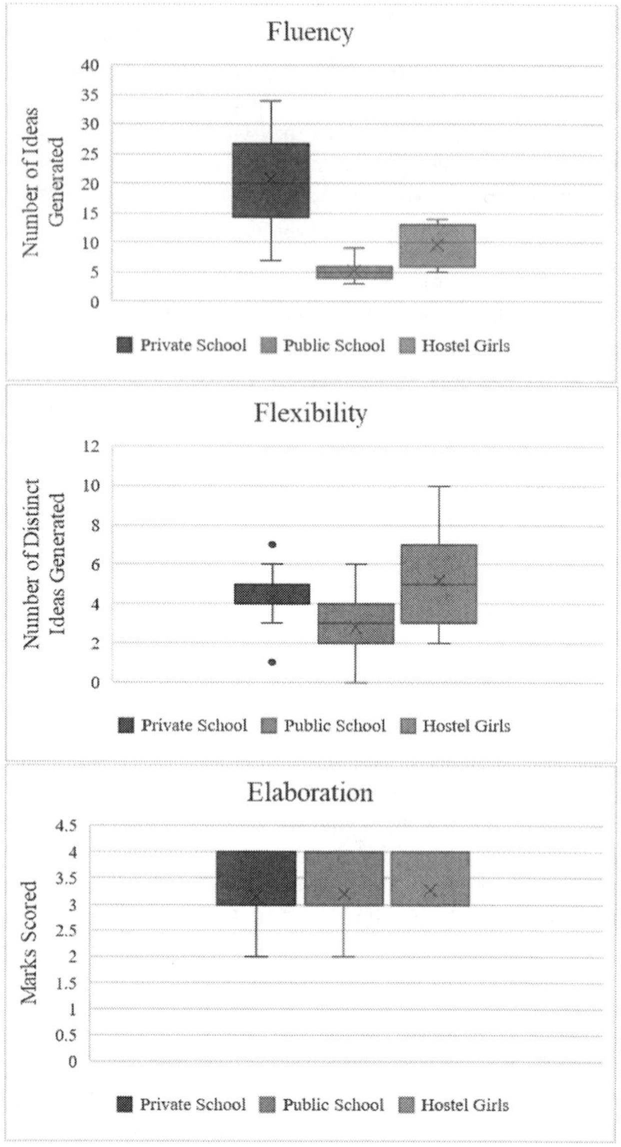

Fig. 2. Creativity dimension scores for each cohort

Pre-Post Changes:

Table II captures the results of the pre-post change analysis. Of the four metrics covered, interest in engineering was consistently the one that produced the most significant change across all groups. A large effect size was measured in all cases shown here. In aggregate, the interest in science also increased significantly. This was, however, largely driven by the public-school students. The private-school students experienced minimal change (p = 2.04E-01), which could be attributed to

https://doi.org/10.52202/073963-0026

TABLE II

RESULTS FROM T-TESTS AND HEDGE'S G TESTS OF THE PRE-POST CHANGE IN INTEREST, CREATIVITY, AND CONFIDENCE LEVELS PERCEIVED BY THE PARTICIPANTS AS A RESULT OF THE INTERVENTION

ALL STUDENTS

		Mean	N	S.D.	Sig. (2-tailed)	|g|	Effect Size
Interest in Science	Pre	7.886	36	1.510	1.77E-03**	0.67	Medium
	Post	8.771	36	1.087			
Interest in Engineering	Pre	6.844	35	2.112	1.11E-06***	1.08	Large
	Post	8.743	36	1.336			
Creativity	Pre	7.939	35	1.767	1.87E-01	-	-
	Post	8.314	36	1.255			
Confidence	Pre	8.294	36	1.426	7.73E-01	-	-
	Post	8.257	36	1.633			

PUBLIC SCHOOL STUDENTS

		Mean	N	S.D.	Sig. (2-tailed)	|g|	Effect Size
Interest in Science	Pre	7.400	15	1.957	2.88E-03**	1.12	Large
	Post	9.200	15	1.146			
Interest in Engineering	Pre	6.833	14	1.337	6.17E-04***	1.93	Large
	Post	9.067	15	0.961			
Creativity	Pre	6.929	14	2.018	6.40E-04***	1.15	Large
	Post	8.667	15	0.816			
Confidence	Pre	8.200	15	1.373	8.00E-01	-	-
	Post	8.333	15	1.543			

SIDDI STUDENTS

		Mean	N	S.D.	Sig. (2-tailed)	|g|	Effect Size
Interest in Science	Pre	6.800	5	1.643	1.21E-01	-	-
	Post	8.667	6	1.506			
Interest in Engineering	Pre	6.800	5	1.643	1.61E-02*	2.02	Large
	Post	9.333	6	0.816			
Creativity	Pre	4.800	5	1.924	1.32E-02*	2.21	Large
	Post	8.000	6	0.894			
Confidence	Pre	7.167	6	1.169	2.86E-01	-	-
	Post	7.833	6	1.472			

*$p \leq 0.05$

**$p \leq 0.01$

***$p \leq 0.001$

the fact that they reported high interest (average 8.250) at the beginning, so there was not much upward progress to make.

The public-school students increased significantly and with large effect size for both interest in science and engineering. If this was one of their first experiences with hands-on STEM work, it is reasonable that this interest would have increased significantly. The public-school students also increased significantly and with large effect size in their perceived creativity levels. Although the public-school students in aggregate did not experience a statistically significant increase in confidence, the public-school females did (p = 7.00E-03).

The Siddi students similarly experienced significant gains from this training, especially in terms of interest in engineering and perceived creativity levels – both increased with statistical significance and large effect size. In fact, one of the Siddi males, who was likely considered a "slow learner" by typical school standards, is the one who reported the largest increases of all.

Although the gender divide appeared to be significant in the team dynamics for the private-school, no difference in statistical significance of pre-post changes were noted for the male and female private-school students. In the public-school, the males did increase in interest for engineering (p = 2.00E-05) whereas the females did not (p = 9.25E-02), and the females did increase in confidence (p = 7.00E-03) whereas the males did not (p = 2.88E-01).

As mentioned earlier, the public-school students objectively performed better than the private-school students in the training. Their pre-survey scores were also generally lower than what the private-school students reported. Hence it is reasonable that the pre-post changes experienced by the public-school students were generally more significant.

Open-Ended Questions:

In general, there was a strong sense of joy and excitement in the rooms as all groups engaged in the training sessions. This is also the main thing that came across in their open-ended responses to "How do you feel after this program?". For the private-school students, the top codes that emerged were "Happy" (8/21) and "Learned new things" (7/21). A few also mentioned "Empathy for skilled laborers" (3/21). One female student said, *"I feel very happy. All things that we did was interesting and new for us. We learnt many things and did hard work to complete these works."* For the public-school students, the top code that emerged was a mention of the item(s) they built (11/15). This suggests that they strongly associated their experience with the specific product created. The second most common code that emerged was "Enjoyment" (8/15) – similar to what the private-school students mentioned. One male student in this group said, *"I learn[ed] how to make a lamp and I enjoyed [it a] lot."*

3) Focus Groups

In their responses, the hostel females expressed significant joy, which is similar to what was captured on the surveys for the other two groups. They noted that they now had ideas for the science projects they'll have to do at school. And they noted that they enjoyed the process of doing this project in the teams - that doing it that way made the process easier.

When asked from multiple angles whether they thought they could replicate this or do other projects on their own later, they responded with a resounding *"yes"*. Most interestingly, the girls brought up the word "confidence" on their own, without being prompted. They self-described as having more confidence in their ability to do projects. One of the participants expanded on this by explaining that *"when you guided us, then we know that we also have skill, that we have to think outside, and we got the feeling that we also can do something."*

IV. DISCUSSION/CONCLUSION

A. Constructs/Framing

A number of observations were made on the constructs being employed. For one, the word "confidence" was interpreted differently than the authors expected from their own context. Confidence didn't seem to be a word that the interviewees readily used to describe their experience prior to the intervention. In fact, all translators struggled to convey the intention behind this in a standardized manner. However, the word organically appeared from the participants when they described the change they experience from the training. Hence there is value in continuing to probe around the language and construct of confidence. It may be worth distinguishing intrinsic and extrinsic confidence. There is a difference between someone being confident because they've been told for their whole life that they are good at whatever and someone who although they may not have received this messaging still knows they can access a "toolbox" to address challenges they face. Work is ongoing to explore what constructs are relevant in the context of the first-year African engineering students (Beem, Ampomah, Takyi & Adomdza, 2023), of which self-efficacy may be one.

Interest is the construct that organically emerged as stakeholders described the starting point in their narratives. From the open-ended responses on the survey, joy, excitement, and pride all emerged. Hence these constructs can also be considered for further use in tool development. The role of community should also be considered in this exploration. The development of desired skills and mindsets does not happen in isolation. The students consistently conferred with each other before jumping into any task, no matter how small. Capturing the role of community and/or socialization may be an important element as well.

B. *Role of stereotypes in the observed dynamics*

The results revealed several interesting dynamics at play for the various student groups. Participants at the private-school tended to rate themselves higher than the public-school students prior to the training, but they performed worse. This suggests that due to societal messaging, they generally exist with a higher self-perception. One of the staff at the public-school used the term "slow learners" to describe their students. With such messaging, the public-school students may think less of their innate abilities.

Although all students increased significantly in their interest in engineering from this intervention, the public-school students are the ones who benefited in other dimensions as well – specifically in interest in science and creativity. The Siddi students specifically also increased significantly in their interest in engineering and perceived creativity levels. The female public-school students are the only disaggregated group that increased significantly in their confidence levels.

One of the Siddi males who appeared to often do work at his own pace and by himself, reported large increases in all questions. Such students may easily be categorized into the "slow learner" bucket but could easily benefit from small amounts of creative outlets for exploration.

Stereotype threat may be at play in inhibiting Siddi youth self-perception, based on societal messaging. Evidence for this could be drawn from the fact that Siddi youth tend to be comfortable when interacting in groups fully composed of Siddis. These results suggest that engineering skills-building activities can have a meaningful impact in breaking misperception and boosting self-perceptions.

Gender dynamics appeared significantly in the study- in the team formation stage, students self-selected along gender lines, but the nature of the activities appeared to break those dynamics down over time.

C. *Limitations*

A few factors should be borne in mind in interpreting the results here. There were slight differences in the grades, ages, and training implementation length between the three groups who received the training. This limits the directness of comparison between the groups. Also, small sample sizes, especially for the disaggregated groupings, limits the robustness of the respective statistical analyses.

D. *Future Work*

This mixed-method exploratory study has revealed significant insight to inform a refined implementation. Future work should consider other constructs to aptly align with the context and describe the phenomenon at hand. The role of gender can be examined more carefully as well. As further work is done to uncover the effects that engineering skills-building activities have on marginalized youth, learnings can be extended across contexts. Three specific goals are outlined here for next steps.

Goal 1: Assess the effects of this first training. Assess which, if any, of the tools that the participants have used post-training. And which, if any, additional fabrication projects have been carried out which may not have required the same tools. Which factors enabled them to do this (support staff, structure of school project/competitions, pressing need, interest, etc.)? Assess the extent to which their self-reported confidence still lives within them, versus being limited to the specific product they built and/or presence of support personnel to guide them.

Goal 2: Identify contextually-relevant language that captures something related to confidence. Conduct deeper interviews with community members to determine their levels of self-efficacy, talk through examples of people who they view as successful and determine what vocabulary they use to describe their actions, decisions and self-perceptions. This can be conducted for both students and adults.

Goal 3: Carry out a next stage of training with select participants. Provide additional skills-building training to select repeat participants. This training can cover additional aspects of the full CCB training which were omitted here, such as learning about the design process itself and then selecting a community need and collectively designing a solution to it.

Once more nuances of the local context are understood as well as an idea of which aspects of the training people most respond to, a control study can be carried out to elicit specific subareas of impact. This can be followed up by a longitudinal study to determine longer-term effects. Results should be considered by researchers seeking to support marginalized

https://doi.org/10.52202/073963-0026

youth globally.

ACKNOWLEDGMENT

The authors thank all the stakeholders on the ground who we had the pleasure of working with. Thanks to Rachel Campbell-Baier for assisting with the fieldwork. Finally, the authors also thank Ashesi University's Engineering Department for the travel support provided to conduct this field research.

REFERENCES

Andrew Drain, Aruna Shekar & Nigel Grigg (2019). 'Involve me and I'll understand': creative capacity building for participatory design with rural Cambodian farmers, CoDesign, 15:2, 110-127, DOI: 10.1080/15710882.2017.1399147

Beem, H., Ampomah, C. O. A., Takyi, J. P. K., & Adomdza, G. (2023). A preliminary exploration of the relevance of self-efficacy, self-determination, and agency in describing the first-year African engineering students' experience. In 2023 ASEE Annual Conference & Exposition. https://peer.asee.org/42268

Beem, H. R. (2021a). The role of machine shop sessions in building confidence amongst first year engineering students in Ghana. In 2021 IEEE Global Engineering Education Conference (EDUCON), pp. 1469-1474.

Beem, H. R. (2021b). Exploring the Role of Project-Based Learning in Building Self-Efficacy in First-Year African Engineering Students. In 2021 ASEE Virtual Annual Conference Content Access. https://peer.asee.org/37161

Beem, H. R., & Jones, G. (2016). Harnessing creativity for design: tests in an introduction to engineering class. Procedia Engineering, 159, 322-325.

Childs, T. (2017). Creative Capacity Building in Uganda: Qualitative case research into the impact of CCB on individuals and communities. *Massachusetts Inst. Technol. D-Lab,* 1-13.

McWilliam, E., Dawson, S., & Tan, J. (2008). From vaporousness to visibility: What might evidence of creative capacity building actually look like? UNESCO Observatory Journal: Multi-disciplinary Research in the Arts, 1(3), 1-19.

MIT D-Lab, "Creative Capacity Building (CCB)", Online https://d-lab.mit.edu/innovation-practice/approach/creative-capacity-building-ccb Accessed on 04 September, 2023

Steele, C. M. (1997). A threat in the air: How stereotypes shape intellectual identity and performance. American Psychologist, 52(6), 613-629. https://doi.org/10.1037/0003-066X.52.6.613

Taha, K. A. (2011). Creative capacity building in post-conflict Uganda (Master's Thesis, Massachusetts Institute of Technology, Department of Urban Studies and Planning). http://hdl.handle.net/1721.1/63238

Copyright statement

 https://doi.org/10.52202/073963-0026

Research in Engineering Education Symposium-2024
Jan 4-6, 2024 - Hubballi, India

Do students' resource-usage patterns vary across institutions? Implementing a pedagogical innovation in an undergraduate engineering course

Mohamed Aziz Dridi[a]; Edward Berger[ab], Jeffrey F. Rhoads[b].
MEERCat Purdue, School of Engineering Education, Purdue University[a],
School of Mechanical Engineering, Purdue University[b]
Corresponding Author Email: mdridi@purdue.edu

Abstract

Context

Effective reform of engineering education necessitates the widespread implementation and dissemination of pedagogical innovations globally. However, to ensure the successful propagation of these innovations, we need to better understand the adaptations that they undergo when adopted at a new institution, and the extent to which they differ from the original innovation. This includes understanding the student experience with the innovation.

Purpose or Goal

This study examines the propagation and adaptation of Freeform, a learning environment for teaching an undergraduate dynamics course developed at a large Midwestern university in the United States. Specifically, our goal is to understand how students at an adopting institution used Freeform's learning resources. Our research questions are: 1) What are the students' archetypical patterns of resource usage at the adopting institution? 2) In what ways do those patterns differ from those of students at the original institution of Freeform?

Methods

We conducted a model-based clustering analysis to answer our two research questions. The analysis was conducted on survey data from 50 engineering students at the Freeform adopting institution. This data articulated how frequently students used nine different resources of the Freeform ecosystem.

Outcomes

Our analysis identified 4 resource-usage patterns in the Freeform adopting institution in comparison to 9 patterns for students at the institution where Freeform originated. In the Freeform adopting institution, the most frequent resources that students utilized were Teaching Assistants (TAs) and other students who were not enrolled in the course. This contrasts with the original institution where students relied mostly on the course lecturebook and their classmates.

Conclusion

This study highlights the importance of taking into consideration the differences across institutions when propagating pedagogical innovations such as Freeform. Our results suggest that instructors should anticipate those differences so that the adoption and onboarding process can be optimized for success.

Keywords— pedagogical innovation, help seeking behavior, resource usage.

I. INTRODUCTION

Pedagogical innovations, when successful, are often propagated outside of their original institutions. During such process, they undergo various transformations to adapt to their new implementation environment. Researchers have investigated the fidelity of implementation of an innovation, (Borrego et al., 2013), with the observation (O'Donnell, 2008) that 'efficacy' of an implementation (the extent to which is resembles the original innovation) is different from its 'effectiveness' (the extent to which it achieves a desired outcome*)*. However, little is known about the efficacy of pedagogical innovations from students' perspectives. That is, to what extent students' experiences of a pedagogical innovation differs across institutions.

In this paper, we examined the propagation efficacy of a pedagogical innovation called Freeform with a focus on students' experiences. Specifically, we ask two research questions 1) What are the students' archetypical patterns of resource usage at the Freeform adopting institution? 2) In what ways do those patterns differ from those of students at the institution in which Freeform originated?

II. BACKGROUND

Freeform, the pedagogical innovation that we are studying in this paper, is a pioneering approach to teaching Dynamics. It started in 2008 within Purdue University's School of Mechanical Engineering as an instructional environment integrating elements of Active, Blended, and Collaborative (ABC) pedagogical techniques. The goal of developing the Freeform ecosystem was to enhance both conceptual understanding and problem-solving abilities in the field of engineering mechanics. To this end, Freeform designers curated a range of both in-person and digital learning activities and resources grounded in ABC learning research (Rhoads et al., 2014).

The debate around the value of these ABC approaches and resources seems to have settled with the general conclusion being that each adds value over a more traditional lecture-based format. In fact, a consensus in the literature demonstrates the effectiveness of active learning practices in the engineering classroom (Freeman et al., 2014). Similarly, blended learning environments, which combine in-class and online learning elements, have been proven to be more beneficial than both in-class and online learning environments (Freeman et al., 2014; Means, 2014). In addition to active and blended learning, collaborative learning has also been demonstrated to have a positive influence on student success (Means, 2014) in traditional, online, and blended instructional settings (Fatos et al., 2006; Jeong & Chi, 2007; Means, 2014). Taken together, Freeform's evidence-based ABC strategies offer a powerful set of instructional tools to support and enable student success.

The name 'Freeform' embodies the educational philosophy underpinning this instructional innovation, granting both educators and learners the latitude to tailor resources according to their specific needs. For example, the Freeform platform includes a bespoke 'lecturebook' (Rhoads & Krousgrill, 2013), conceived to not only support but also integrate active learning activities within the very fabric of the course curriculum. Each Freeform course is further enriched by a dedicated online blog, serving as a centralized portal for academic materials. This virtual platform augments blended learning experiences by offering threaded conversations on coursework and incorporating illustrative video examples. As found in previous studies (Kandakatla et al., 2020; Zadoks et al., 2017), this comprehensive array of methodologies and resources offers considerable decision-making autonomy to both instructors and students in their interactions with the Freeform educational framework. For instance, in our previous work (Evenhouse et al., 2023), we found that many students used the lecture example videos to clarify questions that arose during lecture. Others reported reading the lecturebook as a means of better preparing for class, or using the online discussion forums to further clarify dynamics concepts or problem-solving processes.

III. METHODS

This mixed-method study employed an explanatory sequential design where student survey data was collected first followed by student interview data to further explain the quantitative results.

A. Data collection

The participants to this study were sophomore engineering students enrolled in Dynamics at a large public university in southeast United States. The sampling frame for this study was 107 students enrolled in Dynamics in Spring 2022 out of which 57 consented to our research study. These 57 students were asked to complete a survey at the end of the semester about their study habits, help-seeking behaviors, and resource usage in the class. This study focuses on a subsection of the survey that probed students about their resource usage in Dynamics. The subsection asked students how frequently they used each of the class resources. The response options were verbatim, and in the order in which they appear on the survey): at least once per day, 3–6 times per week, 1–2 times per week, 1–3 times per month, 1-3 times per semester, and never. The nine resources included in the survey question are listed in Table 1 along with their descriptions and their median responses for both the adopting and original institution. Seven students submitted incomplete responses to the survey and their responses were, therefore, discarded rendering our final sample to 50 students.

In addition, we conducted semi-structured interviews with seven junior students in mechanical engineering at the adopting institution who took Dynamics. Each interview lasted around 90 minutes and consisted of several questions organized in five categories: Experiences in Dynamics, relationship, and sense of community in class, use of technology, course structure, course comparison to others.

B. Data analysis

We conducted a model-based cluster analysis using the *mclust* package in R (version 3.3.2) to determine the students' archetypical patterns of resource usage across all the nine resources described in Table 1. Model-based clustering offers two main advantages over frequently employed techniques such as K-means. First, it evaluates multiple shapes for the clusters, not only spherical or circular like in K-means. Second, model-based clusters can overlap since they are calculated based on a vector of probabilities corresponding to the alignment of a student's behavior with that of the other students' behavior in that cluster. In the absence of any a priori knowledge about the shape of resource-usage clusters, we opted for model-based clustering. Following Stites et al., (2019) methodology, we conducted a parametric analysis on 14 different clustering shapes with a number of clusters ranging from one to 10. Our selection of the best fitting model was based on the Bayesian Information Criterion (BIC), a likelihood criterion that penalized models with increased complexity. The combinations of shape and cluster-number that recorded the top three BIC values were identified as the most plausible models. The variations in BIC values among these three top cluster models were marginal, less than 0.5% apart, which prompted us to evaluate the three models based on their differences in the number of distinct, qualitative patterns of resource usage. The most parsimonious model which corroborated the qualitatively unique resource-usage patterns was the four-cluster model and was therefore chosen as the final one.

To gain deeper insights into the reasons behind students' specific resource utilization patterns, we undertook what Merriam (2009) described as a fundamental qualitative investigation using data from student interviews. The primary objective of this qualitative research was to discern the distinctive behaviors that characterized each cluster's resource usage. We used the emergent themes of our qualitative analysis not as findings per se but rather as supporting elements to better characterize the different clusters.

Following this methodology, we were able to identify the archetypical resource usage behaviors of students at the adopting institution. We then compare those behaviors to those of students at the original institution based on results previously published (Dridi et al., 2022). This study was conducted in

https://doi.org/10.52202/073963-0027

accordance with the ethical standards of Purdue University and was approved by its Institutional Review Board (IRB). All study participants provided informed consent prior to participation.

IV. RESULTS

A. Macro-level comparison

Table 1 shows a comparison of the resource usage median frequencies between the adopting institution and the institution where Freeform originated. A comparative analysis reveals noteworthy patterns among students' resources-usage preferences. Students at the original institution demonstrate a proclivity for structured, formal resources, notably the lecturebook and the course blog, with median usage frequencies of 3–6 times per week and 1–2 times per week, respectively. These resources appear to be integral components of their learning strategy. Conversely, students at the adopting institution are less frequent users of these core resources, with the lecturebook being accessed 1–2 times per week and the course blog only 1–3 times per semester.

The usage of online solution videos is consistently moderate across both institutions, indicating that students adopted blended learning, which aligns with Freeform ethos. However, the most striking difference emerges in the realm of interpersonal interactions. Students at the adopting institution are more likely to consult both peers outside of class (i.e. not enrolled in Dynamics) and instructors during office hours, with median frequencies of 1–2 times per week and 1–3 times per semester, respectively. This contrasts sharply with the original institution, where students seldom consult peers outside of class (i.e. not enrolled in Dynamics) and never attend instructor office hours.

Interestingly, while the original institution shows a higher frequency of collaboration with classmates—a core Freeform resource—students at the adopting institution engage less frequently with classmates, reporting a median frequency of 1–3 times per semester.

TABLE I

A DESCRIPTION OF THE NINE RESOURCES INCLUDED ON THE END-OF-SEMESTER SURVEY AND THE MEDIAN FREQUENCY WITH WHICH STUDENTS USED THE RESOURCE FOR ADOPTING AND ORIGINAL INSTITUTIONS

Learning Resource	Description	Median Frequency Adopting institution	Median Frequency Original institution
The lecturebook	Combination of a workbook and concise textbook; students write notes and solve problems directly in book.	1-2 times/week	3–6 times/week
Online solution videos	Screencasts of the instructor solving a problem; every lecturebook example and homework problem has a solution video.	1–2 times/week	1–2 times/week
Peers outside of class	Peers who are not currently enrolled in Dynamics but have taken it previously	1–2 times/week	Never
Help room	A dedicated help room staffed over 40hours/week with undergraduate- and graduate-student TAs	1–2 times/week	1–3 times/semester
The course blog	"Blog" most often refers to the discussion forum but could also be interpreted as the course website.	1-3 times/semester	1–2 times/week
Instructor – In class	Could include questions before, during, or after class.	1–3 times/semester	1–3 times/semester
Instructor- Office hours	Office hours were usually 1 hour long, 2–3 days/week.	1–3 times/semester	Never
Non -course online resources	Could include online videos, online example problems, or online tutoring websites.	1–3 times/semester	1–3 times/semester
Classmates	Group quizzes in class; virtual or in-person collaboration outside of class.	1-3 times/semester	1–2 times/week

B. Cluster analysis of survey data

1) Model selection

The cluster model that had the highest BIC (-14,330) had four clusters, and the models with the second and third highest BIC values (-14, 451 and -14. 490) had three and five clusters respectively. Thus, the four-cluster model was chosen as the most parsimonious model. To measure the goodness of fit of the four-cluster model we examined the uncertainty associated with the cluster classification of each student. For the four-cluster model, almost half the students had an uncertainty of

https://doi.org/10.52202/073963-0027

less than 4% and approximately 80% of the students had an uncertainty of less than 30%. In the institution where Freeform originated, our previous work showed the existence of nine clusters displaying unique archetypical resource usage patterns (Dridi et al., 2022). The difference in number of clusters suggest that students in the adopting institution resorted to a more targeted usage of the Freeform resources in comparison to a more diffuse usage in the original institution. Such difference in usage pattern might be reasonably attributable to cultural features specific to each context that shaped students' resources-usage behaviors.

2) Characteristics of resource-usage patterns

Figure 1 describes how students in each of the four clusters used Freeform resources. As expected, students did not use one specific resource over the others but instead combined multiple resources into an academic plan that they thought would meet their learning needs. All students across the four clusters, used consistently at least one of Freeform core resources, i.e., classmates, lecturebook, online videos, and the course blog. In contrast, students at the institution where Freeform originated (figure 2) consistently used at least two of the core resources. In the adopting institution, we note a consistent pattern across all students consisting of frequently using the help room (1-2 times/ week or more) while rarely resorting to classmates (less than 1-3 times per month). In stark contrast, students at the original institution frequently relied on their classmates (at least 1-2 times/week) and barely used the help room (less than 1-3 times/ month).

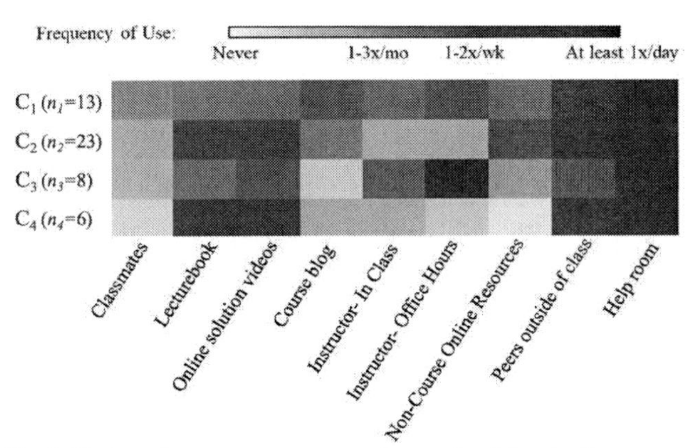

Fig. 1. Clusters of average resource usage frequencies at the adopting institution

A closer look at each of the four clusters at the adopting institution reveals distinctive features among clusters. Students of cluster C_1 used on average the course resources the most frequently. In addition, they displayed a more diversified usage pattern compared to the three other clusters. Conversely, students in cluster C_4 used the course resources the least frequently while displaying a concentrated resource usage pattern around the lecturebook, watching online solution videos, reaching out to peers outside of class and finally using the help room. Students of cluster C_2 displayed a similar resource-usage pattern but with a more focused usage on additional online resources such as the course blog and non-course online resources such as Youtube videos. Students of cluster C_3 seem to prefer interactions with the instructor team over their peers. In fact, students in that cluster are the only ones who used instructor office hours frequently. Conversely, they are the ones who used the least their peers whether in class or outside.

These archetypical resource-usage patterns in the adopting institution differ notably from those of the original institution. In our previous work focused on the original institution of Freeform, we established that collaboration was a distinctive trait in students' resource-usage strategies (Dridi et al., 2022). Such collaboration was particularly salient in students' reliance on their peers in class as a major resource. However, students in the adopting institution approached collaboration differently as they relied more heavily on their peers outside of class. It seems though that the survey respondents might have interpreted "peers outside of class" not as students who previously took Dynamics but rather as interacting with their classmates outside of class. In fact, our interviews with the students at the adopting institution revealed that using the mobile group messaging app GroupMe was a constitutive feature of the students' culture at the adopting institution. This might explain why we see low usage of the course blog across the four clusters. Keeping in mind that the survey was distributed post pandemic, it is also reasonable to assume that a new culture of out-of-class online communication between students had become the norm. In addition, the students whom we interviewed made note of low attendance in class which was also confirmed by the course instructor. In such case, students might not have known each other and therefore had to turn to students who had previously taken the course for support.

Another notable difference between the two institutions resides in how frequently students at the adopting institution sought help from the instruction team (both the instructor and the TA team). In the original institution, our previous research revealed an avoidance mechanism from students towards the instructor and a perceived hierarchal power differential that might contribute to such avoidance (Dridi et al., 2022). In contrast, students from the adopting institution appear to be more comfortable reaching out to the instructor either in class or during office hours as well as to the TA team. This is reflected in our interviews with the students of the adopting institution who expressed a strong alignment with the instructor teaching and pedagogy. In addition, the interviewed students described the TA team as being a key factor in their success in class and recommended future students of the course to abundantly use that resource.

https://doi.org/10.52202/073963-0027

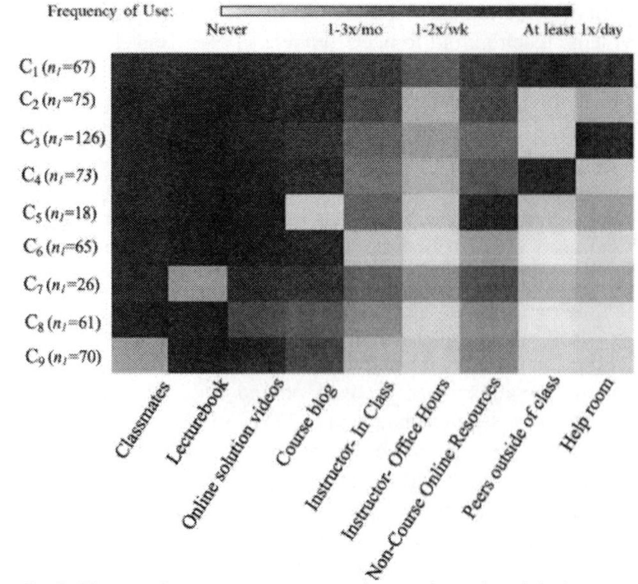

Fig. 2. Clusters of average resource usage frequencies at the original institution.

V. Discussion and Future work

In summary, our analysis uncovered four main findings related to the archetypical resource-usage across both institutions. Students at the adopting institution 1) displayed fewer archetypical resource-usage patterns (i.e., a smaller number of clusters), 2) used Freeform core resources less frequently across all the clusters (i.e., average usage frequencies, 3) collaborated more with their peers who had previously taken the class, and 4) interacted more frequently on the instruction team (i.e., instructor and TAs). We offer two types of hypotheses to explain these four findings, namely contextual and cultural potential factors.

In terms of contextual factors, we noted that students mentioned in the interviews not being aware of the availability of online videos as one of the resources at their disposal. Since these videos offered solutions to typical homework solutions, it is fair to assume that students resorted to the help room more often to understand how to solve problems. Furthermore, it is also important to recognize that the Freeform implementation and associated data collection at the adopting institution was conducted post-pandemic, while our prior research study at the original institution happened pre-pandemic. This element might help contextualize the observed shift in students' resource-usage pattern in the context of the Freeform ecosystem.

In terms of cultural factors, we noted that students at the adopting institution felt closely supported by the instruction team and found the office hours with the instructor particularly helpful. This could be indicative of student-centric culture among faculty with an orientation towards in-person consultation with members of the instructional team. In addition, students' orientation towards help-seeking and collaboration from students who previously took the course in the adopting institution suggest a distinct collaboration culture compared to the original institution where students had notable preference towards interactions with co-enrolled students.

National or societal cultures play a pivotal role in determining the dynamics of the classroom. One of the most significant aspects is the relationship between students and authority figures, such as instructors. In many Western cultures, there is a focus on egalitarianism, promoting open dialogue between students and teachers. Conversely, in many Asian or African cultures, there is a pronounced hierarchical structure, where teachers are seen more as figures of reverence, and challenging their perspectives might be viewed as culturally inappropriate. This cultural gradient can have profound implications for pedagogical innovations. For example, an innovation that encourages students to openly critique and question instructional content may thrive in a culture that values open dialogue. However, the same innovation may require substantial adaptation in a culture where such behaviors might be viewed as confrontational or disrespectful.

Both contextual and cultural factors point to the inevitable adaptation of pedagogical innovations to local characteristics of the implementation institution. Therefore, it is important for both pedagogical innovation designers and implementers to consider the idiosyncrasies of propagation settings before rolling out those innovations.

In previous work, we described how cultural consensus theory (CCT) can be used to characterize the unique cultural characteristics of both faculty and students. We also explored how those cultural dimensions converge or diverge with the design ethos of the Freeform ecosystem. In future work, we will use a CCT-based analysis to explore how student cultural characteristics explain their resource-usage patterns. Specifically, we will conduct a clustering analysis of the same students' sample using the cultural consensus analytical framework we developed previously (Berger et al., 2021) and assess to what extent the cultural clusters overlap with the resource-usage ones.

VI. Acknowledgment

We gratefully acknowledge the support and collaboration of the full MEERCat Purdue team, as well as the administration, faculty, and staff at the adopting institution and Purdue University. This research was supported by the U.S. National Science Foundation under Grant No. 1915574. Any opinions, findings, and conclusions expressed in this material are those of the authors and do not necessarily reflect the views of the National Science Foundation

References

Berger, E. J., Wu, C., Briody, E. K., Wirtz, E., & Rodríguez-Mejía, F. (2021). Faculty subcultures in engineering

https://doi.org/10.52202/073963-0027

and their implications for organizational change. *Journal of Engineering Education*, *110*(1), 230–251.

Borrego, M., Cutler, S., Prince, M., Henderson, C., & Froyd, J. E. (2013). Fidelity of Implementation of Research-Based Instructional Strategies (RBIS) in Engineering Science Courses. *Journal of Engineering Education*, *102*(3), 394–425. https://doi.org/10.1002/jee.20020

Evenhouse, D., Lee, Y., Berger, E., Rhoads, J. F., & DeBoer, J. (2023). Engineering student experience and self-direction in implementations of blended learning: A cross-institutional analysis. *International Journal of STEM Education*, *10*(1), 19. https://doi.org/10.1186/s40594-023-00406-x

Fatos, X., Daradoumis, T., & Martínez-monés, A. (2006). A Layered Framework For Evaluating On-line Collaborative Learning Interactions. *International Journal of Human-Computer Studies*, *7*, 622–635.

Freeman, S., Eddy, S. L., McDonough, M., Smith, M. K., Okoroafor, N., Jordt, H., & Wenderoth, M. P. (2014). Active learning increases student performance in science, engineering, and mathematics. *Proceedings of the National Academy of Sciences*, *111*(23), 8410–8415. https://doi.org/10.1073/pnas.1319030111

Jeong, H., & Chi, M. T. H. (2007). Knowledge Convergence And Collaborative Learning. *Instructional Science*, *35*(4), 287–315. https://doi.org/10.1007/s11251-006-9008-z

Kandakatla, R., Berger, E. J., Rhoads, J. F., & DeBoer, J. (2020). Student Perspectives On The Learning Resources In An Active, Blended, and Collaborative (ABC) Pedagogical Environment. *International Journal of Engineering Pedagogy (iJEP)*, *10*(2), Article 2. https://doi.org/10.3991/ijep.v10i2.11606

M. A. Dridi, E. Berger, J. F. Rhoads, & J. DeBoer. (2022). Characterizing students' subcultures in engineering and their alignment with the adoption of the Freeform pedagogical system. *2022 IEEE Frontiers in Education Conference (FIE)*, 1–7. https://doi.org/10.1109/FIE56618.2022.9962509

Means, B. (2014). *Learning Online: What Research Tells Us About Whether, When and How*. Routledge.

Merriam, S. B. (2009). *Qualitative Research: A Guide to Design and Implementation*. John Wiley & Sons.

O'Donnell, C. L. (2008). Defining, Conceptualizing, and Measuring Fidelity of Implementation and Its Relationship to Outcomes in K–12 Curriculum Intervention Research. *Review of Educational Research*, *78*(1), 33–84. https://doi.org/10.3102/0034654307313793

Rhoads, J. F., & Krousgrill, C. (2013). *Dynamics: A lecturebook* (1.1). AcademicPub.

Rhoads, J., Nauman, E., Holloway, B., & Krousgrill, C. (2014). The Purdue Mechanics Freeform Classroom: A New Approach to Engineering Mechanics Education. *121st ASEE Annual Conference & Exposition, Indianapolis, IN. June 15-18, 2014*. https://peer.asee.org/23174

Stites, N. A., Berger, E., Deboer, J., & Rhoads, J. F. (2019). *A Cluster-Based Approach to Understanding Students' Resource-Usage Patterns in an Active, Blended, and Collaborative Learning Environment*.

Zadoks, A., Stites, N. A., Evenhouse, D., Patel, N., Kandakatla, R., Berger, E., Rhoads, J. F., & DeBoer, J. (2017). Longitudinal analysis of instructor actions in an active, blended, and collaborative classroom environment. *2017 IEEE Frontiers in Education Conference (FIE)*, 1–6. https://doi.org/10.1109/FIE.2017.8190521

Research in Engineering Education Symposium-2024
Jan 4-6, 2024 - Hubballi, India

Svadhyaya: A Story Telling Model for Discovery Oriented Learning

Prakash Hegade, Ashok Shettar, Vijayalakshmi M and Preeti Patil
KLE Technological University
prakash.hegade@kletech.ac.in

Abstract

Context

Developing problem solving skills has been an emphasis from schooling to degree programs and beyond. Problem based learning and discovery oriented learning share several common philosophies of learning. One can be integrated into another for an effective classroom delivery. On other hand, NEP 2020 aims to bring and develop the essential problem solving skills rooting from our arts and literature.

Purpose or Goal

Considering the importance of problem based learning, discovery oriented learning and storytelling, this work proposes the following research questions: How can we integrate the story telling experience from arts and literature into the discovery oriented learning model integrated with problem based learning? How can a case study help to build an inventory of use-cases that can be connected to computer science concepts and principles?

Methods

With a pragmatic philosophical assumption, qualitative research method was adapted where in two cycles of coding: structural, vivo and focused coding were used for data analysis. Data was collected using case study approach. Self-selection was used for sampling and a total of 26 participants were part of the study in three iterations excluding the researcher. Case study templates were designed and improved with iterations starting from the seven jump problem based learning model.

Outcomes

Themes were generated for each of the iteration and the template of problem analysis was improved over the process. A structural phase-wise associative template was developed for the storytelling model. A deductive tree was designed to generate inventory of use cases as a further closure addressing the two formulated research questions.

Conclusion

Objectives, culture and deductions being identified as the three major elements of the case study, further evolved to a template that can assist in storytelling case study design. Storytelling promises to be one of the means to support the discovery oriented learning.

Keywords—discovery oriented learning; problem based learning; story telling; svadhyaya

I. INTRODUCTION

ONE of the essential life skills that an individual must hone to excel in professional life is the problem solving capability. From schooling to a degree program, and also beyond, problem solving skill is emphasized via various aspects. Several models have been designed and deliberated to enhance the problem solving skills. The ability to solve problems is articulated as one of the most important manifestations of human thinking (Holyoak, 1990). The theories of problem solving have been discussed giving prominence to the problem space and providing a framework for understanding the process of solving them (Newell & Simon, 1972). Though the problem solving process can be evaluated (Charles, 1987), the context and methodology usually varies from the domain and the result required.

The experiences of problem solving can help one learn the content and also develop the thinking strategies (Hmelo-Silver, 2004). There is copious literature to prove the effectiveness of problem based learning in numerous learning environments. From problem space analysis to writing objectives and building a project, the approach has been realized in various formats and synergies. The goal four of the United Nations sustainable goals talks about 'quality education' (Biermann et al., 2017). In the regards, the Union Cabinet of India rolled out India's new education system policy through National Education Policy, approved on 29th July 2020.

The National Education Policy 2020 document accentuates the below mentioned opinions (Govinda, 2020):

- The need for a pedagogy that makes education more experiential, holistic, discovery-oriented, flexible and enjoyable.
- The curriculum must be also based on arts, literature, culture, values, etc.
- The course delivery must emphasize conceptual understanding.
- The pedagogy must have story-telling and art-integration.
- The new skilled professionals must have holistic development with respect to humanities, art, social science, problem solving etc.

There can always be alternatives and varied schools of thought to realize the culture and art-form based models in the education system. However, the end learning objective remains the same: pedagogy to enhance the learning via story telling. Through life experiences and events though one develops basic problem solving capability, modern complex problems need more than routine thinking. Understanding the basic principles from the classics can strengthen ones comprehension skills and can serve as major criteria for applied skill enhancement.

Discovery oriented learning emphasizes learning experiences as against the memorization. It attempts to provide meaningful experiences (Castronova, 2002) in the process of learning. The research part of finding out solutions to the problem allows students to analyze information and integrate the concepts for an improved educational experience. Problem solving, active engagement, critical thinking, self-directed learning, inquiry and exploration and application of knowledge are some of the common characteristics between problem based learning (De Graaf and Kolmos, 2003) and discovery oriented learning. It has also been compared with other learning strategies (Savery, 2006). Using discovery oriented learning for problems in problem based learning can help in initial exploration, promote self-directed inquiry, help to address open ended questions and provide a space for reflection and synthesis (Efedni et al., 2020). Hence this work integrates both of them to design a story-telling model.

The paper is further divided into following sections: Section 2 presents the literature survey. Section 3 presents the research design and methodology. Section 4 presents the data analysis and findings. Section 5 initiates the discussion with conclusion in section 6.

II. Literature Survey

This section presents the literature survey majorly on Problem Based Learning (PBL), Discovery-Oriented Learning (DOL) and storytelling. DOL emphasizes open-ended exploration. It is often initiated by the student's curiosity. DOL study often beings with no pre-defined problem to solve. Though there are several common characteristics, both of them also differ on numerous aspects. DOL can be an effective tool to realize PBL. This section hence reviews both the domains.

PBL finds its roots from the medical domain (Barrows, 1986), giving a thick literature of its usage (Barrows and Tamblyn, 1980), and today being used by all other domains. It has also been integrated with other learning styles and frameworks, for example its experimentation in a constructivist learning environment (Savery and Duffy, 1995). If not used for entire curriculum, PBL has been integrated in the classrooms on need basis (Stepien and Gallagher, 1993) as well used by administrators since ages (Bridges, 1992).

PBL is effective on several fronts and usually works well for smaller classrooms (Schmidt, 2011). Though there is no one common definition of PBL, it has common characteristics like critical thinking, self-directed learning, collaboration etc. When the learning is experience based, students learn through the thinking strategies (Hmelo-Silver, 2004). A meta-analysis on the method states that the process is proven to be effective in the learning process (Dochy et al., 2003).

The benefits of PBL have been psychologically analyzed (Norman and Schmidt, 1992). The method has been employed from schools (Achilles and Hoover, 1996) to start-ups (San and Ng, 2006) and the traditional methods have been critiqued in comparison (Mills and Treagust, 2003). PBL helps in knowledge construction process (Colliver, 2000). Merits and demerits of the process have been analyzed (Schmidt et. all., 2011).

In DOL students vigorously explore topics, concepts, or phenomena independently or via guided inquiry (Hammer, 1997). It encourages self-directed learning, critical thinking, and practical exploration, allowing deeper understanding of concepts by letting learners make detections and connections through their own efforts and inquisitiveness. It's a student-centered method (Wolfe, 1992). The theoretical foundations of the method have been discussed (Svinicki, 1998). The method also encourages creative thinking (Rahman, 2017). The model's effectiveness has been studied with respect to critical and cognitive thinking (Martaida et al., 2017). Innovative studies like using moon for DOL have been carried out (Cummins et al., 1992). From mathematics to English, it has been used in several domains as a teaching-learning strategy. We can however comprehend that PBL and DOL are based on the common theories of active learning, inquiry and exploration, critical thinking, collaborative learning. DOL can be integrated with PBL as a mode to solve the problems. DOL and PBL models have been compared and deliberated for similarities and differences and the study concluded that in terms of high-level thinking ability both the models provide the same effectiveness (Setyaningrum et al., 2020).

Story telling in education research has been explored on theoretical basis with story as method (Gallagher, 2011). It has been used as a pedagogical tool in higher education (Abrahamson, 1998). With the advent of technology, story-telling has gone digital and has been integrated in education pedagogies (Alismail, 2015).

The literature survey provides a scope to combine all the methods and develop a model to use it as pedagogy in the classroom teaching and learning. The model can also enable a student to construct a knowledge base which can further assist as per the NEP 2020 guidelines. Using the culture and folklore can also spike the interest in students on how the classic theories and stories are relevant in the contemporary world. With the common characteristics, storytelling and DOL can be integrated with PBL as a delivery model for the problem scenarios.

https://doi.org/10.52202/073963-0028

III. METHODOLOGY

This section presents the research design that was followed for the proposed work. Philosophical assumptions, research question, model, context, data collection process, sampling methods and case study selection process are discussed.

A. Philosophical Assumptions

Pragmatic philosophical assumptions (Creswell & Poth, 2013) are made for this work as it allows the researcher to follow dynamic and innovative ways to explore the research problem at hand (Morgan, 2014). It allows a researcher to take operational decisions in the best interest of research study domain (James, 1975). In the context of research study, observes its inherently defined limitations and biases that could be reflected in the study participants and researchers describing the axiological beliefs. The reality for this work, being multiple, was constructed based the research participants and from the researcher perspective. The knowledge construction has happened with researcher being an insider and by comprehending subjective evidences from the research participants. Every participant constructed their own reality with their experiences constituting the ontology. The methodology followed is qualitative approach.

B. Research Question

Keeping the motivation of the study in mind, in consideration to PBL and DOL, two research questions were formulated as listed below:

RQ1: How can we integrate the story telling experience from arts and literature into the discovery oriented learning model integrated with problem based learning?

RQ2: How can a case study help to build an inventory of use-cases that can be connected to computer science concepts and principles?

C. Context

The context of the study was students who had completed their second year of engineering from KLE Technological University from School of Computer Science and Engineering. The study was carried out over two years with two iterations and two batches.

D. Initial Model Elements

The initial model design work was carried out by using the seven jump model (Murwantini, 2015) which was applied on painting Guernica painted by Pablo Picasso. The painting features imagery and symbolism depicting the horrors of the war, particularly the bombing of the Spanish town of Guernica during the Spanish Civil War (Ray, 2006). The reason to select this case study is because of its color scheme, the figures used, the emotional space and the conflicts it gives rise to (Patterson, 2007). Seven jump is one of the prominent PBL models used by several universities and domains (Harimurti, 2023). The different steps of seven jump model was listed out and analysed with the identified case study. As a researcher, then the steps were used on the painting Guernica to arrive at the initial model elements of Svadhyaya. The term Svadhyaya means self-study derived from the language Sanskrit. The components of the model developed can be seen in Figure 1.

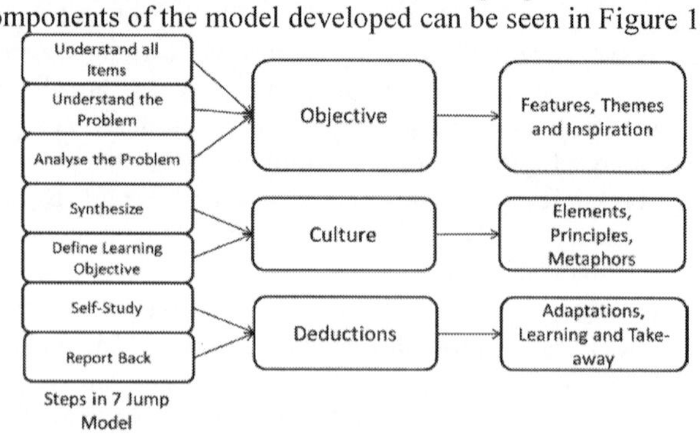

Fig 1. Components of Svadhyaya

The major components identified for the model were Objectives, Culture and Deductions. The mapping of the different phases of the seven jump model is presented in the figure. Each element is defined by three listed properties as shown in the Figure 1. As an example, Culture was identified by the three features namely: elements that constituted it, principles that defined it and metaphors related to it. The features were identified by the detailed qualitative analysis on the developed case study by using structured and in-vivo coding.

E. Data Collection Process and Sampling

Students were provided with case study and templates for the completion and submission with 3 day time. Students were free to use any available resources for the study. No specific training was provided to complete the case studies. Student submitted sheets were used for analysis. Cases studies were employed are they can be used to connect and understand complex issues of real-life context (Cousin, 2005). Case studies can be used for critical thinking and reflections on the real-life scenarios (Merriam, 1988). Cases from arts and literature were selected as they naturally assimilate into the storytelling (Meretoja & Davis, 2017) and provide a power for narration (Lothe, 2017). Reflections on the arts and storytelling have been discussed in the literature (Johnson, 2016).

As per the university guidelines a consent form was shared detailing the research question and process to all the participants. Only upon the agreement, students were handed over the task. The participation was optional. Self-selection was used for the data collection (Sharma, 2017). A call-for was made for the students to participate in the research during summer holidays. Everyone who agreed was provided with a case study and a template to complete. Snowballing process was further used to enroll if anyone else from their contact would be interested in the process (Goodman, 1961). In snowball sampling participants assist in identifying the potential participants for the study.

https://doi.org/10.52202/073963-0028

F. Participant Details

The initial design was made by the researcher in the first iteration. For second and third iterations, research participants were students who had completed their second year. The numbers are presented in Table 1. The total process took over in three iterations of refinement.

TABLE I
PARTICIPANTS

SI. No.	Iteration	Number of Participants
1	01	01
2	02	07
3	03	19

G. Objectives

The objectives of the work are as listed below in Table 2. The objectives were designed based on the research questions and they were further used to design the case study using pieces from arts and literature. The objectives were used to break down the research questions into smaller achievable tasks.

TABLE II
OBJECTIVES

ID.	Objective
OB1	To identify the works from arts and literature for the case study design
OB2	To comprehend the parameters of storytelling via a template design and iteratively improve it.
OB3	To realize the story telling elements into the engineering course delivery using the metaphors and inventory of use-cases

H. Case Study Selection

In the first round the case study selected was the painting Guernica as it has ample metaphors connected to it (Picasso et al., 1956). The painting has also been adapted into several other forms. The painting is basically a story of war and its sufferings. It was a suitable case study to apply the seven steps on. In the second round, as there were seven students, seven new case studies were selected. The selection was made from the possible entire genre list that could be thought of. Books, movies and variety of domains were covered based on popularity and its social awareness from the arts and literature. The selection was guided by the first version of template designed. Cases were selected that had objectives, culture and deductions. The seven selected case studies are presented in Table 3. Every student was assigned with one case study. In the second round, three more studies were included as there were more students and each student was asked to submit two case study solutions. The additional list of three is presented in Table 4.

TABLE III
CASE STUDY LIST 1

No.	Case Study
1	The destruction of Tripura
2	Horcrux from Happy Potter
3	Infinity Stones from Avengers
4	The Seven Ravens
5	The Game of Dice from Mahabaratha
6	The Lord of the Flies
7	Schindler's List

TABLE IV
CASE STUDY LIST 2

No.	Interview Question
1	The Old Man and the Sea
2	Rigveda
3	Aboriginal Australians

IV. DATA ANALYSIS AND FINDINGS

For case study completion, students were provided with a table template to fill the data. The template provided is presented in Table 5 below. This template was designed by the researcher with the model as described in the Figure 1.

TABLE V
CASE STUDY TEMPLATE 1

Criteria	Description
Case Study Metadata	
Name	Name here
Case Study	Cast study name
Prominence	What is the prominence of the case study?
Synopsis	Write a short synopsis
Model Analysis	
Phase 1: Objective	
Theme	What theme do you see in this art?
Inspiration	Does it have an inspiration?
Features	What are the prominent features?
Phase 2: Culture	
Elements	What are the major elements of the art?
Metaphors	Does it stand as a metaphor for something?
Principles	What principles does it observe?
Phase 3: Deductions	
Adaptations	Has there been an adaptation of it?
Learning and Take-away	What did you learn from this? Can you write its applications?

On the case study data, two rounds of coding were carried out. In the first round a mixture of structured and in-vivo coding was used. In structured coding, we code the passages according to the research question or topics (Lampert & Ervin-Tripp, 1993). In In-vivo coding we use the exact phrases and words that are collected from description and perspectives

https://doi.org/10.52202/073963-0028

(Manning, 2017). Focused coding was employed for the second round where we categorize the related and merge the information at need (Stuckey, 2015).

A next level template was generated for iteration 3 students based on the data analysis carried out from iteration 2. The template is presented in Table 6.

TABLE VI
CASE STUDY TEMPLATE 2

Criteria	Description
Case Study Metadata	
Name	Your Name Here
Case Study	Cast study name
Prominence	What is the prominence of the case study?
Synopsis	Write a short synopsis
Model Analysis	
Phase 1: Init	
Theme	What theme do you see in this case study? What does it represent?
Inspiration	Does it have an inspiration?
Features	What are the prominent features?
Objective	What objectives do you observe?
Phase 2: Operational	
Culture	Does it bring out any cultural elements?
Elements	What are the major elements of the case study?
Metaphors	Does it stand as a metaphor for something? Do you connect to some other work of similar kind? Does it have an abstraction?
Principles	What principles does it observe?
Phase 3: Deductions	
Adaptations	Has there been an adaptation of it? Movie? Song? Why?
Analysis	What is your analysis? Did it amaze you?
Inventory of Use-Cases	Write the applications and context you observe. Where can this case study be used? Can we use it to teach something?

Some of the coding samples are described along with how the themes were generated. A student had written inspiration as 'Good and evil, war and peace' for destruction of Tripura. These clearly indicated the THEME of the story. The dice game unfolded later to a war. The theme here was PROBABILITY GAME. With such code generations, a new element THEME was generated with a description of 'what does this represent'.

Students had listed out all the characteristics and parts that constitute a case study. In Seven Ravens, a complete list of characters, sun, moon, grief, death, baptism etc. were listed out. Hence a new theme of ELEMENTS was coded. USE CASES theme was generated in a similar way so to extend a case study to other domains and list its applications.

Horcrux was to protect the one who created it. Tripura wanted to bring order in the universe. Such statements represented the OBJECTIVE of the work and hence a theme was designed for the same. Students had listed out varied thoughts of objectives of each work and they were all relevant. This led to the generation of next level template.

The second template being the learning from the first one was clear with its objectives. Students were given a clear set of indicators on how to work on the case study and what to discover further. They were asked to build upon an inventory of case studies in the end for the case study they were exploring. Trigger points were provided for each element in order to gain a new perspective from the case study. The Table 7 below lists some of the use cases listed by students.

TABLE VII
USE CASES SAMPLES

Sl. No.	Case Study	Use Case
1	Infinity Stones	The six stones are six ingredients of preparation
2	Horcrux	Once when something is on internet, we don't know how many forms it has taken. It may never be deleted and copy might always exist.
3	Game of Dice	Marketing strategies
4	Destruction of Tripura	The strength of doing together, time loop, radicalism of group
5	The Old man and The Sea	When process is more important that results

The case studies have contributed in the learning process and to discover new perceptions about the world we live in. On asking if we could use these case studies as learning materials, 18 students said 'yes'. The numbers are in Table 8. 'Not sure' could also mean a polite no. However, it could also mean they need more clarity on where and how to use it.

TABLE VIII
CASE STUDY USAGE

Attribute	Number of Agreement
Yes	18
No	0
Not sure	8

V. DISCUSSION

Considering the first research question, our analysis indicates that Table 6, case study template 2 is the means by which we can integrate the story telling experience from arts and literature into the discovery oriented learning model integrated with problem based learning. A faculty or a center that generates case studies, has to select a work from arts and literature, apply and analyze with the template and with the uses cases and metaphors generated, can decided to use it on the allied principles and concepts. The table can also help to deduce the cultural elements from the study.

Story telling is an art. The different adaptations of each case study indicate that the same story can take different forms based on the domain. This also means that they can be used to understand principles of foundational courses in engineering (Refer Table 7). Resources optimization is the major theme for operating system course. Process management is what software engineering all about. The different elements constitute in the schema design of the database. These were the points that came from the student inventory use-case samples. This answers the research question number two (Refer Figure 3).

A detailed analysis of these case students has led to a path that they can be connected to the engineering concepts. The strategies could also motivate students to come up with new algorithms design techniques which is the current need of the hour in data science. Most important of all, it makes students curious and excited. Students were asked how much time they spent on the case study and the numbers can be seen in the Figure 2 below.

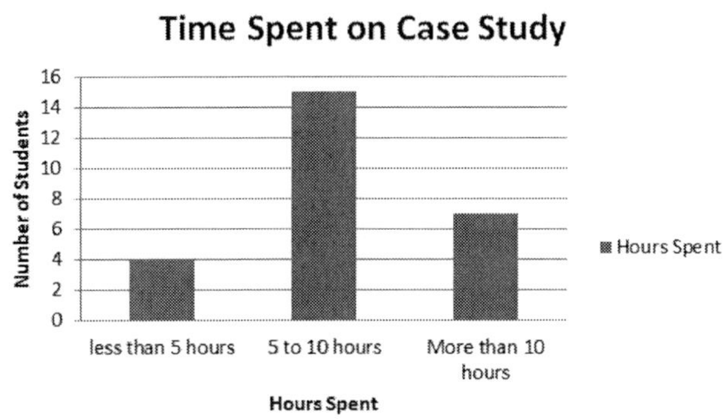

Fig 2. Time spent on case study

Majority (15) of the students spent 5 to 10 hours on each case study. 7 students expressed that they spent more than 10 hours.

A deductive tree can be seen in Figure 3 below.

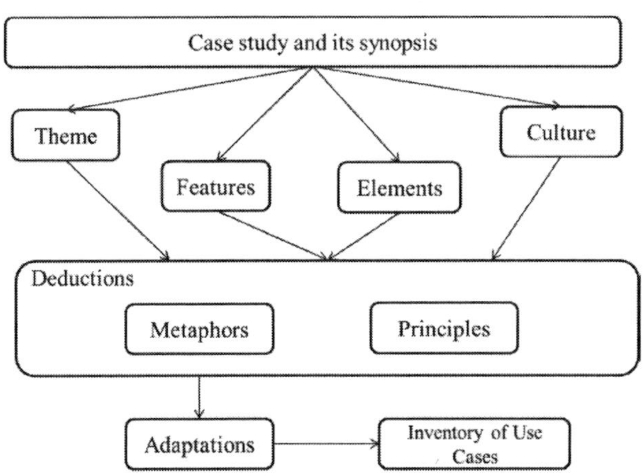

Fig. 3: Case study deductive tree

The tree represents, drawn from the Table 6 elements, as we believe, on how the knowledge of a student grows with respect to the case study. This hierarchy can lead to discovery of use-cases that they can connect with the course concepts. Before arriving at metaphors and principles, they need to identify the four mentioned components on the synopsis. This tree can be used as a template for the storytelling and discovery oriented learning. The tree emerged during the second level coding in iteration 3. The themes, features, elements and culture in the case study help in arriving at deductions. From the deduction, we can classify the metaphors and principles. The different adaptation of the study can be analyzed using these and they lead to the inventory of use cases that can be used as the computer science study materials.

Following are the other significant conclusive discussion points that we can arrive at from the case study research. Each case study had several metaphors to unravel (Barcelona, 2001). The model helped students to get a bigger picture of a classic tale. A lot of our ancient stories have symbolism (Feldman, 1990). They are not just stories, but they capture the social orders and dynamics of a society and they can be used in the case study and storytelling (Goodwin, 1982).

The analysis part connected to the course concepts that students have studied in the past. The examples can be used to connect to the concepts from all the domains. As most of the research is influenced by the sociology, the arts and literature can be used to convey not only the morale but can also stand as a foundational concepts of theories (Landrum et al., 2019). The deductions can help one to build a frame work with respect to individual domain of consideration. The principles can be used to give larger and basic meanings of life principles, to arrive at, and their applications.

VI. Conclusion

Arts and literature are also the means of dissemination of morale and life principles into the social order. An artistic tells a story with his work. Since the ancient times they have been a medium to preserve and forward the cultural and demographic characteristics. Svadhyaya model attempts to bring them systematically into the classroom case studies. Svadhyaya model encompasses the critical elements: objectives, culture and deductions for a case study. The model promises to be one way of bringing arts and literature into the classrooms via storytelling. NEP 2020 aims to build a knowledge base to use such works from arts and literature into the class room delivery. With the template designed, Svadhyaya keeps the first step towards such a knowledge base.

The discussion section also provides a scope to design future research questions, design instruments and validate the metaphors, principles, adaptations and inventory use-cases, effectively and probably with discovery-oriented learning.

https://doi.org/10.52202/073963-0028

References

Achilles, C. M., & Hoover, S. P. (1996). Exploring Problem-Based Learning (PBL) in Grades 6-12.

Abrahamson, C. E. (1998). Storytelling as a pedagogical tool in higher education. Education, 118(3), 440-452.

Alismail, H. A. (2015). Integrate digital storytelling in education. Journal of Education and Practice, 6(9), 126-129.

Barrows, H. S., & Tamblyn, R. M. (1980). Problem-based learning: An approach to medical education. Springer Publishing Company.

Barrows, H. S. (1986). A taxonomy of problem-based learning methods. Medical education, 20(6), 481-486.

Barcelona, A. (2001). On the systematic contrastive analysis of conceptual metaphors: Case studies and proposed methodology. Applied cognitive linguistics II: Language pedagogy, 117-146.

Biermann, F., Kanie, N., & Kim, R. E. (2017). Global governance by goal-setting: the novel approach of the UN Sustainable Development Goals. Current Opinion in Environmental Sustainability, 26, 26-31.

Bridges, E. M. (1992). Problem Based Learning for Administrators. ERIC Clearinghouse on Educational Management, University of Oregon, 1787 Agate Street, Eugene, OR 97403-5207.

Castronova, J. A. (2002). Discovery learning for the 21st century: What is it and how does it compare to traditional learning in effectiveness in the 21st century. Action research exchange, 1(1), 1-12.

Charles, R. (1987). How To Evaluate Progress in Problem Solving. National Council of Teachers of Mathematics, 1906 Association Drive, Reston, VA 22901.

Colliver, J. A. (2000). Effectiveness of problem-based learning curricula: research and theory. Academic medicine, 75(3), 259-266.

Cousin, G. (2005). Case study research. Journal of geography in higher education, 29(3), 421-427.

Creswell, J. W., & Poth, C. N. (2013). Philosophical assumptions and interpretive frameworks. Qualitative inquiry and research design: choosing among five approaches. Los Angeles: Sage Publications, 15-41.

Cummins, R. H., Ritger, S. D., & Myers, C. A. (1992). Using the Moon as a tool for discovery-oriented learning. Journal of Geological Education, 40(2), 142-146.

Dochy, F., Segers, M., Van den Bossche, P., & Gijbels, D. (2003). Effects of problem-based learning: A meta-analysis. Learning and instruction, 13(5), 533-568.

De Graaf, E., & Kolmos, A. (2003). Characteristics of problem-based learning. International Journal of Engineering Education, 19(5), 657-662.

Efendi, F., Fitria, Y., Farida, F., & Hadiyanto, H. (2020). Perbedaan Model Problem Based Learning dengan Discovery Learning terhadap Higher Order Thingking Skills dan Self Directed Learning di Sekolah Dasar. Jurnal Basicedu, 5(1), 301-309.

Feldman, S. P. (1990). Stories as cultural creativity: On the relation between symbolism and politics in organizational change. Human relations, 43(9), 809-828.

Gallagher, K. M. (2011). In search of a theoretical basis for storytelling in education research: Story as method. International Journal of Research & Method in Education, 34(1), 49-61.

Govinda, R. (2020). NEP 2020: A critical examination.

Goodman, L. A. (1961). Snowball sampling. The annals of mathematical statistics, 148-170.

Goodwin, M. H. (1982). "Instigating": Storytelling as social process. American ethnologist, 9(4), 799-816.

Hammer, D. (1997). Discovery learning and discovery teaching. Cognition and instruction, 15(4), 485-529.

Harimurti, S. M. (2023). Seven Jump PBL to Connect Modern Problems to Sciences. Asia Pacific Journal of Management and Education (APJME), 6(1), 1-15.

Hmelo-Silver, C. E. (2004). Problem-based learning: What and how do students learn?. Educational psychology review, 16(3), 235-266.

Holyoak, K. J. (1990). Problem solving. Thinking: An invitation to cognitive science, 3, 117-146.

James, W. (1975). Pragmatism (Vol. 1). Harvard University Press.

Johnson, C. (2016). The Way of the Writer: Reflections on the Art and Craft of Storytelling. Simon and Schuster.

Lampert, M. D., & Ervin-Tripp, S. M. (1993). Structured coding for the study of language and social interaction. Talking data: Transcription and coding in discourse research, 169-206.

Landrum, R. E., Brakke, K., & McCarthy, M. A. (2019). The pedagogical power of storytelling. Scholarship of Teaching and Learning in Psychology, 5(3), 247.

Lothe, J. (2017). Storytelling and Ethics: Literature, Visual Arts and the Power of Narrative.

Martaida, T., Bukit, N., & Ginting, E. M. (2017). The effect of discovery learning model on student's critical thinking and cognitive ability in junior high school. IOSR Journal of Research & Method in Education, 7(6), 1-8.

Manning, K. (2017). Organizational theory in higher education. Routledge.

Meretoja, H., & Davis, C. (Eds.). (2017). Storytelling and ethics: Literature, visual arts and the power of narrative. Routledge.

Morgan, D. L. (2014). Pragmatism as a paradigm for social research. Qualitative inquiry, 20(8), 1045-1053.

Mills, J. E., & Treagust, D. F. (2003). Engineering education—Is problem-based or project-based learning the answer. Australasian journal of engineering education, 3(2), 2-16.

Merriam, S. B. (1988). Case study research in education: A qualitative approach. Jossey-Bass.

Murwantini, S. (2015). Application of Seven Jump Method with Approach to Problem Based Learning on Students in Program Studi of Mechanical Engineering Education Fkip Palangkaraya University Academic Year 2014/2015. BALANGA: Jurnal Pendidikan Teknologi dan Kejuruan, 3(1), 29-37.

Newell, A., & Simon, H. A. (1972). Human problem solving (Vol. 104, No. 9). Englewood Cliffs, NJ: Prentice-hall.

Norman, G. T., & Schmidt, H. G. (1992). The psychological basis of problem-based learning: A review of the evidence. Academic medicine, 67(9), 557-565.

Patterson, I. (2007). Guernica and total war (Vol. 4). Harvard University Press.

Picasso, P., Larrea, J., & Barr, A. H. (1956). Guernica. Hammarby Tryckeri.

Ray, B. (2006). Analyzing political art to get at historical fact: Guernica and the Spanish civil war. The Social Studies, 97(4), 168-171.

Rahman, M. H. (2017). Using discovery learning to encourage creative thinking. International Journal of Social Sciences & Educational Studies, 4(2), 98.

Savery, J. R. (2006). Overview of problem-based learning: definition and distinctions, the interdisciplinary. In Journal of Problem-based learning.

San Tan, S., & Ng, C. F. (2006). A problem-based learning approach to entrepreneurship education. Education+ Training, 48(6), 416-428.

Schmidt, H. G., Rotgans, J. I., & Yew, E. H. (2011). The process of problem-based learning: what works and why. Medical education, 45(8), 792-806.

Setyaningrum, W., Riani, A. L., & Wardani, D. K. (2020). Comparison of Problem-Based Learning and Discovery Learning Model. International Journal of Multicultural and Multireligious Understanding, 7(3), 305-313.

Sharma, G. (2017). Pros and cons of different sampling techniques. International journal of applied research, 3(7), 749-752.

Stepien, W., & Gallagher, S. (1993). Problem-based learning: As authentic as it gets. Educational leadership, 50, 25-25.

Stuckey, H. L. (2015). The second step in data analysis: Coding qualitative research data. Journal of Social Health and Diabetes, 3(01), 007-010.

Svinicki, M. D. (1998). A theoretical foundation for discovery learning. Advances in physiology education, 275(6), S4.

Wood, D. F. (2003). ABC of learning and teaching in medicine: Problem based learning. BMJ. 326 (7384): 328–330.

Wolfe, C. R. (1992). Using an authoring system to facilitate student-centered discovery oriented learning. Computers & Education, 19(4), 335-340.

 https://doi.org/10.52202/073963-0028

Path to SDG 2030: Fostering Sustainable Development at the School of Civil Engineering, KLE Tech - A Case Study

Anand M. Hunashyal[a]; Shivaraj Halyal[a]; Roopa A.K[a];

[a] School of Civil Engineering, KLE TECH, Hubli, India-5000830
Corresponding Author Email: roopa.kuri@kletech.ac.in

Abstract

Context

The 2030 United Nations' Agenda for Sustainable Development highlights the need to integrate Education for Sustainable Development (ESD) principles across all educational levels. By equipping individuals with knowledge and skills to address environmental and social challenges, ESD contributes to achieving the Sustainable Development Goals (SDGs) and building a sustainable future.

Purpose or Goal

Current civil engineering graduates lack awareness sustainability development, resource depletion, environmental pollution, rapid population growth, and ecosystem degradation during their undergraduate studies. It is crucial to address this gap by reevaluating the curriculum and incorporating courses that specifically educate students about the importance of sustainability.

Methods

The approach of this study emphasizes understanding global perspectives on sustainability and Sustainable Development Goals (SDGs) integration especially with civil engineering curriculum. Identifying gaps in the current curriculum is essential and subsequently a four phased implementation approach is proposed to address these gaps effectively. The study further analyzes various models to determine the most suitable training program for students during their academic journey.

Outcomes

By adopting a precise and context-specific approach, current study aims to provide insights into how sustainable development goals can be systematically integrated into the existing civil engineering curriculum and make more aligned with SDGs, ensuring that graduates are well-prepared to contribute to sustainable practices in their professional careers.

Conclusion

Eventually instilling sustainability principles early in their education, future civil engineers will be equipped with the knowledge and skills necessary to integrate sustainable practices into their professional work, mitigating the environmental impact associated with the overuse of materials

Keywords— Sustainable development, SDG Goals, quadrant approach curriculum, curriculum matrix

I. INTRODUCTION

Sustainable development is of paramount importance as it addresses critical global challenges and ensures a better future for current and future generations. It emphases on preserving the environment, eradicating poverty, promoting social inclusivity, fostering economic resilience, and enhancing health and well-being. Sustainable development calls for global cooperation, responsible consumption, and long-term prosperity. By embracing sustainable practices, we can create a more equitable, resilient, and prosperous world where people and the planet thrive together.

In response to this, the Sustainable Development Goals (SDGs) for 2030, introduced by the United Nations (UN) in 2015, encompass an extensive and ambitious global agenda focused on addressing urgent social, economic, and environmental challenges. Consisting of 17 interconnected goals, the SDGs aim to eradicate poverty, hunger, and inequality, while ensuring universal access to quality education, healthcare, and clean water. They advocate for sustainable economic growth, decent work, and responsible industrialization, fostering resilient infrastructure and innovation. Furthermore, the SDGs underscore the importance of gender equality, climate action, and the conservation of marine and terrestrial ecosystems, striving to establish sustainable patterns of consumption and production. Collaboration among governments, civil society, and the private sector is encouraged to accomplish these transformative objectives, aiming to create a world where no one is marginalized, and our planet is safeguarded for future generations. Achieving these goals requires a collective global effort, instilling hope for a more equitable, inclusive, and sustainable world by the year 2030 (Gutierrez-Bucheli et al., 2022).

Amongst various sectors, education sector plays a vital role in promoting and encouraging to accomplish the SDGs by 2030 (Álvarez et al., 2021). It raises awareness about the interconnected global challenges and empowers individuals to contribute to sustainable development efforts. Education equips students with the knowledge and skills to adopt sustainable practices, advocate for change, and drive policy development. Every student as sustainability change agents, can upsurge mindfulness and have the potential to create impacts vis-à-vis the SDGs at the individual, organizational and institutional levels (Hubscher et al., 2022).

Especially this is more factual in the engineering education through engineering students. Throughout the four years of innovation and problem-solving skills, engineering students can develop sustainable solutions that contribute to various SDGs, such as clean energy, sustainable infrastructure, and climate action. Their cross-disciplinary collaboration and ethical engineering practices enable comprehensive approaches to address issues related to cleaner environment, poverty, health, education, and more (Terrón-López., 2020). Engaging with communities and advocating for sustainable practices, engineering students have the potential to make a global impact and drive progress towards achieving the SDGs.

The implementation of sustainability into engineering is feasible by emphasizing analytical methods for analyzing the effects of technology, such as lifecycle evaluations of products or seeking for positive system transformation through technology and social change management and innovation (Mulder, 2017). Many writers outline learning objectives that includes broadly cognizant of global issues and changes, having the ability to comprehend competing norms and values regarding sustainability issues and to think in terms of general structures with continuous feedback, as well as being able to work across disciplinary boundaries (Dlouhá et al, 2017). In contrast, all of these approaches could fail to reach their objectives; as all engineering education needs to be refocused on confronting global concerns. Therefor national and international academic bodies have propelled the sustainable development through educational programs. (Akyazi et al., 2020). Although many universities have begun to implement sustainability into their curricula, it is still challenging to execute a systematic global approach and to assess progress and outcomes.

Civil engineering, a vital branch of engineering, assumes the responsibility of designing, constructing, and maintaining essential infrastructures such as buildings, transportation, hydraulic, and energy supply systems (Beagon., 2023). Transport infrastructures and the construction of buildings have a great impact on the environment, require a high consumption of energy and raw materials, and produce a large volume of waste. Hence it is essential to train the civil engineering graduates on order to develop the sustainable environment by introducing the reuse of materials, the manufacture of ecofriendly materials to minimize the ecological impact of newly constructed infrastructure and adopting the recycled materials, and global assessment of projects from the perspectives of social, environmental, and economic sustainability. Therefore, curriculum designed by incorporating many courses related to the environment and public policy, to elevate the focus on sustainability. By seamlessly integrating most of the Sustainable Development Goals (SDGs) into the civil engineering curriculum, this sector can play a pivotal role in addressing global challenges and creating eco-friendly infrastructures and services. Current study is dedicated to aligning the curriculum with the UN's 2030 sustainability goals, aiming to empower students to be more focused and effectively contribute to the SDGs throughout their educational journey and make a lasting positive impact on the world. The transformation of civil engineering education towards a sustainability-driven approach holds immense promise in creating a more resilient and prosperous future for all.

Current study considers Civil Engineering curriculum (2023-24 Board of Studies Approved) of KLE Technological University as case study. The approach of this study emphasizes understanding global perspectives on sustainability and Sustainable Development Goals especially integration with civil engineering curriculum. Identifying gaps in the current curriculum is essential and subsequently a four phased implementation approach is proposed to address these gaps effectively. By adopting a precise and context-specific approach, current study aims to provide insights into how sustainable development goals can be systematically integrated into the existing civil engineering curriculum and make more aligned with SDGs, ensuring that graduates are well-prepared to contribute to sustainable practices in their professional careers.

II. Understanding the SDG Landscape

The Sustainable Development Goals (SDGs) were established through a comprehensive and inclusive process initiated by the United Nations. The journey towards the SDGs began with the United Nations Conference on Sustainable Development, also known as Rio+20, held in Rio de Janeiro, Brazil, in June 2012. During Rio+20, member states recognized the need to address global challenges and promote sustainable development on a global scale. While the Millennium Development Goals (MDGs) had made significant progress since their inception in 2000, they were seen as having limitations in addressing all dimensions of sustainability and development.

In response to this, the UN General Assembly created an Open Working Group (OWG) on Sustainable Development Goals in January 2013. The OWG's objective was to develop a proposal for the post-2015 development agenda, which would include a set of sustainable development goals and targets. In September 2014, the OWG presented its final report, which contained a set of 17 proposed goals and 169 targets covering a wide range of social, economic, and environmental issues. The proposed goals were designed to be universal and applicable to all countries, recognizing the interconnectedness of global challenges and the shared responsibility for sustainable development. Subsequently, in August 2015, a UN summit was convened in New York, where world leaders formally adopted the 2030 Agenda for Sustainable Development. The 2030 Agenda outlines the 17 SDGs, each with specific targets and indicators, to be achieved by 2030. The SDGs officially came into existence on January 1, 2016 (UN; The Goals 2030 , 2017).

https://doi.org/10.52202/073963-0029

The adoption of the SDGs represents a significant milestone in international cooperation, signifying a collective commitment to address the world's most pressing challenges, including poverty, hunger, inequality, climate change, environmental degradation, and social injustice. The SDGs provide a global roadmap for sustainable development, encouraging collaboration, innovation, and concerted action towards building a more equitable, inclusive, and sustainable future for all (Hák T et al., 2016). By incorporating the SDGs into the curriculum, education empowers students to understand the interconnectedness of global issues and the importance of responsible decision-making. It encourages active participation in community-based initiatives, nurturing a sense of responsibility towards society and the environment. Additionally, integrating the SDGs in education empowers teachers to serve as change agents, fostering a culture of sustainability and inspiring students to become advocates for positive change. Through education, we can create a transformative impact, equipping future generations with the knowledge and values needed to build a more sustainable and prosperous world for all (Cebrián et al., 2020). The 17 SDGs along with their purpose are as follows in the Table 1.

TABLE 1 : SUSTAINABLE DEVELOPMENT GOALS AS PER UN

	Sustainable Development Goals	Purpose
SDG1:	No Poverty	End poverty in all forms.
SDG 2:	Zero Hunger	Achieve food security and improved nutrition.
SDG 3:	Good Health and Well-being	Ensure healthy lives for all.
SDG 4:	Quality Education	Ensure inclusive and equitable education.
SDG 5:	Gender Equality	Achieve gender equality and empower women.
SDG 6:	Clean Water and Sanitation	Ensure access to clean water and sanitation.
SDG 7:	Affordable and Clean Energy	Provide sustainable energy for all.
SDG 8:	Decent Work and Economic Growth	Promote inclusive economic growth and decent work.
SDG 9:	Industry, Innovation, and Infrastructure.	Build sustainable infrastructure and foster innovation
SDG 10:	Reduced Inequalities	Reduce inequalities within and among countries.
SDG 11:	Sustainable Cities and Communities	Create sustainable cities and communities.
SDG 12:	Responsible Consumption and Production	Promote sustainable consumption and production.
SDG 13:	Climate Action	Take urgent action to combat climate change and its impacts.
SDG 14:	Life Below Water	Conserve and sustainably use marine resources.
SDG 15:	Life on Land	Protect and restore terrestrial ecosystems and biodiversity.
SDG 16:	Peace, Justice, and Strong Institutions	Promote peace, justice, and strong institutions.
SDG 17:	Partnerships for the Goals	Strengthen global partnerships for sustainable development.

III. Systematic Literature Reviews on the Integrating SDGs in Education

Current section of the article deals with literature reviews on integrating SDGs in education which have been on the rise in recent years. This section mainly serves the purpose of synthesizing existing studies and evidence related to the implementation of SDGs in educational practices and policies. Most of the studies conducted to gain a comprehensive understanding of the challenges, opportunities, and best practices involved in integrating SDGs into educational systems.

The paper by Osofero et al. (2014) examines the civil engineering program at two European universities, aiming to develop graduates with sustainability awareness and meet stakeholder aspirations. Case studies highlight efforts to integrate sustainability but underscore the need for urgent interventions to enhance students' knowledge and promote responsible actions aligned with sustainability principles.

Fenner et al. (2014) draws on 12 years of experience leading sustainability-focused engineering programs at a UK university. It highlights the essential skills young civil engineers need, such as handling complexity, uncertainty, and environmental constraints. The study explores what education is required for the next generation of civil engineers to act sustainably in their professional practice. It discusses fundamental principles, effective teaching strategies, and examples of linking sustainability to civil engineering practice. Furthermore, it reviews UK and international best practices showcasing progress towards sustainable engineering goals.

Stock et al. (2018) highlights the need to train young engineers for sustainability challenges in a dynamic global environment. It advocates for new perspectives in higher engineering education, emphasizing transnational and project-oriented teaching. The "European Engineering Team" master course exemplifies this approach, fostering sustainable start-up development and enhancing students' key competencies.

The article by Zamora-Polo et al. (2019) explores the evolving concept of sustainability, encompassing various fields like ecology, politics, ethics, and spirituality. It proposes a framework for teaching the UN's Sustainable Development Goals in Higher Education, benefiting students personally and professionally. The framework is applied in a case study for Primary Teacher Degree, aiming to build a change-maker University.

Mansell et al. (2019) emphasizes linking project sustainability to the UN's 2030 goals. It identifies a "golden thread" between sustainability reporting at project and organizational levels, allowing for embedding sustainable development goals into infrastructure project design. This strengthens investment appraisal and promotes success across economic, social, and environmental outcomes. Critical questions raised need resolution within the infrastructure sector for better alignment with global sustainability objectives.

Perpignan et al. (2020) addresses the lack of sustainable engineering teaching in French schools. It proposes a skill crossover matrix derived from a literature review and a survey to aid in eco-design skill development. The matrix guides teachers in creating sustainable engineering curricula and helps students monitor their skill growth. Companies seek engineers with eco-design skills to align with their evolving strategies.

The study by Cebrián et al. (2020), highlights the urgency of embedding Education for Sustainable Development (ESD) principles into all educational levels. The Special Issue titled "Competencies in Education for Sustainable Development" presents recent developments in ESD competencies, including curriculum developments, evaluation tools, and conceptual models. The study also emphasizes the growing importance of sustainability competencies and the need for further research on operationalizing and evaluating competency development among students and educators.

Amidst the climate emergency, organizations globally integrate sustainability and SDGs in their actions. In education, steps have been taken to include sustainability at all levels, including universities. However, challenges like rigid structures and time constraints hinder integration. Civil engineering's role in creating environmentally impactful infrastructures makes it vital for promoting sustainability. Álvarez et al. (2021) presents a multidisciplinary approach utilizing problem-based and project-based learning to foster sustainability in civil engineering education. Positive outcomes and increased SDG integration in final projects indicate potential adoption in other disciplines.

Study by Gutierrez-Bucheli et al. (2022) focuses on sustainability-initiatives in engineering education to achieve the UN's SDGs. It utilizes a realist scoping review to analyze approaches since the 1990s, identifying gaps in learning outcomes. While there is a desire for integrative sustainability education, it requires additional administrative resources. The study emphasizes re-evaluating the engineer's role and social responsibilities to empower students as change agents. Implications for practice and curriculum development are highlighted.

Beagon et al. (2023) examines key competences for engineering students to address sustainability challenges and achieve SDGs. Stakeholders' views (Academics, Employers, and Students) from four countries are compared. Normative, strategic, and systems thinking competences are prioritized, but anticipatory competence, crucial for future-oriented sustainability, is lacking. Educators can use the findings to develop programs and provide opportunities for students to acquire the necessary competences for supporting sustainable development and SDGs.

In conclusion, the rise of literature reviews on integrating SDGs in education reflects the growing awareness of sustainability's importance in various academic fields. However, there is a need to address the gaps in the civil engineering curriculum to effectively prepare students for sustainability challenges. While the existing civil engineering curriculum includes many courses related to the environment or sustainability, they often lack integration with the specific goals outlined in the SDGs. Hence, there is urgent interventions are required to enhance civil engineering students' knowledge and promote responsible actions aligned with sustainability principles. Emphasizing anticipatory competence is crucial for a future-oriented perspective in achieving sustainable development in the construction industry. By incorporating these insights into civil engineering education, we can also bridge the gap and equip the next generation of engineers to contribute meaningfully to the global pursuit of sustainable development and SDGs. With this current study emphasizes understanding global perspectives on sustainability and Sustainable Development Goals (SDGs) integration with existing civil engineering curriculum through a four phased implementation approach.

IV. SDGs Embedding Methodology for Curriculum of Civil Engineering Program

To proactively address the crucial task of integrating a global approach to sustainability, particularly the SDGs, into the existing civil engineering curriculum, a strategic adoption approach based on four distinct phases or quadrants has been devised as shown in the Figure 1.The proposed curriculum focuses on the demand that initiatives are required to integrate a global practical approach to sustainability for current civil engineering degrees at KLE Technological university, Hubballi. This structured method serves as a preliminary step, allowing educators to thoroughly examine and evaluate the opportunities for effectively embedding SDGs in the curriculum. By employing this systematic and comprehensive approach, civil engineering programs can align their educational objectives more cohesively with the broader goals of sustainable development. This empowers future engineers to tackle real-world challenges with a more holistic and responsible approach, contributing to a sustainable and prosperous global future.

Planned strategic four quadrant approach for integrating a global approach to sustainability into the existing civil engineering curriculum encompasses four key quadrants such as courses, laboratories, student projects, and internships. In theoretical courses, sustainability concepts, ethical considerations, and best practices aligned with SDGs are introduced. This fosters critical thinking and analytical skills, essential for sustainable engineering practices. Laboratory courses offer hands-on experience in implementing sustainable solutions, exploring renewable energy, and sustainable materials. Through practical projects, students develop innovative design solutions, promoting creativity and problem-solving abilities.

Student projects offer a unique opportunity to connect classroom learning with real-world applications. By encouraging projects aligned with specific SDGs, students can directly contribute to sustainability goals, fostering a sense of

https://doi.org/10.52202/073963-0029

responsibility and impact. Moreover, internships with sustainability-focused companies provide students with practical exposure to sustainable engineering in a professional setting. Working on SDG-related projects during internships enhances their understanding and experience in sustainable practices, while mentorship supports their professional growth.

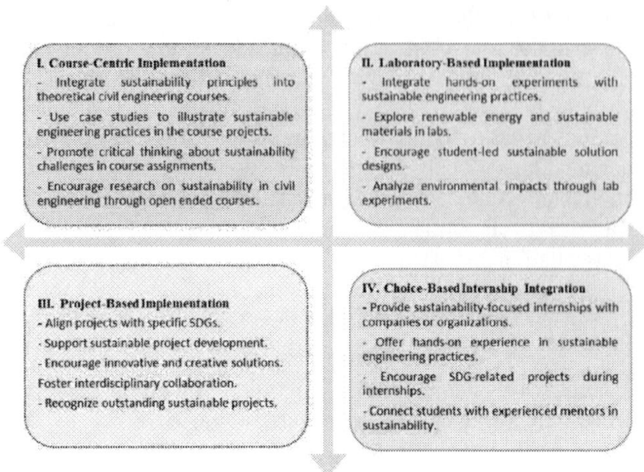

Fig.1: Strategic four quadrant approach for integrating a global approach to sustainability into the existing civil engineering curriculum

A. *Integration of curriculum with SDGs*

Based on strategic four-quadrant approach for integrating Sustainable Development Goals (SDGs) in the civil engineering curriculum involves a meticulous selection of subjects, semester-wise, based on their impact on the final degree. A comprehensive matrix is formulated, encompassing all courses chosen, including theory, design, laboratory, projects, and internship training. The courses in the matrix are then color-coded according to the SDGs (as per Table 1) that has all the possibility to directly integrated with it. This matrix serves as a roadmap in upcoming days to guide the curriculum design and delivery trajectory, ensuring the seamless implementation of sustainability principles throughout the civil engineering degree. The curriculum matrix for the SDG roadmap is shown in the Figure 2.

Subsequently in-depth discussions are conducted to explore the potential and possibilities of integrating possible SDGs with identified courses is made. The aim is to identify innovative ways to embed sustainability principles, making the curriculum more effective and impactful. This process allows educators to develop a holistic approach to sustainability education, preparing civil engineering students to tackle real-world sustainability challenges.

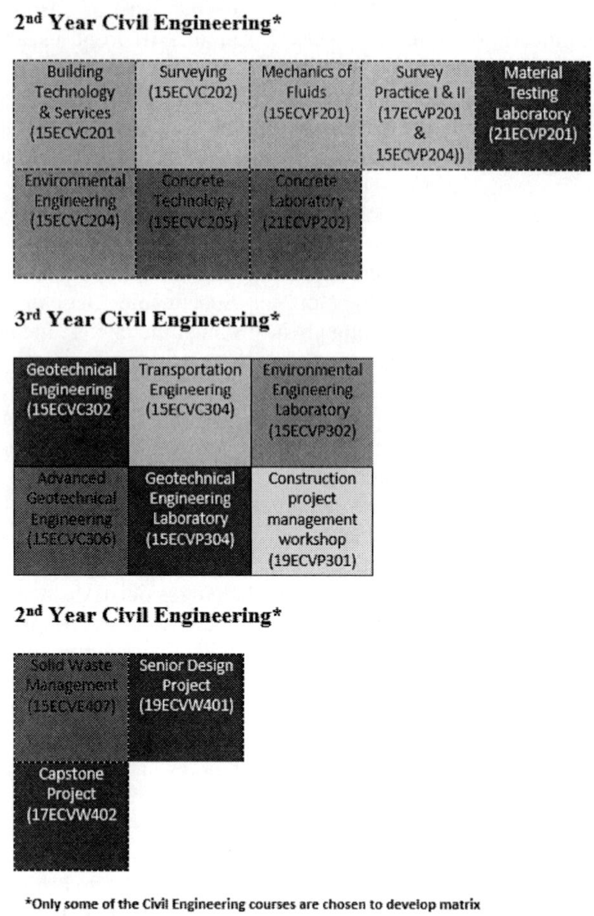

Fig. 2: The curriculum matrix for the SDG roadmap

As shown in the Figure 2, Civil Engineering curriculum has Building Technology & Services during 2nd year of curriculum as one of the basic theory courses but yet important one. This course has predominant opportunity to integrate some of the SDG's goals in the syllabus where it detailed with fundamental of buildings such as categories of buildings, building components, building materials etc. SDG 12 (Responsible Consumption and Production) can be incorporated by focusing on sustainable building materials and construction methods with lower environmental impacts. Concepts such as waste reduction strategies can be taught effectively to promote responsible consumption and production in building projects.

Current civil curriculum of civil engineering one theory course and two laboratory course related surveying practices in infrastructure industry. As SDG 11 clearly address on (Sustainable Cities and Communities), hence in these course and laboratory students can learn about the critical role of accurate land surveys in urban planning and development. Even they can made to understand how Geographic Information Systems (GIS) can be utilized to analyze spatial data for sustainable city planning. Surveying plays a crucial role in

https://doi.org/10.52202/073963-0029

assessing land use, infrastructure development, and resource management within cities. By integrating SDG 11 into the course, students recognize the importance of promoting inclusive and sustainable cities through their surveying work. They also explore how their expertise can contribute to improving urban living conditions, enhancing accessibility, and ensuring environmental sustainability in urban areas

The Mechanics of Fluids one important course during 2nd year of the curriculum, also presents an opportunity to address SDG 13 (Climate Action), which focuses on taking urgent action to combat climate change and its impacts. Students can learn about fluid dynamics in the context of environmental applications, such as studying the behavior of air and water currents, ocean circulation patterns, and weather systems. They explore how climate change can influence fluid behaviors, leading to extreme weather events, sea-level rise, and other climate-related challenges. By integrating SDG 13, civil engineering students become more aware of their role in mitigating the effects of climate change through sustainable engineering practices.

In the Material Testing Laboratory course, civil engineering students are exposed to various testing methods for materials, particularly steel, which is a crucial component in building infrastructure. By integrating SDG 9 (Industry, Innovation, and Infrastructure), students explore how innovations in material testing techniques can enhance the quality and durability of infrastructure projects. They learn about non-destructive testing methods, advanced technologies, and automation in material testing, enabling them to contribute to the development of safer and more efficient construction practices.

Environmental Engineering course during 2nd year and Environmental Engineering Laboratory during 3rd year of the curriculum are significant course those can unswervingly integrate with SDG 6 (Clean Water and Sanitation) by emphasizing the importance of providing clean and safe water to communities and managing wastewater effectively. Students here learn about various water treatment processes, including filtration, disinfection, and desalination, to ensure access to clean drinking water. Additionally, they explore techniques for treating and reusing wastewater, reducing water pollution, and protecting water resources. By understanding the principles of sustainable water management, civil engineering students can efficiently contribute to achieving SDG 6 and ensuring that all individuals have access to clean water and adequate sanitation, which is essential for human health and well-being.

In the Concrete Technology theory and subsequent laboratory course of civil engineering curriculum, the integration of SDG 15 (Life on Land) can be made essential to promote sustainable land use and biodiversity conservation. Concrete production is associated with significant environmental impacts, including land degradation and habitat loss. By focusing on eco-friendly concrete production techniques, such as using recycled aggregates and supplementary cementitious materials, students can reduce the demand for natural resources and minimize the environmental footprint.

In the Geotechnical Theory and Lab Course of civil engineering, the integration of SDG 9 (Industry, Innovation, and Infrastructure) is play essential role to ensure the development of resilient and sustainable infrastructure. Geotechnical engineering plays a crucial role in the design and construction of foundations, retaining walls, and underground structures. By incorporating innovative techniques in the syllabus integrating with SDG 9, such as geosynthetics and soil stabilization methods etc, students can contribute to the advancement of infrastructure that withstands natural hazards and environmental challenges. Where as in the case of advance geotechnical course along with conventional aspects by integrating SDG 14 (Life Below Water), the course can also focus on the significance of geotechnical investigations for marine environments and the protection of marine ecosystems. Students can learn about sustainable solutions for coastal engineering, which involve preserving marine habitats and mitigating potential impacts on marine life. Emphasizing SDG 14 course can raise awareness about the need for advanced geotechnical practices that safeguard life below water and contribute to ocean conservation efforts.

Again, Transportation Engineering course can integrate with SDG 11 (Sustainable Cities and Communities) very effectively by focusing on the development of sustainable and efficient transportation systems in urban areas. Students can learn about innovative transportation planning and design practices that prioritize public transport, non-motorized modes, and integrated mobility solutions. Emphasis can be placed on creating pedestrian-friendly streets, dedicated cycling lanes, and efficient public transit networks. By integrating SDG 11, the course can highlight the role of transportation engineers in promoting sustainable urban development, reducing congestion, and enhancing accessibility for all residents.

The Solid Waste Management course can contribute to SDG 8 (Decent Work and Economic Growth) by focusing on the economic aspects of waste management. Students can be trained in the course to explore waste-to-resource opportunities, such as recycling and composting, which can create employment and contribute to economic growth. The course can also examine the role of civil engineers in designing cost-effective waste management systems that benefit local economies and communities.

During final year of curriculum students will experience the long-term projects. Most of the projects are either field or laboratory based. Ultimately most of the projects try to address the current industry needs. Hence in the Project Courses, SDG 9 (Industry, Innovation, and Infrastructure) can be integrated by focusing on sustainable and innovative infrastructure development. Some of the student teams can work on projects that promote energy-efficient technologies, smart solutions, and inclusive infrastructure. By considering environmental impact and social inclusion, they contribute to SDG 9 goal of promoting sustainable industrialization and resilient

 https://doi.org/10.52202/073963-0029

infrastructure. Through this integration, civil engineering students gain valuable skills in designing projects that align with global sustainability objectives, fostering economic growth, and supporting sustainable development.

V. SDGs Implementation Through Pedagogical Approaches

A. Research-Based Learning

A significant flaw in higher education has been noted as being excessively theoretical instruction with little practical application and inadequate research preparation. Thus, it requires the new holistic paradigm approach of sustainability compels science to become more inclusive and receptive to taking a holistic perspective to society. There are various ways of integrating teaching with research; in some of these ways, students play a more passive part, while in others, they play a more active role. RBL, or research-based learning, is the strategy that is most widely used in higher education. The present curriculum proposed the Research investigation-based course project in 4th semester to enhance students research skills and critical thinking. Research projects are carried out in material testing laboratory and concrete technology laboratory course. Students make attempt to develop the sustainable and ecofriendly cement composite by using cement-based demolition waste, industrial waste, granite waste and recycled aggregate for and municipal solid waste incineration fly ash to produce eco-friendly binders for building construction. The samples prepared using above materials and evaluate the mechanical behavior of cement. The strategy presented by RBL aims to achieving the SDG 9 and SDG 12 objectives through the revaluation and reuse of industrial waste and municipal solid waste incineration and concrete debris produced during construction. The outcomes of research projects promote waste management and reduces the impact on environmental. Additionally, this new strategy would enable the industrial sector to strengthen its production infrastructure through the development of new, cutting-edge manufacturing processes built on eco-friendly models and utilizing resources more effectively.

B. Project Based Learning

PBL aids in the development of students problem-solving abilities. Students typically excel at closely specified textbook problems but have lack of knowledge with poorly structured, open-ended problems, which they frequently face in the real world. This is particularly true for concerns about sustainability, which seek for multidisciplinary, integrative, flexible problem-solving methods rather than closed approach. Students often learn that dealing with problems frequently entails more than just a cognitive training. Implementing solutions to problems is necessary, however, there may be difficulties in the way of progress. Through PBL, students acquire new skills along with social and political awareness which is required for implementing solutions into actions. Hence, In the Fifth Sem, the 2-credit course titled "Construction project management workshop" is introduced where students are exposed to methodology to construct projects from the initial design to the detailed construction project. In this subject, students apply all the knowledge acquired during the degree in the development of a project, and it is the prelude to the Final Degree Project. This course serves as a prerequisite for the real time construction project and requires students to apply all of the knowledge they have learned from academic curriculum. In the proposed work, students carry out case study of complex real problem related to structural, geotechnical and transportation field. The objective of the course is understanding the multiple stages involved in execution. The students collect the data regarding structural conditions, performance and environmental aspect which impact the different criteria such as sustainability, economy, and society. Students are groups into 6 numbers, and they have analyzed the above data and prepare the report about methodology adopted to solve of the case study problem which helps them understand the reality of large civil engineering projects. In addition, it has been helpful to have a better understanding of how decision-making is carried out in complex environments, the usefulness of multi criteria evaluation, and the need to consider sustainability criteria in this type of large-scale work. It has also been beneficial to gain a deeper comprehension of the way of decision-making is carried out in complicated situations, the value of multi-criteria evaluation, and the necessity of taking sustainability criteria into account in this kind of large-scale activity. As a result, this kind of work appears to be enabling civil engineering students to consider additional criteria (sustainability, society, and economy) in projects that are related to their own profession, beyond the simple application of technical and functional criteria.

VI. Discussion and Conclusions

The integration of Sustainable Development Goals (SDGs) in the Civil Engineering curriculum is of utmost importance to prepare future engineers to address global sustainability challenges. Through systematic literature review, the current study has emphasized understanding the global perspectives on sustainability and identified gaps in the existing curriculum. By adopting a strategic four-quadrant approach, the study has provided insights into how SDGs can be systematically integrated into the Civil Engineering curriculum at KLE Technological University through some of courses.

The study has highlighted the opportunities to link SDGs in theoretical courses, laboratory work, student projects, and internships. By incorporating SDGs into theoretical courses such as Building Technology & Services, Mechanics of Fluids, Environmental Engineering, etc students gain a deep understanding of how civil engineering practices can contribute to sustainability. Laboratory courses in Surveying, Material Testing, and Environmental Engineering etc provide hands-on

https://doi.org/10.52202/073963-0029

experience in implementing sustainable solutions and exploring innovative technologies.

Furthermore, the study recognizes the significance of student projects and internships in promoting sustainability. By encouraging projects aligned with specific SDGs, students can actively contribute to sustainable development goals and understand the broader impact of their work. Internships with sustainability-focused companies offer practical exposure to sustainable engineering practices and mentorship to support professional growth.

In conclusion, the integration of SDGs in the Civil Engineering curriculum is a critical step towards producing responsible and future-oriented engineers. The strategic approach outlined in this study enables educators to develop a holistic approach to sustainability education, preparing students to tackle real-world sustainability challenges.

Current study acts as a base for the SDGs integration planning in the curriculum considering only few courses. For future work, it is essential to evaluate the actual effectiveness through integrating SDGs in the curriculum using innovative teaching methods and continuously monitoring and taking feedback from students, if required by industry professionals to as third party.

Overall, by adopting a comprehensive approach to integrate SDGs in the curriculum, civil engineering programs can play a vital role in achieving sustainable development goals and shaping a sustainable and prosperous future for generations to come.

REFERENCES

Hubscher, C., Hensel-Borner, S., & Henseler, J. (2022). Social marketing and higher education: partnering to achieve sustainable development goals. Journal of Social Marketing, 12(1), 76-104.

Terrón-López, M. J., Velasco-Quintana, P. J., Lavado-Anguera, S., & Espinosa-Elvira, M. D. C. (2020). Preparing sustainable engineers: A project-based learning experience in logistics with refugee camps. Sustainability, 12(12), 4817.

Álvarez, I., Etxeberria, P., Alberdi, E., Pérez-Acebo, H., Eguia, I., & García, M. J. (2021). Sustainable civil engineering: Incorporating sustainable development goals in higher education curricula. Sustainability, 13(16), 8967.

Beagon, U., Kövesi, K., Tabas, B., Nørgaard, B., Lehtinen, R., Bowe, B., ... & Spliid, C. M. (2023). Preparing engineering students for the challenges of the SDGs: what competences are required?. European Journal of Engineering Education, 48(1), 1-23.

Cebrián, G., Junyent, M., & Mulà, I. (2020). Competencies in education for sustainable development: Emerging teaching and research developments. Sustainability, 12(2), 579.

Fenner, R. A., Cruickshank, H. J., & Ainger, C. (2014, October). Sustainability in civil engineering education: why, what, when, where and how. In Proceedings of the Institution of Civil Engineers Engineering Sustainability (Vol. 167, No. 5, pp. 228-237). Thomas Telford Ltd.

Gutierrez-Bucheli, L., Kidman, G., & Reid, A. (2022). Sustainability in engineering education: A review of learning outcomes. Journal of Cleaner Production, 330, 129734.

Hák, T., Janoušková, S., & Moldan, B. (2016). Sustainable Development Goals: A need for relevant indicators. Ecological indicators, 60, 565-573.

Mansell, P., Philbin, S. P., Broyd, T., & Nicholson, I. (2019). Assessing the impact of infrastructure projects on global sustainable development goals. In Proceedings of the Institution of Civil Engineers-Engineering Sustainability (Vol. 173, No. 4, pp. 196-212). Thomas Telford Ltd.

Osofero, A. I., Udeaja, C., Corradi, M., & Vo, T. (2014). Sustainability Challenges in Civil Engineering Education. In Computing in Civil and Building Engineering (2014) (pp. 2047-2054).

Perpignan, C., Baouch, Y., Robin, V., & Eynard, B. (2020). Engineering education perspective for sustainable development: A maturity assessment of cross-disciplinary and advanced technical skills in eco-design. Procedia CIRP, 90, 748-753.

R. Bali Swain & F. Yang-Wallentin (2020) Achieving sustainable development goals: predicaments and strategies, International Journal of Sustainable Development & World Ecology, 27:2, 96-106.

Stock, T., & Kohl, H. (2018). Perspectives for international engineering education: Sustainable-oriented and transnational teaching and learning. Procedia Manufacturing, 21, 10-17.

United Nations (n.d.). Sustainable development goals. THE 17 GOALS | Transforming Our World: The 2030 Agenda for Sustainable Development. UN General Assembly, 21 October 2015. Retrieved August 3, 2023, from

Zamora-Polo, F., & Sánchez-Martín, J. (2019). Teaching for a better world. Sustainability and sustainable development goals in the construction of a change-maker university. Sustainability, 11(15), 4224.

https://doi.org/10.52202/073963-0029

How can a Holistic Approach to Practice, Research and Policy for Sustainable Engineering Education be Developed? An Investigation

Nikita Hari, Abel Nyamapfene, and John Mitchell
University College London, Centre for Engineering Education, London, UK
nikita.hari@eng.ox.ac.uk

Abstract

Context
In the last two decades, significant pedagogical advances aimed at enhancing the UK's engineering education have emerged. However, there's a noted absence of an integrated approach linking policy and practice to evidence-based research in this field.

Purpose or Goal
This study seeks to bridge this gap by exploring the lived experiences and perspectives of engineering education leaders who have been at the forefront of pioneering education reforms in the UK. It aims to understand the challenges they face and identify potential solutions and models that could address the evolving needs of undergraduate engineering education.

Methods
The study utilizes a qualitative research design with purposeful sampling. Qualitative data is obtained through interviews to gain in-depth insights into the experiences and perspectives of stakeholders. Thematic analysis of the collected data is used to identify common themes, patterns, and relationships.

Outcomes
The findings of this qualitative preliminary investigation are to develop research questions to inform and frame a more comprehensive quantitative study aimed at shedding light on the development of curriculum frameworks for implementing effective and scalable engineering education models in the UK.

Conclusion
This study reveals the conflicting and complementary factors in the UK engineering education landscape. This highlights the need for a systems-based approach connecting policy and practice, informed by evidence-based research for developing a sustainable engineering education framework in the UK.

Keywords—Engineering Education Reform; Sustainable Curriculum Innovation and Design; Integrated Engineering Education; Holistic Engineering Education, Systems Approach to Engineering Education.

I. INTRODUCTION

Undergraduate engineering education in the UK faces significant challenges in adapting to the 21st century (Jones et al., 2000 and Spinks et al., 2006). Since the turn of the century, there has been widespread agreement that reforms are urgently needed to prepare students for increasingly complex global issues (Graham, 2012 and Haghighi, 2005). The UK government, on its part, has implemented various policies to attract young people into engineering (Clark, 2011). Additionally, the accreditation of engineering degrees and chartered engineers by the Engineering Council and the various professional engineering institutions has ensured global standards and quality control (Levy, 2000). However, coordinated efforts through partnerships, policy, and research are required from universities, professional engineering institutions, government, and other stakeholders to guarantee meaningful and sustainable reforms in engineering education (Graham, 2012).

There have been ongoing reforms in UK engineering education over the past decade. These reforms have been highlighted by the Engineering Professors Council (EPC) and the Institution of Engineering Technology (IET) (EPC and IET, 2017), and in a report highlighting innovation and good practice in engineering education across the UK (EPC and IET, 2019). Collectively the reforms highlight how engineering educators in the UK are responding to the challenges and opportunities of the 21st century through innovation and reforms to engineering education practices and pedagogies. Examples of approaches discussed in these papers include reforms to curriculum design, assessment, teaching methods, student engagement, industry collaboration, use of state-of-the-art technology, and sharing good practices and lessons learned across the sector (Fowler et al, 2023).

Reaffirming the international standing of engineering education reforms by UK universities, University College London (UCL) and the University of Cambridge, have been identified as global leaders in engineering education alongside ten other universities, with UCL being further identified as an emerging leader alongside four other universities (Graham, 2018). Additionally, policy changes in higher education enacted in the Higher Education and Research Act 2017 (HERA) have enabled new providers such as the New Model in Technology and Engineering (NMITE), Dyson Institute of Engineering and Technology (DIET) and the Engineering & Design Institute London (TEDI-London) to establish new universities that focus on modern engineering pedagogies.

However, UK research in engineering education practice, although it is growing, is still limited and does not reflect the reform work that is being implemented. There is low engagement in engineering education research (EER) in the UK, evidenced by few publications, mostly single-author or single-institution (Nyamapfene, 2017), suggesting lack of collaboration between engineering education researchers and practitioners. With respect to policy, Cooper et al. (2023), have argued that UK engineering policy, unlike science policy, is rarely discussed or scrutinised in the academy or in public governance. The authors concluded that engineering is largely absent or marginalised in government ministries, committees, agencies, and public bodies.

It could be argued that one of the main reasons for this shortcoming is the lack of a holistic system in the UK that connects policy, and practice in undergraduate engineering education to evidence-based research, and that encompasses all stakeholders.By examining the interplay between these three elements of policy, practice and research, the study aims to identify strategies and frameworks that promote effective and scalable engineering education models. In this study we explore the lived experiences and perceptions of one group of stakeholders, namely engineering education changemakers, who have been at the forefront of pioneering engineering education reforms in the UK.

The goal is to help engineering educators enhance the adaptability, innovation, and preparedness of engineering graduates, enabling them to tackle complex challenges and contribute to societal development. However, limited research exists on the integration of policy, practice, and research as a cohesive paradigm to cultivate future-ready engineers. This study seeks to bridge that gap by exploring the synergistic relationship between policy, practice, and research and its impact on engineering education outcomes for designing effective learning ecosystems.

This preliminary investigative study is a component of a broader project focused on the following research questions.:

RQ1: What are the strategies and frameworks that promote effective and scalable engineering education models in the UK?

RQ2: How have current engineering education models enhanced the adaptability, innovative thinking, and work-readiness of engineering graduates in the UK?

RQ3: How has policy, practice and research in engineering education collectively influenced and contributed to the development of current engineering education models in the UK?

The objective of this qualitative preliminary investigation is to use these research questions to inform and frame a more comprehensive quantitative study that addresses these proposed questions in depth.

A. RQ1

Delved into the engineering educational model at the participants' respective Universities/Departments/Institutions. It sought to get an overview of the model, exploring its foundational principles, pedagogical approaches, and key components. Additionally, it sought to understand the driving forces behind the model's adoption or development, shedding light on the motivations that led to its implementation within participant's educational context. Finally, this research question sought to provide a comprehensive foundation for examining the subsequent research questions, offering insight into the model's impact and purpose within the participants' academic institutions.

B. RQ2

Focused on the tangible impacts of the engineering education models at participants' institutions. It examined how the model had positively influenced academic and employment outcomes for students. Furthermore, this research question investigated the evolution of engineering graduates in terms of their work-readiness, aptitude, and skills compared to earlier cohorts before the model's implementation, supported and justified by empirical evidence.

C. RQ3

Delved into the past influence of government policies and evolving engineering practices on the development of new engineering education models in the UK. It examined the extent to which these external factors influenced curriculum design and delivery, as well as the development of innovative educational models. Additionally, it investigated the degree to which these models were shaped by evidence-based research, shedding light on the research-informed nature of the educational approaches.

II METHODS

Qualitative data for this initial study is obtained through interviews with engineering leaders, educators, and industry professionals. The collected data is thematically analyzed to identify common themes, patterns, and relationships.

https://doi.org/10.52202/073963-0030

A. Study Approach

Semi-structured, online interviews with engineering leaders, educators, and industry professionals were conducted on UCL MS Teams and/or Zoom, and timings were chosen to fit in with participants' availability. Online interviews were found to be most appropriate to this study, as our potential research participants are from different UK higher education institutions and arranging face-to-face interviews would be financially expensive and time-consuming.

B. Participant Selection

Following the approach by Graham (2018) in her study on the global state of the art in engineering education, we selectively engaged individuals known for their contributions to engineering education reforms in the UK. This purposeful sampling was guided by their contributions to their institutions and/or their impact in scholarship and research in engineering education. In selecting participants, we also considered diverse factors such as gender, ethnicity, institution types, leadership roles, personal and professional experiences, and the nature and scope of reforms they have implemented.

These individuals have on average 20 to 30 years' experience within the UK engineering higher education sector, and hence, collectively these interviews give us almost 200 to 300 years of experience and perspectives of the UK engineering higher education landscape. The research participants also have varied lengths of experience within the sector, and are at different hierarchies of leadership, which again, gives us insights from different hierarchical perspectives. The participants have all followed different career trajectories to their current positions, with some having non-engineering backgrounds such as Economics, Entrepreneurship and Design and Innovation. Some of the participants are established academics with strong technical research credentials, whilst others have come up through the education and management routes, and others have progressed from professional engineering practice to academia.

C. Positionality Statement

Regarding positionality, three of us have a technical engineering background, are currently education-focused academics within engineering, and collectively, we have led engineering educational reforms at six UK universities, five of which are research intensive institutions, and one a start-up engineering higher education institution. As educators and practitioners at the forefront of leading education reform, our positionality and background significantly shape our perspectives and approaches to this research. Collectively our breadth of experiences have exposed us to the complexities and challenges of implementing engineering education reforms in different higher education contexts.

Furthermore, we acknowledge and are aware that our background and training can impact the way we interpret data, engage with participants, and frame research questions. We are also cognisant of our positionality, potential biases, personal values, and beliefs, which include a strong commitment to embedding values and ethics in engineering education research and practise. While our values drive our passion for this research, we are self-conscious of the need to maintain objectivity and consider multiple perspectives throughout the research process.

To mitigate potential biases, we employ reflexivity and engage in continuous self-examination to incorporate diverse voices and viewpoints in our research, without seeking to impose our own voices and interpretations. We are committed to conducting a rigorous and ethical study that contributes to the ongoing dialogue on engineering education transformation in the UK.

D. Data Collection and Analysis

This study adheres to the ethical research guidelines established by the British Education Research Association (BERA, 2018). The research process is characterized by rigorous ethical considerations, starting with the acquisition of informed consent, which was obtained from all participants prior to conducting interviews. Participants were provided with detailed information about voice recordings and the assurance of their anonymity within the research. The interview sessions were conducted in a conversational manner, fostering a relaxed and open atmosphere, with participants displaying no signs of apprehension. Our approach employed semi-structured interviews, designed to delve deeply into the authentic experiences of the participants in relation to the research questions.

The interviews underwent thematic analysis. Each conversation was recorded and subsequently transcribed. The two researchers collectively analysed the transcripts to identify thematic categories that emerged organically from the participants' discussions. By posing similar questions to various participants across diverse data samples, we were able to shed light on common thematic concerns. These identified themes were then subjected to further examination through a review of existing literature, providing insights into how previous scholarship has addressed these issues.

III RESULTS AND DISCUSSIONS

All participants displayed a tendency to intermingle their responses, deviating from the specific research questions provided. They frequently initiated their answers to one question and proceeded to address additional questions from various research areas without prompting. This behavior was influenced by our interview style, characterized by an open-ended discussion approach. From the analysis of the interview transcripts, we identified ten emerging themes. Consequently, we have restructured this section, organizing the collective findings into subtopics that align with the emerging themes.

A. Driving Forces Behind Curriculum Transformation in Engineering Education: (RQ1, RQ2 and RQ3)

We identified several drivers and variables for engineering education reform as listed here:

1) Individual Visionaries vs. Institutional Initiatives

The study data suggests that curriculum transformation initiatives are primarily led by visionary individuals within most institutions who are passionate about change but face limited institutional buy-in.

2) Successful Funding Attraction

A participant from one of the engineering institutions succeeded in attracting significant funding for curriculum reform, and this helped to capture c the institution's leadership's attention and support.

3) Industry Influence

Industry-driven curriculum models aimed to shape engineers based on specific needs, feeling universities didn't align with their requirements.

4) Positive Impact of Policy Changes

Policy shifts permitting private institutions for targeted engineering needs benefited two of the engineering institutions by providing justification and funding for the desired engineering education reforms.

5) Institutional Imperative for Entrepreneurship

Some of the institutions integrated entrepreneurships into their engineering curricula with philanthropic or commercial support.

6) Bottom-Up, Passion-Driven Initiatives

Curriculum reforms frequently start with individual drive at the grassroots, as evidenced by one of the engineering institutions whose departmental evolution was spurred by the CDIO framework.

7) Unique Nature of Engineering Institutional Transformation

One of the engineering institution curriculum growths stemmed from its founder's vision and charisma. With funding and institutional backing, its distinct journey highlights a blend of factors not easily duplicated elsewhere.

In the landscape of engineering education, visionary individuals within institutions spearhead curriculum transformation, driven by passion and innovation. However, they often grapple with limited institutional support. This illustrates the dynamic interplay between individual visionaries and institutional initiatives in shaping engineering education.

B. Strategic Considerations for the Timely Implementation of Educational Reforms

Timely integration of educational reforms into an institution's culture is vital for enduring, significant changes:

1) Sustainability

Ensuring reforms' long-term viability is paramount. Extending implementation time helps deeply root new practices and minimizes superficial changes.

2) Cultural Shift

Educational modifications entail cultural transitions, necessitating sustained alignment efforts with the new vision.

3) Leadership Role

Leadership significantly influences reform pace, with figures like Engineering Institution 3's founder vital in championing change.

4) Experience Insights

Comparing institutions illuminates the diverse reform paths followed by individual institutions and offers insights for specific contexts.

5) Urgency vs. Sustainability

Balancing swift change with sustainable integration is key, considering potential resistance and momentum loss.

In conclusion, the timeframe for educational reforms should be carefully considered to ensure both sustainability and effective cultural integration. It's a delicate balance that requires leadership, adaptability, and a keen understanding of the institution's unique context. Learning from the experiences of different institutions can help inform the best approach for successful reform initiatives.

C. The Complex Challenges of Educational Reforms on Engineering Education Culture

The impact of educational reforms on the culture of engineering education is complex and multifaceted:

1) Sustained Recognition of Education-Focused Academics

Educational reforms have created pathways for the recognition and promotion of academics who excel in the domain of education. This recognition is a positive

 https://doi.org/10.52202/073963-0030

development, as it values teaching and pedagogical expertise alongside research, but there is no systemic and sustained implementation.

2) *Challenges in Cultural Shift*
Despite these positive changes, challenges remain in achieving a broader cultural shift in engineering education. Regulatory bodies, professional institutions, and councils may continue to prioritize traditional practices, which can hinder the full realization of a culture that values education as much as research.

3) *Need for Alignment*
Achieving a cultural shift requires alignment across all stakeholders in the field. This includes regulatory bodies, educational institutions, professional organizations, and industry. A collective effort is necessary to bring about a comprehensive transformation in the culture of engineering education.

4) Work in Progress
Cultural change often takes time and persistence. While progress has been made, it's important to recognize that the transformation of a long-established culture is an ongoing journey.

In summary, the impact of educational reforms on the culture of engineering education is a mixed bag of positive recognition and the persistence of traditional practices. To fully realize the desired cultural shift, it's essential for all stakeholders to work collaboratively and align their priorities with the evolving needs of engineering education in the modern era.

D. Multifaceted Outcomes of Educational Reforms

The insights from two participants', in particular offer a comprehensive view of the multifaceted outcomes of educational reforms. Here's a closer look at the key takeaways from their perspectives:

1) *Employability and Industry Relevance*
The study highlights the practical and industry-oriented aspects of educational reforms. For example, the reforms implemented at one of the institutions led to success in collaborating with industry in aligning education with workforce needs. This enhanced graduates' employability by making them more attractive to employers and better prepared for real-world challenges.

2) *Learning Experience and Student Satisfaction*
The study highlights that in addition to improving employability, engineering education reforms contribute to student contentment and educational quality. This, in turn, leads to improved student engagement and understanding, which can lead to more effective learning outcomes.

These two perspectives demonstrate that educational reforms can have a multi-dimensional impact, encompassing both employability and the overall educational experience. Successful reforms aim to strike a balance between preparing students for the workforce and providing them with a rewarding and effective learning journey. Ultimately, a well-rounded education aligns with the needs of both students and employers, creating a win-win scenario for all stakeholders.

E. Evolving Career Perspectives of Academics and the Value of Diverse Experiences

The recognition of leaders and engineering educators' expertise, as well as their mobility between institutions, can indeed be a positive outcome of educational reforms. When institutions value and acknowledge the contributions of these individuals, it can lead to increased opportunities for them to share their expertise and insights across various academic settings. This mobility not only benefits the educators but also enhances the exchange of innovative teaching methods, curriculum designs, and pedagogical approaches, ultimately contributing to the broader improvement of engineering education on a larger scale.

The career journeys of many participants' underscore the evolving perspective on career fluidity and the value of diverse experiences in academia and beyond. Here are some key takeaways from their trajectories:

1) *Experience Diversity*
Their careers emphasize the richness of diverse experiences, moving between academia, industry, and varied institutions.

2) *Challenging Stereotypes*
Their paths debunk the myth that startups harm careers, spotlighting how entrepreneurial ventures add value to academia and other fields.

3) *Intersectional Career Recognition*
The rise in acceptance of careers spanning sectors underscores that varied experiences fuel innovation.

4) *Skill Transferability*
Their moves highlight how skills from one setting can be valuable in another, leading to versatile academic and professional realms.

5) *Change Adoption*
Their trajectories underscore the essence of embracing varied career opportunities in today's fast-paced world.

In conclusion, the career paths of most of the participants' highlight the evolving nature of careers and the growing appreciation for interdisciplinary experiences. Their

https://doi.org/10.52202/073963-0030

willingness to navigate between academia and other sectors enriches both their own professional development and the broader workforce and academic landscape.

F. Educational Leaders' Industry Background and Collaboration with Industry: Complex Dynamics

The observation that changes in educational institutions, apart from few institutions, do not have direct input from industry, despite the leaders often having a rich industry background, raises several important points:

1) Leaders' Lived Experiences

It's noted that the changes driven by these leaders are influenced by their personal experiences and insights gained from their industry backgrounds. Their understanding of industry needs and practices inform their decisions regarding educational reforms, even if industry is not formally involved.

2) Challenges of Non-Industry Experience

While industry experience are valuable, leaders without direct industry backgrounds often face challenges in working with industry partners. Bridging the gap between academia and industry can be complex, and leaders with industry experience may have an advantage in navigating this terrain.

3) Importance of Collaboration

Collaboration with industry is a crucial aspect of aligning education with workforce needs. While industry may not always be the driving force behind change, their input and collaboration can help ensure that educational reforms are relevant and responsive to industry demands.

The relationship between educational leaders' industry backgrounds and their ability to drive change in collaboration with industry is complex. While industry experience can be beneficial, the success of educational reforms often depends on effective collaboration between academia and industry, regardless of who initiates the changes.

G. Employer thoughts about current graduates

The feedback loop between employers and educational institutions regarding current graduates are challenging to establish and maintain consistently. Several factors contribute to this challenge:

1) Resource Constraints

Many educational institutions lack the necessary resources to implement and sustain continuous tracking and monitoring of graduates in the workforce. This includes financial constraints, limitations in data collection and analysis tools, and insufficient staff dedicated to alumni relations and career tracking.

2)Diverse Employer Views

Employers have varied expectations and criteria for evaluating graduates. This diversity can make it difficult to create a standardized feedback system that effectively captures the full range of employer perspectives.

3) Graduate Mobility

Graduates often move between jobs and even industries over their careers, making it challenging for institutions to track their progress consistently.

4) Time Lag

There can be a significant time lag between a graduate entering the workforce and any potential feedback from their employer, making it challenging to provide timely insights to educators.

Despite these challenges, establishing a feedback loop between educational institutions and employers is valuable for improving educational programs and ensuring graduates are well-prepared for the workforce. Efforts to overcome these challenges involve developing better data collection and analysis systems, fostering strong alumni networks, and building collaborative relationships with employers to facilitate ongoing communication and feedback.

H. Elevating Student Satisfaction: A Catalyst for Educational Reform Success

The participant perspective underscores the significance of student satisfaction as a pivotal factor in the success and widespread adoption of educational reforms. Here, we can further elaborate on the importance of this aspect:

1) Student-Centric Approach

Placing students at the centre of educational reforms is essential. By actively seeking their feedback and addressing their needs, institutions can tailor their reforms to provide a more engaging and effective learning experience.

2) Holistic Assessment

Student assessments should go beyond just measuring academic outcomes. Evaluating the broader educational goals, such as critical thinking and problem-solving skills, is crucial. This ensures that reforms align with the overarching objectives of education.

3) Continuous Improvement

Educational reforms should be viewed as an ongoing process. Regularly collecting and analysing student feedback allows institutions to identify areas for improvement and make necessary adjustments to enhance the learning experience continually.

4) Model for Adoption

When educational reforms prioritize student satisfaction and holistic learning experiences, they can serve as a model for other departments or institutions seeking to implement similar changes. A successful approach in one context can inspire and guide reforms elsewhere.

5) Enhancing Competencies

Ultimately, the aim of educational reforms is not only to impart knowledge but also to equip students with essential skills and competencies that are valuable in their future careers and in life.

Focusing on student satisfaction and evaluating educational reforms from a generic learning perspective can lead to more effective, adaptable, and widely adopted improvements in education, benefiting both students and the institutions themselves.

I. Navigating the Transition from Startup to Operational Stability

The transition from the startup phase to the growth and operational stability phase is a critical juncture for many institutions and new departments, and it often requires a change in leadership and a shift in focus. Here's a deeper look at this transformation using Engineering Institution 3 as an example:

1) Changing Skill Requirements

In the early phase of growth, changemakers and/or founders prioritize entrepreneurial and innovative skills, driven by vision and a hands-on approach. As the institution expands, the focus shifts to management, scalability, and efficiency.

2) Leadership Transition

Knowing when to transition leadership is crucial. New leaders introduce expertise in strategic planning, fundraising, and organization. For many Engineering Institutions, a leadership change ushered in a new educational program perspective.

3) Reducing Dependency on Individuals

For institutional sustainability, reducing reliance on specific individuals is key. Establishing strong systems and processes ensures continuity and growth, independent of specific leaders, especially during transitions.

4) Promoting Vision Continuity

While leadership may change, it's essential to maintain the core vision and mission of the institution. A clear and shared vision can guide the institution through transitions and changes, ensuring that it stays true to its founding principles.

Transitioning from startup to growth requires understanding changing skill needs, introducing new leadership, and building systems for sustainability and continuity. This phase is pivotal for an institution's long-term success.

J. Navigating Complex Challenges in Engineering Education Policies

The issues highlighted regarding the policies and challenges in engineering and education in some regions are indeed complex and multifaceted:

1) Policy Fragmentation

The absence of cohesive and comprehensive policies can hinder the development and growth of education and engineering sectors. Fragmented policies make it difficult to establish a clear direction for educational institutions and can lead to inconsistency in quality and focus.

2) Political Use

Education and engineering are sometimes leveraged for political gain rather than being guided by a long-term strategic vision for the country. This can lead to policy decisions that prioritize short-term political interests over the broader needs of the education and engineering sectors.

3) Lack of Apex Body

The absence of a central governing body for engineering education can result in challenges related to standardization, quality control, and research coordination. A well-defined apex body can help set standards and drive improvements.

4) Funding Challenges

Adequate funding is crucial for research, development, and maintaining high-quality educational programs. The lack of consistent funding can create a "chicken and egg" situation where universities struggle to invest in research and innovation.

5) Policy Disconnect

There appears to be a gap between policy decisions and the real-world needs and challenges faced by educational institutions. The disconnect between policy and practice can hinder progress.

6) Lack of Collaboration

Collaborative efforts between universities and the government are essential for addressing these challenges. A coordinated approach can help advocate for change and drive policy reforms.

7) Respect for Universities

Universities and research institutions play a vital role in societal progress. A lack of political respect for these institutions can undermine their ability to contribute effectively to national development.

https://doi.org/10.52202/073963-0030

Addressing these issues requires a multi-faceted approach, including the development of comprehensive policies, fostering collaboration, and advocating for the importance of education and engineering in societal progress. It's crucial for stakeholders to work together to overcome these challenges and create a more conducive environment for education and engineering to thrive.

IV CONCLUSIONS

Mapping the above themes to a loop diagram as depicted in Fig. 1, the conflicting and complementary factors in this landscape operate in complex ways in the UK engineering education landscape and hence, there is a need for a concentrated systems-based approach to solve the issues. Findings from this study concur with Cooper (2023) that engineering policy has always been sidelined with respect to government policy. The issues raised by the engineering academics in this study echo findings from earlier papers (Davis, et al 2002) which indicated that an apex body had been set up to drive change within Higher Education, including engineering higher education. This was in the form of the Learning and Teaching Support Network (LTSN), which was set up in 2000, and which had an engineering subject centre that sought to address and drive change within engineering education across the UK. LTSN subsequently transformed into the Higher Education Academy, which, in turn merged with the Leadership Foundation for Higher Education and the Equality Challenge Unit in 2018 to form Advance Higher Education (HE), and the subject centre initiative was dismantled, negatively impacting collaboration on engineering education across the UK.

Therefore, we need a comprehensive agenda on advancing Engineering Education through critical exploration of engineering education policy, involving development of comparative data, rich descriptions of engineering education, research, and policy intersections, leveraging recent progress in engineering education, practice, sustainability, Equality, Diversity and Inclusion (EDI),and ethics, and establishing a central hub for engineering education policy-focused research. This framework, we hope would propel the comprehension and influence of engineering educations' role in policy realms.

Recommendations highlight critical areas for improvement in engineering education:

A. Consistent and Clear Policy

Establishing a consistent and clear policy framework for engineering education is crucial. This can provide guidance, standards, and a unified vision for the sector, ensuring that educational reforms align with broader national goals.

B. Adaptation of Regulatory Bodies and PEIs

Regulatory bodies and professional engineering institutions (PEIs) should be flexible and adaptable in response to new models and approaches to delivering education. This can help facilitate innovation and responsiveness to changing industry needs.

C. Structured Promotional Pathway for Educators

Creating a well-defined promotional pathway and recognition system for engineering educators can incentivize excellence in teaching and research. This can help attract and retain talented educators in the field.

D. Investment in Pedagogical Research

Increased investment in engineering education research is essential for advancing teaching methodologies, curriculum development, and educational outcomes. Research can drive evidence-based improvements in education.

E. Collaboration and Partnerships

Collaboration among universities, educators, and relevant stakeholders is crucial. Collaborative efforts can enhance the ability to secure funding, lobby for policy changes, and collectively address challenges in engineering education.

Implementing these recommendations can contribute to the development of a more robust and responsive engineering education ecosystem that aligns with the needs of industry and society while fostering excellence in teaching and research.

V FUTURE WORK

Considering the initial findings that correspond to complicated push-pull factors in the field, this work will build on the current investigation and expand by delving deeper into UK engineering education policy and the interplay between pedagogy, research, and practice. Studying the vital role of collaboration among diverse stakeholders will be key for the expansion and success of this project.

In addition to academia and government bodies, this initiative will seek to engage key stakeholders like Engineering Professional Councils (EPC), Royal Academy of Engineering, IEEE, IET and other relevant organizations to structurally address the challenging landscape of UK engineering education.

We aspire for this work to spark a robust discourse within the critical engineering community. In tandem with the authors' ongoing efforts to foster collaboration among engineering communities, we aim to cultivate a dynamic platform for the

 https://doi.org/10.52202/073963-0030

productive exchange and evolution of ideas in this space. By fostering partnerships and cooperation, this approach will aim to create a unified front in reshaping engineering education policy and practices, ensuring a sustainable and effective model that benefits both students and society at large.

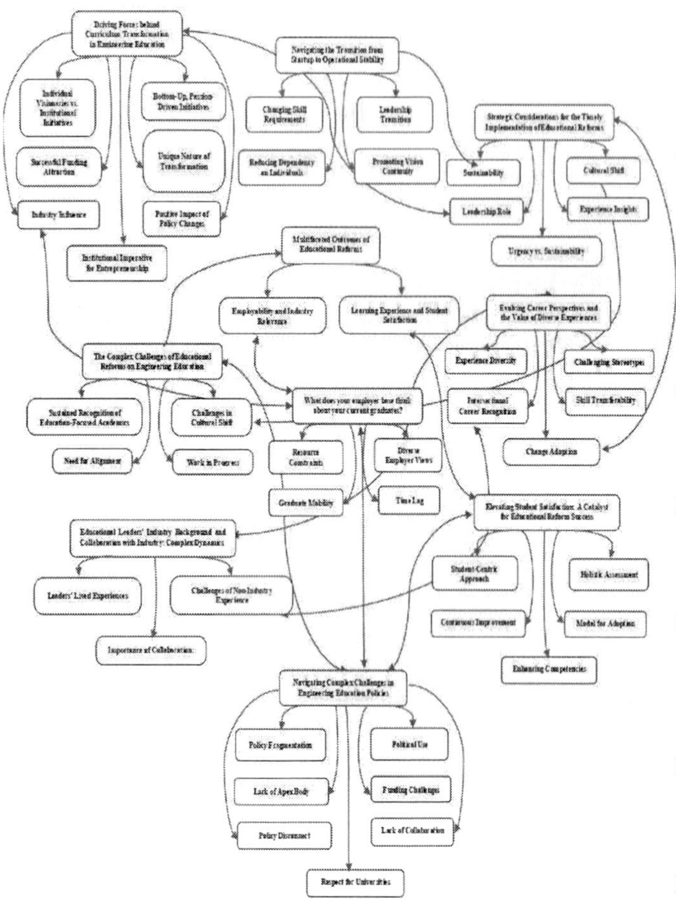

Fig. 1. Causal Loop Diagram Mapping Interactions at Play Between the Various Factors of Pedagogy, Policy, And Practice (red line signifies interconnection between the themes and blue lines within the themes) (adapted from An introductory systems thinking toolkit for civil servants - GOV.UK (www.gov.uk))

ACKNOWLEDGMENT

We acknowledge all the participants for their generous time and appreciate them sharing their valuable insights with us. We thank Dr. Janet Hanson, visiting research fellow at University of Winchester for her advice and guidance for this project.

REFERENCES

Jones, B., Scott, P., Bolton, B., Bramley, A., & Manske, F. (2000). The British engineer problem: A comparison of careers, employment, and skills. *Policy Studies, 21*(1), 5-23.

Spinks, N., Silburn, N., & Birchall, D. W. (2006, July). Making it all work: the engineering graduate of the future, a UK perspective. In *2006 Technology Management for the Global Future-PICMET 2006 Conference* (Vol. 3, pp. 1124-1132). IEEE.

Graham, R. H. (2012). Achieving excellence in engineering education: the ingredients of successful change. London: Royal Academy of Engineering.

Haghighi, K. (2005). Systematic and sustainable reform in engineering education. *Journal of Environmental Engineering, 131*(4), 501-502.

Clark, R. (2011). Today's pupils, tomorrow's engineers! Pedagogy and policy: a UK perspective. *Journal of Engineering, Design and Technology, 9*(2), 227-241.

Levy, J. (2000). Engineering education in the United Kingdom: standards, quality assurance and accreditation. *International Journal of Engineering Education, 16*(2), 136-145.

Engineering Professors Council. (2022, October 10). Proceedings of the conference: New Approaches to Engineering in Higher Education - Engineering Professors Council. Retrieved from link

Engineering Professors Council. (2017). *New Approaches to Engineering in Higher Education- Engineering Professors Council.* Retrieved from link

Fowler, S., Direito, I., Bellingham, K., & Mitchell, J. (2023). Emerging trends, approaches, and challenges in engineering education in the UK, in *SEFI 2023*, Dublin, Ireland.

Higher Education and Research Act 2017 c29. Retrieved from link

Graham, R. (2018). The global state of the art in engineering education. Massachusetts Institute of Technology (MIT) Report, Massachusetts, USA.

Nyamapfene, A., & Williams, B. (2017). Evolution of Engineering Education Research as a Field of Inquiry in the UK.

Davis, L. E., Eley, R. M., & Lamb, F. M. (2002). Engineering education issues in the UK.

Cooper, A. C., Lioté, L., & Colomer, C. (2023). We need to talk about engineering policy. *Technology in Society, 72*, 102196.

Government Office for Science (2023). An introductory system thinking toolkit for civil servants. Retrieved from link

Copyright Statement

https://doi.org/10.52202/073963-0030

Appendix A: List of Reviewers

(In alphabetical order of first name)

Sl. No	Reviewer Name	Affiliation
1	Adam Carberry	The Ohio State University
2	Adarsh Patil	KLE Technological University
3	Aida Guerra	Aalborg University
4	Alexa Ray Fernando	National University, Philippines
5	Alexander Kist	University of Southern Queensland
6	Am Abirami	Thiagarajar College of Engineering
7	Anish Nair	Kalasalingam Academy of Research and Education
8	Anita Campbell	University of Cape Town
9	Anitha D	Thiagarajar College of Engineering
10	Anu Singh	University of Nebraska Lincoln
11	Ashish Agrawal	College of Engineering Technology, Rochester Institute of Technology, NY, USA.
12	Avneet Hira	Boston College
13	Balachandra Halemani	KLE Technological University, Hubballi
14	Basawaraj Patil	KLE Technological University, Hubballi
15	Binu K.G.	St Joseph Engineering College - Vamanjoor, Mangaluru
16	Brian P. Self	California Polytechnic State University
17	Bruce Kloot	University of Cape Town
18	Carlo Gabriel	Southern Institute of Technology
19	Carola Hernandez	Facultad de Educación, Universidad de los Andes
20	Catherine Hills	University of Southern Queensland
21	Cynthia Finelli	University of Michigan
22	David Asante-Asare	Ashesi University
23	Devlin Montfort	Oregon State University
24	Diana Martin	TU Eindhoven
25	Faris Tarlochan	Qatar University

Sl. No	Reviewer Name	Affiliation
26	Gopalkrishna Joshi	MIT Vishwaprayag University, Solapur
27	Gururaj Bhadri	KLE Technological University
28	Hadley Perkins	Arizona State University
29	Helen Inglis	University of Pretoria
30	Hemlata Gaikwad	Rajarambapu Institute of Technology
31	Hiyam Al-Kilidar	University of Technology, Sydney
32	Homero Murzi	Virginia Tech
33	Ieshya Anderson	Arizona State University
34	Javeed Kittur	The University of Oklahoma
35	Jenni Case	University of Cape Town
36	Jens Myrup Pedersen	Aalborg University, Dept. of Electronic Systems
37	Jessica Leeker	University of Colorado at Boulder
38	Jeyamala	Thiagarajar College of Engineering
39	Jiabin Zhu	Shanghai Jiao Tong University
40	Jillian Seniuk Cicek	University of Manitoba
41	John Mitchell	University College London
42	Juebei Chen	Aalborg University
43	Julius Fusic	TCE
44	Kanmani Buddhi	BMS College of Engineering
45	Karthik Shastry	RV College of Engineering
46	Kaushik Mallibhat	KLE Technological University
47	Kavitha D	Thiagarajar College of Engineering
48	Keith Raymond Fernandes	St Joseph Engineering College Mangaluru
49	Kim Blackmore	Australian National University
50	Kirsten Davis	Purdue University
51	Krishnaraja Kodancha	KLE Technological University, Hubballi
52	Lauren Jansen	University of Pretoria

Sl. No	Reviewer Name	Affiliation
53	Laxmikant Patil	KLE Technological University Hubballi
54	Medha Dalal	Arizona State University
55	Mitra Addi	Universiti Teknologi Malaysia
56	Muhibul Haque Bhuyan	American International University Bangladesh (AIUB), Bangladesh
57	Nadeyah Alreiahi	Kuwait University
58	Nalini Iyer	KLE Technological University
59	Ousman Badjie	Amrita Vishwa Vidyapeetham University
60	Paresh Tanna	RK University
61	Prakash Hegade	KLE Technological University
62	Prashant Revankar	KLE Technological University
63	Prathiba Nagabhushan	ACU
64	Preethi Baligar	KLE Technological university
65	Preeti Satish	Dayananda Sagar College of Engineering
66	Preeti Patil	KLE Technological University
67	Rachel Figard	Arizona State University
68	Radhika Devi	MLR Institute of Technology
69	Radhika Amashi	KLE Technological University
70	Rajlaxmi Chouhan	Indian Institute of Technology Jodhpur
71	Raju G U	KLE Technological University, Hubli
72	Roger Hadgraft	University of Technology, Sydney
73	Rohit Kandakatla	KG Reddy College of Engineering and Technology
74	Rucha Joshi	Plaksha University
75	Saeed Shaeri	Charles Sturt University
76	Sanjeev Kavale	Arizona State University
77	Santhosh Pokala	JNTUH
78	Santosh Naik	Hyderabad Institute of Technology and Management
79	Sarah Dart	Queensland University of Technology

80	Scott Daniel	University of Technology Sydney
81	Shannon Chance	Technological University (TU) Dublin
82	Shivalingsarj Desai	KLE Technological University
83	Shreeshail M L	KLE Technological University
84	Siddhant Joshi	Purdue University
85	Siddharthsinh Jadeja	B H Gardi College of Engineering and Technology
86	Sivakumar Krishnan	Vishnu Educational Development and Innovation Centre
87	Sohum Sohoni	Milwaukee School of Engineering
88	Soumya Narayanan	IIT Bombay
89	Srinivas Dustker	Purdue University
90	Stefanija Klaric	Charles Darwin University
91	Surendra Bandi	HITAM
92	Syakirah Samsudin	Universiti Pendidikan Sultan Idris
93	Teresa Hattingh	University of Johannesburg
94	Thanikachalam Vedhathiri	NITTTR
95	Tony Vo	Monash University
96	Vijayalakshmi M	KLE technological University
97	Vinayak Kulkarni	KLE Technological University
98	William Oakes	Purdue University
99	Xiaofeng Tang	Tsinghua University
100	Yogesh Velankar	Amrita Vishwa Vidyapeetham
101	Zach Simpson	University of Johannesburg

Appendix B- Sponsors

Institutional Sponsors

KLE Technological University | Creating Value, Leveraging Knowledge

https://www.kletech.ac.in/

Hyderabad Institute of Technology and Management
https://hitam.org

RK University
https://rku.ac.in/

Kalasalingam Academy for Research and Education
https://kalasalingam.ac.in/

Model Institute of Engineering and Technology
https://www.mietjmu.in/

Nalla Malla Reddy Engineering College
https://www.nmrec.edu.in/

Thiagarajar College of Engineering
https://www.tce.edu/

Rajarambapu Institute of Technology
https://www.ritindia.edu/

St Joseph Engineering College
https://www.sjec.ac.in/

Saintgits Group of Institution
https://saintgits.org/

RV College of Engineering
https://www.rvce.edu.in/

Ramco Institute of Technology
https://www.ritrjpm.ac.in/

Hindustan University
https://hindustanuniv.ac.in/

Marwadi University
https://www.marwadiuniversity.ac.in/

Vasireddy Venkatadri Institute of Technology
https://www.vvitguntur.com/

PLATINUM SPONSOR

GOLD SPONSOR

GOLD SPONSOR

SILVER SPONSOR

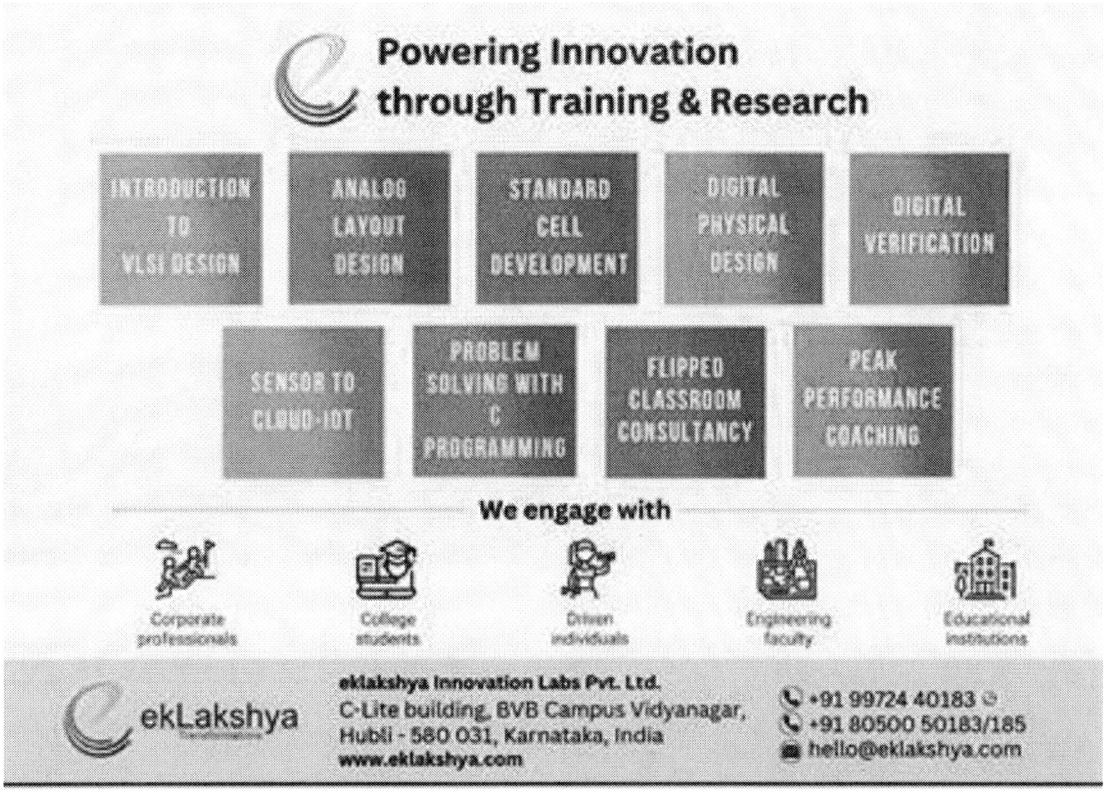

SILVER SPONSOR

BEST COMPLEMENTS FROM

IGIP

BRONZE SPONSOR

BRONZE SPONSOR

Learn the basics of practical signal processing techniques in MATLAB®. Use spectral analysis and filtering techniques to preprocess, analyze, and extract information from signal data.

Get an e-certificate on completion of this 2-hour course.

Link and QR Code to start the course: https://matlabacademy.mathworks.com/details/signal-processing-onramp/signalprocessing?s_eid=PEP_32264

MathWorks®

BRONZE SPONSOR

Best Compliments From

CTI

Consulting,
Technology,
Information

AUTHOR INDEX